The Ants of Fiji

University of California Press, one of the most distinguished university presses in the United States, enriches lives around the world by advancing scholarship in the humanities, social sciences, and natural sciences. Its activities are supported by the UC Press Foundation and by philanthropic contributions from individuals and institutions. For more information, visit www.ucpress.edu.

University of California Publications in Entomology, Volume 132

University of California Press
Berkeley and Los Angeles, California

University of California Press, Ltd.
London, England

Cataloging-in-Publication data for this title is on file with the Library of Congress.

ISBN 978-0-520-09888-6 (pbk. : alk. paper)

The paper used in this publication meets the minimum requirements of ANSI/NISO Z39.48-1992 (R 1997) (Permanence of Paper).

The Ants of Fiji

ELI M. SARNAT
University of Illinois, Urbana-Champaign

EVAN P. ECONOMO
University of Michigan

UNIVERSITY OF CALIFORNIA PRESS
Berkeley Los Angeles London

CONTENTS

List of Plates ix
Abstract xi
Acknowledgments xiii
Introduction 1
  Geography of Fiji 1
  Geologic history 4
  Scientific exploration and study of the Fijian ant fauna 4
  Overview of the Fijian ant fauna 6
Methods 6
  Specimen data 6
  Sources of material 7
  Undescribed and undetermined species 7
  Material examined 7
  Collection localities 8
  Geographic and ecological distributions 8
  Specimen images 8
  The species concept as applied to the Fijian ant fauna 9
  Standard measurements and indices 9
  Abbreviations of depositories 9
Taxonomic synopsis of the Fijian ant fauna 10
  Key to the ant genera of Fiji 10
  Subfamily Amblyoponinae 22
    Genus Amblyopone 22
      Amblyopone zwaluwenburgi (Williams) 22
    Genus Prionopelta 23
      Prionopelta kraepelini Forel 23
  Subfamily Cerapachyinae 24
    Genus Cerapachys 24
      Cerapachys cryptus Mann 26
      Cerapachys fuscior Mann 26
      Cerapachys sp. FJ06 27
      Cerapachys lindrothi Wilson 28
      Cerapachys zimmermani Wilson 28
      Cerapachys sp. FJ01 29
      Cerapachys majusculus Mann 30
      Cerapachys sculpturatus Mann 30
      Cerapachys vitiensis Mann 30
      Cerapachys sp. FJ07 31
      Cerapachys sp. FJ05 32
      Cerapachys sp. FJ04 32
      Cerapachys sp. FJ08 and *Cerapachys* sp. FJ10 32
  Subfamily Dolichoderinae 33
    Genus Iridomyrmex 33
      Iridomyrmex anceps (Roger) 33
    Genus Ochetellus 34
      Ochetellus sororis (Mann) 34

Genus Philidris 35
Philidris nagasau (Mann) 36
Genus Tapinoma 37
Tapinoma melanocephalum (Fabricius) 37
Tapinoma minutum Mayr 38
Tapinoma sp. FJ01 38
Tapinoma sp. FJ02 39
Genus Technomyrmex 39
Technomyrmex vitiensis Mann 39
Subfamily Ectatomminae 41
Genus Gnamptogenys 41
Gnamptogenys aterrima (Mann) 41
Subfamily Formicinae 42
Genus Acropyga 42
Acropyga lauta Mann 42
Acropyga sp. FJ02 43
Genus Anoplolepis 43
Anoplolepis gracilipes (Smith, F.) 43
Genus Camponotus 44
Camponotus chloroticus Emery 46
Camponotus oceanicus (Mayr) 47
Camponotus polynesicus Emery 48
Camponotus vitiensis Mann 50
Camponotus sp. FJ04 51
Camponotus fijianus Özdikmen, NEW STATUS 52
Camponotus dentatus (Mayr) 53
Camponotus bryani Santschi 54
Camponotus manni Wheeler 55
Camponotus umbratilis Wheeler, NEW STATUS 55
Camponotus sp. FJ02 56
Camponotus sp. FJ03 56
Camponotus cristatus Mayr 57
Camponotus laminatus Mayr 58
Camponotus levuanus Mann, NEW STATUS 59
Camponotus maafui Mann 60
Camponotus sadinus Mann, NEW STATUS 60
Camponotus kadi Mann, NEW STATUS 61
Camponotus lauensis Mann 62
Camponotus schmeltzi Mayr 63
Genus Nylanderia 63
Nylanderia glabrior (Forel) 64
Nylanderia vaga (Forel) 64
Nylanderia vitiensis (Mann), REVISED STATUS 65
Nylanderia sp. FJ03 66
Genus Paraparatrechina 66
Paraparatrechina oceanica Mann, REVISED STATUS 67
Genus Paratrechina 68
Paratrechina longicornis (Latreille) 68

Genus Plagiolepis 69
Plagiolepis alluaudi Emery 69
Subfamily Myrmicinae 70
Genus Adelomyrmex 70
Adelomyrmex hirsutus Mann 70
Adelomyrmex samoanus Wilson & Taylor 71
Genus Cardiocondyla 71
Cardiocondyla emeryi Forel 72
Cardiocondyla kagutsuchi Terayama 72
Cardiocondyla minutior Forel 73
Cardiocondyla nuda (Mayr) 73
Cardiocondyla obscurior Wheeler 74
Genus Carebara 74
Carebara atoma (Emery) 74
Genus Eurhopalothrix 75
Eurhopalothrix emeryi (Forel) 76
Eurhopalothrix insidiatrix Taylor 77
Eurhopalothrix sp. FJ52 77
Genus Lordomyrma 77
Lordomyrma curvata Sarnat 79
Lordomyrma desupra Sarnat 80
Lordomyrma levifrons (Mann) 80
Lordomyrma polita (Mann) 81
Lordomyrma rugosa (Mann) 81
Lordomyrma stoneri (Mann) 82
Lordomyrma striatella (Mann) 82
Lordomyrma sukuna Sarnat 83
Lordomyrma tortuosa (Mann) 83
Lordomyrma vanua Sarnat 84
Lordomyrma vuda Sarnat 84
Genus Metapone 85
Metapone sp. FJ01 85
Genus Monomorium 86
Monomorium destructor (Jerdon) 87
Monomorium floricola (Jerdon) 88
Monomorium pharaonis (Linnaeus) 88
Monomorium sechellense Emery 89
Monomorium vitiense Mann, REVISED STATUS 89
Monomorium sp. FJ02 90
Genus Myrmecina 90
Myrmecina cacabau (Mann) 91
Myrmecina sp. FJ01 92
Genus Pheidole 92
Pheidole caldwelli Mann 96
Pheidole fervens Smith, F. 97
Pheidole knowlesi Mann 98
Pheidole megacephala (Fabricius) 99
Pheidole oceanica Mayr 100

Pheidole onifera Mann 101
Pheidole sexspinosa Mayr 101
Pheidole umbonata Mayr 102
Pheidole vatu Mann 102
Pheidole wilsoni Mann 103
Pheidole sp. FJ05 104
Pheidole sp. FJ09 105
Pheidole bula Sarnat 106
Pheidole colaensis Mann 106
Pheidole furcata Sarnat 107
Pheidole pegasus Sarnat 107
Pheidole roosevelti Mann 108
Pheidole simplispinosa Sarnat 109
Pheidole uncagena Sarnat 109
Genus Poecilomyrma 110
Poecilomyrma myrmecodiae Mann, NEW STATUS 112
Poecilomyrma senirewae Mann 113
Poecilomyrma sp. FJ03 113
Poecilomyrma sp. FJ05 113
Poecilomyrma sp. FJ06 114
Poecilomyrma sp. FJ07 114
Poecilomyrma sp. FJ08 114
Genus Pristomyrmex 115
Pristomyrmex mandibularis Mann 115
Pristomyrmex sp. FJ02 116
Genus Pyramica 117
Pyramica membranifera (Emery) 118
Pyramica trauma Bolton 118
Pyramica sp. FJ02 119
Genus Rogeria 119
Rogeria stigmatica Emery 120
Genus Romblonella 121
Romblonella liogaster (Santschi), NEW STATUS 121
Genus Solenopsis 122
Solenopsis geminata (Fabricius) 122
Solenopsis papuana Emery 123
Genus Strumigenys 124
Strumigenys basiliska Bolton 127
Strumigenys chernovi Dlussky 128
Strumigenys daithma Bolton 128
Strumigenys ekasura Bolton 129
Strumigenys frivola Bolton 129
Strumigenys godeffroyi Mayr 130
Strumigenys jepsoni Mann 130
Strumigenys mailei Wilson & Taylor 130
Strumigenys nidifex Mann 131
Strumigenys panaulax Bolton 132
Strumigenys praefecta Bolton 132

Strumigenys rogeri Emery 133
Strumigenys scelesta Mann 133
Strumigenys sulcata Bolton 134
Strumigenys tumida Bolton 135
Strumigenys sp. FJ01 135
Strumigenys sp. FJ13 135
Strumigenys sp. FJ14 136
Strumigenys sp. FJ17 136
Strumigenys sp. FJ18 136
Strumigenys sp. FJ19 136
Genus Tetramorium 137
Tetramorium bicarinatum (Nylander) 138
Tetramorium caldarium (Roger) 139
Tetramorium insolens (Smith, F.) 140
Tetramorium lanuginosum Mayr 140
Tetramorium manni Bolton 141
Tetramorium pacificum Mayr 141
Tetramorium simillimum (Smith, F.) 142
Tetramorium tonganum Mayr 143
Genus Vollenhovia 143
Vollenhovia denticulata Emery 144
Vollenhovia sp. FJ01 145
Vollenhovia sp. FJ03 145
Vollenhovia sp. FJ04 146
Vollenhovia sp. FJ05 146
Subfamily Ponerinae 146
Genus Anochetus 146
Anochetus graeffei Mayr 147
Genus Hypoponera 147
Hypoponera confinis (Roger) 149
Hypoponera eutrepta (Wilson), REVISED STATUS 149
Hypoponera monticola (Mann) 150
Hypoponera opaciceps (Mayr) 150
Hypoponera pruinosa (Emery) 151
Hypoponera punctatissima (Roger) 152
Hypoponera turaga (Mann) 152
Hypoponera vitiensis (Mann), REVISED STATUS 153
Hypoponera sp. FJ16 153
Genus Leptogenys 153
Leptogenys foveopunctata Mann 155
Leptogenys fugax Mann 155
Leptogenys humiliata Mann 156
Leptogenys letilae Mann 156
Leptogenys navua Mann 157
Leptogenys vitiensis Mann 157
Leptogenys sp. FJ01 157
Genus Odontomachus 158
Odontomachus angulatus Mayr 158

Odontomachus simillimus Smith, F. 159
Genus Pachycondyla 160
Pachycondyla stigma (Fabricius) 160
Genus Platythyrea 161
Platythyrea parallela (F. Smith) 161
Genus Ponera 161
Ponera colaensis Mann 162
Ponera manni Taylor 163
Ponera swezeyi (Wheeler) 163
Ponera sp. FJ02 163
Subfamily Proceratiinae 164
Genus Discothyrea 164
Discothyrea sp. FJ01 165
Discothyrea sp. FJ02 165
Discothyrea sp. FJ04 165
Genus Proceratium 166
Proceratium oceanicum De Andrade 166
Proceratium relictum Mann 167
Proceratium sp. FJ01 167
Omitted Taxa 167
Dubious Records 168
Camponotus rufifrons (Smith, F.) 168
Linepithema humile (Mayr) 168
Tetramorium tenuicrine (Emery) 169
Odontomachus haematodus (Linnaeus) 169
Hypoponera ragusai (Emery) 169
Nomina Nuda 169
Species plates 170
Literature cited 357
Appendix A. Collection localities 369
Appendix B. Checklist of the ants of Fiji 380

LIST OF PLATES

Plate 1........*Amblyopone zwaluwenburgi*
Plate 2........*Prionopelta kraepelini*
Plate 3........*Cerapachys cryptus*
Plate 4........*Cerapachys fuscior*
Plate 5........*Cerapachys* sp. FJ06
Plate 6........*Cerapachys lindrothi*
Plate 7........*Cerapachys zimmermani*
Plate 8........*Cerapachys* sp. FJ01
Plate 9........*Cerapachys majusculus*
Plate 10......*Cerapachys sculpturatus*
Plate 11......*Cerapachys vitiensis*
Plate 12......*Cerapachys* sp. FJ07
Plate 13......*Cerapachys* sp. FJ05
Plate 14......*Cerapachys* sp. FJ04
Plate 15......*Cerapachys* sp. FJ08
Plate 16......*Cerapachys* sp. FJ10
Plate 17......*Iridomyrmex anceps*
Plate 18......*Ochetellus sororis*
Plate 19......*Philidris nagasau*
Plate 20......*Tapinoma melanocephalum*
Plate 21......*Tapinoma minutum*
Plate 22......*Tapinoma* sp. FJ01
Plate 23......*Tapinoma* sp. FJ02
Plate 24......*Technomyrmex vitiensis*
Plate 25......*Gnamptogenys aterrima*
Plate 26......*Acropyga lauta*
Plate 27......*Acropyga* sp. FJ02
Plate 28......*Anoplolepis gracilipes*
Plate 29......*Camponotus chloroticus*
Plate 30......*Camponotus polynesicus*
Plate 31......*Camponotus vitiensis*
Plate 32......*Camponotus* sp. FJ04
Plate 33......*Camponotus fijianus*
Plate 34......*Camponotus dentatus*
Plate 35......*Camponotus bryani*
Plate 36......*Camponotus manni*
Plate 37......*Camponotus umbratilis*
Plate 38......*Camponotus* sp. FJ02
Plate 39......*Camponotus* sp. FJ03
Plate 40......*Camponotus cristatus*
Plate 41......*Camponotus laminatus*
Plate 42......*Camponotus levuanus*
Plate 43......*Camponotus maafui*
Plate 44......*Camponotus sadinus*
Plate 45......*Camponotus kadi*
Plate 46......*Camponotus lauensis*
Plate 47......*Camponotus schmeltzi*
Plate 48......*Nylanderia glabrior*
Plate 49......*Nylanderia vaga*
Plate 50......*Nylanderia vitiensis*
Plate 51......*Nylanderia* sp. FJ03
Plate 52......*Paraparatrechina oceanica*
Plate 53......*Paratrechina longicornis*
Plate 54......*Plagiolepis alluaudi*
Plate 55......*Adelomyrmex hirsutus*
Plate 56......*Adelomyrmex samoanus*
Plate 57......*Cardiocondyla emeryi*
Plate 58......*Cardiocondyla kagutsuchi*
Plate 59......*Cardiocondyla minutior*
Plate 60......*Cardiocondyla nuda*
Plate 61......*Cardiocondyla obscurior*
Plate 62......*Carebara atoma*
Plate 63......*Eurhopalothrix emeryi*
Plate 64......*Eurhopalothrix insidiatrix*
Plate 65......*Eurhopalothrix* sp. FJ52
Plate 66......*Lordomyrma curvata*
Plate 67......*Lordomyrma desupra*
Plate 68......*Lordomyrma levifrons*
Plate 69......*Lordomyrma polita*
Plate 70......*Lordomyrma rugosa*
Plate 71......*Lordomyrma stoneri*
Plate 72......*Lordomyrma striatella*
Plate 73......*Lordomyrma sukuna*
Plate 74......*Lordomyrma tortuosa*
Plate 75......*Lordomyrma vanua*
Plate 76......*Lordomyrma vuda*
Plate 77......*Metapone* sp. FJ01
Plate 78......*Monomorium destructor*
Plate 79......*Monomorium floricola*
Plate 80......*Monomorium pharaonis*
Plate 81......*Monomorium sechellense*
Plate 82......*Monomorium vitiense*
Plate 83......*Monomorium* sp. FJ02
Plate 84......*Myrmecina cacabau*
Plate 85......*Myrmecina* sp. FJ01
Plate 86......*Pheidole caldwelli*
Plate 87......*Pheidole fervens*
Plate 88......*Pheidole knowlesi*
Plate 89......*Pheidole megacephala*
Plate 90......*Pheidole oceanica*

Plate 91......*Pheidole onifera*
Plate 92......*Pheidole sexspinosa*
Plate 93......*Pheidole umbonata*
Plate 94......*Pheidole vatu*
Plate 95......*Pheidole wilsoni*
Plate 96......*Pheidole* sp. FJ05
Plate 97......*Pheidole* sp. FJ09
Plate 98......*Pheidole bula*
Plate 99......*Pheidole colaensis*
Plate 100....*Pheidole furcata*
Plate 101....*Pheidole pegasus*
Plate 102....*Pheidole roosevelti*
Plate 103....*Pheidole simplispinosa*
Plate 104....*Pheidole uncagena*
Plate 105....*Poecilomyrma myrmecodiae*
Plate 106....*Poecilomyrma senirewae*
Plate 107....*Poecilomyrma* sp. FJ03
Plate 108....*Poecilomyrma* sp. FJ05
Plate 109....*Poecilomyrma* sp. FJ06
Plate 110....*Poecilomyrma* sp. FJ07
Plate 111....*Poecilomyrma* sp. FJ08
Plate 112....*Pristomyrmex mandibularis*
Plate 113....*Pristomyrmex* sp. FJ02
Plate 114....*Pyramica membranifera*
Plate 115....*Pyramica trauma*
Plate 116....*Pyramica* sp. FJ02
Plate 117....*Rogeria stigmatica*
Plate 118....*Romblonella liogaster*
Plate 119....*Solenopsis geminata*
Plate 120....*Solenopsis papuana*
Plate 121....*Strumigenys basiliska*
Plate 122....*Strumigenys chernovi*
Plate 123....*Strumigenys daithma*
Plate 124....*Strumigenys ekasura*
Plate 125....*Strumigenys frivola*
Plate 126....*Strumigenys godeffroyi*
Plate 127....*Strumigenys jepsoni*
Plate 128....*Strumigenys mailei*
Plate 129....*Strumigenys nidifex*
Plate 130....*Strumigenys panaulax*
Plate 131....*Strumigenys praefecta*
Plate 132....*Strumigenys rogeri*
Plate 133....*Strumigenys scelesta*
Plate 134....*Strumigenys sulcata*
Plate 135....*Strumigenys tumida*
Plate 136....*Strumigenys* sp. FJ01
Plate 137....*Strumigenys* sp. FJ13
Plate 138....*Strumigenys* sp. FJ14
Plate 139....*Strumigenys* sp. FJ17
Plate 140....*Strumigenys* sp. FJ18
Plate 141....*Strumigenys* sp. FJ19
Plate 142....*Tetramorium bicarinatum*
Plate 143....*Tetramorium caldarium*
Plate 144....*Tetramorium insolens*
Plate 145....*Tetramorium lanuginosum*
Plate 146....*Tetramorium manni*
Plate 147....*Tetramorium pacificum*
Plate 148....*Tetramorium simillimum*
Plate 149....*Tetramorium tonganum*
Plate 150....*Vollenhovia denticulata*
Plate 151....*Vollenhovia* sp. FJ01
Plate 152....*Vollenhovia* sp. FJ03
Plate 153....*Vollenhovia* sp. FJ04
Plate 154....*Vollenhovia* sp. FJ05
Plate 155....*Anochetus graeffei*
Plate 156....*Hypoponera confinis*
Plate 157....*Hypoponera eutrepta*
Plate 158....*Hypoponera monticola*
Plate 159....*Hypoponera opaciceps*
Plate 160....*Hypoponera pruinosa*
Plate 161....*Hypoponera punctatissima*
Plate 162....*Hypoponera turaga*
Plate 163....*Hypoponera vitiensis*
Plate 164....*Hypoponera* sp. FJ16
Plate 165....*Leptogenys foveopunctata*
Plate 166....*Leptogenys fugax*
Plate 167....*Leptogenys humiliata*
Plate 168....*Leptogenys letilae*
Plate 169....*Leptogenys navua*
Plate 170....*Leptogenys vitiensis*
Plate 171....*Leptogenys* sp. FJ01
Plate 172....*Odontomachus angulatus*
Plate 173....*Odontomachus simillimus*
Plate 174....*Pachycondyla stigma*
Plate 175....*Platythyrea parallela*
Plate 176....*Ponera colaensis*
Plate 177....*Ponera manni*
Plate 178....*Ponera swezeyi*
Plate 179....*Ponera* sp. FJ02
Plate 180....*Discothyrea* sp. FJ01
Plate 181....*Discothyrea* sp. FJ02
Plate 182....*Discothyrea* sp. FJ04
Plate 183....*Proceratium oceanicum*
Plate 184....*Proceratium relictum*
Plate 185....*Proceratium* sp. FJ01

ABSTRACT

The ant fauna of the Fijian archipelago is a diverse assemblage of endemic radiations, pan-Pacific species, and exotics introduced from around the world. Here we provide a taxonomic synopsis of the entire Fijian ant fauna by incorporating previously published information with the results of a recently completed, archipelago-wide biodiversity inventory. This synopsis updates the first and only other treatment of the fauna, W. M. Mann's 1921 monograph, *The Ants of the Fiji Islands*. A total of 187 ant species representing 43 genera are recognized here. Of these species, 88% are native to the Pacific region, 70% are endemic to Fiji, and 12% are introduced into the Pacific region. Approximately 45 ant species in Fiji are undescribed, and are identified here by assigned code names. An illustrated key to genera, synopses of each species, keys to species of all genera, and a species list is provided. The work is further illustrated with specimen images, distribution maps, and habitat-elevation charts for all species. Seven taxa are promoted to full species status: *Camponotus fijianus* Özdikmen, **stat. n.**, *Camponotus kadi* Mann **stat. n.**, *C. levuanus* Mann **stat. n.**, *C. sadinus* Mann **stat. n.**, *C. umbratilis* Wheeler **stat. n.**, *Poecilomyrma myrmecodiae* Mann **stat. n.**, *Romblonella liogaster* (Santschi) **stat. n.** The following five taxa are revived from synonymy: *Hypoponera eutrepta* (Wilson) **stat. rev.**, *H. vitiensis* (Mann) **stat. rev.**, *Monomorium vitiense* (Mann) **stat. rev.**, *Paraparatrechina oceanica* (Mann) **stat. rev.**, *N. vitiensis* (Mann) **stat. rev.** The following **new synonymies** are proposed (senior synonym listed first): *Camponotus cristatus* Mann = *C. cristatus nagasau* Mann; *C. kadi* Mann = *C. loloma* Mann = *C. trotteri* Mann; *C. polynesicus* Emery = *C. maudella* Mann = *C. maudella seemanni* Mann = *C. janussus* Bolton; *Pheidole knowlesi* Mann = *P. extensa* Mann; *Philidris nagasau* (Mann) = *P. alticola* (Mann) = *P. agnatus* (Mann); *Romblonella liogaster* (Santschi) = *R. vitiensis* Smith, M. **Lectotypes** are designated for the following species: *Camponotus vitiensis* Mann, *Gnamptogenys aterrima* (Mann), *Poecilomyrma senirewae* Mann.

ACKNOWLEDGMENTS

Vinaka vakalevu to the Fiji government, landowners and people of Fiji for their kind hospitality and for allowing specimen collection and exportation. Special thanks to David Olson and Linda Farley for encouraging EMS to take on this project and for their considerable support and advice. Neal Evenhuis, Dan Bickel, Evert Schlinger and Leah Brorstrom provided specimens from the Fiji Terrestrial Arthropod Survey. Akanisi Caginitoba, Moala Tokota'a, Seta and Adi, and the Wildlife Conservation Society staff provided logistical support and field collections. John Heraty, Jack Longino, and an anonymous reviewer made substantial improvements to the manuscript. Phil Ward provided invaluable mentorship. Peter Cranston and Brian Fisher offered insightful suggestions regarding the pursuit and publication of this research. Marek Boroweic provided images and review for the *Cerapachys* section. Steve Shattuck provided images and review for the *Discothyrea* section. Hilda Waqa provided specimens. Ted Shultz, Eugenia Okonski, Stefan Cover and Suzanne Ryder assisted with museum material. Scott Solomon and Sasha Mikheyev provided specimens and participated in a field expedition. Darren Ward provided data on geographic distributions. Matthew Martinez and Benjamin Blanchard assisted with specimen curation. Jasmine Joseph, Anna Lam, Will Ericson and Shannon Hartman assisted with specimen imaging. Thanks to the U. of Michigan Museum of Zoology for funding specimen imaging, and Lacey Knowles, Huijie Gan and Liane Racelis for assisting with the imaging at UM. Special thanks to Michele Esposito for specimen and imaging help, and for assistance integrating our content with Antweb. Liem Tran assisted with database entry. Aaron King assisted with ArcGIS formatting. Julia Schreiber assisted with database entry, field collections and manuscript review. This research was supported by the Ernst Mayr Grant (EMS), the Pacific Rim Research Program (EMS), the Encyclopedia of Life Rubenstein Fellowship Program (to EMS), the National Science Foundation (NSF), DEB-0425970 "Fiji Terrestrial Arthropod Survey", the NSF Graduate Research Fellowship (to EMS and EPE), UMMZ Insect Division (EPE), the U. of Texas graduate program (EPE), NSF IGERT fellowship (EPE) and the UC Davis Entomology Department (EMS).

INTRODUCTION

Biologists have long sought to document and understand the unique evolution and ecology of island biotas. Oceanic archipelagoes are often adorned with spectacular evolutionary radiations and unique ecosystems. These distinctive faunas, however, are highly vulnerable to human activities, climate change, and introduction of exotic species. Among island ant faunas, perhaps nowhere are these themes so prominently on display as in the Fijian archipelago.

The Fijian terrestrial biota was assembled during approximately 20 million years of over-water colonization, *in situ* evolution and speciation, and more recently through the arrival of species as stowaways on canoes, galleys and battleships (Figure 1). Today's Fijian ant fauna is characterized by extreme geographic isolation from source areas, differentiation and pattern formation among islands, and contemporary invasions. The list of species occurring in Fiji, which continues to grow, includes both widespread dominant species and rare taxonomic oddities.

The motivation of this study is to provide an update to W. M. Mann's (1921) monograph *The Ants of the Fiji Islands*, published 89 years ago. At the time, Mann lamented that the insect fauna of Fiji had been almost entirely neglected, and the limited knowledge accrued in the years since his publication is even more lamentable. With the recent collection of a large number of ant specimens in recent years, and a surge of interest in biodiversity research and conservation in Fiji, the opportunity has arrived to synthesize the taxonomy of the Fijian ant fauna for new generations of biologists. Our goal is to provide a resource that will allow a scientist to collect an ant specimen anywhere in Fiji and connect it to information on its taxonomy, geographic distribution, habitat distribution and natural history. With 187 species distributed over seven islands of moderate size, and hundreds of smaller islands, the system represents a diverse yet tractable fauna that can be useful for testing hypotheses in evolutionary biology, island biogeography, community ecology, invasion biology and other disciplines.

It would be remiss to conclude this introduction without a note of recognition and thanks to the people of Fiji. Mann (1921) wrote in the introduction of his own treatise on Fijian ants, "I shall remember the native Fijians...as the kindliest, most hospitable folk I have known." Eighty-nine years later, we both share those sentiments and add our admiration for the Fijians' thoughtful stewardship of their native lands. We hope this small study will be useful for scientific discovery and conservation of Fiji's fascinating natural heritage in the generations to come.

Geography of Fiji

Fiji is a tropical archipelago located between Melanesia and Polynesia. The archipelago is composed of 332 islands exceeding 2.6 km^2 and 512 islets (Ash, 1992; Neall & Trewick, 2008). Of the islands, two are considered large (Viti Levu and Vanua Levu), and five are considered medium sized (Taveuni, Kadavu, Ovalau, Gau and Koro). The total land area of the archipelago is approximately 18,270 km^2 and the majority of the islands occur between 16°–19°S and 176°E–178°W. The archipelago is approximately 2,600 km east of Australia, and approximately

Figure 1. Photographs of Fijian ants. A) *Acropyga* sp. FJ02 (endemic) carrying mealybug. B) *Hypoponera eutrepta* (endemic) carrying larva. C) *Tetramorium lanuginosum* (introduced). D) *Camponotus dentatus* (endemic). E) *Odontomachus simillimus* (Pacific native). F) *Camponotus polynesicus* (endemic).

3,000 km southeast of New Guinea. The highest peak in Fiji reaches 1324 m. A map showing Fiji's location in the southwest Pacific relative to Australia and nearby islands is presented in Figure 2. A reference map of the Fijian archipelago, including all the localities from which ant specimens were collected and examined is presented in the Species Plates section.

Figure 2. Map of the southwest Pacific showing the relative position and distance of Fiji to Australia and nearby islands.

Climate

Fiji has an oceanic climate with mean temperatures varying within the narrow range of 22°C in July and 26°C in January (Ash, 1992). Daily temperatures fluctuate approximately 6°C. The high mountain ranges of the larger islands cause a strong orthographic pattern in which the windward (south-eastern) sides of the islands are typically wetter and cloudier than the drier and sunnier leeward (north-western) sides. Annual rainfall increases from about 200–300 cm on the southeast coasts and smaller islands up to 500–1000 cm on the windward mountain ranges. The northwest coasts of the larger islands receive as little as 150 cm. The seasonality of Fiji is divided into the warm and wet months between November and April, and the cooler and drier months between May and October. The warm and wet season can be punctuated by tropical cyclones that bring tremendous winds and rainfall through the island system, and often results in pronounced flooding within river valleys.

Habitats

There are three dominant ecological gradients in Fiji's terrestrial environment: elevation, disturbance and precipitation. All of these drive patterns of vegetation (Evenhuis & Bickel, 2005) and are relevant for ant distributions. Lowland rain forests occur from sea-level to 600 m. Montane rain forests occur above 600 m and are characterized by temperatures that are 4–6°C lower than in lowland forests. Cloud forests occur at the highest peaks and ridges of Viti Levu, Vanua Levu, Taveuni and Kadavu. Fiji's dry zones were historically sclerophyllous forests and scrublands found on the leeward sides of islands from sea level to 450 m, although most of this habitat is now converted to farmland. Humans have transformed the landscape through a long history of logging, deforestation, agriculture and urbanization. Today, as with much of the tropics, there is a strong disturbance gradient from the most urbanized areas to the forest interior. This gradient is highly correlated with elevation, as little native forest remains adjacent to the coast. Plates 1–185 depict the distribution of each ant species across the elevation and disturbance gradient.

Geologic history

Fiji's geologic history is rich and complex. While many studies pertain to particular geologic periods or regions of the Fijian archipelago (Auzende, 1995; Colley & Hindle, 1984; Davey, 1982; Falvey, 1978; Hathway & Colley, 1994; Kumar, 2005; Musgrave & Firth, 1999; Nunn, 1990; Nunn, 1999; Stratford & Rodda, 2000; Whelan et al., 1985) comprehensive reviews of geologic processes and formations within the archipelago are more limited (Chase, 1971; Ewart, 1988; Rodda, 1994; Rodda & Kroenke, 1984).

The primary rocks of the Fijian islands are composed of intruded and extruded volcanics, uplifted marine sediments, and limestones (Dickinson & Shutler, 2000; Rodda, 1994; Rodda & Kroenke, 1984). Viti Levu is the oldest island, and formed as the easternmost extent of the ancient Vitiaz Arc (Rodda, 1994) (Table 1). Although Viti Levu's oldest rocks date back as far as the Late Eocene, it is unlikely the island emerged above sea-level until the Late Oligocene to earliest Miocene (Whelan et al., 1985), and the bulk of the island did not emerge until the Early Pliocene when volcanism and uplift raised the center of Viti Levu above 1000 m (Rodda, 1994; Stratford & Rodda, 2000). More detailed discussions of Fijian geology and its impact on the biogeographic history of Fijian ant fauna are discussed in Lucky and Sarnat (2010) and Sarnat and Moreau (2011).

Scientific exploration and study of the Fijian ant fauna

With the exception of Mann's 1921 monograph, the ant fauna of Fiji has remained relatively neglected since scientific exploration of the archipelago began in the early 19th century. While

Table 1. Approximated geologic ages of formation and emergence above sea-level for islands of the Fijian archipelago arranged from oldest to youngest. The paucity of emergence dates published in the geologic literature is reflected by the number of missing values in the tables.

Island	*Oldest rocks*	*Earliest emergence*	*Reference*
Viti Levu	40 Ma	25 Ma	Rodda, 1994; Whelan, et al., 1985
Lau Group	15–3.0 Ma	—	Rodda, 1994
Yasawa Islands	8.0–6.0 Ma	—	Rodda, 1994
Vanua Levu	8.0–6.5 Ma	4.0 Ma	Kroenke & Yan, 1993; Rodda, 1994; Rodda & Kroenke, 1984
Moala Group	7.0–5.0 Ma	—	Rodda, 1994
Lomaiviti Group	5.0 Ma	—	Rodda, 1994
Kadavu	3.2 Ma	2.0 Ma	Rodda, 1994
Taveuni	0.8 Ma	—	Rodda, 1994; Rodda & Kroenke, 1984

the lack of collections has limited the scientific community's appreciation for the rich endemism of the Fijian ant fauna, it has also served to keep the literature uncluttered by the nomenclatural morass that tends to plague more intensively surveyed regions. A comprehensive review of published reports of Fiji's ant species is compiled in Ward & Wetterer (2006), and a brief history of entomological expeditions in Fiji is presented in Evenhuis and Bickel (2005).

The first significant scientific exploration of the Fijian insect fauna was initiated in the 1860's by Eduard Graeffe, founder of the Godeffroy Museum in Hamburg. Although Graeffe ventured into the then hostile interior of Viti Levu, the only records of his ant collections are those from Ovalau described and catalogued by Gustav Mayr (1866; 1870), which included seven endemic species and the more widespread *Cardiocondyla nuda*. A more detailed account of the expedition in Fiji is available from the recent translation of his personal travelogue (Graeffe, 1986).

The first and only thorough survey of the archipelago's ant fauna, prior to our own, was conducted during a 10-month expedition from 1915–1916 by William M. Mann, who published his monograph on the ants of Fiji in 1921. Soon after receiving his doctorate degree from Harvard University under the tutelage of W. M. Wheeler, Mann was offered a position at the National Museum of Natural History and was sent on a collecting expedition to Fiji and the Solomon Islands. An autobiographical account of these expeditions is published in Ant Hill Odyssey (Mann, 1948). Mann collected on Ovalau, Viti Levu, Taveuni, Vanua Levu, Kadavu, and across the Lau group. He described 50 of Fiji's currently recognized ant species from his own expedition (Mann, 1921), and one from a collection of D. Stoner (Mann, 1925), along with descriptions of the myrmecophilous guest species from Fiji and the Solomons (Mann, 1920).

Subsequent myrmecological work on Fijian ants has been scattered. The young naturalist E. H. Bryan collected *Romblonella liogaster* Santschi and *Camponotus bryani* Santschi during the *Whitney South Seas Expedition* of 1924. W. M. Wheeler (1934) described three additional species of *Camponotus* from Mann's collections that had all been treated as *C. dentatus* Mann, and reviewed the region's ant fauna in his checklist of the ants of Oceania (1935). Edward O. Wilson collected on Viti Levu for several days in December of 1954, and included Fiji as the easternmost extension of Melanesia in a series of regional revisions that included *Hypoponera* (1957), *Leptogenys* (1958b), *Cerapachys* (1959c), and the Odontomachini genera *Odontomachus* and *Anochetus* (1959b). These works on the Melanesian ant fauna led Wilson to develop several influential publications concerning the evolution, ecology and biogeography of island species (MacArthur & Wilson, 1967; Wilson, 1959a;1961). Wilson continued his work in the Pacific with Robert W. Taylor, and together they published *The Ants of Polynesia* (Wilson & Taylor, 1967) which remains an authoritative taxonomic resource for Pacific ants. Taylor, who collected briefly on Viti Levu in 1962, described and reviewed Fijian ant species in the genera *Ponera* (1967) and *Eurhopalothrix* (1980a), and published the rediscovery of *Myrmecina cacabau* Mann (1980b). Noel H. N. Krauss, the Hawaii State exploratory entomologist, made trips to Fiji in the 1970's in search of insects and other arthropods to be used in biological control of agricultural pests in Hawaii (Evenhuis & Bickel, 2005), and collected ant specimens from several different islands. The description of many new species of dacetines from Fiji were included in Barry Bolton's (2000) global revision of the tribe. Following a Soviet expedition to the region, G. M. Dlussky published a series papers (Dlussky, 1993a;b;1994) discussing zoogeography in the southwestern Pacific which included analyses of Fiji's ant fauna, described several species, and synonymized a handful of others. Darren Ward and James Wetterer (2006) published an exhaustive account of literature pertaining to the Fijian ant taxa, new records of ants introduced to the archipelago, and a checklist of 138 species and subspecies of Fijian ants. Ward and Beggs (2007) published a study examining the ecological patterns of primarily invasive ant communities in the Yasawa islands of Fiji. Sarnat revised the species-level taxonomy of two clades endemic to the archipelago—the Fijian *Lordomyrma* (Lucky & Sarnat, 2008; Sarnat, 2006) and the *Pheidole roosevelti* group (Sarnat, 2008b; Sarnat & Moreau, 2011).

Overview of the Fijian ant fauna

Unlike typical remote oceanic islands, Fiji boasts a diverse and distinctive ant fauna with 43 genera, 187 known species, and endemism rates of over 70% (see Appendix B for checklist). A number of Fiji's ant genera (e.g. *Cerapachys*, *Leptogenys*, *Proceratium*) are not expected to occur on a remote oceanic island, and are more typically associated with continental faunas. Study of these taxa in relation to their relatives throughout the Pacific may prove useful for elucidating the biogeographical origin for much of Fiji's enigmatic biota. Additionally, many genera (e.g. *Camponotus*, *Pheidole*, *Strumigenys*, *Lordomyrma*) have diversified within the archipelago into a radiation of endemic species. A detailed study of these genera combined with a broad study of the entire ant fauna is sure to reveal a wealth of information concerning the ecological and evolutionary assembly of Fiji's biota over geologic time.

The following four genera are reported here from Fiji for the first time: *Amblyopone*, *Acropyga*, *Discothyrea* and *Metapone*. One species of *Amblyopone* (Amblyoponinae) was collected from leaf litter sampling, but the species has been collected elsewhere in the Pacific. Two species of *Acropyga* (Formicinae), at least one of which is endemic, were found nesting and foraging in the soil with associated mealybugs. Several species of *Discothyrea* (Proceratiinae) are thus far known only from Fiji. At least one endemic species of *Metapone* (Myrmicinae) was collected from across the archipelago in malaise traps, though no workers have been captured.

The only genus endemic to the archipelago is the elegant *Poecilomyrma* (Myrmicinae). Although the workers are known only from a handful of field collections, the genus is morphologically very diverse in the archipelago. Nests are arboreal and very rarely encountered. Queens are ergatoid and wingless. Whether the genus evolved within Fiji, or whether Fiji is merely the last refuge of an older and more widespread lineage, remains unknown.

METHODS

Specimen data

As of this publication 10,447 unique specimen records based on material we collected and examined are databased and available online through Antweb <www.antweb.org/fiji.jsp> or from the authors upon request. Records include taxonomic, locality, collection and specimen information. The full database records are more detailed than the abbreviated information presented in this publication. Readers seeking additional information are referred to the online database and are encouraged to use the advanced search and mapping tools available on Antweb.

Sources of material

The following review of Fiji's ant fauna is based primarily on the examination of W. M. Mann's (1921) original collections and the material collected from 2002–2008 as part of an archipelago-wide survey. The recent survey of 298 unique localities on twelve islands yielded over 10,000 pinned specimens of ants representing 175 species. Of these species, 45 are believed to be undescribed. There are an additional nine species that are published as occurring in Fiji, but were not recovered during the recent survey. In total there are 187 described and undescribed species known from Fiji representing 43 genera in eight subfamilies. The species that occur on Rotuma are excluded, because the island is not considered here to be geographically part of the Fijian archipelago.

Undescribed and undetermined species

Undescribed species, in addition to potentially previously described species whose names we were unable to determine, are included in the following sections. These morphospecies are designated by code names using the abbreviation "sp." (species), "FJ" (Fiji), and a two-digit number. Images of voucher specimens for each morphospecies are available on Antweb so that future changes in their taxonomic status can be readily updated. Readers are referred to Appendix B for a list of voucher specimens associated with each species.

Material examined

The majority of the material examined was collected from 2002–2008 during several surveys of Fiji's terrestrial arthropod fauna. Ants were collected during an extensive survey of leaf litter arthropods conducted by the Wildlife Conservation Society (WCS) from 2002–2003 (Evenhuis & Bickel, 2005; Sarnat & Moreau, 2011). Malaise trapping of arboreal and winged arthropods was conducted by the Schlinger Fiji Bioinventory of Arthropods and the NSF Fiji Terrestrial Arthropod Survey with the help of WCS and local villagers (Evenhuis & Bickel, 2005). Ants were collected by hand and litter sifting by the authors during numerous expeditions to different islands of the archipelago between 2004–2008. Finally, ant specimens collected from an elevational canopy fogging transect were made available by Hilda Waqa (University of the South Pacific).

The second significant source of specimens is that originally collected by W. M. Mann during his expedition of 1915–1916. These collections include many type specimens that are currently deposited at the USNM (Washington D.C., USA) and the MCZC (Cambridge, MA, USA). Mann designated cotypes, which in most cases can be considered conspecific syntype series as they represent single nest collections from single localities. However, there are several cotype series, such as those originally designated for *Camponotus laminatus*, *Gnamptogenys aterrima*, and *Poecilomyrma senirewae* that include specimens from different species collected at the same locality. In cases such as these, lectotypes are designated.

Although the aforementioned collections represent the most extensive survey undertaken for Fijian ants, a several other collectors have visited the archipelago, and most of that material is deposited at the ANIC (Canberra, Australia). A review of the ANIC material from Fiji (aside from *Lordomyrma* and the *Pheidole roosevelti* group) was beyond the scope of the current study, but we hope this contribution will be useful in organizing those collections.

Collection localities

The collection localities listed for the material examined of each species are organized by island and abbreviated to a unique locality code in the interest of reducing redundancy. The locality codes consist of a locality abbreviation often followed by an elevation record expressed in meters (e.g., 'Mt. Kuitarua 485'). For localities with no elevation records, such as island records (e.g., 'Ovalau') and Mann's old collection records (e.g., 'Suene'), only the locality abbreviation is listed. In cases where multiple localities differ only by latitude and longitude, a lowercase letter is used to distinguish them (e.g., 'Devo Peak 1187 b' vs. 'Devo Peak 1187 c'). A more detailed list of the locality data, including the full locality name, elevation and geocoordinates is presented in Appendix A. For all data associated with a particular locality code, readers can enter the code on Antweb using the site's search tools.

Geographic and ecological distributions

The recent inventory produced extensive distribution data across geography and habitat for many Fijian ant species. For each species, geographic distribution derived from type localities and examined material is plotted on a map of the Fiji archipelago (Plates 1–185). The distribution records for introduced ants reported in Ward and Wetterer (2006) are also included in the locality maps for the respective species. Ecological data was recorded from specimen records and plotted onto charts to show distribution across elevation and habitats (Plates 1 –185). Habitats are classified into five categories of human disturbance, based on ecological field notes and in a few ambiguous cases, corroborated by satellite imagery. Definitions of the classifications are presented at the beginning of the Species Plates section.

Specimen images

Digital color specimen photographs were taken using four primary imaging systems. Images credited to Eli Sarnat were taken using the Auto-Montage software package (Syncroscopy) in combination with a JVC KY-F7U digital camera mounted on a Leica MZ16 dissecting scope. Images credited to Evan Economo were taken using the Image-Pro Plus and Helicon Focus software packages in combination with a Leica DC300 digital camera mounted on a Leica MCZ16 dissecting scope. Images credited to Marek Boroweic were taken using the software package Zerene Stacker in combination with a JVC KY-F7U digital camera mounted on a Leica MZ16 dissecting scope. Images credited to Shannon Hartman and Will Ericson were taken using the LAS v3.8 software package (Leica) in combination with a DVC425 digital camera mounted on a Leica MSV266 dissecting scope. All images credited to Sarnat, Economo, Hartman and Ericson were further edited by Michele Esposito (California Academy of Sciences) in Adobe Photoshop in accordance with the guidelines available in the documentation section of Antweb. Credits for all images are available on Antweb with the exception of images for *Strumigenys daithma*, *Discothyrea* sp. FJ02 and *Discothyrea* sp. FJ04 which were supplied by Steve Shattuck.

The species concept as applied to the Fijian ant fauna

Species are defined here under the premise of the biological species concept (Mayr, 1942) and are presumed to be reproductively isolated from all other populations. Inference of these species boundaries is based on gross comparative morphology and aided by known geographic distributions. Where multiple populations exhibit consistently different phenotypes in sympatry, they are considered different species. Where such populations are not found in sympatry, more caution is taken in ascribing them to separate species. Demarcation of ant species within the Fijian archipelago is challenging due to the geographical complications introduced by the island landscape. When closely related populations occur on multiple islands, and each island population has a distinctive morphology, it is difficult to discern what level of gene flow is occurring among them. Although there is a temptation to split each morphologically distinct island population into its own species, we generally consider them differentiated populations of a single species until more careful examination can be made on a case-by-case basis. Where possible, descriptions of the local variation observed across populations are given in the species accounts. An attempt is also made to reconcile the nomenclature of the Fijian ant fauna with the modern trend in ant systematics towards elimination of subspecies, varieties and races. Although these terms can be useful—and especially so for a fauna that occurs across an island archipelago—they are not supported by the concepts of evolutionary biology and are better treated as distinct but unnamed subpopulations.

Standard measurements and indices

HW *Head width*: maximum width of head not including the eyes.
HL *Head length*: maximum length of head from the posterior margin to the tip of the anterior clypeal margin measured along the midline. For species in which the posterior margin of the head is impressed, the length is measured along the midline to the point perpendicular to the posteriormost extent of the posterolateral lobes.
CI *Cephalic index:* HW/HL.
FL *Metafemur length*: length of metafemur measured along its long axis.
FI *Metafemur length index*: FL/HL.
SL *Scape length*: length of first antennal segment excluding the radicle.
SI *Scape index*: SL/HL.

Abbreviations of depositories

MHNG Natural History Museum of Geneva (Geneva, Switzerland).
USNM United States National Museum of Natural History (Washington DC, USA).
BPBM Bernice Pauahi Bishop Museum (Honolulu, Hawaii, USA).
MCZC Harvard Museum of Comparative Zoology (Cambridge, Massachusetts, USA).
PSWC Philip S. Ward Collection (Davis, California, USA).
FNIC Fiji National Insect Collection.[1]

[1] At the time of publication, the FNIC was not built and not yet ready to accept type material. The BPBM is therefore acting as custodian of holotypes described from material collected during the recent survey, namely from *Lordomyrma* and the *Pheidole roosevelti* group.

TAXONOMIC SYNOPSIS OF THE FIJIAN ANT FAUNA

The following section provides taxonomic synopses, identification characters, and general information about each of the 187 ant species and 43 genera currently recognized in Fiji.

Key to the ant genera of Fiji based on the worker caste.

1 Mesosoma attached to gaster by a 1-segmented waist (abdominal segment 2) (Fig. 3). 2

– Mesosoma attached to gaster by a 2-segmented waist (abdominal segments 2+3) (Fig. 4, Fig. 5)........ 25

2(1) Tip of gaster armed with sting (Fig. 6). Gaster broken in outline by a distinct constriction between first and second gaster segments (abdominal segments 3+4) (Fig. 9); if no constriction present then the gaster is curled beneath itself so that the sting points anteriorly........ 3

– Tip of gaster armed with an acidopore (Fig. 7) or ventral slit (Fig. 8), but never with a sting. Gaster uniform in outline without a constriction between first and second gaster segments (abdominal segments 3+4) (Fig. 10) 14

3(2) Gaster strongly arched so that it curls beneath itself and the sting points anteriorly (Fig. 11). Eyes minute (< 4 facets). Rarely encountered leaf litter inhabitants (Proceratiinae)........ 4

– Gaster weakly arched so that the sting points posteriorly (Fig. 12); or if the gaster is strongly arched then the eyes are not minute (> 4 facets)........ 5

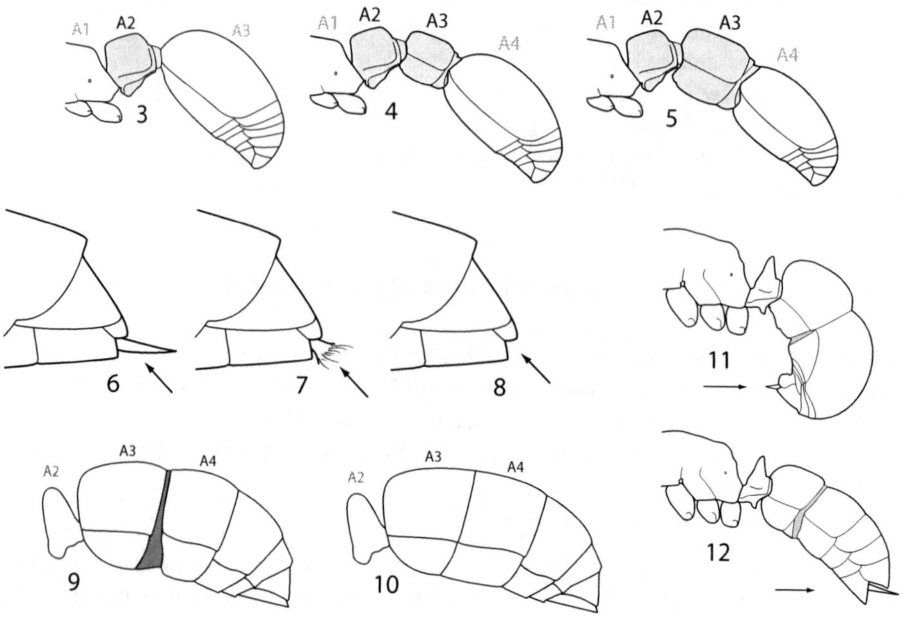

4(3) Last antennal segment large and bulbous (Fig. 13), approaching length of the remaining funicular segments combined. Clypeus projecting strongly over mandibles. Head and body lacking long flexous hairs. Antenna with fewer than 10 segments. .... ***Discothyrea***

– Last antennal segment not large and bulbous, distinctly shorter than the length of the remaining funicular segments combined (Fig. 14). Clypeus not projecting strongly over mandibles. Head and body covered with long flexous hairs. Antenna 12-segmented .... ***Proceratium***

5(3) Petiole broadly attached to gaster and lacking a posterior face (Fig. 15). Anterior margin of clypeus with a row of small peg-like teeth (Fig. 17). Eyes minute (< 4 facets) or absent. Rarely encountered leaf litter inhabitants (Amblyoponinae)........................ 6

– Petiole with a distinct posterior face and attached to gaster by a narrow constriction (Fig. 16). Anterior margin of clypeus lacking a row of small peg-like teeth. Eyes of various sizes 7

6(5) Mandibles long and linear with 7 teeth (1 apical, 1 sub-apical, and 5 basal) (Fig. 18). Head square-shaped, approximately as long as broad. Subpetiolar process narrowly developed and lacking fenestra (Fig. 20) .... ***Amblyopone***

– Mandibles tridentate (Fig. 19). Head rectangular, distinctly longer than broad. Subpetiolar process broadly developed and with fenestra present (Fig. 21). .... ***Prionopelta***

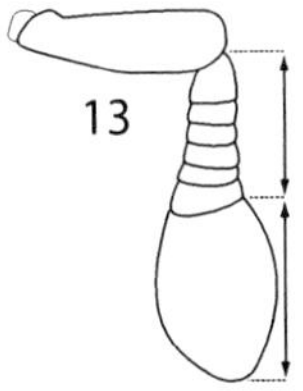

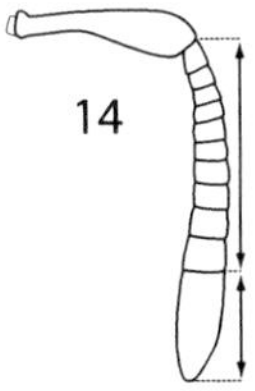

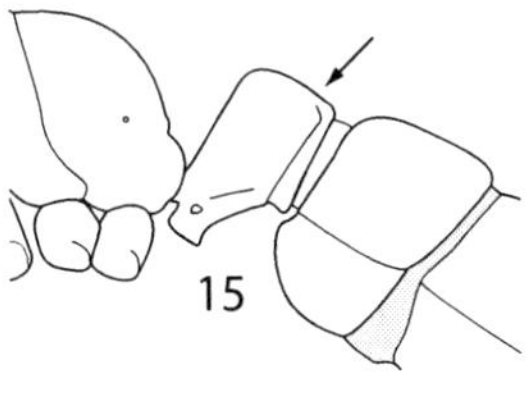

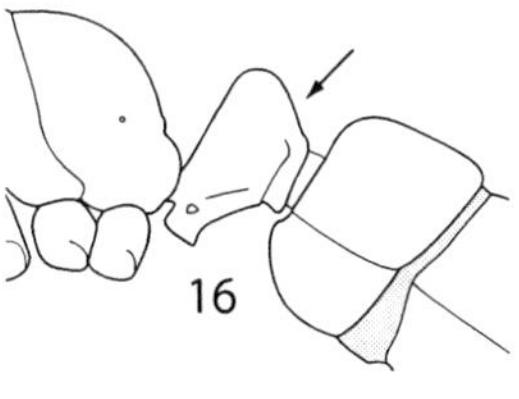

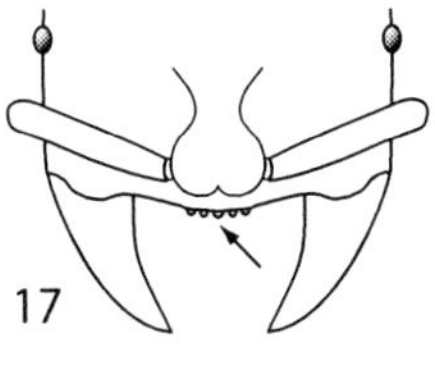

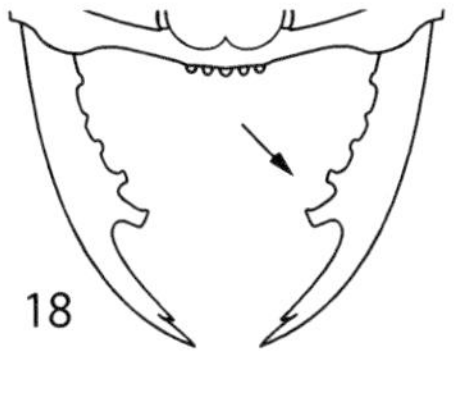

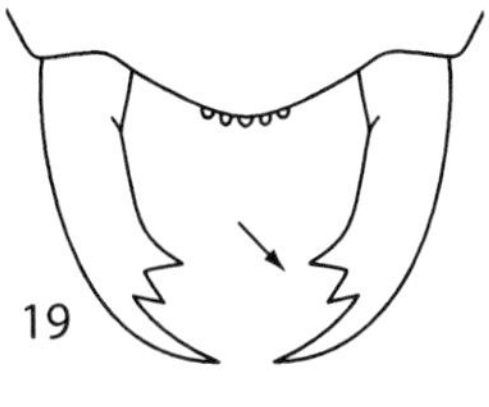

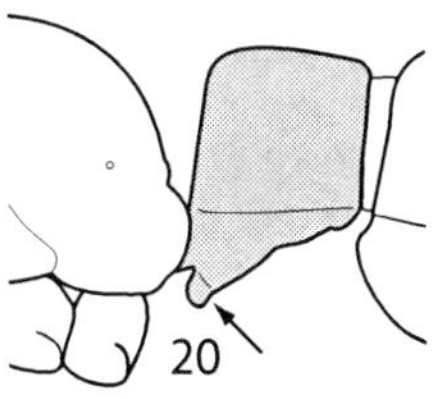

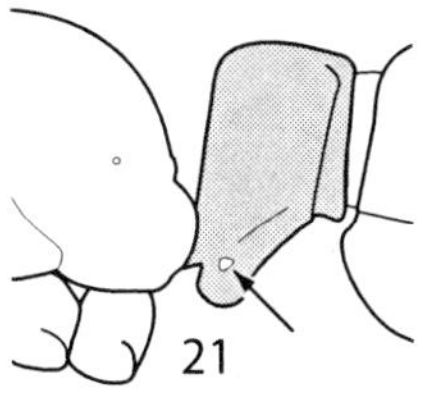

7(5) Head covered in large, widely spaced, piligerous foveae (Fig. 22). Mesosoma and gaster smooth and shiny. Subpetiolar process approximately half node height (Fig. 23). Eyes large (> 10 facets) and located at head midline (Fig. 25) (Ectatomminae). .... ***Gnamptogenys***

– Head variously sculptured, but if foveate then mesosoma and gaster not smooth and shiny. Subpetiolar process distinctly less than half node height (Fig. 24). Eyes variously sized but located distinctly anterior to head midline (Fig. 26) (Ponerinae). 8

8(7) Mandibles linear and armed with an apical fork (Fig. 27a); mandibles inserted towards the middle of the anterior head margin (Fig. 27b). Head constricted posterior to eye level. Clypeus with anterior margin flat to convex but never forming a distinct triangle that projects anteriorly beyond the base of the mandibles.......... 9

– Mandibles triangular (Fig. 28a), or if linear then lacking an apical fork (Fig. 29a); mandibles inserted towards lateral corners of the of the anterior head margin (Fig. 28b, Fig. 29b). Head not constricted posterior to eye level. Clypeus variously shaped 10

9(8) Posterior of head interrupted by a median groove (Fig. 31). Petiolar node armed with an apical spine (Fig. 33).......... ***Odontomachus***

– Posterior of head not interrupted by a median groove (Fig. 32). Petiolar node convex and lacking an apical spine (Fig. 34) .... ***Anochetus***

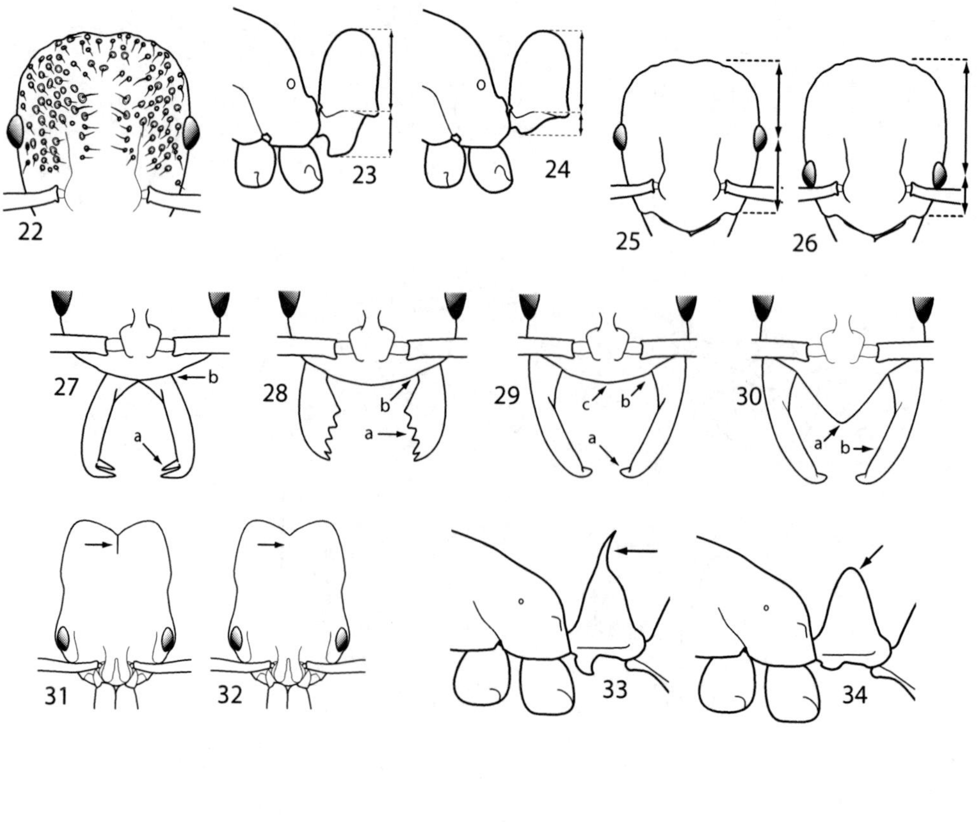

10(8) Clypeus distinctly triangular and projects anteriorly well beyond the base of the mandibles (Fig. 30a). Mandibles elongate and edentate (Fig. 30b). Hind legs with pectinate tarsal claws (Fig. 35). .... ***Leptogenys***

– Clypeus with anterior margin flat to convex but never forming a distinct triangle that projects anteriorly beyond the base of the mandibles (Fig. 29c). Mandibles not elongated and with at least 5 teeth or denticles (Fig. 28a). Hind legs lacking pectinate tarsal claws (Fig. 36) 11

11(10) Petiolar node rectangular in profile (approximately as tall as long) and posterior face distinctly concave (Fig. 37a). Frontal lobes separated by a longitudinal median groove (Fig. 39), not by the posterior extension of the clypeus. Hind tibia with two pectinate spurs (Fig. 41)... ***Platythyrea***

– Petiolar node cuneiform (wedge-shaped) in profile, distinctly taller than long with posterior face convex to flat (Fig. 38a). Frontal lobes separated by the posterior extension of the clypeus (Fig. 40), not by a longitudinal median groove. Hind tibia with either one pectinate spur (Fig. 42) or one pectinate spur (Fig. 43a) and one simple spur (Fig. 43b) 12

12(11) Mandibles with 5 well-developed teeth (Fig. 44). Hind tibia with one pectinate spur and one simple spur (Fig. 43). Eyes minute (< 4 facets) .... ***Pachycondyla***

– Mandibles with more than 5 teeth and denticles (Fig. 45). Hind tibia with pectinate spur but lacking simple spur (Fig. 42). Eyes of various sizes 13

13(12) Subpetiolar process with fenestra (Fig. 46)... ***Ponera***

– Subpetiolar process lacking fenestra (Fig. 47) .... ***Hypoponera***

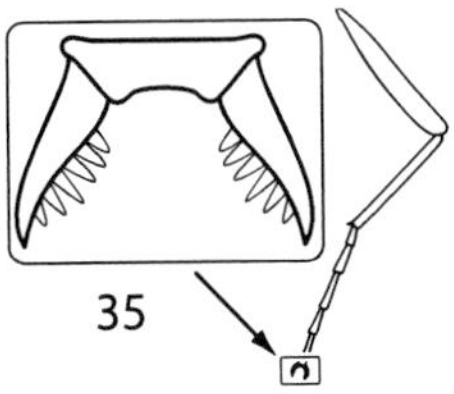

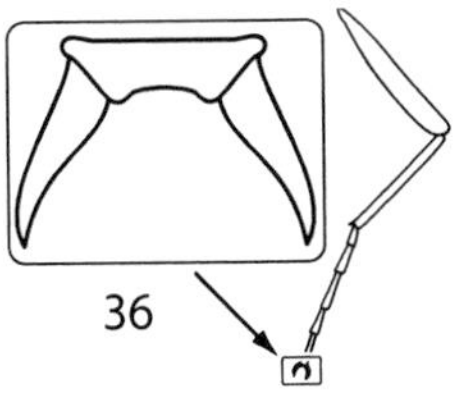

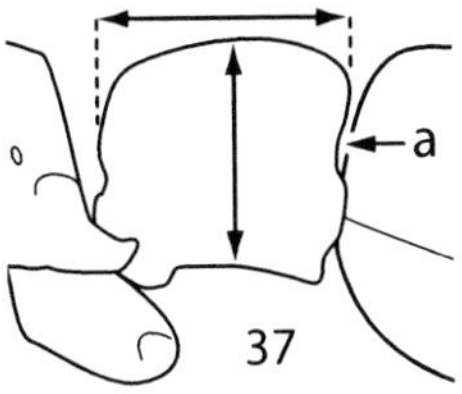

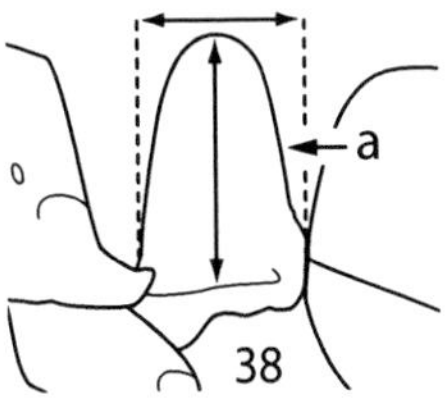

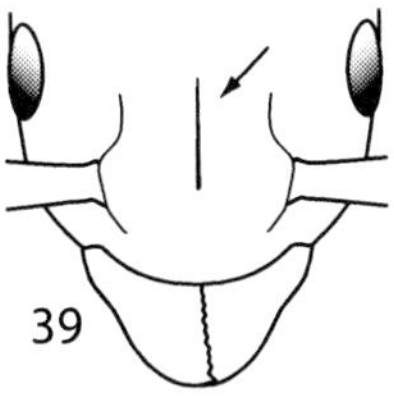

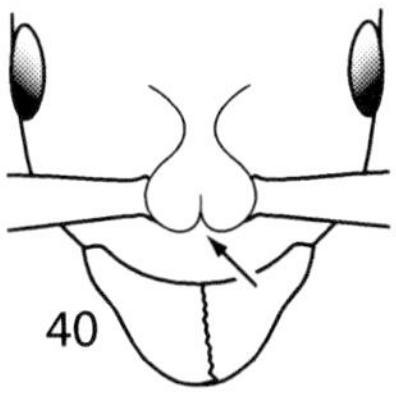

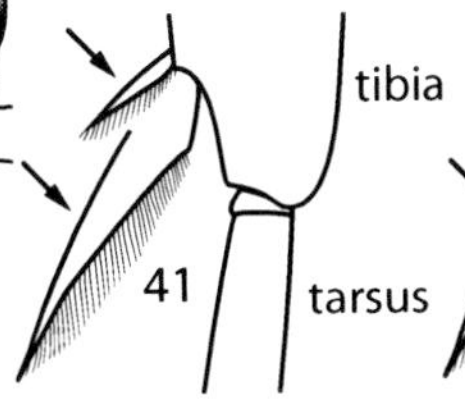

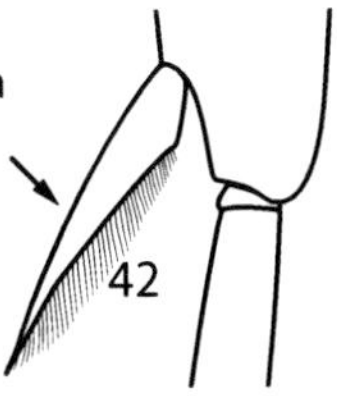

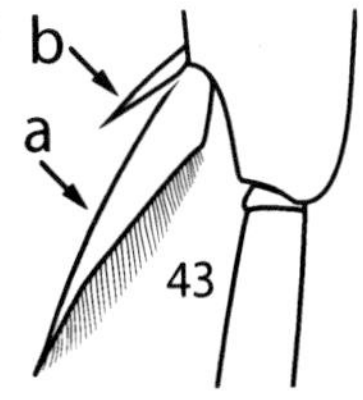

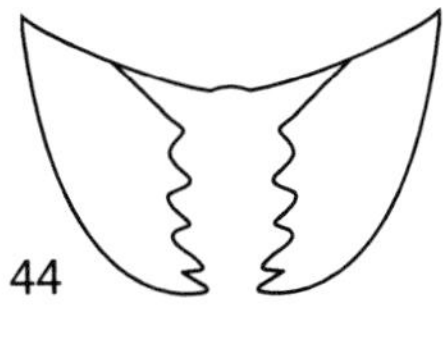

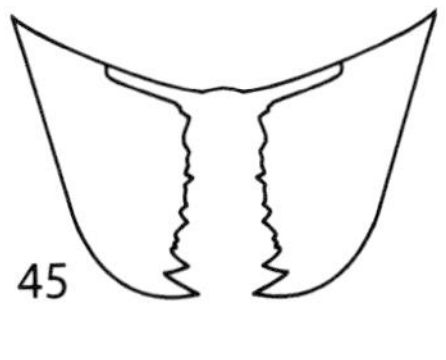

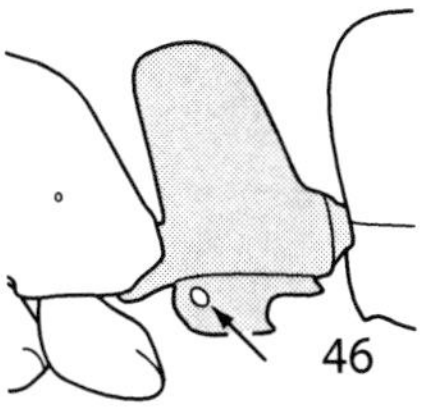

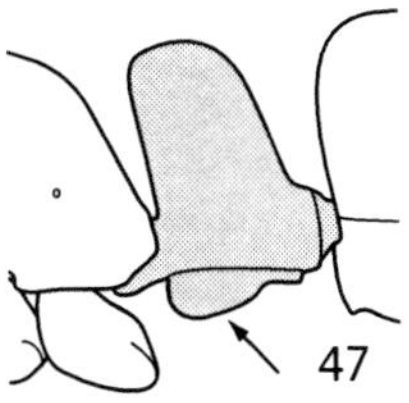

14 (2) Tip of gaster armed with a circular opening (acidopore) that is often fringed with short hairs (Fig. 7) (Formicinae) 15

– Tip of gaster armed with a ventral slit that is never fringed with short hairs (Dolichoderinae) (Fig. 8) 21

15(14) Legs and antennae very elongated, with antennal scapes 1.5 times head length (Fig. 48). Eyes distinctly break head outline (Fig. 50) 16

– Legs and antennae of various proportions, but antennal scapes never 1.5 times head length (Fig. 49). Eyes break outline of head, or do not break outline of head (Fig. 51) 17

16(15) Dorsum of mesosoma with long thick hairs (Fig. 52). Antenna 12-segmented (Fig. 54). Dark brown to black .... ***Paratrechina***

– Dorsum of mesosoma lacking erect hairs (Fig. 53). Antenna 11-segmented. Yellow to reddish-brown .... ***Anoplolepis***

17(15) Eyes minute (< 4 facets) (Fig. 55). Antenna 8- or 9-segmented. Antennal scapes do not reach posterior head margin (Fig. 57). Pale yellow. Rarely encountered leaf-litter species .... ***Acropyga***

– Eyes variously sized, but always composed of more than 4 facets (Fig. 56). Antenna 11- or 12-segmented. Antennal scapes do not reach (Fig. 57), just reach, or extend beyond (Fig. 58) posterior margin of head. Variously colored 18

18(17) Head, mesosoma and gaster with pairs of long thick hairs (Fig. 52). Antenna 12-segmented. Mesosoma impressed between promesonotum and propodeum (Fig. 59). Eyes never breaking outline of head (Fig. 51) 19

– Head, mesosoma and gaster lacking pairs of long thick hairs; if long hairs are abundant then they are flexous, not thick (Fig. 53). Antenna 11- or 12-segmented. Profile of mesosoma forming an unbroken, continuously convex curve (Fig. 60), or impressed between promesonotum and propodeum (Fig. 59). Eyes variously located relative to head outline 20

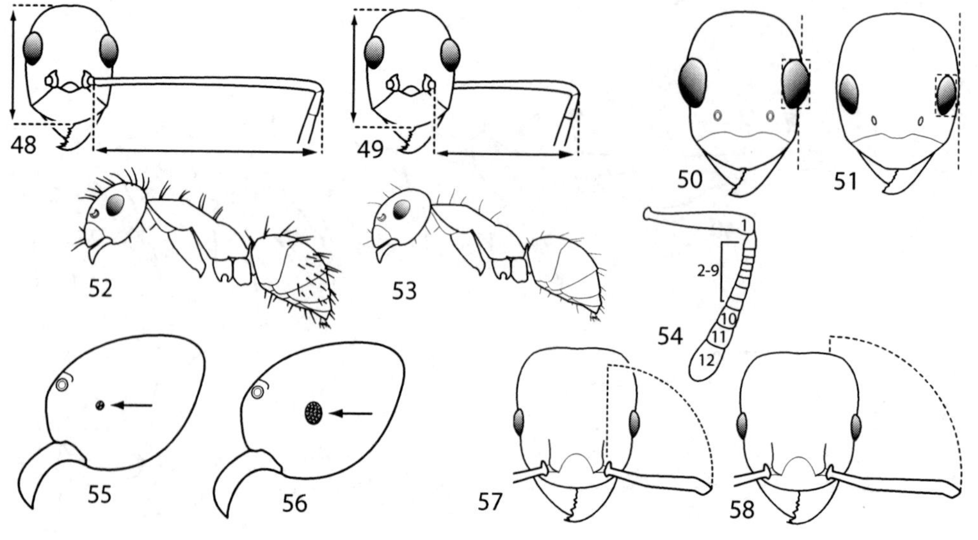

48 49 50 51 52 53 54 55 56 57 58

19(18) Antennal scapes with erect hairs (Fig. 61) .... ***Nylanderia***

– Antennal scapes lacking erect hairs (Fig. 62)..................................... ***Paraparatrechina***

20 (18) Antenna 12-segmented. Metapleuron with a distinct gland orifice fringed by short hairs Fig. 63). Head longer than broad (Fig. 65), but if not then cuticle is thick and propodeum armed with spines (Fig. 67). Medium to large species (HW > 0.8 mm). .... ***Camponotus***

– Antenna 11-segmented. Metapleuron lacking a distinct gland orifice fringed by short hairs (Fig. 64). Head as broad as long (Fig. 66). Minute species (HW < 0.4 mm). .... ***Plagiolepis***

21(14) Petiolar node greatly reduced or absent, the anterior face absent or at most indistinct (Fig. 68). Propodeum with dorsal surface distinctly shorter than posterior face (Fig. 70) 22

– Petiolar node present and with distinct anterior and posterior faces (Fig. 69). Propodeum with dorsal surface longer or approximately equal to posterior face (Fig. 71)...... 23

22(21) Gaster with 4 plates on its dorsal surface (fifth gastral tergite ventral) (Fig. 72a). First gastral segment lacking erect hairs (Fig. 72b) .... ***Tapinoma***

– Gaster with 5 plates on its dorsal surface (fifth gastral tergite dorsal) (Fig. 73a). First gastral segment with erect hairs (Fig. 73b)...................................... ***Technomyrmex***

23(21) Propodeal dorsum with posterior protrusion (Fig. 74a). Propodeal declivity concave (Fig. 74b). Dorsum of mesosoma lacking erect hairs .... ***Ochetellus***

– Propodeal dorsum lacking a posteriorly projecting protrusion (Fig. 75a). Propodeal declivity flat to convex (Fig. 75b). Dorsum of mesosoma with erect hairs 24

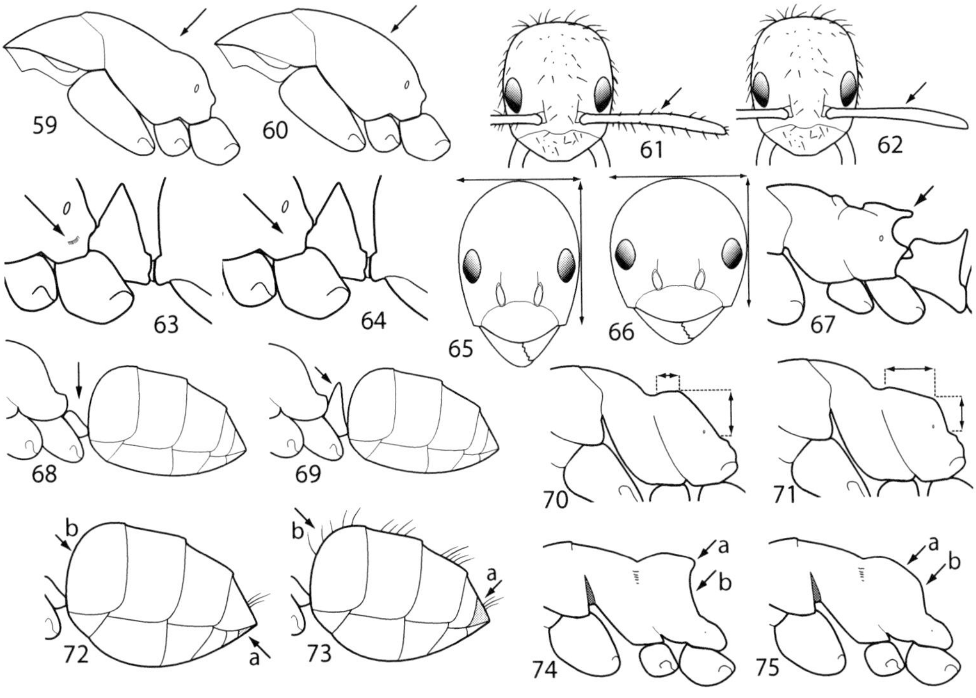

24(23) Head heart-shaped with strongly concave posterior margin (Fig. 76a). Eyes located distinctly below midline of face (Fig. 76b). Anterior clypeal margin weakly convex (Fig. 78) and never with 1 median and 2 lateral rounded projections .... ***Philidris***

– Head ovoid, not heart-shaped, with posterior margin flat to weakly convex (Fig. 78a). Eyes located at or above, but never below midline of face (Fig. 78b). Anterior clypeal margin with 1 median and 2 lateral rounded projections (Fig. 79). .... ***Iridomyrmex***

25(1) Antennal club 1-segmented (Fig. 80). Antennal insertions never covered by frontal lobes (Fig. 85). Pygidium flattened and armed with a row of small peg-like teeth (Fig. 87). (Cerapachyinae). .... ***Cerapachys***

– Antennal club variously produced (Fig. 81–83) or indistinct (Fig. 84), but never composed of only one segment. Antennal insertions at least partly covered by frontal lobes (Fig. 86). Pygidium convex and not armed with a row of small peg-like teeth (Fig. 88). (Myrmicinae). 26

26(25) Head shape triangular with broad posterior margin tapering to narrow anterior margin (Fig. 89). Antenna 6- or 7-segmented (Fig. 90–91) 27

– Head shape rectangular to ovoid (Fig. 90), but never triangular. Antenna 11- or 12-segmented (Fig. 54) 29

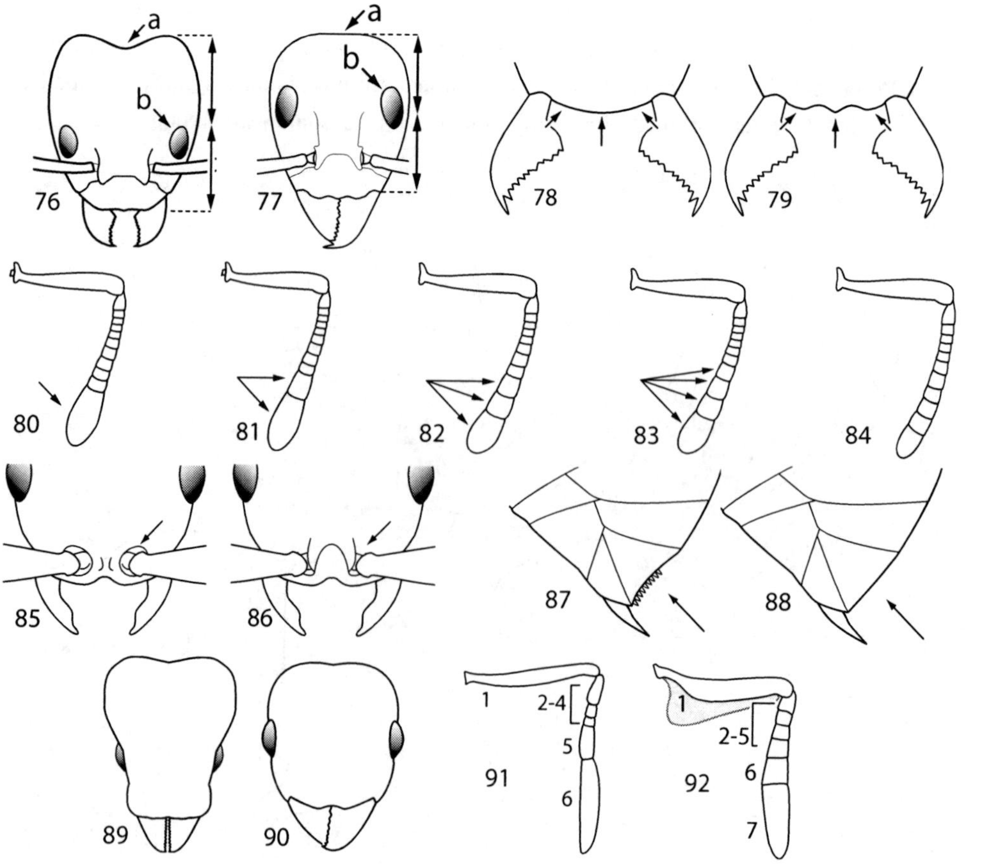

27(26) Eyes located on upper margin of antennal scrobes (Fig. 93). Antenna 7-segmented (Fig. 92). Waist lacking spongiform tissue (Fig. 94). Mandibles triangular (Fig. 95a). .... ***Eurhopalothrix***

– Eyes located on lower margin of antennal scrobes (Fig. 96). Antenna 6-segmented (Fig. 90). Waist at least partially covered by spongiform tissue (Fig. 97). Mandibles triangular (Fig. 95) or linear (Fig. 98a) 28

28(27) Mandibles linear and armed with an apical fork (Fig. 98a, b) .... ***Strumigenys***

– Mandibles triangular and armed with denticles (Fig. 95a, b) .... ***Pyramica***

29(26) Petiole with distinct peduncle (Fig. 99a) and node (Fig. 99b) 30

– Petiole lacking a distinct peduncle (Fig. 100a) and lacking a distinct node (Fig. 100b). 40

30(29) Antennal club 2-segmented (Fig. 81). Propodeum unarmed (Fig. 101), or armed at most with denticles (Fig. 102), but never with substantial teeth or spines (Figs. 103–104). Antenna 9-, 10-, or 12-segmented 31

– Antennal club 3-segmented (Fig. 82). Propodeum variously armed, but usually with distinct teeth (Fig. 103) or spines (Fig. 104). Antenna 11- or 12-segmented 33

31(30) Distinct basal tooth present near base of mandible that is separated from masticatory teeth by a considerable distance (Fig. 105). Antenna 12-segmented. Petiole weakly pedunculate. Propodeum armed with denticles (Fig. 102) .... ***Adelomyrmex***

– Basal tooth absent (Fig. 106). Antenna 9- or 10-segmented. Petiole with distinct peduncle. Propodeum either unarmed (Fig. 100) or armed with small denticles (Fig. 101) 32

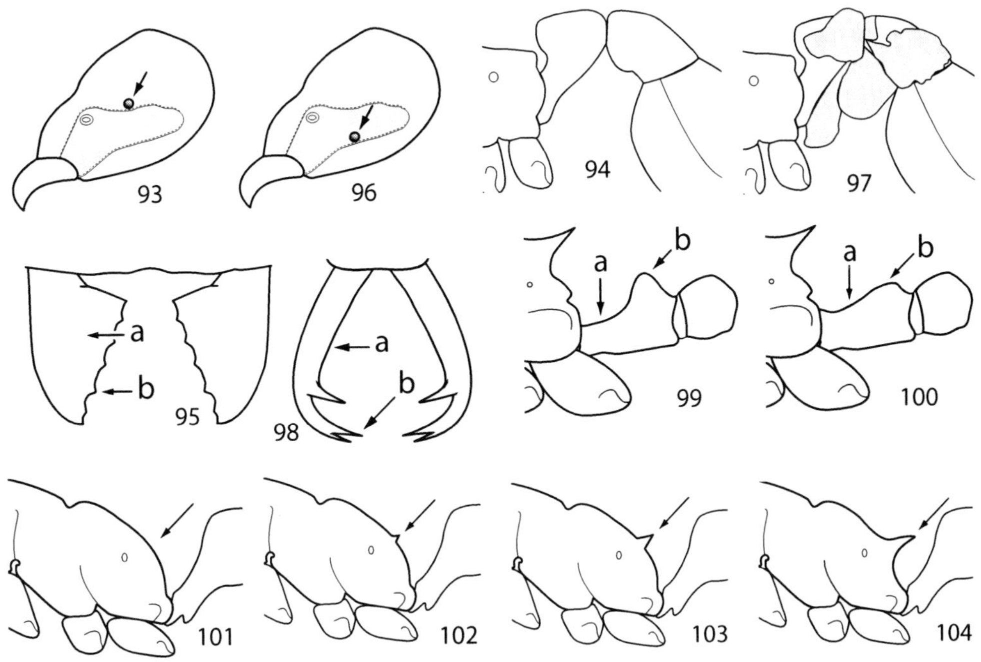

32(31) Antenna 10-segmented. Propodeum unarmed (Fig. 101). Workers either monomorphic or continuously polymorphic, but not never bimorphic with discrete minor and major subcastes. Minute to medium sized .... ***Solenopsis***

– Antenna 9-segmented. Propodeum armed with small denticles (Fig. 102). Workers bimorphic with discrete minor and major subcastes. Minor workers minute. .... ***Carebara***

33(30) Postpetiole swollen, in dorsal view wider than long and much broader than petiole (Fig. 107). Mesosoma lacking erect hairs (Fig. 109). Antennal scapes not reaching posterior margin of head (Fig. 57). Antennal scrobes lacking (Fig. 111). .... ***Cardiocondyla***

– Postpetiole not swollen relative to petiole (Fig. 108). Mesosoma with at least some erect hairs (Fig. 110). Antennal scapes do not reach (Fig. 57) or surpass (Fig. 58) posterior margin of head. Antennal scrobes lacking (Fig. 111) or present (Fig. 112)........... 34

34(33) Anterior clypeal margin armed with 3 broad and blunt teeth (Fig. 113). Mandible with 4 distinct teeth on the masticatory margin and 1 distinct basal tooth on the basal margin (Fig. 115). Antenna 11-segmented. Face circular (Fig. 66).... ***Pristomyrmex***

– Anterior clypeal margin unarmed (Fig. 114). Mandibles with more than 4 teeth on the masticatory margin and never with a distinct tooth on the basal margin (Fig. 116). Antenna 12-segmented. Face ovoid (Fig. 65).. 35

35(34) Propodeum either unarmed (Fig. 101) or armed with small denticles (Fig. 102); if small denticles are present then the eyes are minute (< 4 facets) (Fig. 55). Antennal scrobes absent (Fig. 111)... ***Monomorium***

– Propodeum armed with spines (Fig. 104), teeth (Fig. 103) or denticles (Fig. 102); if armed with denticles then the eyes are composed of more than 4 facets (Fig. 56). Antennal scrobes present (Fig. 112) or absent (Fig. 111)..................................... 36

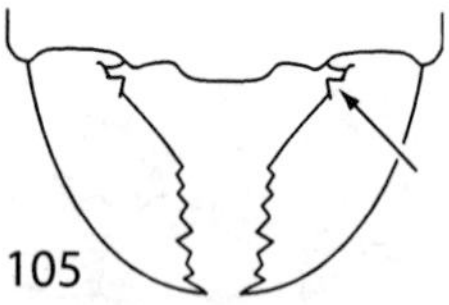

105 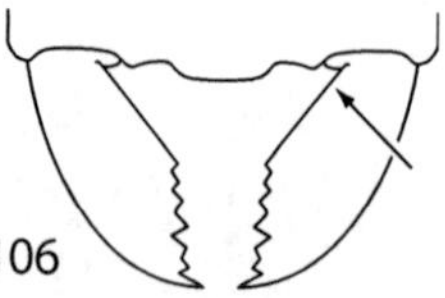 106 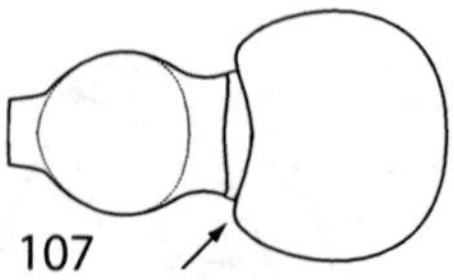 107 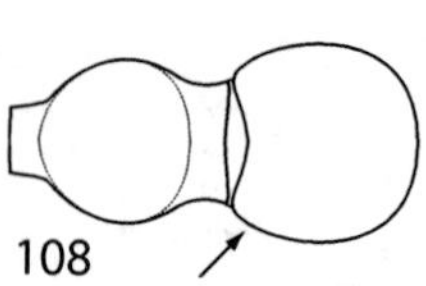 108

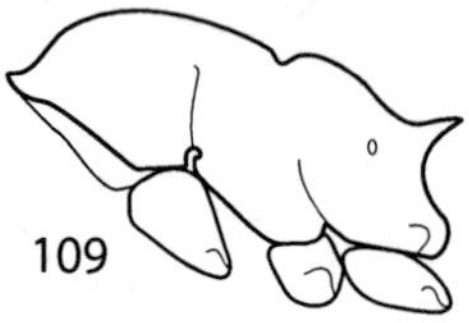

109 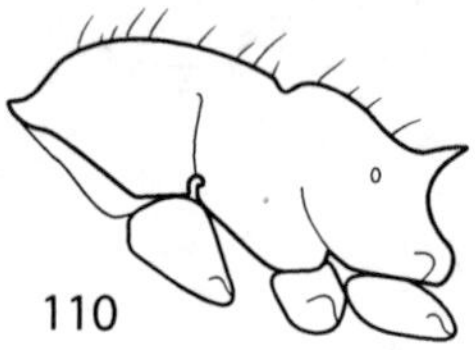 110 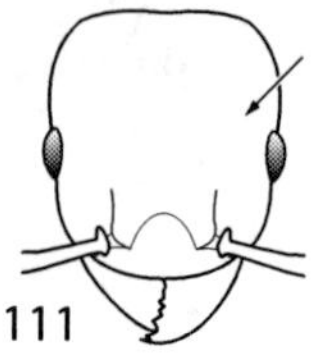 111 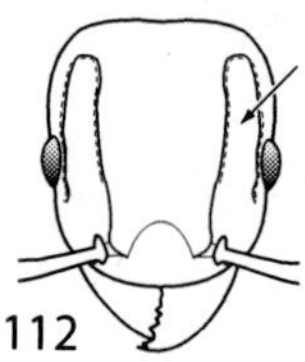 112

36(35) Propodeal lobes (Fig. 117a) modified into elongated spines that are longer than propodeal spines (Fig. 117b). Petiolar peduncle distinctly elongated (Fig. 117c). Head and mesosoma sculptured with broad rugae and deep furrows. Often bicolored red and black, but occasionally uniform black .... ***Poecilomyrma***

– Propodeal lobes (Fig. 118a) not modified into elongated spines that are longer than propodeal spines (Fig. 118b). Petiolar peduncle variously shaped, but not distinctly elongated (Fig. 118c). Head and mesosoma variously sculptured. Variously colored... 37

37(36) Frontal lobes relatively close together so that the posteromedian portion of the clypeus, where it projects between the frontal lobes, is at most only slightly broader than one of the lobes (Fig. 119). Mesosoma with depression distinctly separating promesonotum from propodeum (Fig. 121). Workers bimorphic with discrete minor and major subcastes, and the antennal scapes of the minor workers reach or exceed posterior margin of head (Fig. 58). Antennal insertions never surrounded by a raised sharp-edged ridge (Fig. 124).. ***Pheidole***

– Frontal lobes relatively far apart so that the posteromedian portion of the clypeus, where it projects between the frontal lobes, is much broader than one of the lobes (Fig. 120). Mesosoma with (Fig. 121) or lacking (Fig. 122) depression separating promesonotum from propodeum. Workers monomorphic. Antennal scapes do not reach the posterior margin of the head (Fig. 57). Raised sharp-edged ridge surrounding antennal insertions either absent (Fig. 124) or present (Fig. 123)..... 38

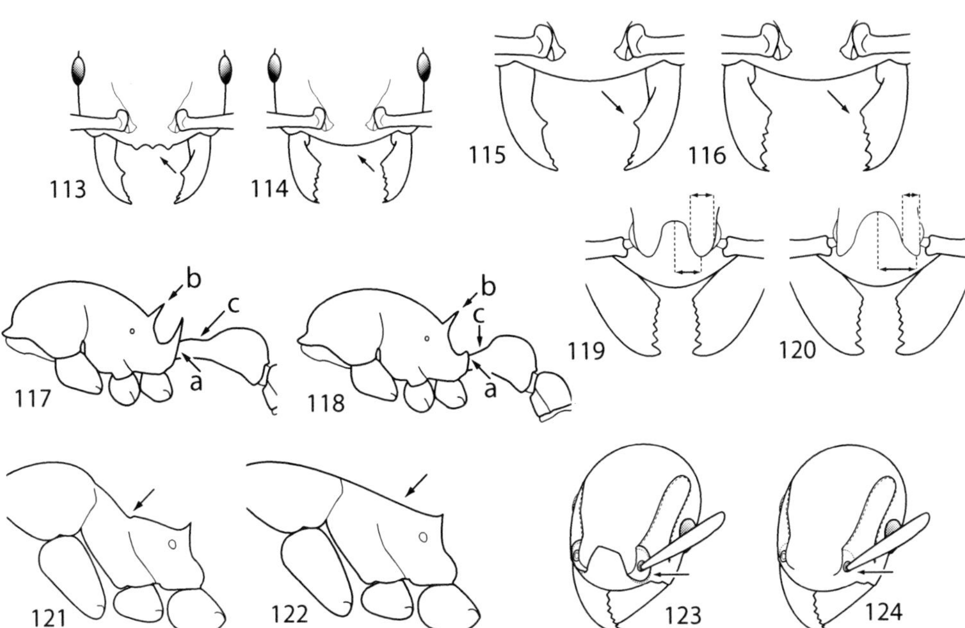

38(37) Antennal insertions surrounded by a raised sharp-edged ridge (Fig. 123). Tip of sting with a triangular to pendant-shaped extension (Fig. 125). Mesosoma evenly convex without depression separating promesonotum from propodeum (Fig. 122). Propodeal spines distinctly longer than diameter of the propodeal spiracle (Fig. 127). Antennal scrobes distinct (Fig. 112) .... ***Tetramorium***

– Antennal insertions not surrounded by a raised sharp-edged ridge (Fig. 124). Tip of sting lacking a triangular to pendant-shaped extension (Fig. 126). Mesosoma with (Fig. 121) or lacking (Fig. 122) depression separating promesonotum from propodeum. Propodeal spines of various lengths. Antennal scrobes present (Fig. 112) or absent (Fig. 111).. 39

39(38) Propodeal spines distinctly longer than diameter of the propodeal spiracle (Fig. 127). Mesosoma with depression separating promesonotum from propodeum (Fig. 121). Antennal scrobes usually distinct (Fig. 112). Face variously sculptured. .... ***Lordomyrma***

– Propodeal spines short, approximately equal to diameter of the propodeal spiracle (Fig. 128). Mesosoma evenly convex without depression separating promesonotum from propodeum (Fig. 122). Antennal scrobes absent (Fig. 111). Face strongly rugoreticulate .... ***Rogeria***

40(29) Petiole with large anteroventral subpetiolar process (Fig. 129). Propodeum either unarmed (Fig. 101) or armed at most with one pair of small denticles (Fig. 102). 41

– Petiole lacking large anteroventral subpetiolar process (Fig. 130). Propodeum armed with at least one pair of spines (Fig. 103–104)... 42

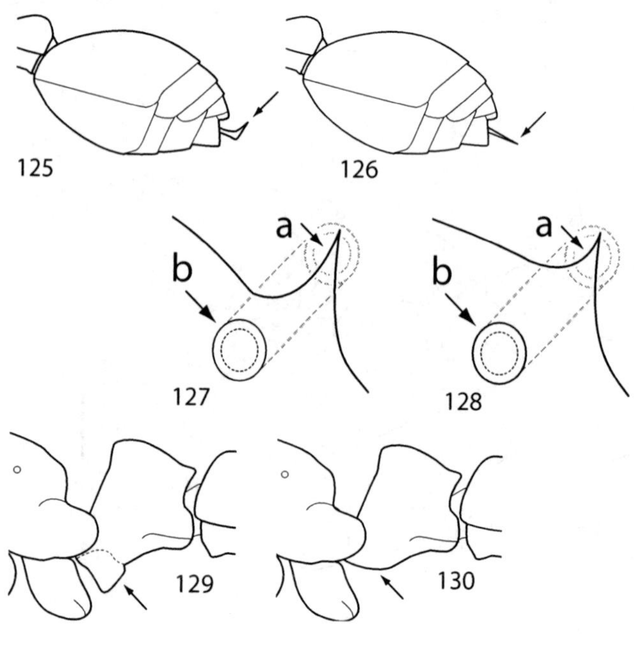

41(40) Anterior margin of clypeus with a rectangular projection extending over base of mandibles (Fig. 131). Antennal scrobes deeply impressed (Fig. 112). Antenna 11-segmented. Body highly modified into heavily armored cylindrical shape. Very rarely encountered..***Metapone***

– Anterior margin of clypeus evenly convex and lacking a rectangular projection extending over base of mandibles (Fig. 132). Antennal scrobes absent (Fig. 111). Antenna 12-segmented. Body not highly modified into heavily armored cylindrical shape. Uncommonly encountered..***Vollenhovia***

42(40) Propodeum armed with two pairs of spines (Fig. 133). Sides of head with a carinate ridge extending below eye-level from mandibular insertions to posterolateral head margin (Fig. 135). ....***Myrmecina***

– Propodeum armed with one pairs of spines (Fig. 134). Sides of head lacking a carinate ridge extending below eye-level from mandibular insertions to posterolateral head margin (Fig. 136). ....***Romblonella***

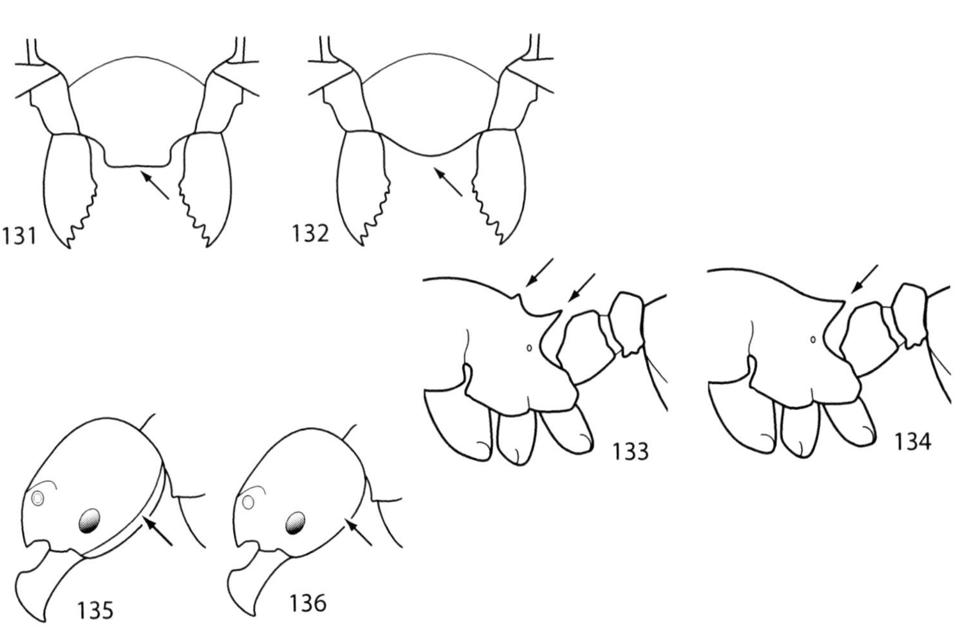

Subfamily Amblyoponinae

Brown (1960), in a treatise on the subfamily, described the members of this group as obligatory predators of arthropods and predominately cryptic foragers in leaf litter and rotting logs. The amblyoponines are a moisture-loving group that is most commonly encountered in the tropics, though they are also known from wet habitats in temperate zones. There are two genera of Amblyoponinae present in Fiji, each represented by a single species.

Key to the Amblyoponinae workers of Fiji.

1 Mandibles short and narrow with three teeth. Head distinctly longer than broad. Eyes present as a single facet. Subpetiolar process broadly developed and with fenestra present. .... ***Prionopelta kraepelini***

– Mandibles very long and linear with six or more teeth (preapical and basal teeth small to minute). Head as long as broad. Eyes entirely absent. Subpetiolar process narrowly developed and lacking fenestra. ....***Amblyopone zwaluwenburgi***

Genus Amblyopone

Diagnosis of worker caste in Fiji. Antenna 11-segmented, last antennal segment not large and bulbous, distinctly shorter than remaining funicular segments combined. Head square-shaped, approximately as long as broad. Eyes entirely absent. Anterior margin of clypeus flat and armed with five conical peg-like teeth. Mandibles long and linear with seven teeth (one apical, one sub-apical, and five basal). Waist 1-segmented; petiole (abdominal segment 2) broadly attached to the first gastral segment (abdominal segment 3). Subpetiolar process narrowly developed; lacking fenestra. Gaster armed with sting; distinct constriction between abdominal segments 3+4; tip pointing posteriorly or straight down, but never anteriorly.

Amblyopone is represented in Fiji by a single species, *A. zwaluwenburgi*. *Amblyopone* species are distributed across the world both in the tropics and to a lesser extent in the temperate regions. Most species of *Amblyopone* are known from moist, forested areas where they nest in rotten wood, leaf litter, or in the soil under stones or logs, but some also occur in arid habitats across the globe (Brown, 1960). All species are believed to feed exclusively on arthropods, and many are known to specialize on chilopods.

Amblyopone zwaluwenburgi (Williams)

(Plate 1)

Stigmatomma (*Fulakora*) *zwaluwenburgi* Williams, 1946: 639; worker described. Type locality: HAWAII [not examined]. Holotype: worker [headless] (BPBM, not examined). Paratype: 1 worker (BPBM, not examined). Combined in *Amblyopone* (Brown, 1960: 122); figured and discussed (Wilson & Taylor, 1967: 19, fig. 3); discussed (Taylor, 1978); redescribed and figured (Onoyama, 1999: 158).

Amblyopone zwaluwenburgi is a small brownish-yellow species with a square head, no eye facets and long slender mandibles. The mandibles of the examined specimen appear armed with a long pointed apical tooth, a minute subapical tooth, a strong third tooth and three minute basal teeth. In his redescription of the species, based on a paratype worker, Onoyama (1999) mentioned that there are four teeth basal to the third.

The single species of *Amblyopone* known from Fiji is represented by a single specimen collected from leaf litter. *Amblyopone zwaluwenburgi* is otherwise reported only from Christmas Island (Framenau & Thomas, 2008; Taylor, 1990) and the type locality in Hawaii. Wilson and Taylor (1967) concurred with Brown's (1960) prediction that, "...the species has been introduced into Hawaii from Melanesia or the East Indies," and the discovery of the species in Fiji serves as some validation. We do not know whether the species is native to Fiji or was recently introduced from elsewhere. However, the collection locality of the single specimen in Fiji is from a coastal area with abundant populations of introduced and widespread species.

Material examined. **Viti Levu:** Naboutini 300.

Genus Prionopelta

Diagnosis of worker caste in Fiji. Antenna 12-segmented, last antennal segment not large and bulbous, distinctly shorter than remaining funicular segments combined. Head rectangular, head length longer than head width. Eyes minute (≤3 facets) but distinct. Anterior margin of clypeus denticulate. Mandibles tridentate. Waist 1-segmented; petiole (abdominal segment 2) broadly attached to the first gastral segment (abdominal segment 3). Subpetiolar process broadly developed; fenestra present. Gaster armed with sting; distinct constriction between abdominal segments 3+4; tip pointing posteriorly or straight down, but never anteriorly.

Prionopelta was only recently reported from Fiji (Ward & Wetterer, 2006). Six of the world's 15 species are currently known to occur in Melanesia, where they are widespread but rare, and largely restricted to rainforest habitats where they nest in the soil, leaf litter or rotting logs (Shattuck, 2008).

Prionopelta kraepelini Forel

(Plate 2)

Prionopelta kraepelini Forel, 1905: 3; worker, queen described. Type locality: INDONESIA [not examined]. Syntypes: workers, queen (MHNG, not examined). Discussed (Brown, 1960: 122). In Polynesia (Wilson & Taylor, 1967: 19, fig. 4); in Micronesia (Clouse, 2007a); in Indo-Pacific revision (Shattuck, 2008).

Prionopelta kraepelini is a pale yellow shiny species with the face covered densely by small punctures, the pronotum covered more lightly by small punctures, and a mostly smooth propodeal surface marked occasionally by larger punctures. Short, suberect pilosity is abundant.

Shattuck (2008) reported *P. kraepelini* as being one of the most widespread species in the genus, citing its range as extending from Sumatra and Peninsular Malaysia east through the Philippines and Micronesia to Samoa. The discovery of the species in Fiji helps reconcile the geographical disjunction represented by the previously outlying population found in Samoa. The possibility that the species was brought to Polynesia by humans, as suggested by Wilson and Taylor (1967), is supported by the collection of this species in Fiji beneath a roadside stone.

Material examined. **Viti Levu:** Colo-i-Suva Forest Park 220.

Subfamily Cerapachyinae

Genus Cerapachys

Diagnosis of worker caste in Fiji. Antenna 9- or 12-segmented; antennal club 1-segmented. Antennal insertions not covered by frontal lobes. Waist 2-segmented, although postpetiole (abdominal segment 3) can be enlarged and broadly attached to first gastral segment (abdominal segment 4). Petiole quadrate and apedunculate. Gaster armed with sting; pygidium flattened and armed with a row of small peg-like teeth. Body shape cylindrical.

The Melanesian *Cerapachys* were revised by Wilson (1959c), and the worldwide fauna was revised by Brown (1975). Fiji hosts seven described species (and approximately the same number of undescribed species) that descend from at least two different lineages: the *typhlus* group and the *dohertyi* group. The *typhlus* group workers have 9-segmented antennae, strongly constricted postpetioles, absent or vestigial eyes, and strongly foveate to foveoreticulate sculpture. The *dohertyi* group workers have 12-segmented antennae, an enlarged postpetiole that is attached broadly to the first gastral segment, large multifaceted eyes, and a general reduction of foveate sculpture. The one Fijian species that does not fit comfortably into either group is *Cerapachys* sp. FJ04. Interpretation of species boundaries is difficult in Fiji and other regions because of the considerable variation exhibited by different colonies and populations with respect to size and sculpture.

Cerapachys are known for being specialized predators of other ants. Although most of the Fijian species are found in the leaf litter, there is one species (*C. zimmermani*) that is apparently adapted for arboreal nesting and foraging.

Key to the *Cerapachys* workers of Fiji.[2]

1 Antenna 9-segmented. Postpetiole (abdominal segment 3) subequal in size to petiole and narrowly attached to first gastral segment (abdominal segment 4). Eyes vestigial to absent. Sculpture strongly foveate to foveoreticulate. (*C. typhlus* group) 2
– Antenna 12-segmented. Postpetiole (abdominal segment 3) distinctly larger than petiole and broadly attached to first gastral segment (abdominal segment 4). Eyes variable. Mostly smooth and shiny 4
2 Spaces between foveolae smooth, giving the integument a shiny appearance. Relatively small in size (HW < 0.60 mm) .... ***Cerapachys* sp. FJ06**
– Spaces between foveolae shagreened, giving the integument a rough appearance. Relatively large in size (HW > 0.70 mm) 3
3 Sides of propodeum and sides of petiole with dense deeply impressed foveae that are discernibly wider in diameter than propodeal spiracle or foveae on head. Dark species.

[2] Additional taxonomic work is required before members of the *vitiensis* complex can be reliably diagnosed.

.... ***C. fuscior***

– Sides of propodeum and sides of petiole with sparse, lightly impressed foveae that are not discernibly wider in diameter than propodeal spiracle or foveae on head. Pale species. .... ***C. cryptus***

4 Eyes minute (< 5 facets) to absent. Front of face covered with well-defined foveae. Subpetiolar process with fenestra. Minute species (HW < 0.40 mm). .... ***Cerapachys* sp. FJ04**

– Eyes large (> 10 facets). Front of face smooth or with small punctures but never covered with well-defined foveae. Subpetiolar process lacking fenestra. Small to medium sized species (HW > 0.50 mm). (*C. dohertyi* group)... 5

5 Petiole broader than long. Pronotum with an anterodorsal carinula. Mesosoma with or lacking well-defined foveae... ***C. vitiensis*** complex

– Petiole longer than broad. Pronotum lacking an anterodorsal carinula. Mesosoma smooth and shiny and lacking well-defined foveae 6

6 Thin carinula separating anepisternum from katepisternum. Postpetiole sternum almost entirely smooth. Large species (HW > 0.80 mm) .... ***C. zimmermani***

– Lacking thin carinula separating anepisternum from katepisternum. Postpetiole sternum foveate to lightly rugose. Medium to small species (HW < 0.70 mm) 7

7 Petiole and mesosoma smooth and shiny. Medium sized species (HW > 0.65 mm). .... ***C. lindrothi***

– Petiole and mesosoma smooth and shiny to lightly rugose. Small species (HW < 0.65 mm). .... ***Cerapachys* sp. FJ01**

typhlus group

The *typhlus* group of *Cerapachys*, as defined by Brown (1975), is represented in Fiji by at least three species. Two species (*C. cryptus* and *C. fuscior*) were described by Mann (1921), and the third species (*Cerapachys* sp. FJ06) is currently undescribed. The most obvious characters separating *Cerapachys* sp. FJ06 from *C. cryptus* and *C. fuscior* are its substantially smaller size (HW < 0.60 mm) and more shiny appearance caused by finer and shallower foveae. *Cerapachys fuscior* can be distinguished from *C. cryptus* by its darker reddish brown color, larger diameter foveae, and more smooth and shiny sculpture between the foveae. Additionally, the subpetiolar process of *C. fuscior* (and *Cerapachys* sp. FJ06) is composed of two parts: a posterior rounded lamella and an anterior narrowly attached tooth. The subpetiolar process of *C. cryptus* is a single rounded lamella with a fenestra. The eyes of the examined *C. fuscior* material were composed of a single but distinct ommatidium. The eyes of the examined *C. cryptus* material were most often entirely absent, although a faint hint of an ommatidium were visible in several specimens.

Although no Fijian nest series of the *typhlus* group have been collected with males, we attempt here to associate male specimens collected in malaise traps with the worker specimens collected by hand and litter sifting. While these hypotheses require the collection of nest series of both castes or use of genetic markers for validation, the low specific diversity of this group in Fiji, combined with strong geographic patterns and consistent morphological similarities adds confidence to the associations. In addition to collecting males that match the morphology and distributions of the three *typhlus* group species for which workers are known, male specimens from several localities on Viti Levu appear to represent a fourth species (*Cerapachys* sp. FJ52) for which the worker caste is unknown.

Cerapachys cryptus Mann
(Plate 3)

Cerapachys (*Syscia*) *cryptus* Mann, 1921: 408 fig. 1; worker, ergatogyne described. Type locality: FIJI, Viti Levu, Nadarivatu (W. M. Mann). Syntypes: 9 workers (MCZC, examined); 67 workers, 1 ergatogyne (USNM, examined). Larva described (Wheeler, 1950). In revision of Melanesian fauna (Wilson, 1959c).

See the discussion under the *typhlus* group and the key to *Cerapachys* workers for characters used to diagnose *C. cryptus* from its Fijian congeners. Although no collections of *C. cryptus* workers are known subsequent to Mann's expedition, malaise trapping has produced a dozen male specimens from across Viti Levu that closely match the morphology predicted by the worker caste, and which are included in the material examined and distribution map.

Material examined. **Viti Levu:** Naikorokoro 300, Waimoque 850, Nasoqo 800 a, Nasoqo 800 b, Veisari 300 (3.5 km N).

Cerapachys fuscior Mann
(Plate 4)

Cerapachys (*Syscia*) *cryptus*, subsp. *fuscior* Mann, 1921: 410; worker described. Type locality: FIJI, Taveuni, Somosomo (W. M. Mann). Syntypes: 7 workers (MCZC, examined); 8 workers (USNM, examined). Raised to species (Wilson, 1959c).

See the discussion under the *typhlus* group and the key to *Cerapachys* workers for characters used to diagnose *C. fuscior* from its Fijian congeners. Wilson (1959c), recognizing the close similarity between *C. fuscior* and *C. cryptus*, offered that further collections of these species may demonstrate that the two are insular variants of the same species. The morphology of the single *C. fuscior* worker (CASENT0171152, Mt. Delaikoro, Vanua Levu) collected since Mann's expedition to Taveuni shares the features that distinguish the type specimen from *C. cryptus*. The Vanua Levu specimen, despite coming from a different island, strongly resembles the type series and further validates the consideration of *C. fuscior* being a distinct species from *C. cryptus*. Furthermore, a large number of male specimens collected from Taveuni, Vanua Levu, and Kadavu are close matches to the morphology predicted by the worker caste of *C. fuscior*, and are readily distinguishable from the males predicted to belong to *C. cryptus*. The inferred *C. fuscior* males are included in the material examined and distribution map, and extend the range from Vanua Levu to the islands of Taveuni and Kadavu. The sympatry of both forms at several sites on Viti Levu adds additional support for the current taxonomy.

Mann (1921) obtained several small colonies of this species beneath stones after rains, and described the workers as rolling up and feigning death when disturbed. The only other known worker specimen was collected from sifted litter.

Material examined. **Kadavu:** Moanakaka 60. **Taveuni:** Lavena 235, Lavena 234, Lavena 217, Lavena 219, Lavena 229, Mt. Devo 1064, Somosomo 200. **Vanua Levu:** Kilaka 61, Mt. Delaikoro 699, Rokosalase

150, Rokosalase 97, Lomaloma 587, Lomaloma 630.

Cerapachys sp. FJ06

(Plate 5)

See the discussion under the *typhlus* group and the key to *Cerapachys* workers for characters used to diagnose *Cerapachys* sp. FJ06 from its Fijian congeners. The single ommatidium of this species varies from distinct to absent, and there is some range in the strength of sculpture and color across its distribution. *Cerapachys* sp. FJ06 bears close resemblance to *C. biroi* Forel. Both species are similar in size, shape and sculpture. The most striking difference separating the two involves the subpetiolar process. In *C. biroi*, the subpetiolar process takes the shape of a rounded lamella with an anterior fenestra. In *Cerapachys* sp. FJ06, it appears that the dorsal portion of the process has been removed, causing the disappearance of the fenestra and leaving a narrowly attached anterior tooth and a more broadly attached posterior lamella. A similar modification of the subpetiolar process is observed in *C. fuscior.*

Cerapachys sp. FJ06 has a relatively wide distribution across the archipelago. Putative males of *Cerapachys* sp. FJ06 (included in the material examined and distribution map) expand the distribution to include Vanua Levu, Kadavu, Koro, Lakeba, and Gau. A hand collection from Monasavu Road in Viti Levu included several ergatogynes nesting under a stone with workers.

Material examined. **Gau:** Navukailagi 387, Navukailagi 496, Navukailagi 564. **Kadavu:** Moanakaka 60. **Koro:** Mt. Kuitarua 500, Mt. Kuitarua 505, Mt. Kuitarua 485, Nasau 465 a, Mt. Kuitarua 380. **Lakeba:** Tubou 100 a, Tubou 100 b, Tubou 100 c. **Macuata:** Vunitogoloa 4. **Moala:** Naroi 75. **Taveuni:** Lavena 235, Lavena 234, Lavena 219, Lavena 229, Mt. Devo 1064. **Vanua Levu:** Wainibeqa 87, Wainibeqa 150, Kilaka 98, Rokosalase 180. **Viti Levu:** Nabukavesi 40, Mt. Evans 800, Mt. Evans 800, Mt. Evans 800, Vaturu Dam 620, Nakobalevu 340, Waimoque 850, Colo-i-Suva 460, Colo-i-Suva 325, Colo-i-Suva 372, Nabukavesi 300, Mt. Rama 300, Naikorokoro 300, Veisari 300 (3.8 km N), Waivudawa 300, Veisari 300 (3.5 km N), Nausori.

dohertyi group

The *dohertyi* group is represented in Fiji by five described species (*C. lindrothi*, *C. majusculus*, *C. sculpturatus*, *C. vitiensis* and *C. zimmermani*) and what appear to be several undescribed species. The *dohertyi* group is taxonomically challenging because it is difficult to determine confidently whether observed morphological variation represents different populations of a single species or variation among different species. Much of the morphological variation involves size and sculpture, both of which tend to form continuous states. Males collected from malaise sampling reflect taxonomic difficulties similar to the workers, and it is beyond the scope of this study to associate male specimens with the workers of the *dohertyi* group.

One clear divide among the *dohertyi* group in Fiji separates the *lindrothi* and *vitiensis* complexes. The *lindrothi* complex is readily identified by the shape of the petiole, which is longer than broad, and the lack of a carinula running transversely across the anterodorsal pronotum. The *vitiensis* complex is more sculptured, possesses a petiole that is as broad or broader than long, and has a carinula running transversely across the anterodorsal pronotum.

The members of the *dohertyi* group are presumed to have originated in Asia and reached Fiji by way of New Guinea and the Solomons (Wilson, 1959c). Wilson reviewed two competing hypotheses for the invasion of Fiji from western Melanesia. In one scenario, the Fijian species form a clade derived from a single colonizing population. In the second scenario, the Fijian species form two clades with *C. majusculus*, *C. lindrothi*, and *C. zimmermani* descending from a first wave of colonists, and *C. sculpturatus* and *C. vitiensis* descending from a second wave of colonists.

lindrothi complex

The *lindrothi* complex is represented in Fiji by two described species (*C. lindrothi* and *C. zimmermani*) and at least one undescribed species (*Cerapachys* sp. FJ01). These species form a tentative group that is characterized by a combination of the following characters: (1) 12-segmented antennae, (2) large eyes, (3) petiole longer than broad, (4) anterodorsal pronotum lacking a transverse carinula, and (5) mostly smooth and shiny integument that lacks deep, well-defined foveolae. Compared to their close relatives in the *vitiensis* complex, the three species are relatively easy to separate. It remains to be seen whether the distinct characters separating these species are genuine or the artifacts of poor sampling.

Cerapachys lindrothi Wilson
(Plate 6)

Cerapachys lindrothi Wilson, 1959c: 53; worker described. Type locality: FIJI, Viti Levu, Nadala, near Nadarivatu (E. O. Wilson, acc. no. 28). Holotype: worker (MCZC type no. 30133, examined). Paratypes (same data as holotype): 10 workers, 2 ergatogynes (MCZC, examined).

Cerapachys lindrothi is a long-limbed species of moderate size lacquered with an almost entirely smooth and shiny black integument. This species is most similar to *C. zimmermani*, but the latter is larger, entirely lacks foveae, and bears a thin but distinct carinula separating the anepisternum from the katepisternum. The most obvious difference between *C. lindrothi* and *Cerapachys* sp. FJ01 is the much smaller size of the latter species. Finding other characters to distinguish these two taxa is made more difficult by the considerable diversity of sculpturing (ranging from entirely smooth to bearing shallow foveolae) observed in *Cerapachys* sp. FJ01. The type series of *C. lindrothi* from Viti Levu and both collections from Koro were taken by hand from colonies nesting in rotting logs.

Material examined. **Koro:** Mt. Kuitarua 440 b, Nasau 420 b. **Viti Levu:** Nadala 300.

Cerapachys zimmermani Wilson
(Plate 7)

Cerapachys zimmermani Wilson, 1959c: 54; worker described. Type locality: FIJI, Viti Levu, Mt. Korombamba [= Mt. Korobaba] (E.C. Zimmerman). Holotype: worker (BPBM 2827, examined). Paratypes (same data as holotype): 1 worker (BPBM, examined); 1 worker

(MCZC, examined).

Cerapachys zimmermani is the largest and least sculptured of Fijian *Cerapachys* known to date, and can be distinguished from its Fijian congeners using the key to workers and notes under *C. lindrothi*. The Kadavu specimens are a darker red than the type material, but otherwise match well. Interestingly, *C. zimmermani* may have at least a partial arboreal habit. In his original description of the species, Wilson (1959c) reported that two of Zimmerman's collections were made by beating shrubs leading him to speculate, "If *zimmermani* is indeed a low arboreal forager…it is exceptional in this respect among the Melanesian cerapachyines whose habits are known." Further evidence of arboreal tendencies comes from a discovery during the recent survey of this species nesting in a hollow twig of a tree near the summit of Mt. Washington (Kadavu).

Material examined. **Kadavu:** Mt. Washington 760. **Viti Levu:** Korobaba.

Cerapachys sp. FJ01

(Plate 8)

The specimens collectively assigned to *Cerapachys* sp. FJ01 appear most closely related to *C. lindrothi* and are grouped together based on their small size and longer than broad petiole. As mentioned in the discussion of *C. lindrothi*, the specimens vary slightly in their sculpture and size, but too few specimens (and nest series, in particular) are available to determine whether the variation is reflective of population-level differences or discrete species. While all specimens of *Cerapachys* sp. FJ01 bear foveolae on the postpetiolar sternum, those from Viti Levu and Taveuni also bear shallow but distinct foveolae on the postpetiole dorsum, and the sides of the petiole and lateral portions of the mesosoma are weakly rugose. The single specimen from Koro, aside from the postpetiolar sterna, is entirely smooth. However, the Koro specimen is more similar in size to the Viti Levu specimens than to the larger-bodied Taveuni series. All of the collections of *Cerapachys* sp. FJ01 were from sifted litter.

Material examined. **Koro:** Nasau 465 a. **Taveuni:** Lavena 300, Tavuki 734. **Viti Levu:** Naikorokoro 300.

vitiensis complex

The *Cerapachys vitiensis* complex exhibits a gradation of intermediate morphologies that makes separation of distinct species a significant challenge. This study recognizes seven taxa belonging to the complex: *C. majusculus*, *C. sculpturatus*, *C. vitiensis*, *Cerapachys* sp. FJ05, *Cerapachys* sp. FJ07, *Cerapachys* sp. FJ08, and *Cerapachys* sp. FJ10. A wide range of sculpture and size variation as exhibited by the limited specimens and fewer nest series makes it difficult to determine what character states are indicators of different species and what states constitute minor geographic, population and individual variation.

Mann (1921) originally described two species (*C. majusculus*, *C. vitiensis*) and one subspecies (*C. vitiensis* subsp. *sculpturatus*), which Wilson (1959c) later elevated to species rank. There are clear differences among these species when the type material is examined. The type specimens of both *C. vitiensis* and *C. sculpturatus* bear broad, well-defined foveae on the postpetiolar dorsum and anterior portions of the pronotum, while *C. majusculus* (in addition to being distinctly larger) bears only small punctures on the postpetiolar dorsum and pronotum. The syntypes of *C. vitiensis* can be separated from the holotype of *C. sculpturatus* again by the size and density of foveae. The postpetiolar foveae of *C. sculpturatus* are so broad and densely distributed as to form a foveoreticulate sculpture composed of overlapping foveae. The postpetiolar foveae of *C. vitiensis*, in contrast, are mostly smaller and more sparsely distributed such that only a few of the posterior foveae overlap each other and the postpetiole does not appear foveoreticulate. Additionally, the anterior portion of the *C. sculpturatus* pronotum is more coarsely sculptured than that of *C. vitiensis.* With the addition of new collections, however, the relatively clear morphological lines separating the described species become increasingly blurred.

Cerapachys majusculus Mann
(Plate 9)

Cerapachys (Cerapachys) majusculus Mann, 1921: 408; worker described. Type locality: FIJI, Viti Levu, Nadarivatu (W. M. Mann). Syntypes: 8 workers, 3 pupae (MCZC, examined); 27 workers (USNM, examined).

Cerapachys majusculus is both the largest and least sculptured species of the *vitiensis* complex. See the notes under the *C. vitiensis* complex for a discussion of how it differs from *C. sculpturatus* and *C. vitiensis*. *Cerapachys* sp. FJ07 is significantly smaller than *C. majusculus*, but otherwise quite similar in shape and form. See notes under the former species for a more detailed discussion of differences separating the two. No additional specimens of this species have been collected since Mann's collection of workers from several colonies beneath stones in the forests of Nadarivatu.

Cerapachys sculpturatus Mann
(Plate 10)

Cerapachys (Cerapachys) vitiensis subsp. *sculpturatus* Mann, 1921: 407; worker described. Type locality: FIJI, Viti Levu, Nasoqo (W. M. Mann). Holotype [single specimen]: worker (USNM, examined). Raised to species (Wilson, 1959c).

Cerapachys sculpturatus is the most heavily sculptured member of the *vitiensis* complex. It is the only species in which the majority of the postpetiolar dorsum is covered by overlapping, broadly formed and deeply impressed foveae, causing a foveoreticulate sculpturing. See the notes under the *vitiensis* complex for further discussion of this species. No additional specimens of this species have been collected since Mann's (1921) collection of a single worker.

Cerapachys vitiensis Mann
(Plate 11)

Cerapachys (Cerapachys) vitiensis Mann, 1921: 406; worker described. Type locality: FIJI, Vanua Levu, Lasema (W. M. Mann). Syntypes: 4 workers (MCZC type no. 8681, examined); 3 workers (USNM, examined).

Although there are a number of specimens that are treated in this study as belonging to *C. vitiensis*, their inclusion requires a broadening of the description and a reworking of the diagnostic key to allow for a less foveate postpetiole. The three specimens (CASENT0175784, CASENT0177204[3], CASENT0177237[4]) from Vanua Levu (the island of the type locality) are quite similar to the syntypes in size, shape and sculpture distribution and strength. The two specimens from Viti Levu differ from the type series in that the foveae of the postpetiolar dorsum are reduced to scattered fine punctures—a condition reminiscent of *C. majusculus*. Despite both being collected from the same locality, one of the Viti Levu specimens (CASENT0175799) is similar in size to the syntypes, while the other (CASENT0175795) is larger and more robust. The series from Gau (CASENT0175834, CASENT0175797, CASENT0175807) differs from the syntypes in the smaller but more uniformly shaped foveae on the dorsum of the petiole and postpetiole.

Mann (1921) collected the type series from beneath a stone in a bog, and the other specimens listed above were all collected from sifted litter.

Material examined. **Gau:** Navukailagi 408. **Vanua Levu:** Drawa 270, Lasema a. **Viti Levu:** Nasoqo 800 b, Nasoqo 800 d.

Cerapachys sp. FJ07

(Plate 12)

The specimens assigned here to *Cerapachys* sp. FJ07 are similar to *C. majusculus*, but substantially smaller in size. Like *C. majusculus*, the sculpture on the mesosoma, petiole and postpetiole is very much reduced compared to that of *C. sculpturatus* or *C. vitiensis*. In addition to the difference in size, *Cerapachys* sp. FJ07 can be distinguished from *C. majusculus* by the lack of well-defined foveae on the lateral portions of the petiole. In *C. majusculus*, the foveae on the sides of the petiole are as well-defined as those on the sternum of the postpetiole. In *Cerapachys* sp. FJ07, however, the foveae on the sides of the petiole are reduced to irregular and feeble rugae that are poorly-defined compared to the foveae on the sternum of the postpetiole.

The two specimens from Viti Levu vary slightly in size and sculpture. The larger specimen (CASENT0175825, near Colo-i-Suva) has a marginally stronger sculpture that includes irregular shaped dents on the posterolateral portion of the head, while the smaller specimen (CASENT0175746, near Mt. Tomanivi) is uniformly smooth and shiny aside from feeble rugae on the sides of the petiole. The series from Vanua Levu includes five specimens taken in a sample of sifted litter that are uniform in appearance, and bear a stronger foveate sculpture (particularly on the sides of the postpetiole) than the Viti Levu specimens. The two specimens from Gau were collected from the same sifted litter sample, but one (CASENT0175826) is distinctly larger than the other (CASENT0175790). Both specimens exhibit a relatively smooth petiole and postpetiole, although they also show a modest amount of sculpturing on the posterolateral portion of the face.

Material examined. **Gau:** Navukailagi 408. **Vanua Levu:** Yasawa 300. **Viti Levu:** Nasoqo 800 c, Nakobalevu 340.

[3] Locality either Mt. Delaikoro or Weisali Reserve (both Vanua Levu).
[4] Locality either Kilaka or Natewa (both Vanua Levu).

Cerapachys sp. FJ05

(Plate 13)

The most obvious feature distinguishing the specimens assigned to *Cerapachys* sp. FJ05 is the abnormally swollen appearance of the antennae, in which the antennomeres between the scape and the apical segment approach a four to one ratio of width to length. Brown (1975) describes a species (*Cerapachys kodecorum*) from Borneo that occasionally will have as many as, "six abnormally thick segments of the funiculus." Another useful feature separating *Cerapachys* sp. FJ05 from *Cerapachys* sp. FJ07 (the most morphologically similar species) is the well-defined, continuous carinae that form a transverse bridge from the frontal lobes to the longitudinal carinae found between the eye and the antennal insertion. All specimens of *Cerapachys* sp. FJ05 were taken from sifted litter.

Material examined. **Gau:** Navukailagi 408. **Moala:** Mt. Korolevu 300.

Additional *Cerapachys* species

Cerapachys sp. FJ04

(Plate 14)

Cerapachys sp. FJ04 is the smallest species of this genus thus far known from Fiji, and it is also the only species of the genus that does not fit comfortably into either the *typhlus* group or the *dohertyi* group. Although *Cerapachys* sp. FJ04 has 12-segmented antennae and a weakly constricted postpetiole, it also bears minute eyes composed of fewer than five ommatidia and a strongly foveate sculpture covering most of its mesosoma and face. Another way to distinguish this species from its other 12-segmented Fijian congeners is the pale coloration, strongly margined propodeal declivity and katepisternum, and sunken anepisternum. This species bears a superficial resemblance to *C. biroi*, but the latter has 9-segmented antennae and no distinct ommatidia.

Workers of *Cerapachys* sp. FJ04 have been collected from several localities in Viti Levu and from a single locality in Koro, all from litter sifting. The male specimen from Colo-i-Suva is tentatively associated with the worker collected in the same litter sample, although there is a possibility that they do not belong to the same nest. Males of the same species are also known from malaise trapping at Koroyanitu (Viti Levu), Vanua Levu, and Taveuni.

Material examined. **Kadavu:** Moanakaka 60. **Koro:** Nasau 465 a. **Taveuni:** Mt. Devo 892. **Vanua Levu:** Lagi 300. **Viti Levu:** Mt. Evans 800, Colo-i-Suva 186 d, Naboutini 300, Veisari 300 (3.5 km N).

Cerapachys sp. FJ08 and *Cerapachys* sp. FJ10

(Plates 15–16)

Two *Cerapachys* specimens (CASENT01755805, *Cerapachys* sp. FJ08; CASENT0177223, *Cerapachys* sp. FJ10) examined are difficult to place within any of the aforementioned described or undescribed species, and are treated here as distinct species for the purpose of analyses and morphological description. Both specimens belong to the *vitiensis* group, and both are the smallest specimens of that group examined from Fiji. While there is a distinct possibility that one or both specimens represent very small workers of a previously discussed species, the lack of nest series or foraging series makes the possibility difficult to test.

Material examined (*Cerapachys* sp. FJ08): **Viti Levu:** Nasoqo 800 d.

Material examined (*Cerapachys* sp. FJ10): **Vanua Levu:** Mt. Delaikoro 391.

Subfamily Dolichoderinae

Genus Iridomyrmex

Diagnosis of worker caste in Fiji. Antenna 12-segmented. Head ovoid, not heart-shaped, with posterior margin flat to weakly convex. Scape length less than 1.5x head length. Eyes do not break outline of head and located distinctly below midline of face. Anterior margin of clypeus with one median and two lateral rounded projections. Mandibles with distinct basal angle and dentition absent from basal margin. Dorsum of mesosoma with erect hairs. Propodeum with dorsal surface longer or approximately equal to posterior face. Propodeum lacking posteriorly projecting protrusion; declivity flat to convex. Waist 1-segmented. Petiolar node present and with distinct anterior and posterior faces. Gaster armed with ventral slit.

There is only one species of *Iridomyrmex* currently recognized from Fiji. The genus is most readily diagnosed in Fiji by the distinctive median projection of the anterior clypeal margin. *Iridomyrmex* is most likely to be confused in Fiji with *Philidris*, which has a more cordate head with eyes closer to the posterior clypeal margin.

Iridomyrmex anceps (Roger)
(Plate 17)

Formica anceps Roger, 1863a: 164; worker described. Type locality: WEST MALAYSIA [not examined]. Combined in *Iridomyrmex*, senior synonym of *excisus* (Dalla Torre, 1893). Senior synonym of *papuana* (and its junior synonym *discoidalis*) (Wilson & Taylor, 1967); of *formosae*, *ignobilis*, *meinerti*, *metallescens*, *sikkimensis*, *watsonii* (Heterick & Shattuck, 2011).

Iridomyrmex anceps is a dull brown ant with dense pubescence, erect pilosity of the head, mesosoma and gaster, and relatively long limbs. The taxonomy of this widespread and morphologically diverse species is somewhat problematic. Six previously described taxa from across the Pacific, including *I. ignobilis* Mann from Fiji, were recently synonymized with *I. anceps* (Heterick & Shattuck, 2011). The only specimen examined from the recent survey that matches Mann's original description is a small worker (CASENT0177568) collected in a malaise trap in Vanua Levu. Although other populations of *I. anceps* examined from Fiji show some polymorphism among nestmates, with smaller workers tending to be relatively more gracile, with more slender appendages and narrower heads, the specimen in question appears qualitatively different even from the smallest *I. anceps* specimens examined. The most obvious difference is the slender antennae that extend approximately one third their length past the posterior margin of the head, compared to the more robust antenna of other Fijian *Iridomyrmex* that exceed the posterior margin of the head by one fourth their length. If the gaster is removed, it is easy to see that in posterior view, the petiole of Vanua Levu specimen is significantly narrower apically than that of the others. Although the compression of the femora is difficult to see, the other differences elucidated by Mann hold true. While we accept the synonymy of *I. ignobilis*, and defer to the broader study undertaken by Heterick and Shattuck, a molecular study of all the taxa now considered *I. anceps* might reveal interesting population structure that could elucidate dispersal patterns across the Pacific.

The synonymy of *I. ignobilis* suggests that *Iridomyrmex anceps* is not as recent a colonist as had previously been assumed (Wilson & Taylor, 1967). However, the presence of *anceps* at the primary airport and Lautoka shipping port reveals that there are many opportunities for its further spread. The species was not collected during baiting surveys of the Nausori airport (Viti Levu) or Savusavu port and surrounding areas (Sarnat, 2008a). Part of the reason for not capturing this species during baiting exercises is that, unlike many of its dolichoderine relatives, *I. anceps* appears to be a poor recruiter to food items. In an attempt to videotape and photograph *I. anceps* at a bait station, we placed sugar solution baits and peanut butter baits in high density surrounding a very active nest made in the sand. Despite frequent encounters with the baits by several individuals, no member of the colony stayed for more than half a minute, and the foragers made little attempt to recruit other colony members to the resource they had found. After fifteen minutes, foragers of *Solenopsis geminata* from a distant nest appeared and took control of the baits without any resistance from *I. anceps*. This result is contradicted by a more formal study of the behavior of *I. anceps* at baits conducted by Darren Ward in Fiji (2007), and the results placed it near the top of bait dominance of the eleven species surveyed.

Material examined. **Naroi:** McDonald's Resort 10 a. **Vanua Levu:** Lagi 300. **Viti Levu:** Sigatoka 30 a, Lautoka Port 5 b, Vaturu Dam 700, Vaturu Dam 530, Vaturu Dam 540, Volivoli 55, Volivoli 25, Nadi town. **Yasawa:** Tamusua 118.

Genus Ochetellus

Diagnosis of worker caste in Fiji. Antenna 12-segmented. Scape length less than 1.5x head length. Eyes do not break outline of head. Dorsum of mesosoma lacking erect hairs. Propodeum with dorsal surface longer or approximately equal to posterior face. Propodeum with distinct and posteriorly projecting protrusion. Propodeal declivity concave. Waist 1-segmented. Petiolar node present and with distinct anterior and posterior faces. Gaster armed with ventral slit.

Ochetellus is represented in Fiji by the single species *O. sororis*. The genus is immediately diagnosable from all other Fijian dolichoderines by the elevated and acutely angled propodeum and strongly concave propodeal declivity.

Ochetellus sororis (Mann)

(Plate 18)

Iridomyrmex sororis Mann, 1921: 469, fig. 26; workers described. Type locality: FIJI, Viti

Levu, Nadarivatu (W. M. Mann). Syntypes: 3 workers (MCZC type no. 8712, examined); 3 workers (USNM, examined). Combined in *Ochetellus* (Shattuck, 1992: 17).

Note: Specimens from the non-type localities of Navai and Waiyanitu housed at the MCZC bear red cotype labels but are not true syntypes.

The only species of the genus known from Fiji, *O. sororis* appears to be a close relative to the more widespread *O. glaber* Mayr. Mann describes the species as differing from the Australian *O. glaber* and New Caledonian *O. someri* Forel by having a head that is broader and shorter with more convex sides. Measurements of eight workers each from Australia, New Caledonia and Fiji failed to isolate the Fijian specimens with respect to cephalic index, although they do tend towards the broader end of the spectrum. Specimens of *Ochetellus* from New Caledonia (in PSWC) generally differ from *O. sororis* by their shinier and less punctate face, and more dorsally flattened petiole. The Australian specimens of *Ochetellus* examined exhibit greater variability, however, and variously approach the Fijian specimens in sculpture, color, shape and size. The New Guinea specimens examined tend to be on the larger side, although some polymorphism in the worker caste is obvious from specimens from all regions.

Brown (1958b), in his discussion of the New Zealand ant fauna, observed that several described species of *Ochetellus*, such as *O. itoi* (Japan), *O. punctatissimus* (Australia), together with *O. sororis* represent geographical variants that all grade into the morphological diversity of *O. glaber.* A thorough review of the *glaber* group across the Pacific may reveal conclusive support for the synonymy of *O. sororis* with *O. glaber*, but in the meantime its status as a distinct species is preserved.

The type series was collected from an ant-plant, and Mann reports finding a clavigerid beetle associated with the nest. Workers were collected during the recent survey from vegetation, malaise traps and sifted litter.

Material examined. **Kadavu:** Moanakaka 60. **Vanua Levu:** Mt. Vatudiri 570. **Viti Levu:** Navai 700, Mt. Tomanivi 700, Narokorokoyawa 700, Mt. Batilamu 1125 b, Colo-i-Suva 372, Navai 1020, Veisari 300 (3.8 km N), Nadarivatu 750, Navai, Waiyanitu.

Genus Philidris

Diagnosis of worker caste in Fiji. Head heart-shaped with strongly concave posterior margin. Antenna 12-segmented. Scape length less than 1.5x head length. Eyes do not break outline of head; located at or above, but never below midline of face. Anterior margin of clypeus weakly concave, and never with one median and two lateral rounded projections. Mandibles with distinct basal angle and dentition absent from basal margin. Dorsum of mesosoma with erect hairs. Propodeum with dorsal surface longer or approximately equal to posterior face. Propodeum lacking posteriorly projecting protrusion; declivity flat to convex. Petiolar node present and with distinct anterior and posterior faces. Waist 1-segmented. Gaster armed with ventral slit.

Philidris is recognizable among dolichoderines in Fiji by the cordate head. It can be separated from *Ochetellus* by the convex propodeal declivity and from *Iridomyrmex* by the eyes that occur closer to the posterior margin of the clypeus. Although three taxa (one species and its two subspecies) are described from Fiji, it is difficult to find good characters that reliably separate them from each other across their respective ranges.

The genus is often associated with myrmecophilous ant plants in the family Rubiaceae and the same holds true for the Fijian species. *Philidris* is quite abundant and aggressive where it occurs, but is thus far unknown from Viti Levu. The absence likely corresponds to the different species of ant plant that occur there.

Philidris nagasau (Mann)

(Plate 19)

Iridomyrmex nagasau Mann, 1921: 470, fig. 27; worker described. Type locality: FIJI, Taveuni, Nagasau (W. M. Mann). Syntypes: 3 workers (USNM, examined).

Iridomyrmex nagasau subsp. *alticola* Mann, 1921: 472. Type locality: Taveuni, Nagasau, near lake [Lake Caginitoba] (W. M. Mann). Syntypes: 3 workers (USNM, examined). NEW SYNONYMY.

Iridomyrmex nagasau subsp. *agnatus* Mann, 1921: 472. Type locality: FIJI, Vanua Levu, Wainunu (W. M. Mann). Syntypes: 3 workers (USNM, examined). NEW SYNONYMY.

Philidris nagasau was described by Mann from numerous workers collected from a single forest of the MacKenzie estate that were all found nesting in the bulbous stems of *Hydnophytum* (Rubiaceae) ant-plants. These workers were described as glossy black with yellowish-white tarsi. Two subspecies were also described that varied from *P. nagasau* in shape and color. *Philidris nagasau alticola* was also described from Nagasau (but closer to Lake Caginitoba) and was said to have a less elevated propodeum, a uniformly reddish brown head and a somewhat lighter thorax. *Philidris nagasau agnatus* was described from several foragers collected on Vanua Levu, and differed from *P. nagasau* in its less elevated propodeum, brown color of head and mesosoma, and slightly darker gaster.

After examining numerous series from hand collections and malaise trapping, the discrete lines of color and shape separating these three taxa bleed into a near seamless continuum. Color appears to be conserved within nest series, but varies without regard to island or locality. When a series of dark black specimens is examined next to a series of light brown specimens from the same locality, it is tempting to make the case for two distinct species living in sympatry. However, series of intermediate workers with darker brown heads and a lighter mesosoma also collected from the same locality suggest that genetic exchange is continuing to occur. Furthermore, no other characters can be found that correlate with color. The shape of the propodeum varies within nest series, and even the worker of *P. nagasau* figured in Mann's description has a propodeum that is less elevated and more rounded than those of certain workers from lighter colored series.

The specimen from Kadavu is a geographic outlier, and it is not known whether it represents a true resident population or an erroneous record.

Material examined. **Kadavu:** Moanakaka 60. **Taveuni:** Devo Peak 1188, Devo Peak 1187 b, Lavena 235, Lavena 234, Lavena 229, Tavuki 734, Mt. Devo 892, Mt. Devo 1064, Tavoro Falls 160, Soqulu Estate 140, Lavena 235. **Vanua Levu:** Mt. Delaikoro 910, Kilaka 61, Kilaka 98, Kasavu 300, Mt. Vatudiri 641, Mt. Delaikoro 699, Rokosalase 118, Lomaloma 587, Lomaloma 630, Lomaloma 630.

Genus **Tapinoma**

Diagnosis of worker caste in Fiji. Antenna 12-segmented. Scape length less than 1.5x head length. Eyes do not break outline of head. Dorsum of mesosoma lacking erect hairs. Propodeum with dorsal surface distinctly shorter than posterior face; lacking posteriorly projecting protrusion and declivity flat to convex. Waist 1-segmented, but can be obscured by gaster. Petiolar node greatly reduced or absent, the anterior face absent or at most indistinct. First gastral segment lacking erect hairs. Gaster with four plates on its dorsal surface, and with the fifth plate on the ventral surface; armed with ventral slit.

The most common species of *Tapinoma* in Fiji is the highly successful pan-tropical tramp *T. melanocephalum*. In addition to the distinctly bicolored *T. melanocephalum*, several uniform brown species and one uniformly pale species are also established on Fiji. The taxonomy of *Tapinoma* is in a poor state. It is difficult to know which of these additional species (if any) remain undescribed, and if any available names can be applied to the others. Clouse (2007a) discusses some of the difficulties involved with the taxonomy of Pacific *Tapinoma*, but as we have not examined the specimens from Micronesia it is not apparent if either of the morphotypes he distinguishes as separate from *T. melanocephalum* and *T. minutum* match either *Tapinoma* sp. FJ01 or *Tapinoma* sp. FJ02.

Key to *Tapinoma* workers of Fiji.

1 Head dark and sharply contrasting with pale gaster, appearing bicolored without magnification. Antennal scapes surpassing posterior margin of head by at least length of first funicular segment. Introduced in Pacific.................................... ***T. melanocephalum***
– Head and gaster not sharply contrasting, appearing uniform in color without magnification. Antennal scapes rarely surpassing posterior margin of head, and never by more than length of first funicular segment.. 2
2 Anterior margin of clypeus distinctly concave. Mandible with distinct break in tooth size after fourth tooth. Small species (HW $\leq$0.45 mm). Pacific native .... ***T. minutum***
– Anterior margin of clypeus either flat or indistinctly concave. Mandible either with teeth progressively diminishing in size or with distinct break in size after the fifth tooth. Larger species (HW $\geq$ 0.5 mm).. 3
3 Head, mesosoma and gaster brown and contrasting with paler brown or yellow antennae and tarsi .... ***Tapinoma* sp. FJ01**
– Entire body, including antennae and tarsi, same shade of yellow............ ***Tapinoma* sp. FJ02**

Tapinoma melanocephalum (Fabricius)
(Plate 20)

Formica melanocephala Fabricius, 1793: 353, worker described. Type locality: FRENCH GUIANA [not examined]. In Fiji (Mann, 1921: 475); in Polynesia (Wilson & Taylor, 1967: 14, fig. 66). Current subspecies: nominal plus *coronatum*, *malesianum*. For additional synoptic history see Bolton et al. (2006).

This common tramp species is easily separated from other *Tapinoma* in Fiji by the striking bicolored appearance, due to its dark head and pale yellow abdomen. Closer inspection will also reveal that unlike other Fijian *Tapinoma*, the scapes of this species surpass the posterior margin of the head by a length greater than the first funicular (third antennal) segment. This latter character is especially useful in distinguishing *T. melanocephalum* from *T. minutum*, which is the other small species with a distinctly concave anterior clypeal margin.

Tapinoma melanocephalum is a common household pest ant in the Pacific and in tropical habitats across the globe. These ants are particularly attracted to sugar, and are also fond of foraging on vegetation and walls. They are nearly ubiquitous in the villages and towns of Fiji, but also sometimes can be found at low abundance in the forests. Specimens were collected from malaise traps, sifted litter, nests in dead branches, and foraging trails on the ground and vegetation.

Material examined. **Gau:** Navukailagi 387, Navukailagi 496, Navukailagi 564. **Kadavu:** Moanakaka 60, Namalata 100. **Koro:** Mt. Kuitarua 500, Mt. Kuitarua 505. **Lakeba:** Tubou 100 a, Tubou 100 b. **Macuata:** Vunitogoloa 10, Vunitogoloa 36. **Moala:** Naroi 75, Mt. Korolevu 375. **Ovalau:** Lovoni valley, Levuka. **Taveuni:** Devo Peak 1187 b, Soqulu Estate 140. **Vanua Levu:** Wainibeqa 53, Vusasivo Village 190, Mt. Vatudiri 641, Rokosalase 150, Rokosalase 94, Lomaloma 630, Mt. Delaikoro 391, Lagi 300. **Viti Levu:** Nabukavesi 40, Mt. Evans 800, Mt. Evans 800, Mt. Evans 800, Mt. Tomanivi 700 b, Navai 700, Ocean Pacific 1, Mt. Naqarababuluti 912, Sigatoka 30 a, Koronivia 10, Savione 750 a, Volivoli 55, Volivoli 25, Nausori, Lami. **Yasawa:** Tamusua 118, Nabukeru 144, Nabukeru 120.

Tapinoma minutum Mayr

(Plate 21)

Tapinoma minutum Mayr, 1862: 703; worker described. Type locality: AUSTRALIA [not examined]. Material referred to by Wilson & Taylor (1967: 82) as *fullawayi* (unavailable name). Current subspecies: nominal plus *broomense*, *cephalicum*, *integrum*.

Tapinoma minutum is a small brown species that is most easily confused with *T. melanocephalum* and *Tapinoma* sp. FJ01 in Fiji. It can be distinguished from the former by notes given under that species, and from *Tapinoma* sp. FJ01 by its smaller size, concave anterior clypeal margin, and mandibular tooth arrangement. Two specimens (CASENT0177208 from Viti Levu, CASENT0177148 from Taveuni) collected from malaise traps are most similar to *T. minutum* in their mandibles and anterior clypeal margins, but are distinctly larger than the other specimens available for comparison. They are included in analyses based on this taxonomic survey as tentatively belonging to *T. minutum*. Most collections were made from malaise trapping, but hand collections were made as well, including several from colonies nesting in *Hydnophytum* (Rubiaceae) ant plants.

Material examined. **Lakeba:** Tubou 100 a. **Taveuni:** Lavena 217, Mt. Devo 1064. **Vanua Levu:** Kilaka 146. **Viti Levu:** Mt. Evans 800, Mt. Evans 800, Mt. Tomanivi 700, Monasavu 800 a, Monasavu Dam 600, Navai 1020, Veisari 300 (3.8 km N).

***Tapinoma* sp. FJ01**

(Plate 22)

Tapinoma sp. FJ01 is most similar to *Tapinoma* sp. FJ02, but whereas the former is mostly brown in color with lighter antennae and tarsi, the latter is uniformly yellow in color. It can be distinguished from other Fijian *Tapinoma* by the characters given in the key and in the discussions of those species. This species is currently known in Fiji only from malaise samples collected on the islands of Kadavu and Gau.

Material examined. **Gau:** Navukailagi 496, Navukailagi 564. **Kadavu:** Moanakaka 60.

Tapinoma sp. FJ02

(Plate 23)

Tapinoma sp. FJ02 appears to be a close relative of the darker and less uniformly colored *Tapinoma* sp. FJ01. Together, these two species have a shared set of characters that separate them from the smaller *T. melanocephalum* and *T. minutum*. The only collection of this species was made from beneath the bark of a tree on Mt. Korobaba (Viti Levu).

Material examined. **Viti Levu:** Lami 304.

Genus Technomyrmex

Diagnosis of worker caste in Fiji. Antenna 12-segmented. Scape length less than 1.5x head length. Eyes do not break outline of head. Mandibles lacking a distinct basal angle such that dentition occurs continuously along mandibular and basal and apical margins. Dorsum of mesosoma with erect hairs, but hairs occasionally broken or rubbed off. Propodeum with dorsal surface distinctly shorter than posterior face; lacking posteriorly projecting protrusion and declivity flat to convex. Waist 1-segmented, but can be obscured by gaster. Petiolar node greatly reduced or absent, the anterior face absent or at most indistinct. Gaster with five plates on its dorsal surface; armed with ventral slit.

Technomyrmex is represented in Fiji by the single species *T. vitiensis*. The genus is most likely to be confused with *Tapinoma*, but can be separated by the larger size and the presence of standing hairs on the mesosomal dorsum. Another character often used to separate it from *Tapinoma* is the number of gastral tergites visible in dorsal view (five in *Technomyrmex*, four in *Tapinoma*), but the character can be difficult to discern depending on the condition of the specimens being examined.

Technomyrmex vitiensis Mann

(Plate 24)

Technomyrmex albipes var. *vitiensis* Mann, 1921: 473; workers described. Type locality: FIJI, Viti Levu, Nadarivatu (W. M. Mann). Syntypes: 1 queen (USNM, examined); 1 queen (MCZC, examined). Synonymized with *T. albipes* (Wilson & Taylor, 1967: 82); revised status, raised to species, senior synonym of *rufescens* (Bolton, 2007: 104, fig. 2).

Note: Specimens from the non-type localities of Waiyanitu, Somosomo, Nasoqo, Suene, Nagasau, and Vesari housed at the MCZC and USNM bear red cotype labels but are not true syntypes.

Technomyrmex vitiensis is the only species of the genus recorded from Fiji, although most of the specimens collected over the past century have no doubt been determined as *T. albipes* (following Wilson and Taylor, 1967). Bolton's (2007) global revision of *Technomyrmex* cleared the confusion surrounding specimens previously identified as *T. albipes*. The resurrection of *T. vitiensis* and elevation to species status is founded on several morphological characters separate from those used by Mann to distinguish it from *T. albipes*. Mann's distinction relied upon a slight difference in the color of the legs, and is most likely unreliable when large series from a wide geographic range of the two species are examined. Bolton (2007) lists several ways to separate the two species. The most reliable character differences are the absolutely and relatively longer antennal scapes and larger eyes of *T. vitiensis*. Both the differences in antennal scape and eye size are obvious when specimens of both species are compared next to each other, but they require careful measurements if only one species is available. Although Bolton places an emphasis on the difference in propodeal shape between the two species, the large series of specimens examined in this study reveals that the propodeum of *T. vitiensis* is often less angulate than the type specimens and those figured and described by Bolton, and the common occurrence of worker-queen intercastes, with the more developed mesoscutum, further adds to the variation in shape.

Technomyrmex vitiensis is one of the most pervasive components of the Fijian ant fauna, and has successfully penetrated into primary and disturbed forest habitats. That the species was collected heavily in malaise trap samples and leaf-litter samples suggests it forages in both the ground and arboreal strata. The species is also frequently encountered during hand collecting, and nests were frequently encountered in dead logs and occasionally in ant-plants.

Material examined. **Beqa:** Mt. Korovou 326, Malovo 182. **Gau:** Navukailagi 387, Navukailagi 415, Navukailagi 364, Navukailagi 408, Navukailagi 475, Navukailagi 356, Navukailagi 496, Navukailagi 564, Navukailagi 575. **Kadavu:** Mt. Washington 800, Moanakaka 60, Moanakaka 60, Mt. Washington 700, Namalata 100, Namalata 120, Namalata 50, Namalata 75, Namalata 139, Buka Levu. **Koro:** Mt. Kuitarua 500, Mt. Kuitarua 505, Mt. Kuitarua 485. **Lakeba:** Tubou 100 a, Tubou 100 b, Tubou 100 c. **Macuata:** Vunitogoloa 4. **Moala:** Maloku 80, Maloku 120, Maloku 1, Mt. Korolevu 375. **Ovalau:** Ovalau. **Taveuni:** Devo Peak 1187 b, Lavena 300, Lavena 235, Lavena 234, Lavena 217, Lavena 219, Lavena 229, Soqulu Estate 140, Qacavulo Point 300, Somosomo 200, Nagasau. **Vanua Levu:** Kilaka 146, Wainibeqa 87, Kilaka 61, Wainibeqa 135, Wainibeqa 53, Wainibeqa 150, Kilaka 98, Mt. Kasi Gold Mine 300, Yasawa 300, Kasavu 300, Vusasivo Village 190, Drawa 270, Vusasivo 50, Vuya 300, Mt. Wainibeqa 152 c, Mt. Vatudiri 641, Vusasivo Village 400 b, Rokosalase 180, Rokosalase 150, Rokosalase 97, Rokosalase 118, Rokosalase 94, Lomaloma 587, Lomaloma 630, Lomaloma 630, Mt. Delaikoro 391, Lagi 300, Suene, Labasa, Lasema a. **Viti Levu:** Nabukavesi 40, Mt. Evans 800, Mt. Evans 800, Mt. Evans 800, Navai 700, Nadala 300, Monasavu 800 a, Abaca 525, Colo-i-Suva Forest Park 220, Ocean Pacific 1, Ocean Pacific 2, Vaturu Dam 700, Vaturu Dam 620, Vaturu Dam 550, Vaturu Dam 530, Narokorokoyawa 700, Vaturu Dam 540, Monasavu Dam 600, Lami 200, Lami 400, Colo-i-Suva 460, Colo-i-Suva 325, Colo-i-Suva 372, Colo-i-Suva 186 d, Volivoli 55, Mt. Evans 700, Veisari 300 (3.8 km N), Veisari 300 (3.5 km N), Nadarivatu 750, Nausori, Nasoqo, Waiyanitu, Tailevu, Vesari. **Yasawa:** Tamusua 118, Nabukeru 144.

Subfamily Ectatomminae

Genus Gnamptogenys

Diagnosis of worker caste in Fiji. Antenna 12-segmented. Last antennal segment not large and bulbous, distinctly shorter than remaining funicular segments combined. Clypeus with anterior margin flat to convex, but never forming a distinct triangle that projects anteriorly beyond the base of the mandibles. Mandibles triangular with at least five distinct teeth or denticles. Head covered in large, widely spaced, piligerous foveae and mesosoma and gaster smooth and shiny. Eyes medium to large (> 15 facets); located at head midline. Waist 1-segmented. Petiole narrowly attached to gaster; bearing large subpetiolar process. Gaster armed with sting; distinct constriction between abdominal segments 3+4; tip pointing posteriorly or straight down, but never anteriorly.

Gnamptogenys aterrima (Mann)

(Plate 25)

Wheeleripone aterrima Mann, 1921: 411, fig. 2; worker, queen, male described. Type locality: FIJI, Viti Levu, Waiyanitu (W. M. Mann). Syntypes: 3 workers (MCZC type no. 8685, examined); 9 workers, 5 queens, 1 male (USNM, examined). USNM worker (unique specimen identifier USNM ENT 00528958) here designated Lectotype. Combined in *Gnamptogenys* (Brown, 1958a: 227). Described, figured, phylogeny inferred, key to Old World species (Lattke, 2004: 60, fig. 12).

Note: Specimens from the non-type localities of Kadavu and Vanua Ava bear red cotype labels, but are not true syntypes.

Gnamptogenys aterrima is the only representative of the Ectatomminae known from Fiji. The combination of a single waist segment bearing a subpetiolar process, a sting, small convex eyes, and the foveate face coupled with and a glassy smooth mesosoma and gaster can easily separate this species from all other ant species in Fiji. A lectotype is designated here to distinguish the type series from other series collected by Mann from Lasema and Suene that erroneously bear red cotype labels. Measurements for workers and a queen from the type series are recorded in Lattke (2004).

Gnamptogenys aterrima belongs to the *albiclava* group, and its occurrence in Fiji represents the most eastern extent of the Asian members of the genus (Lattke, 2003; Lattke, 2004). The other members of the *albiclava* group are restricted to the Solomon Islands, and phylogenetic work suggests that the group descended from a Papuan origin. The alternative hypothesis, also suggested by Lattke, that *G. aterrima* arrived in Fiji via rafting on the 'Eua terrane, is dubious (Lucky & Sarnat, 2010). Besides the geologic evidence that the terrane was submerged subsequent to its sundering from the Norfolk Ridge, there is no evidence that any *Gnamptogenys* occurred in New Caledonia, and the ancient vicariant event predicts *G. aterrima* would occupy a more basal position on the tree than is observed.

There are several circumstantial clues that suggest *G. aterrima* is a relatively recent arrival in Fiji. The species is a good disperser as evidenced by its widespread occurrence throughout the archipelago, even on the more remote island of Moala. Furthermore, both queens and males were readily intercepted by malaise traps, and the species varies little in morphology across Fiji. Unlike some of the lineages of apparently old residence in Fiji, *Gnamptogenys* has not radiated into a suite of locally endemic species or even morphologically distinctive populations. Despite its apparently widespread distribution within Fiji, *G. aterrima* workers are rarely encountered foraging. Aside from the sexual castes in malaise traps, the species is most often collected by litter sifting, or discovery of nests under stones or in dead logs.

Material examined. **Beqa:** Mt. Korovou 326, Malovo 182. **Gau:** Navukailagi 625, Navukailagi 675, Navukailagi 387, Navukailagi 408, Navukailagi 432, Navukailagi 475, Navukailagi 496, Navukailagi 564, Navukailagi 575, Navukailagi 300. **Kadavu:** Moanakaka 60, Moanakaka 60, Mt. Washington 700, Daviqele 300, Vanua Ava b. **Koro:** Tavua 220, Mt. Kuitarua 500, Mt. Kuitarua 485, Mt. Nabukala 520, Mt. Kuitarua 440 b, Nasau 420 b, Nasau 465 a, Nasoqoloa 300, Mt. Kuitarua 380. **Moala:** Mt. Korolevu 375, Mt. Korolevu 300. **Ovalau:** Draiba 300. **Taveuni:** Devo Peak 1187 b, Lavena 217, Lavena 219, Lavena 229, Mt. Devo 892, Mt. Devo 1064. **Vanua Levu:** Kilaka 61, Kilaka 98, Yasawa 300, Kasavu 300, Nakanakana 300, Drawa 270, Vuya 300, Mt. Vatudiri 570, Vusasivo Village 400 b, Rokosalase 150, Lomaloma 630, Mt. Delaikoro 391, Suene, Lasema a. **Viti Levu:** Mt. Evans 800, Mt. Evans 800, Mt. Evans 800, Mt. Evans 700, Vaturu Dam 550, Vaturu Dam 530, Nadakuni 300 b, Lami 200, Nakobalevu 340, Colo-i-Suva 325, Colo-i-Suva 372, Volivoli 25, Nakavu 200, Vunisea 300, Nabukavesi 300, Naikorokoro 300, Waivudawa 300, Veisari 300 (3.5 km N), Nausori Highlands 400, Waiyanitu.

Subfamily Formicinae

Genus Acropyga

Diagnosis of worker caste in Fiji. Antenna 8- to 9-segmented. Antennal club indistinct. Scape length less than 1.5x head length. Eyes minute to absent (≤3 facets). Metapleuron with a distinct gland orifice, propodeum unarmed and with convex declivity. Petiolar node present and with distinct anterior and posterior faces. Gaster armed with acidopore; distinct constriction not visible between abdominal segments 3+4.

The *Acropyga* of Fiji are small, nearly blind pale yellow species with four mandibular teeth. *Acropyga* live beneath the soil tending mealybugs (Schneider & LaPolla, 2011). There are two species in Fiji, both of which are very rarely collected.

Acropyga lauta Mann

(Plate 26)

Acropyga (*Rhizomyrma*) *lauta* Mann, 1919: 365; worker, queen described. Type locality: SOLOMON IS.: San Cristoval, Pamua, Wainoni Bay (W. M. Mann). Lectotype: 1 worker (MCZC, examined). Paralectotypes: 12 workers, 4 males (USNM, examined); 6 workers, 1 queen (MCZC type no. 9183, examined). In generic revision (LaPolla, 2004: 65, fig. 27c).

In the most recent survey, *Acropyga lauta* was only collected from a single locality in Vanua Levu, where workers were found in the soil while excavating a *Pheidole* nest. During review of museum specimens at the NMNH, we found collections from Kadavu made by Mann that were apparently overlooked in his monograph. The Fiji specimens are larger than those from other populations of the species (LaPolla, pers. comm.). *Acropyga lauta* is one of four ant species on Fiji currently known to occur on only a single additional island archipelago in the Pacific.

In the Solomons, *A. lauta* is reported as being trophophoretic (Johnson et al., 2001), and has been associated with the mealybugs *Eumyrmoccus kolobangarae* and *E. kusiacus* (LaPolla, 2004). The species is tentatively placed in the *myops* group by LaPolla, with the caveat that the torulae are more closely set than in the other species of this group. *Acropyga lauta* and *Acropyga* sp. FJ02 are most similar in morphology to *A. kinomurai* (Japan) and *A. sauteri* (Taiwan).

Material examined. **Kadavu:** Vunisea. **Vanua Levu:** Mt. Vatudiri 570, Vusasivo Village 400 b.

Acropyga sp. FJ02

(Plate 27)

Acropyga sp. FJ02 is known from a single nest excavation in the Colo-i-Suva Forestry Park near Suva. Workers and males of this species were found nesting in the soil and actively tending mealybugs belonging to the species *Eumyrmococcus sarnati* Schneider & LaPolla that were feeding upon roots. These specimens are confirmed to represent a previously undescribed species (Schneider & LaPolla, 2011).

Material examined. **Viti Levu:** Colo-i-Suva 186 d.

Genus Anoplolepis

Diagnosis of worker caste in Fiji. Antenna 11-segmented. Antennal club indistinct. Scape length greater than 1.5x head length. Eyes medium to large (> 3 facets) and break outline of head. Metanotum impressed. Metapleuron with a distinct gland orifice. Propodeum unarmed and with a convex declivity. Waist 1-segmented. Petiolar node present and with distinct anterior and posterior faces. Gaster armed with acidopore. Head, mesosoma and gaster lacking pairs of long thick hairs.

Anoplolepis gracilipes (Smith, F.)

(Plate 28)

Formica gracilipes Smith, F., 1857: 55; worker described. Type locality: SINGAPORE [not examined]. Junior synonym of *longipes* Jerdon (Emery, 1887: 247). First available replacement name for *Formica longipes* Jerdon which is the junior primary homonym of *Formica longipes* Latreille (now in *Pheidole*) (Bolton, 1995: 67). In Fiji (Mann, 1921: 474, as *Plagiolepis longipes*). For additional synoptic history see Bolton, et al. (2006).

Anoplolepis gracilipes, known most commonly as the Yellow Crazy Ant, is a relatively large, yellow to orange colored species with long legs, large eyes and extremely long antennal scapes. *Anoplolepis gracilipes* is occasionally confused with species of *Leptomyrmex* and *Oecophylla*, but neither of these genera occur in Fiji.

Considered by the IUCN/SSC Invasive Species Specialist Group (ISSG) to be among the 100 worst invasive species in the world, *A. gracilipes* is widespread across the tropics, and populations are especially dense in the Pacific region. The species is most infamous for causing the "ecological meltdown" of Christmas Island (O'Dowd et al., 2003). While *A. gracilipes* is mostly restricted to low elevations, and its distribution among localities is idiosyncratic, it tends to reach high abundance where it does occur. Mann (1921) reports the species as being abundant throughout the islands, especially in the cultivated districts. During the recent survey, alates were frequently encountered in malaise traps, and workers were encountered in sifted litter and foraging on the ground. Several colonies were found nesting in trees.

Material examined. **Beqa:** Mt. Korovou 326. **Gau:** Navukailagi 408, Navukailagi 356, Navukailagi 300. **Kadavu:** Moanakaka 60, Moanakaka 60. **Koro:** Mt. Kuitarua 505, Nasau 470 (4.4 km). **Lakeba:** Tubou 100 a. **Macuata:** Vunitogoloa 10, Vunitogoloa 36. **Moala:** Maloku 80, Maloku 120, Naroi 75, Maloku 1. **Naroi:** McDonald's Resort 10 a. **Taveuni:** Waiyevo, Somosomo 200. **Vanua Levu:** Wainibeqa 135, Kilaka 98, Rokosalase 180, Rokosalase 150, Lagi 300, Lasema a. **Viti Levu:** Mt. Evans 800, Mt. Evans 800, Sigatoka 30 a, Lautoka Port 5 b, Suva, Vaturu Dam 700, Volivoli 55, Volivoli 25, Mt. Evans 700, Galoa 300, Nausori, Nadi town, Lami. **Yasawa:** Tamusua 118, Nabukeru 144, Nabukeru 120.

Genus Camponotus

Diagnosis of worker caste in Fiji. Antenna 12-segmented. Antennal club indistinct. Scape length less than 1.5x head length. Head length longer than head width. Eyes medium to large (> 3 facets). Metanotum impressed or not impressed. Metapleuron lacking a distinct gland orifice. Propodeum armed with spines and with concave declivity or unarmed with convex declivity. Waist 1-segmented. Petiolar node present and with distinct anterior and posterior faces. Gaster armed with acidopore. Head, mesosoma and gaster lacking pairs of long thick hairs.

Camponotus is one of the most diverse genera in the world and in Fiji. In addition to hosting one more widely distributed species (*C. chloroticus*), Fiji is home to many endemic lineages of *Camponotus*, some of which have diversified into small radiations. The two most diverse groups of *Camponotus* in Fiji are the small, armored and rare *dentatus* group (subgenus *Colobopsis*), and the larger, more soft-bodied and abundant species in the subgenus *Myrmogonia*. Although *Camponotus* is one of the better known genera in Fiji, several species remain undescribed.

Key to the minor workers of Fijian *Camponotus*.[5]

1 Propodeum armed with pair of posteriorly projecting spines. Petiole armed with one pair of upright spines on the posterodorsal margin, and possibly an additional pair of upright spines on the anterodorsal margin. Head and mesosoma heavily armored with thick rugoreticulate sculpture. (*Colobopsis*) *C. dentatus* group.. 2

– Propodeum and petiole lacking pairs of spines. Head and mesosoma lacking thick

[5] *Camponotus oceanicus* is not treated here because the worker caste is unknown.

rugoreticulate sculpture 8

2 Metanotal groove moderately impressed, the depression shallower and narrower than length of eye. Ventral surface of head, mesopleuron and sides of petiole uniformly covered in a rough microreticulate sculpture. Length of petiolar peduncle posterior to node less than or equal to portion anterior to node. 3

– Metanotal groove strongly impressed, the depression deeper and broader than length of eye. Ventral surface of head, mesopleuron and sides of petiole uniformly smooth and shiny. Length of petiolar peduncle posterior to node distinctly greater than portion anterior to node 4

3 Pronotum with humeri dorsoventrally flattened and appearing pinched in dorsal and lateral view. In full face view the posterior margin of head distinctly concave between median portion and posterolateral corners. Tergites of gaster posteriorly broadly banded with yellow. Fiji endemic .... ***C. fijianus***

– Pronotum with humeri obtuse, but not dorsoventrally flattened or appearing pinched in dorsal and lateral view. In full face view the posterior margin of head convex to flat between median portion and posterolateral corners. Tergites of gaster black posteriorly, or at most very narrowly banded with yellow. Fiji endemic .... ***C. dentatus***

4 Mesosoma black in color. Fiji endemic .... ***Camponotus* sp. FJ02**

– Distinctly bicolored with head and gaster black, mesosoma and petiole red 5

5 Sides of propodeum and sides of petiole with long white hairs approaching length of those on the ventral surfaces of petiole and gaster. Lower portion of pronotal sides smooth and shiny. Petiole node longer than broad or broader than long 6

– Sides of propodeum and sides of petiole with long hairs absent. Lower portion of pronotal sides with foveolate sculpture. Petiole node longer than broad. Fiji endemic... ***C. manni***

6 Metapleuron, propodeum and dorsum of petiolar node sculptured. Fiji endemic. .... ***C. umbratilis***

– Metapleuron, propodeum and dorsum of petiolar node smooth and shiny 7

7 In dorsal view petiole node broader than long. Frontal lobes bulging, lateral margins bluntly rounded and with flattened edge. Fiji endemic .... ***Camponotus* sp. FJ03**

– In dorsal view petiole node longer than broad. Frontal lobes not bulging, lateral margins with a carinate and raised edge. Fiji endemic. .... ***C. bryani***

8 Head, scapes, mesosoma, waist and gaster covered by a dense pelt of erect hairs, the majority of which are equal to or greater than eye length. Uniformly black. Fiji endemic. .... ***Camponotus* sp. FJ04**

– Pilosity variable, but never with head, scapes, mesosoma, waist and gaster all covered by a dense pelt of erect hairs, the majority of which are equal to or greater than eye length. Color variable 9

9 Dorsal surfaces of mesosoma, waist, and first gastral segment lacking hairs equal to or greater than eye length. Propodeum with obtuse angle and with dorsum and declivity approximately equal in length. Uniformly dull black with coriaceous microsculpture. Fiji endemic .... ***C. vitiensis***

– Dorsal surfaces of mesosoma, waist, or first gastral segment with at least several hairs equal to or greater than eye length; propodeum variously shaped. Color and sculpture variable 10

10 Head and mesosoma brownish yellow to dark brown, but never black. Mesonotum and propodeum never laterally compressed. Posterolateral corners never angulate 11

– Head black. Mesosoma either black or red, but never yellow or brown. Mesonotum and propodeum laterally compressed. Posterolateral corners either angulate or not angulate.

(*Myrmogonia*) 12
11 Propodeum uniformly curved and lacking angle separating dorsum from declivity. All dorsal surfaces of mesosoma with erect hairs equal to or longer than eye length. Pacific native .... ***C. chloroticus***
– Propodeum with obtuse but distinct angle separating dorsum from declivity. Mesosoma entirely lacking erect hairs equal to or longer than eye length. (*Colobopsis*). Fiji endemic .... ***C. polynesicus***
12 Head with posterolateral corners submarginate and acutely angled, giving them a pinched appearance. Petiole, in profile, longer than tall. Petiolar node with vertex projecting anteriorly over peduncle. Mesonotum and propodeum not laterally compressed into a thin dorsal keel 13
– Head with posterolateral corners evenly rounded and lacking margins, acute angles, or a pinched appearance. Petiole, in profile, cuneiform and distinctly taller than long. Petiolar node with vertex achieved medially and not projecting anteriorly over peduncle. Mesonotum and propodeum laterally compressed into a thin dorsal keel 14
13 Leg with trochanter and distal portion of the coxa a pale yellow that contrasts with the darker brown of the tibia and basal portion of the coxa. Head with posterolateral corners less acutely angled. Fiji endemic .... ***C. kadi***
– Entire leg from coxa to femur is uniformly light brown to black. Head with posterolateral corners more acutely angled.. 15
14 Propodeum with posteriorly projecting tooth overhanging petiole. Declivity between mesonotum and propodeal tooth convex and evenly arched. Fiji endemic .... ***C. lauensis***
– Propodeum lacking posteriorly projecting propodeal tooth, or if present, then declivity between mesonotum and propodeal tooth concave. Fiji endemic................. ***C. schmeltzi***
15 Pilosity on gastral tergites dense and yellow 16
– Pilosity on gastral tergites sparse and dark 18
16 Dorsal keel projecting posteriorly as a distinct acutely angled tooth. Mesosoma usually red and sharply contrasting with black head and gaster (Vanua Levu specimens can be darker). Fiji endemic .... ***C. laminatus***
– Dorsal keel forming an obtuse angle posteriorly, but not projecting as a distinct acute tooth. Mesosoma always same color as head and gaster... 17
17 Integument of gaster light brown red on dorsal portion of first three tergites. Petiolar node convex to pointed apically. Legs dark black. Fiji endemic .... ***C. maafui***
– Integument of gaster uniformly black. Pilosity on gastral tergites variable. Petiolar node scalloped to evenly convex apically. Legs black to reddish black. Fiji endemic.
.... ***C. cristatus***
18 Pronotum with erect hairs present. Gaster lacking short appressed hairs. Petiolar node broadly concave apically. Mesosoma often red and contrasting with black head and gaster. Fiji endemic .... ***C. levuanus***
– Pronotum lacking erect hairs. Gaster with short appressed hairs present. Petiolar node pointed apically. Mesosoma black and not contrasting with head and gaster. Fiji endemic .... ***C. sadinus***

Camponotus chloroticus Emery

(Plate 29)

Camponotus maculatus subsp. *chloroticus* Emery, 1897: 574; worker described. Type locality:

NEW GUINEA [not examined]. Subspecies of *irritans* (Emery, 1920b: 7). Raised to species, senior synonym of *chlorogaster*, *sanctaecrucis*, and material of the unavailable name *samoaensis* (Wilson & Taylor, 1967: 93, fig. 76). For complete synoptic history see Bolton et al. (2006).

Camponotus chloroticus is a Melanesian member of the *C. maculatus* group. The *C. maculatus* group includes a wide diversity of poorly studied forms from the Old World that have been variously referred to as species, subspecies, varieties and races (McArthur & Leys, 2006). It is uncertain which species Mann (1921) refers to, as he listed his original Fiji material as belonging to an unnamed variety of *C. maculatus* subsp. *pallidus* F. Smith. Although Mann described the specimens as being, "exceedingly close to, if not identical with var. *samoaensis* Santschi," he did not formally list them under that name. Wilson and Taylor (1967: 93) placed Santschi's material (*samoaensis* is an unavailable name because of its infrasubspecific status) under *C. chloroticus* Emery. Suffice it to say, whatever the true identity of this species, it is widespread across the Pacific and is in need of significant revision. Without a thorough review of the *maculatus* group across the Pacific, it is beyond the scope of this study to determine whether *pallidus*, *chloroticus* or another of the myriad available names most appropriately applies to Fijian material.

There is some variation of this species across the archipelago with respect to size and color, but none of the specimens examined match the descriptions of its presumably close Pacific relatives, the darker headed *C. navigator* Wilson & Taylor or *C. variegatus* Fr. Smith with its more banded gaster. The species was collected most frequently in the smaller islands and the coastal sections of the larger islands. Colonies were found nesting in dead branches, in live mangroves, and under tree bark.

Material examined. **Kadavu:** Moanakaka 60. **Koro:** Nabuna 115. **Lakeba:** Tubou 100 b, Tubou 100 c. **Macuata:** Vunitogoloa 4, Vunitogoloa 36. **Mago:** Mango. **Naroi:** McDonald's Resort 15. **Ovalau:** Cawaci, Wainiloca, Andubagenda. **Taveuni:** Lavena 235. **Vanua Levu:** Rokosalase 180, Rokosalase 97, Rokosalase 118, Savusavu, Eavatu. **Viti Levu:** Nabukavesi 40, Mt. Evans 800, Mt. Evans 800, Mosquito Island 1, Ocean Pacific 1, Ocean Pacific 2, Colo-i-Suva 325, Volivoli 55, Volivoli 25.

Camponotus (*Colobopsis*)

Camponotus oceanicus (Mayr)

Colobopsis oceanica Mayr, 1870: 943; queen described. Type locality: FIJI, Ovalau [not examined]. Combined in *Camponotus* (Dalla Torre, 1893: 245); in *C.* (*Myrmamblys*) (Forel, 1914: 272); in *C.* (*Myrmotemnus*) (Emery, 1920b: 258); in *C.* (*Colobopsis*) (Emery, 1925: 146).

Camponotus oceanicus is known only from the single queen described by Mayr in 1870. The distinguishing feature of the specimen, according to Mayr, is the particularly broad clypeus and the particularly low petiolar node. It is one of the very few species that Mann was not able to find, and the recent survey also failed to recover any specimens matching the description. It is possible that the queen belongs to what is considered here as *C. polynesicus*, but examination of the type would be required to confirm its current status.

Camponotus polynesicus Emery
(Plate 30)

Camponotus polynesicus Emery, 1896: 374. Replacement name for *Colobopsis carinata* (Mayr, 1870: 943). Type locality: FIJI, Ovalau [not examined]. Junior secondary homonym of *Formica carinata* (Brullé, 1840: 84). Replacement name of *mayriella* (Mann, 1921: 494). Combination in *C.* (*Colobopsis*) as *mayriella* (Emery, 1925: 146). Senior synonym of *mayriella* (unnecessary second replacement name for *carinata* Mayr) (Bolton, 1995).

Camponotus (Colobopsis) janus Mann, 1921: 498 (s.w.) FIJI, Kadavu, Buke Levu [Junior primary homonym of *janus* Forel]. Replacement name: *janussus* Bolton.

Camponotus maudella Mann, 1921: 496. Type locality: FIJI, Viti Levu, Waiyanitu. Syntypes: 12 minor workers, 12 major workers, 2 queens, 3 males (MCZC type no. 8725, examined); 13 minor workers, 20 major workers, 7 queens, 4 males (USNM, examined). NEW SYNONYMY.

Camponotus maudella var. *seemanni* Mann, 1921: 498. Type locality: FIJI, Viti Levu, Nadarivatu. Syntypes: 4 minor workers, 1 major worker, 2 males, 1 queen (MCZC type no. 8726, examined); 3 minor workers, 3 major workers, 2 males, 2 queens (USNM, examined) [specimens at MCZC from non-type locality of Waiyanitu bearing red cotype labels are not true syntypes]. NEW SYNONYMY.

Camponotus janussus Bolton, 1995: 106. Type locality: FIJI, Kadavu, Buke Levu. Syntypes: 2 minor workers, 1 major worker (MCZC type no. 21553, examined); 4 minor workers, 2 major workers (USNM, examined). [Replacement name for *janus* Mann]. NEW SYNONYMY.

Note: When Mann gave the replacement name of *Camponotus mayriella* to *Colobopsis carinata* Mayr, he assigned cotype status to several series from his own collections that are housed at the MCZC (type no. 8724). The replacement name was unnecessary because Emery had previously assigned one. The designation of type status was erroneous because Mayr had already designated type specimens in his original description. Therefore, none of the material collected by Mann retains type status.

Camponotus polynesicus, as defined here, is a highly variable species both with regard to color and sculpturing. The color of the major workers varies from uniformly dark reddish black to yellow brown with variegated gasters. Intermediate specimens may be bicolored with a paler head and mesosoma and a strongly contrasting dark gaster. The other character showing considerable variation is the sculpture of the clypeus and cheeks.

The large variation exhibited by these ants is almost entirely among colonies; there is virtually no variation among nestmates. It is fascinating to conjecture what types of genetic characters, gene flow and evolutionary processes have led to such a diverse array of forms. A population-level study of the complex would no doubt improve upon the inferences made with only the aids of morphology and geography. A more thorough study may well reveal that *C. polynesicus*, as defined here, is actually a complex of multiple species, but we are considering it a single species with highly structured populations and introgression occurring across geographical boundaries.

The type series described by Mayr was collected from Ovalau. The majors examined from Ovalau for this study have a reddish brown head and antennae, a yellow brown mesosoma and matching legs, and a contrasting dark brown gaster. The clypeus bears a well-defined median carinae that terminates before reaching the anterior border. The clypeus and the cheeks are both marked with broken longitudinal carinae. The minor workers have a yellow brown head, a pale yellow mesosoma and a contrasting dark brown gaster.

The situation on Viti Levu is quite confusing. For example, the full spectrum of color variation occurs sympatrically in Vaturu Dam area on the western part of the island. Nest series of dark specimens (e.g., CASENT0187042), strongly bicolored specimens (lighter mesosoma and head, darker gaster, e.g., CASENT0187099), and uniformly light specimens (e.g., CASENT0187262) were all collected from the exact same locality (Vaturu Dam 575 b). Although the strength of the sculpture is weakly associated with color (darker specimens with heavier sculpture), majors of all the aforementioned series exhibit a strongly-defined median carinae that extends to the apical margin of the clypeus, and all specimens exhibit longitudinal rugae on their cheeks.

The trend of color and sculpture found at the Vaturu Dam area is broken by the major workers collected from the Koroyanitu area less than ten kilometers to the west. There, the majors (e.g., CASENT0187078) are darker than any from the Vaturu Dam area, but their facial sculpture is as weak as any specimens of this group collected from the archipelago. These specimens correspond well to *C. mayriella* reported by Mann. The clypeus is entirely smooth except for a weak median carina that terminates well before attaining the anterior margin, and the cheeks are marked by only one or two very weak and short carinulae near their anterior margins. Although no majors of other color types were collected, the minors show the same color variation as in Vaturu Dam.

In the Nadarivatu area, the major workers (e.g., CASENT0187185, CASENT0187149) closely approximate those of the Vaturu Dam site with respect to facial sculpture, and exhibit the bicolored appearance with the lighter head and mesosoma contrasting with a darker gaster. This combination of sculpture and color approximates the description of *C. maudella* described by Mann from Waiyanitu, and also reported from Nadarivatu and Taveuni. An aberrant single major worker (CASENT0187269) of relatively uniform yellow brown color was collected from the Monasavu Dam region. The head is larger than those of the Nadarivatu majors, and the clypeus cheeks and mandibles are covered in a dense, thickly developed rugoreticulum from which individual rugae are impossible to separate. Minor workers from the Nadarivatu area also include the uniformly light individuals that correspond to *C. maudella* var. *seemanni* Mann, described from Nadarivatu.

In Kadavu, two varieties of majors were collected. One variety [e.g., CASENT0187304] has the lighter head and mesosoma contrasting sharply with a dark gaster. The facial sculpture is defined by strong median carinae running the length of the clypeus, which, like the cheeks, is otherwise marked by weak and short rugae. The other variety [e.g., CASENT0187350] is uniformly light in color. In addition to the strong median carinae, the clypeus and the cheeks are both covered in regular, long well-defined rugae. Although dark majors are absent from the collection, there are many uniformly dark minor workers. The other minor workers collected from the island are all strongly bicolored, with the yellow brown head and mesosoma and the dark gaster. The major worker of *Camponotus janussus* Bolton from Kadavu was described as having a particularly long head. However, the material examined shows considerable variation in the heads of majors with respect to both the size and shape, and the differences are not deemed significant enough to warrant status as a species distinct from *C. polynesicus* as defined here.

On Vanua Levu, the entire spectrum of variety is expressed. There are uniformly light colored majors (e.g., CASENT0187242) and majors with contrasting dark gasters (e.g., CASENT018712), both with well-defined carinae on the clypeus and cheeks. There are also strongly bicolored majors (e.g., CASENT018723) and uniformly dark majors (e.g.,

CASENT0177854) that both have a smooth clypeus (with the exception of the median carinae), and smooth cheeks. One uniformly light major from the Natewa Peninsula (CASENT0187376) even approaches the Monasavu Dam major in the amount of rugoreticulate sculpture on its clypeus and cheeks, though distinct rugae can be identified.

The only major examined from Taveuni has a reddish brown head and mesosoma contrasting with a darker gaster, and moderately developed facial sculpturing. Again, minor workers of all varieties are present. On Koro, nest series of a very dark form and a paler form with a dark abdomen were collected. The face of the dark form major is marked only by a median clypeal carina. The face of the paler form is much reduced in sculpture. Finally, on Gau, the darkest majors (e.g., CASENT0187106) have the most sculptured faces, while the uniformly light major (e.g., CASENT0187494) and the more bicolored major (e.g., CASENT0187940) have strongly reduced facial sculpture.

Camponotus polynesicus occurs across a wide elevation gradient, and was most often encountered in forested habitat, where it nests in dead branches, sticks, logs, and occasionally ant-plants. The species was also consistently collected by both malaise trapping and litter sifting, and several collections were made from human-dominated landscapes.

Material examined. **Beqa:** Mt. Korovou 326. **Gau:** Navukailagi 325, Navukailagi 387, Navukailagi 415, Navukailagi 408, Navukailagi 432, Navukailagi 356, Navukailagi 496, Navukailagi 564. **Kadavu:** Moanakaka 60, Moanakaka 60, Lomaji 580, Namalata 100, Namalata 120, Namalata 50, Namalata 139, Buka Levu. **Koro:** Nabuna 115, Mt. Kuitarua 500, Mt. Kuitarua 505, Kuitarua 480, Mt. Kuitarua 485, Koro. **Lakeba:** Tubou 100 a, Tubou 100 c. **Macuata:** Vunitogoloa 10, Vunitogoloa 36. **Mago:** Mango. **Moala:** Maloku 80, Maloku 120, Maloku 1, Mt. Korolevu 375, Naroi. **Munia:** Munia. **Ovalau:** Levuka 400, Draiba 300, Wainiloca, Andubagenda, Cawaci, Vuma, Ovalau, Levuka. **Taveuni:** Devo Peak 1188, Devo Peak 1187 b, Lavena 300, Lavena 235, Lavena 217, Lavena 219, Lavena 229, Mt. Devo 892, Mt. Devo 1064, Soqulu Estate 140, Lavena 235, Nagasau. **Vanua Levu:** Mt. Delaikoro 910, Kilaka 146, Wainibeqa 53, Wainibeqa 150, Kilaka 98, Vusasivo Village 190, Drawa 270, Vusasivo 50, Vuya 300, Mt. Vatudiri 641, Mt. Delaikoro 699, Vusasivo Village 400 b, Rokosalase 180, Rokosalase 150, Rokosalase 97, Rokosalase 118, Rokosalase 94, Lomaloma 587, Lagi 300, Wainunu, Suene, Labasa, Lasema a. **Viti Levu:** Nabukavesi 40, Mt. Evans 800, Mt. Evans 800, Mt. Evans 800, Mt. Evans 700, Mt. Tomanivi 700 b, Navai 700, Mt. Tomanivi 700, Vaturu Dam 575 b, Colo-i-Suva Forest Park 220, Mt. Naqarababuluti 912, Naqaranabuluti 860, Suva, Vaturu Dam 700, Vaturu Dam 620, Vaturu Dam 550, Vaturu Dam 530, Mt. Batilamu 840 c, Savione 750 a, Monasavu 800, Monasavu Dam 600, Lami 200, Nakobalevu 340, Colo-i-Suva 200, Korobaba, Colo-i-Suva 460, Colo-i-Suva 325, Colo-i-Suva 372, Lami 171, Colo-i-Suva 186 d, Mt. Evans 700, Naboutini 300, Navai 1020, Nakavu 200, Nabukavesi 300, Naikorokoro 300, Veisari 300 (3.8 km N), Waivudawa 300, Nadarivatu 750, Nausori, Waiyanitu, Colo-i-Suva, Belt Road. **Yasawa:** Tamusua 118, Nabukeru 120.

Camponotus vitiensis Mann

(Plate 31)

Camponotus (*Colobopsis*?) *vitiensis* Mann, 1921: 490, fig. 36; worker described. Type locality: FIJI, Viti Levu, Nadarivatu (W. M. Mann). Syntypes: 1 worker (MCZC type no. 8723, examined); 12 workers (USNM, examined). One USNM worker here designated Lectotype. Soldier, queen, male described (Donisthorpe, 1946: 69).

Note: Specimens from the non-type locality of Waiyanitu bear red cotype labels but are not true syntypes.

Camponotus vitiensis is one of the more easily recognizable species in the forests of Fiji owing to its behavior of cocking the gaster above the mesosoma while running in large and aggressive foraging lines. It is a large uniformly dull black species of robust build. The dull appearance is caused by the coriaceous microsculpture covering the head, mesosoma, petiole and legs. The microsculpture of the gaster, however, runs in neat transverse lines. Mann, in his description, made note of the slender and compressed legs of this species. Although there is a small but noticeable and consistent difference in size among *C. vitiensis* from the different islands, all are recognized here as belonging to the species described by Mann (1921). In particular, the Viti Levu specimens tend to be larger than specimens taken from any of the other islands. However, no corresponding differences in morphology can be discerned.

Mann proposed that this species nests high in trees, and the absence of any nest collections gives credence to his conjecture. Moreover, none of the dozens of specimens collected from malaise traps and by hand are soldiers. The queen, male and major described by Donisthorpe (1946) were taken off a tree by R. A. Lever in the Mt. Evans range. The description by Donisthorpe of the soldier's head being flattened anteriorly suggests that *C. vitiensis* does belong to the *Colobopsis* lineage. The geographic range of *Camponotus vitiensis* is much greater than previously recorded (Ward & Wetterer, 2006), and it is now known from all the large and mid-sized islands in Fiji. Malaise trapping proves to be particularly effective for collecting this species, while it was recorded from only one sifter-litter sample.

Material examined. **Gau:** Navukailagi 632, Navukailagi 387, Navukailagi 408, Navukailagi 496, Navukailagi 557, Navukailagi 564. **Kadavu:** Moanakaka 60. **Koro:** Mt. Kuitarua 500, Mt. Kuitarua 505, Kuitarua 480, Mt. Kuitarua 485, Mt. Nabukala 520. **Taveuni:** Devo Peak 1188, Devo Peak 1187 b, Mt. Devo 734, Mt. Devo 892, Mt. Devo 1064, Soqulu Estate 140. **Vanua Levu:** Kilaka 98, Drawa 270. **Viti Levu:** Mt. Evans 800, Mt. Evans 800, Mt. Naqarababuluti 864, Mt. Batilamu 840 c, Savione 750 a, Monasavu Dam 600, Lami 260, Nakobalevu 340, Colo-i-Suva 372, Veisari 300 (3.8 km N), Veisari 300 (3.5 km N), Mt. Nakobalevu 200, Nadarivatu 750, Waisoi 300, Waiyanitu, Tamavua.

Camponotus sp. FJ04

(Plate 32)

Camponotus sp. FJ04 is a large black ant unique among all Fijian species on account of the long, light, thin pilosity covering nearly all dorsal surfaces, including the antennal scapes. The species is most similar in appearance to the nearly hairless *C. vitiensis*, and is provisionally placed in the *Colobopsis* section. The petiole is similar to *C. vitiensis*, but the node forms a rounded apex and is not indented apically, and the legs are redder. The species is known only a few malaise traps and hand collections on Kadavu, Vanua Levu and Taveuni.

Material examined. **Kadavu:** Moanakaka 60. **Taveuni:** Lavena 217. **Vanua Levu:** Drawa 270.

dentatus group

The *dentatus* group is one of the world's most peculiar radiations of *Camponotus* species. The armored cuticle, heavy sculpturing and conspicuous spines make them appear more ponerine or myrmicine than formicine. In the field they move slowly and deliberately without the timid or frenetic gestures that characterize most of their congeners. Although workers are encountered rarely, they can occasionally be found foraging on vegetation or tree trunks. Nests have avoided discovery by collectors and are thus likely to occur either high in trees or deep inside them.

Although the morphological differences among species are recognizable upon close scrutiny, the large gaps between these and typical *Colobopsis* perhaps caused Mann to treat them as a single species. Upon receiving specimens of *C. dentatus* from Mann, W. M. Wheeler (1934) decided that a number of them diverged so strongly from the type that they deserved species status (as in the case of *C. manni*) or subspecies status (as in the case of *C. fijianus* and *C. umbratilis*). The vast majority of specimens examined for this study are of minor workers collected either by malaise trapping or, less frequently, hand collection of individual foragers. No nest collections of any of the following species were made during field expeditions of the recent survey. Fortunately, the minor workers are rich in morphological characters, while the major workers appear more difficult to distinguish among species. Although no nest series are examined, several alates and major workers collected by malaise can be tentatively associated with one or another recognized species owing to their restrictive and exclusive known ranges in addition to educated guesses based on shared morphological characters.

Camponotus fijianus Özdikmen, NEW STATUS

(Plate 33)

Camponotus dentatus fijianus Özdikmen 2010: 524. Replacement name for *Camponotus* (*Colobopsis*) *dentatus* subsp. *humeralis* Wheeler, 1934: 416; worker described. Type locality: FIJI, Nadarivatu (W. M. Mann). Syntypes: 5 workers (MCZC, examined). Junior primary homonym of *humeralis* Emery.

Camponotus fijianus was first described by W. M. Wheeler upon inspection of Mann's *C. dentatus* material. Wheeler proposed that the specimens collectively identified by Mann as *C. dentatus* could be split into a number of different species. He noticed that the humeri of some workers from Nadarivatu varied considerably from the other specimens. Indeed, the humeri of *C. fijianus* appear pinched so as to form broad concave depressions whose submarginate apical borders project anteriorly over the neck. The species can also be distinguished from *C. dentatus* by the concave posterior margin of the head between the median portion and posterolateral corners, and by the broad yellow posterior bands on gastral tergites.

The collections of *C. fijianus* are limited to a few workers collected from malaise traps on Macuata just off the north shore of Viti Levu, the Yasawa Islands and the western mountains of Viti Levu, in addition to the type material described from Nadarivatu. Although the closely related *C. dentatus* is quite widespread across the archipelago, Nadarivatu is the only site where it occurs sympatrically with *C. fijianus*. The co-occurrence in Nadarivatu suggests that despite their close similarities, the two species are reproductively isolated. It is possible that the difference in range reflects a preference for drier habitats in the case of *C. fijianus*, but little is known about the ecologies of either species.

Material examined. **Macuata:** Vunitogoloa 10, Vunitogoloa 36. **Viti Levu:** Mt. Evans 800, Mt. Evans 800, Nadarivatu 750. **Yasawa:** Tamusua 118.

Camponotus dentatus (Mayr)
(Plate 34)

Colobopsis dentata Mayr, 1866: 492, fig. 5; worker, soldier, queen, male described. Type locality: FIJI, Ovalau. Syntypes: 4 minor workers, 2 major workers, 1 queen, 2 males (MCZC, examined). Combined in *Camponotus* (Dalla Torre, 1893: 228); in *C.* (*Myrmamblys*) (Forel, 1914: 271); in *C.* (*Myrmotemnus*) (Emery, 1920b: 258). Combined in *C.* (*Colobopsis*), description of minor worker, major worker, queen and male (Mann, 1921: 491, fig. 37a-c). Discussed (Wheeler, 1934: 415).

Camponotus dentatus displays a wide range of variation across the Fijian archipelago. Although there is a temptation to interpret the variation as evidence for multiple species, especially when the more extreme forms are compared, specimens of intermediate morphologies are frequent enough to warrant a more conservative taxonomic approach. The four primary axes of variation are the propodeum shape and armament, petiole shape, facial sculpture, and color. All four primary characters vary independently from each other, which further hampers any attempt to neatly separate the island populations into multiple species.

The differences in propodeal shape and armament are the most striking among the different islands. In general, there is a trend among the northern islands of Vanua Levu and Taveuni for the propodeum to be more evenly flattened along its length, and for the propodeal projections to be broadly attached basally. There are many specimens on Vanua Levu in which the propodeum is concavely sloped between its lateral margins, and the lateral margins continue posteriorly on an unbroken trajectory to form the convergent propodeal projections. A nearly equal number of specimens exhibit similarly shaped propodeal projections, but the propodeum itself is either flat or forms a weakly convex slope between its lateral margins. These differences are unrelated to locality. There is a single specimen (CASENT0177338) in which the propodeum is flat, but the projections are narrowly attached to its posterior margin (a condition more often found among specimens from Viti Levu and the southern islands). This specimen is also sympatric with one in which the propodeum has a concave slope and broadly attached projections.

Most of the Taveuni specimens bear strong resemblance to those from Vanua Levu. In eight of the ten specimens examined, the propodeum is weakly raised medially, and its projections are broadly attached and convergent distally. In two specimens (CASENT0175838, CASENT0177512) the propodeum is elevated medially with narrowly attached projections that diverge distally. These two specimens, even more so than the aberrant CASENT0177338 from Vanua Levu, bear a closer resemblance to specimens from Viti Levu than to other Taveuni specimens, including those from the same locality.

Unlike many other members of the group, *C. dentatus* can be found in marginal habitats in addition to forest. Mann (1921) found them living in scrubby vegetation near the coasts and in cultivated areas. Wheeler (1934) found them running on tree trunks in Suva. They were also observed by the authors in the streets of Suva among a heap of rubbish and vegetative debris, and they readily recruited to sugar water baits placed on the ground. Specimens were also collected from sifted litter.

Material examined. **Gau:** Navukailagi 325. **Kadavu:** Moanakaka 60, Vanua Ava b. **Lakeba:** Tubou 100 a, Tubou 100 b, Tubou 100 c. **Ovalau:** Levuka 450, Ovalau, Levuka. **Taveuni:** Lavena 235, Lavena 234, Lavena 217, Lavena 229, Soqulu Estate 140, Somosomo 200, Nagasau. **Vanua Levu:** Kilaka 146, Wainibeqa 87, Wainibeqa 53, Kilaka 98, Vusasivo Village 400 b, Rokosalase 180, Rokosalase 94,

Rokosalase 97, Rokosalase 118, Lagi 300, Labasa. **Viti Levu:** Nabukavesi 40, Ocean Pacific 1, Ocean Pacific 2, Lautoka Port 5 b, Suva, Colo-i-Suva 460, Colo-i-Suva 325, Colo-i-Suva 372, Volivoli 55, Volivoli 25, Waivudawa 300, Veisari 300 (3.5 km N), Nadarivatu 750, Waiyanitu.

The *bryani* complex

The workers of the *bryani* complex are most readily distinguished from other *dentatus* group species by a deeper metanotal groove and more shiny surfaces. This study recognizes five largely allopatric species within the *bryani* complex. The morphological differences among the recognized species seemed too great to consider them all a single species with geographic variants. Three species (*C. bryani*, *C. umbratilis*, and *Camponotus* sp. FJ03) all occur on Viti Levu. The other two species are known only from the islands of Taveuni (*C. manni*) and Gau (*Camponotus* sp. FJ02). Only *Camponotus* sp. FJ03 is known from more than one island (Viti Levu and Vanua Levu).

The only case of strict sympatry among the complex is that of *C. bryani* and *Camponotus* sp. FJ03 at Nakobalevu (Viti Levu). The suite of morphological characters discussed in the key and in the following accounts demonstrates the relatively strong differences between the two taxa and their shared range validates them as distinct species. Furthermore, the specimens of *Camponotus* sp. FJ03 from Nakobalevu are virtually indistinguishable from the Vanua Levu specimens. Therefore, we argue that if gene flow is possible between southeastern Viti Levu and several localities on Vanua Levu, gene flow should also be possible between southeastern Viti Levu and central Viti Levu. However, *Camponotus umbratilis* is morphologically as different from *Camponotus* sp. FJ03 and *C. bryani* as these two are from each other, which, in turn, offers support for its status as a distinct species.

Camponotus manni and *Camponotus* sp. FJ02 are considered here to be allospecies, isolated from all other close relatives on their respective medium-sized islands. Their distinction as reproductively isolated species is, therefore, a more difficult case to argue. The rationale presented here is to use the magnitude of morphological difference observed among the Viti Levu species as a scale by which to measure the distinctiveness of the allospecies. By this measure, the differences separating *Camponotus manni* and *Camponotus* sp. FJ02 from each other and the Viti Levu species, constitutes enough morphological divergence to suggest that given the opportunity for crossbreeding, the island endemic populations would remain reproductively isolated.

Camponotus bryani Santschi

(Plate 35)

Camponotus (*Condylomyrma*) *bryani* Santschi, 1928a: 72, fig. 2a–d; worker described. Type Locality: FIJI, Viti Levu, Golo-i-Sova [Colo-i-Suva], 20.vi.1924 (E. H. Bryan). Syntypes: 3 workers (BPBM, not examined). Combination in *C.* (*Colobopsis*) (Wheeler, 1934: 421).

Camponotus bryani is a close relative of *C. dentatus*. The most distinguishing features of this species are the long, narrow, low and smooth petiole together with the smooth and shiny metapleuron and sides of the propodeum. The most obvious difference between *C. bryani* and *Camponotus* sp. FJ02 is the bicolored appearance of the former and the uniform black appearance of the latter. *Camponotus bryani* differs from *C. manni* by the possession of suberect hairs on the sides of the propodeum and petiole. *Camponotus bryani* is separated from *Camponotus* sp. FJ03 by the petiolar node, which is longer than broad in the former and broader than long in the latter, and by the network of rugulae on the face which is coarser and more regular. *Camponotus bryani* is quite similar to *C. umbratilis*, but the former is smooth and shiny on the metapleuron and sides of the propodeum, whereas those of the latter are strongly sculptured.

The species was described by Santschi (1928a) from specimens collected by the Bishop Museum entomologist, E. H. Bryan Jr., during the 1924 Whitney South Seas Expedition. It is known from the mountains surrounding Fiji's capital city, Suva, on eastern Viti Levu. The few specimens collected during the recent survey were obtained by hand collection and litter sifting.

Material examined. **Viti Levu:** Colo-i-Suva Forest Park 220, Nakobalevu 340, Nausori.

Camponotus manni Wheeler
(Plate 36)

Camponotus (*Colobopsis*) *manni* Wheeler, 1934: 418, Fig. 4 a, c, h; worker described. Type locality: FIJI, Taveuni, Somosomo (W. M. Mann). Syntypes: 4 minor workers (MCZC type no. 23342, examined).

Camponotus manni and *C. manni* subsp. *umbratilis* were described by W. M. Wheeler from specimens originally assigned to *C. dentatus* by Mann (Wheeler, 1934). Wheeler discusses a number of character differences between of *C. manni* and *C. umbratilis* that, upon inspection of additional material, do not appear to consistently separate the two. A more reliable way to separate *C. manni* from the other species of the *bryani* complex is the reduced pilosity, which is discussed in more detail under the notes for *C. umbratilis*. *Camponotus manni* was only collected during the recent survey from the island of Taveuni where it was caught in malaise traps and foraging on vegetation. Previous collectors took it from several localities in Vanua Levu.

Material examined. **Taveuni:** Devo Peak 1188, Lavena 234, Lavena 217, Lavena 219, Lavena 229, Lavena 235, Somosomo 200. **Vanua Levu:** Wainunu, Suene.

Camponotus umbratilis Wheeler, NEW STATUS
(Plate 37)

Camponotus (*Colobopsis*) *manni* subsp. *umbratilis* Wheeler, 1934: 420, Fig. 4, d, e; worker described. Type locality: FIJI, Viti Levu, Nadarivatu (W. M. Mann). Syntypes: 3 workers (MCZC type no. 23343, examined).

Whereas *Camponotus umbratilis* is rugose on the metapleuron, sides of propodeum and dorsum of petiolar node, those regions are smooth and shiny in both *C. bryani* and *Camponotus* sp. FJ03. Although the differences separating the species of the *C. bryani* complex are subtle, the presence of both *C. umbratilis* and *C. bryani* at the same locality Nakobalevu (Viti Levu) suggests the two are reproductively isolated. Furthermore, the presence of specimens matching the description of *C. umbratilis* on both eastern Viti Levu and several localities on Vanua Levu argues that the morphological differences separating *bryani* complex populations are more than intraspecific geographical variation. The species is known only from three specimens collected independently from the mountains of western Viti Levu. Specimens were collected in malaise traps, in sifted litter, and foraging on vegetation.

Material examined. **Vanua Levu:** Kilaka 98, Mt. Vatudiri 570. **Viti Levu:** Nakobalevu 340, Colo-i-Suva 460, Colo-i-Suva 372, Nadarivatu 750.

Camponotus sp. FJ02

(Plate 38)

Known only from the island of Gau, *Camponotus* sp. FJ02 is also the only member of the *bryani* complex with a black mesosoma. Additional features of the species include standing pilosity on the sides of the propodeum and petiole (shared by *C. bryani*, *C. umbratilis* and *Camponotus* sp. FJ03), and a glossy smooth integument on the sides of the propodeum and dorsum of the petiolar node. The petiolar node is longer than broad, but not as narrow as *C. umbratilis*. Without the test of sympatry with other members of the *bryani* complex, the decision to treat *Camponotus* sp. FJ02 as a distinct species must be taken as a tentative guess based upon a review of the available material. All the collections of *Camponotus* sp. FJ02 were made from malaise traps on Gau.

Material examined. **Gau:** Navukailagi 387, Navukailagi 496, Navukailagi 564.

Camponotus sp. FJ03

(Plate 39)

Camponotus sp. FJ03 is the only member of the *bryani* complex with a petiolar node that is broader than long, and nearly as tall as it is long. The petiole and propodeal sides both have standing pilosity and are smooth and shiny without rugae. The rugulae on the face of this species are finer than those of the other species in the complex, and tend to run transversely above the frontal carinae. *Camponotus* sp. FJ03 is known from three unique collections of single workers from the mountains of western Viti Levu in Koroyanitu National Park. The workers were collected from malaise traps and foraging on stick piles.

Material examined. **Viti Levu:** Mt. Evans 800, Mt. Batilamu 840 c.

Camponotus (Myrmogonia)

"The subgenus *Myrmegonia*[6] [*Myrmogonia*] is made up of a complex of species, with such closely connecting forms that one might consider them all a single species or divide them into many," Mann (1921). While the same dilemma applies to many of the ants dispersed across the Fijian archipelago, it is true that this group presents a significant challenge to the taxonomist. *Myrmogonia* Forel was erected for the three Fijian species, *C. laminatus* (type species), *C. cristatus* and *C. schmeltzi*, first described by Mayr (1866). Subsequently, a number

[6] Mann (1921) consistently misspelled the subgenus *Myrmogonia* Forel as *Myrmegonia*.

of Australian and one New Guinea species were combined in *Myrmogonia* (Forel, 1914), all of which were combined in different subgenera over the following years (Emery, 1920a; Santschi, 1928b). Mann (1921) argued that the Australian species did not belong in *Myrmogonia* because the soldiers did not have the truncate head typical of the Fijian species.

In a recent attempt to address the chaotic taxonomy and nomenclature of Australian *Camponotus*, the species were assigned to several dozen species groups (McArthur, 2003). Of these, the Fijian species bear closest resemblance to the following: *gibbinotus* group, *walkeri* group, *armstrongi* group, *evae* group, *rubiginosus* group, *oetkeri* group, *claripes* group, and the *aureopilus* group. Although this latter group includes a number of New Guinea species (Shattuck, 2005; Shattuck & Janda, 2009), as well, it is interesting to note the high diversity of Australian forms bearing at least superficial resemblance to the Fijian species. The profemur of the Fijian taxa is wider with respect to the meso- and metafemur than it is in the aforementioned non-Fijian groups.

laminatus complex

The *laminatus* complex is composed of five species (*C. cristatus*, *C. laminatus*, *C. levuanus*, *C. maafui*, and *C. sadinus*) and is recognized by the mesonotum and propodeum being laterally compressed into a thin dorsal keel. The distribution of the *laminatus* complex across the archipelago shows some interesting biogeographic patterns, especially with respect to Vanua Levu and Taveuni, as individual species may be replacing each other on different islands.

Camponotus cristatus Mayr
(Plate 40)

Camponotus cristatus Mayr, 1866: 489, fig. 3; minor worker described. Type locality: FIJI, Ovalau [not examined]. Combination in *C.* (*Myrmogonia*) (Forel, 1912b: 92). Major worker, male, queen described (Mann, 1921: 479, fig. 30).

Camponotus cristatus var. *nagasau*, Mann 1921: 482; worker described. Type locality: FIJI, Taveuni, Nagasau. Syntypes: 20 minor workers (MCZC type no. 8719, examined); 2 minor workers, 1 major worker (USNM, examined). NEW SYNONYMY.

Camponotus cristatus is uniform black to black with reddish legs and antennae. The petiole node is scalloped to evenly convex or pointed apically. A mesonotal keel is present, but does not project posteriorly into a defined tooth. The gastral pilosity is light-colored and dense like *C. laminatus*, and erect hairs are present on pronotum. The species is most similar to *C. maafui*, but can be separated by the uniformly black integument on the gaster. *Camponotus cristatus* var. *nagasau* Mann was considered to be a distinct variety based on the dark red femora of the soldiers bearing a consistent difference from the deep black femora of *C. cristatus* Mayr. Color, however, is a weak character for discriminating among the *C. laminatus*-group species. The weak differences observed among the type specimens are used here to justify synonymy of *C. nagasau* with *C. cristatus.* In addition to the differences of femora color between the soldiers of the Taveuni population and the Viti Levu population, the Taveuni minors also tend to exhibit a more uniformly pointed petiole node.

Most collections of *C. cristatus* were made from malaise traps, though they were also taken from sifted litter. Workers were collected foraging on ground, stones, logs, trees and vegetation. The species is absent from Vanua Levu, where it is replaced by the closely related *C. maafui.*

Material examined. **Gau:** Navukailagi 325. **Kadavu:** Moanakaka 60, Vanua Ava b. **Lakeba:** Tubou 100 a, Tubou 100 b, Tubou 100 c. **Ovalau:** Levuka 450, Ovalau, Levuka. **Taveuni:** Lavena 235, Lavena 234, Lavena 217, Lavena 229, Soqulu Estate 140, Somosomo 200, Nagasau. **Vanua Levu:** Kilaka 146, Wainibeqa 87, Wainibeqa 53, Kilaka 98, Vusasivo Village 400 b, Rokosalase 180, Rokosalase 94, Rokosalase 97, Rokosalase 118, Lagi 300, Labasa. **Viti Levu:** Nabukavesi 40, Ocean Pacific 1, Ocean Pacific 2, Lautoka Port 5 b, Suva, Colo-i-Suva 460, Colo-i-Suva 325, Colo-i-Suva 372, Volivoli 55, Volivoli 25, Waivudawa 300, Veisari 300 (3.5 km N), Nadarivatu 750, Waiyanitu.

Camponotus laminatus Mayr

(Plate 41)

Camponotus laminatus Mayr, 1866: 489, fig. 4; minor worker described. Type locality: FIJI, Ovalau [not examined]. Combination in *C.* (*Myrmogonia*) (Forel, 1912b: 92). Designated as type species for *C.* (*Myrmogonia*) (Wheeler, 1913b: 81). Major worker, male, queen described (Mann, 1921: 477, fig. 29).

Camponotus laminatus is a conspicuous member of the arboreal ant community in Fiji. The species is most often recognized by its relatively large size and bicolored black and red appearance (except on the island of Vanua Levu where it takes on a more uniform dark coloration). *Camponotus laminatus* is a close relative of *C. levuanus*, but can be distinguished from the latter by the presence of erect hair on the pronotum, the dense light-colored pilosity of the gaster in combination with the petiolar node shape. Whereas the petiolar node of *C. levuanus* is broadly concave, that of *C. laminatus* is more often flat to convex, occasionally coming to a point at the apex. However, the petiolar nodes of some *C. laminatus* are very narrowly concave apically. These can be separated from *C. levuanus* by the abundant, light-colored pilosity.

There considerable variation in *C. laminatus* among the different islands of the archipelago is mostly evident in the shape of the petiole, density and length of the gastral pilosity, and color. The petiolar nodes of specimens collected from Ovalau, where the type series was collected, are evenly convex at the apex, and the gasters are covered in a dense pelt of light pilosity. On Viti Levu, there appears to be a split between populations from the southeastern Suva region and populations from the central and western forests. The gasters of those specimens from the Suva area are covered in hairs that are generally denser and longer than their more western counterparts. The difference in pilosity between the two regions is not extreme compared to the difference of gastral pilosity of *C. levuanus*, but it is stark enough to suggest a geographical restriction to gene flow.

Specimens from Kadavu, Koro, Beqa and Moala are all similar to those of Viti Levu, with gastral pilosity tending towards the denser spectrum and petiolar nodes ranging from convex to very narrowly concave at the apex.

On Vanua Levu, *C. laminatus* strongly trends towards a darker pigmentation, and the mesosoma is often as dark as the head and gaster. In general, the antennal scapes and hind tibia of the Vanua Levu specimens maintain the lighter, reddish appearance of those from other islands, and in contrast to the dark antennal scapes and hind tibia of *C. levuanus*. The petiole nodes of these specimens attain a pointed peak at the apex. This distinct shape is shared by one of two specimens collected from Taveuni. The gastral pilosity of some specimens is also dark.

Several series of specimens are uniformly darker than even those previously discussed from Vanua Levu, but are in most ways similar to typical *C. laminatus* with respect to pilosity, propodeum shape, and size of majors. These include a large series from Nadarivatu (Viti Levu), a few minors from Nosoqo (Viti Levu), and a few minors and majors from Labasa (Vanua Levu). These specimens do not key out easily, but are treated here with reservation as *C. laminatus*.

Camponotus laminatus is among the most ubiquitous species in Fiji, and is frequently encountered foraging on vegetation and occasionally on the ground.

Material examined. **Beqa:** Mt. Korovou 326, Malovo 182. **Gau:** Navukailagi 336, Navukailagi 387, Navukailagi 364. **Kadavu:** Mt. Washington 800, Moanakaka 60, Mt. Washington 700, Lomaji 580, Namalata 100, Namalata 120, Namalata 50, Namalata 139, Solo taviene. **Koro:** Mt. Kuitarua 505, Nasau 476, Kuitarua 480, Mt. Kuitarua 485. **Moala:** Mt. Korolevu 375. **Ovalau:** Levuka 500, Levuka 450, Cawaci, Wainiloca, Cawaci, Ovalau, Levuka. **Taveuni:** Devo Peak 1187 b, Lavena 217, Nagasau. **Vanua Levu:** Wainibeqa 87, Kilaka 61, Wainibeqa 53, Vusasivo Village 190, Drawa 270, Vusasivo Village 400 b, Rokosalase 180, Rokosalase 94, Rokosalase 150, Rokosalase 97, Rokosalase 118, Rokosalase 94, Labasa. **Viti Levu:** Nabukavesi 40, Mt. Evans 800, Mt. Evans 800, Mt. Evans 800, Mt. Tomanivi 700 b, Navai 700, Vaturu Dam 575 b, Navai 863, Navai 870, Colo-i-Suva Forest Park 220, Vaturu Dam 700, Vaturu Dam 620, Vaturu Dam 550, Vaturu Dam 530, Savione 750 a, Vaturu Dam 540, Monasavu Dam 600, Korobaba 300, Nakobalevu 340, Colo-i-Suva 200, Colo-i-Suva 325, Colo-i-Suva 372, Colo-i-Suva 186 d, Navai 1020, Waivudawa 300, Veisari 300 (3.5 km N), Nadarivatu 750, Waisoi 300, Nausori, Navai, Waiyanitu, Vunidawa, Colo-i-Suva, Saiaro, Tailevu.

Camponotus levuanus Mann, NEW STATUS

(Plate 42)

Camponotus (*Myrmogonia*) *laminatus* var. *levuanus* Mann, 1921: 479; minor worker described. Type locality: Fiji, Vanua Levu, Wainunu (W. M. Mann). Syntypes: 4 workers, 3 males (USNM, examined).

Note: Specimens from the non-type locality of Labasa bear a red cotype labels, but are not true syntypes.

Camponotus levuanus is a dark species of moderate size that is a closely related to the sympatric and more ubiquitous *C. laminatus*. *Camponotus levuanus* can be distinguished from the latter by the darker and sparser pilosity of the gaster, which gives it a more shiny appearance. The gaster is also more elongate and slender in form, and the petiolar node is broad and often deeply concave apically. In addition, the antennal scapes and hind tibia of *C. levuanus* are generally the same dark color as the head and gaster, whereas even those of the *C. laminatus* from Vanua Levu with the dark mesosoma contrast with the head and gaster. The other *laminatus*-group species with dark and sparse gastral pilosity is *C. sadinus*, which can be distinguished by the presence of erect hairs on the pronotum, and the broader, less elongate gaster with very short appressed hairs. A high degree of color variation occurs within *C. levuanus*. Specimens from Vanua Levu and Kadavu tend to have a dark mesosoma that give them a uniform dark appearance, while those from Beqa, Moala and Viti Levu tend to have a reddish mesosoma more similar to the sympatric *C. laminatus*.

Camponotus levuanus was captured most frequently in malaise traps, though specimens were also taken from sifted litter and from workers foraging on vegetation.

Material examined. **Beqa:** Malovo 182. **Gau:** Navukailagi 432. **Kadavu:** Moanakaka 60, Moanakaka 60, Mt. Washington 700, Namalata 120, Vanua Ava b. **Lakeba:** Tubou 100 b. **Moala:** Mt. Korolevu 375. **Ovalau:** Cawaci, Ovalau. **Vanua Levu:** Kilaka 146, Wainibeqa 150, Vusasivo Village 190, Vuya 300, Mt. Wainibeqa 152 c, Vusasivo Village 400 b, Vusasivo Village 400 b, Rokosalase 180, Rokosalase 150, Rokosalase 118, Wainunu, Labasa. **Viti Levu:** Nabukavesi 40, Mt. Evans 800, Mt. Evans 800, Vaturu Dam 575 b, Vaturu Dam 620, Colo-i-Suva 186 d, Vunisea 300, Nabukavesi 300, Nadarivatu 750, Nausori, Vunidawa, Lami, Colo-i-Suva, Tailevu.

Camponotus maafui Mann

(Plate 43)

Camponotus (*Myrmogonia*) *maafui* Mann, 1921: 482, fig. 31; major and minor worker described. Type locality: Fiji, Vanua Levu, Lasema, (W. M. Mann). Syntypes: 1 major worker (MCZC 8721, examined); 7 minor workers, 8 major workers (USNM, examined).

Minor and major workers of *Camponotus maafui* are immediately identifiable both in the field and under the microscope by the golden gaster. The dorsal portions of first three gastral tergites of both majors and minors are a light brown red that contrasts strongly with the black integument of the rest of the body. The effect is exaggerated in the minor worker caste by the dense pelt of long semierect to semidecumbant yellow pilosity that covers the dorsal portion of the gaster. The species is otherwise quite similar in appearance to *C. cristatus* (see notes under that species for distinguishing characters).

Camponotus maafui occurs widely across Vanua Levu, but is unknown from any of the other islands, including nearby Taveuni. Like other *Myrmogonia*, it is most often encountered recruiting among the vegetation, sometimes in large numbers. Mann apparently named the species in honor of Enele Ma'afu'atuitoga, a 19th century Tongan prince who became chief of a large region of northern Fiji that included part of Vanua Levu.

Material examined. **Vanua Levu:** Kilaka 146, Wainibeqa 53, Wainibeqa 150, Kilaka 98, Vusasivo Village 190, Drawa 270, Vusasivo 50, Mt. Wainibeqa 152 c, Mt. Vatudiri 641, Mt. Delaikoro 734, Vusasivo Village 342 b, Vusasivo Village 400 b, Rokosalase 143, Rokosalase 180, Rokosalase 150, Rokosalase 97, Rokosalase 118, Rokosalase 94, Wainunu, Suene, Lasema a.

Camponotus sadinus Mann, NEW STATUS

(Plate 44)

Camponotus (*Myrmogonia*) *cristatus* subsp. *sadina* Mann, 1921: 482, workers described. Type locality: FIJI, Taveuni, Somosomo (W. M. Mann). Syntypes: 1 minor worker, 2 major workers, 1 queen (MCZC type no. 8720, examined).

Note: Specimens at the MCZC and USNM from the non-type locality of Nagasau bear a red cotype label, but are not true syntypes. The type label of syntype minor worker at the MCZC is 8702, but this may be an error in place of 8720. Furthermore, two pins of *C. sadinus* at the MCZC from Nagasau (non-type locality) bear the *C. nagasau* cotype label (type no. 8719), but are not true syntypes.

Camponotus sadinus is a black species with legs that vary from black to reddish. The gastral pilosity is sparse and dark. The mesonotal keel projects posteriorly as an acute tooth of variable size. The petiolar node is even apically, or comes to a point, but is never scalloped. See notes under *C. levuanus* for how to distinguish it from that species. In the Taveuni specimens, the mesonotal keel projects strongly into a tooth. In Kadavu specimens, the mesonotal keel projects posteriorly into a weak and obtuse angle, and the petiolar node tends to be more pointed apically. In Viti Levu, the single examined specimen has a mesonotal keel that projects weakly into a posterior tooth.

Mann originally described *C. sadinus* as a subspecies of *C. cristatus*, noting that the soldiers of the former were smaller in size, exhibit a more angular propodeum, and have an indistinct carina on the basal third of the clypeus. He also correctly noted that although the minor workers of both species are identical in size, the mesonotal keel of *C. sadinus* projects posteriorly into an acute tooth, and that of *C. cristatus* terminates in an obtuse angle. Furthermore, the gastral hairs of *C. sadinus* are sparser and darker than those of *C. cristatus*. Mann also stated that *C. sadinus* is intermediate between *C. cristatus* and *C. laminatus*. In general, morphological characters do not give clear answers regarding species distinctions in the laminatus complex. It is quite possible that limited gene flow continues to persist among species as delimited here, but the sympatry of both *C. sadinus* and *C. cristatus* on Taveuni suggests that reproductive isolation has been achieved on at least one island.

The majority of the *C. sadinus* material was collected from malaise trapping.

Material examined. **Kadavu:** Moanakaka 60, Namalata 100, Namalata 120, Namalata 50, Namalata 75, Namalata 139. **Taveuni:** Devo Peak 1187 b, Lavena 300, Lavena 235, Lavena 217, Lavena 219, Lavena 229, Tavoro Falls 160, Soqulu Estate 140, Lavena 235, Somosomo 200, Nagasau. **Vanua Levu:** Wainibeqa 87, Kilaka 61, Suene. **Viti Levu:** Nabukavesi 40, Vaturu Dam 575 b, Nadarivatu 750.

schmeltzi complex

The *schmeltzi* complex is composed of three species (*C. kadi*, *C. lauensis*, and *C. schmeltzi*), and is recognized by the angled posterolateral corners of the head and laterally compressed mesosoma.

Camponotus kadi Mann, NEW STATUS

(Plate 45)

Camponotus (*Myrmogonia*) *schmeltzi* var. *kadi* Mann, 1921: 485; minor worker described. Type locality: FIJI, Vanua Levu, Labasa (W. M. Mann). Syntypes: 2 minor workers, 1 male (MCZC type no. 8722, examined); 1 minor worker, 1 major worker, 1 male (USNM, examined) [additional material exists from type locality, but does not bear a red cotype label].

Camponotus (*Myrmogonia*) *schmeltzi* subsp. *troterri* Mann, 1921: 486, fig. 34; worker described. Type locality: FIJI, Taveuni, mountains near lake [Lake Tagimaucia] (W. M. Mann). Holotype [single specimen]: 1 minor worker (USNM, examined). NEW SYNONYMY.

Note: Specimens of *C. kadi* from the non-type localities of Wainunu and Suene at the MCZC bear red cotype labels, but are not true syntypes.

Camponotus kadi, as defined here, encompasses a broad range of geographically distinct morphotypes. Although the type locality of the species is in Vanua Levu, no specimens matching the description were collected there during the recent survey. However, the survey did collect a considerable amount of material from Viti Levu and Gau, where leg color and size tend to vary among individuals and populations within islands. The most extreme deviations are observed in the Gau specimens, where the coxae and legs of individuals can approach a color nearly as dark as the body.

Mann also described *C. schmeltzi* var. *loloma* from Lau, and *C. schmeltzi* subsp. *troterri* from Taveuni. Here we are synonymizing *C. loloma* with *C. schmeltzi*, and *C. trotteri* with *C. kadi*. The propodeum of *C. trotteri* forms a distinct posterior tooth that is distinctly different from any of the material examined from Viti Levu or Gau, but the difference in pilosity as discussed by Mann is far more subtle. The propodeum variation found in *C. loloma* varies from being nearly identical to that of *C. kadi* from Viti Levu and Gau to a very close approximation of *C. trotteri* from Taveuni. If the various forms occurred in sympatry, it would be easier to justify raising each of them to species. However, the lack of sympatry combined with a demonstrated tendency towards morphological plasticity makes the matter more complicated, and thus the tentative decision to consider them geographic variants of a single species.

Even with the expanded definition of *C. kadi*, as proposed here, distinct differences can separate both the majors and minors from the larger *C. schmeltzi*. The minor workers of *C. schmeltzi* are relatively large with the posterolateral corners of the head acutely angled, and the entire leg from coxa to femur is uniformly light brown to black. The minor workers of *C. kadi* are relatively small, the posterolateral corners of the head are evenly rounded, and the trochanter and distal portion of the coxa is a pale yellow that contrasts with the darker brown of the tibia and basal portion of the coxa. The major workers of C. *schmeltzi* are also larger, exhibit the same uniform color from coxa to the tibia as in seen in the minors, and series of strong well-defined longitudinal carinae are present on both the cheeks and clypeus. The majors of *C. kadi* (examined only from the Viti Levu and Gau population) are smaller, possess the same contrasting trochanters, and weak poorly defined carinulae are present on the cheeks and clypeus.

Camponotus kadi was most often collected in malaise traps, nesting in dead wood, and foraging on vegetation. The species was also collected from the ground and by litter sifting, however.

Material examined. **Gau:** Navukailagi 387, Navukailagi 408, Navukailagi 432, Navukailagi 490, Navukailagi 496, Navukailagi 557, Navukailagi 564, Navukailagi 575. **Kabara:** Kabara. **Ovalau:** Andubagenda. **Taveuni:** Lake Tagimaucia. **Vanua Levu:** Mt. Delaikoro 910, Wainunu, Suene, Labasa. **Viti Levu:** Mt. Tomanivi 700 b, Navai 700, Mt. Tomanivi 700, Nadala 300, Navai 770, Navai 930, Monasavu Dam 800, Mt. Naqaranabuluti 1050, Mt. Tomanivi 950, Monasavu 800, Monasavu Dam 600, Navai 1020, Nadarivatu 750, Navai.

Camponotus lauensis Mann
(Plate 46)

Camponotus (*Myrmogonia*) *lauensis* Mann, 1921: 488; minor worker described. Type locality: FIJI, Lau, Kabara, Waquava (W. M. Mann). Syntypes: 2 minor workers (MCZC type no.

8720, examined); 1 minor worker (USNM, examined).

Camponotus lauensis is a distinctive species recognizable by the submargined posterolateral corners of the head that appear dorsoventrally pinched, in combination with the robust mesonotal keel that forms an acute posterior tooth overhanging the petiole. The species remains known only from the type series, and is thus far the only ant species that can be considered endemic to the Lau Group.

Camponotus schmeltzi Mayr

(Plate 47)

Camponotus (*Myrmogonia*) *schmeltzi* Mayr, 1866: 490, worker described. Type locality: FIJI, Ovalau [not examined].

Camponotus (*Myrmogonia*) *schmeltzi* var. *loloma* Mann, 1921: 486, fig. 33, worker. Type locality: FIJI, Lau, Kabara, Waquava (W. M. Mann). Syntypes: 1 minor worker (MCZC, examined). NEW SYNONYMY.

Camponotus schmeltzi is known from six of the archipelago's major islands. It can be distinguished from *C. kadi* by its larger size, acutely angled posterolateral corners of the head, and uniform color from coxae to tibia. A more complete discussion on their morphological differences is given in the *C. kadi* section. *Camponotus schmeltzi* was most often collected in malaise traps and nesting in dead and dry branches, and occasionally in more moist logs.

Material examined. **Kadavu:** Mt. Washington 700, Lomaji 580. **Koro:** Nasau 476, Mt. Kuitarua 485, Mt. Nabukala 500, Mt. Kuitarua 440 b. **Ovalau:** Ovalau. **Taveuni:** Devo Peak 1188, Lavena 217. **Vanua Levu:** Kilaka 98, Kasavu 300, Vuya 300, Mt. Vatudiri 641, Mt. Vatudiri 570, Mt. Delaikoro 699, Rokosalase 143, Rokosalase 180, Rokosalase 150, Rokosalase 97, Rokosalase 118, Rokosalase 94, Lomaloma 587, Lomaloma 630. **Viti Levu:** Mt. Evans 800, Vaturu Dam 575 b, Navai 870, Colo-i-Suva Forest Park 220, Vaturu Dam 700, Vaturu Dam 550, Vaturu Dam 530, Mt. Batilamu 840 c, Vaturu Dam 540, Mt. Naqaranabuluti 1050, Lami 200, Nakobalevu 340, Colo-i-Suva 200, Colo-i-Suva 460, Colo-i-Suva 325, Colo-i-Suva 372, Colo-i-Suva 186 d, Mt. Evans 700, Naikorokoro 300, Veisari 300 (3.8 km N), Veisari 300 (3.5 km N), Nausori, Waiyanitu.

Genus Nylanderia

Diagnosis of worker caste in Fiji. Antenna 12-segmented. Antennal club indistinct. Antennal scapes with erect hairs. Scape length less than 1.5x head length. Eyes medium to large (> 3 facets); do not break outline of head. Metanotum impressed. Metapleuron with a distinct gland orifice. Propodeum unarmed and with convex declivity. Waist 1-segmented but can be obscured by gaster. Petiolar node present and with distinct anterior and posterior faces. Gaster armed with acidopore. Head, mesosoma and gaster with pairs of long thick hairs.

Nylanderia is a recently revived generic name into which many species formerly assigned to *Paratrechina* have been placed (LaPolla et al., 2010). *Nylanderia* species in Fiji are recognizable by the rows of thick standing hairs arranged in pairs on the head, mesosoma and gaster. They can be separated from *Paratrechina longicornis* by their shorter appendages and their eyes which do not break the outline of the head. For characters separating *Nylanderia* from *Paraparatrechina*, see the discussion under that genus. *Nylanderia* reaches its highest diversity in the tropics, but in many temperate areas *Nylanderia* form important components of the fauna. In addition to several invasive species and widespread Pacific species, several of the Fijian *Nylanderia* are considered to be endemic. The genus, however, has not radiated on the archipelago.

Key to *Nylanderia* workers of Fiji.

1 Mesopleuron with abundant appressed pilosity (usually > 20 hairs). Appressed pilosity abundant across mesosoma; head and gaster very shiny and appressed pilosity sparsely distributed. Pronotum and mesonotum each with 3 or more pairs of long standing hairs. Eyes large (facets > 50). Head approximately as wide as long. Dark brown in color. Pacific native .... ***Nylanderia* sp. FJ03**

– Mesopleuron shiny with appressed pilosity absent to sparse (usually < 20 hairs). Appressed pilosity sparse to absent across mesosoma. Head and gaster very shiny to dull with appressed pilosity sparse to abundant. Pronotum and mesonotum each with 2 pairs of long standing hairs. Eyes small to large. Head distinctly longer than wide. Pale yellow brown to dark brown in color 2

2 Gaster with first tergite more opaque and covered by a dense pubescence of often overlapping hairs. Eyes larger (facets > 35). Pacific native .... ***N. vaga***

– Gaster with first tergite shiny and free of pubescence. Eyes smaller (facets < 35 3

3 Eyes very small (facets < 8). Uniformly pale yellow brown. Fiji endemic .... ***N. vitiensis***

– Eyes larger (facets > 10). Darker brown. Pacific native .... ***N. glabrior***

Nylanderia glabrior (Forel)

(Plate 48)

Prenolepis braueri var. *glabrior* Forel, 1902: 490; worker, queen, male described. Type locality: AUSTRALIA [not examined]. Combination in *Paratrechina* (*Nylanderia*) (Emery, 1925: 221); in *Nylanderia* (LaPolla et al., 2010).

Nylanderia glabrior is a very shiny dark yellow brown species with small eyes and a first gastral tergite that is free of pubescence. These last two characters separate it from *N. vaga*, and the larger eyes and darker color separate it from *N. vitiensis*. The species occurs throughout the Pacific. *Nylanderia glabrior* was collected in Fiji from a handful of sifted litter samples taken from Viti Levu and Vanua Levu.

Material examined. **Vanua Levu:** Drawa 270, Mt. Vatudiri 641, Mt. Delaikoro 699. **Viti Levu:** Navai 870, Naboutini 300, Nabukavesi 300, Mt. Rama 300, Nausori.

Nylanderia vaga (Forel)

(Plate 49)

Prenolepis obscura r. *vaga* Forel, 1901: 26; worker described. Type locality: NEW BRITAIN, Ralum [not examined]. Senior synonym of *crassipilis* and *irritans* (Wilson & Taylor, 1967: 90). Combination in *Nylanderia* (LaPolla et al., 2010). For complete synoptic history see Bolton, et al. (2006).

Nylanderia vaga, as it is currently defined, encompasses a broad spectrum of morphological variation, and specimens range from pale yellow brown to black with every shade in between. Eye size, body size, head size and the distribution and quantity of appressed pilosity also vary considerably. *P. vaga* is most often confused with *Nylanderia* sp. FJ03, and the two can generally be distinguished by examining the pilosity of the mesopleuron (discussed under *Nylanderia* sp. FJ03). *Nylanderia crassipilis* Santschi was originally described from Fiji as a variety of *vaga*, but was synonymized with *N. vaga* by Wilson and Taylor (1967). The same authors also point ed out that Mann (1921) misidentified *N. vaga* from Fiji was as *N. vividula* Nylander, and (possibly) Cheesman and Crawley (1928) misidentified it as *N. vitiensis* Mann.

Nylanderia vaga is ubiquitous in Fiji, ranging from sea level to over 1000m, foraging on the ground and in the vegetation, and nesting opportunistically. Workers are also known to tend hemipterans. In addition to occurring in even the most disturbed of human landscapes, the species also frequently penetrates into interior forest habitats. It is easily collected by hand, baiting, litter sifting and malaise trapping.

Material examined. **Beqa:** Mt. Korovou 326, Malovo 182. **Gau:** Navukailagi 675, Navukailagi 415, Navukailagi 408, Navukailagi 432, Navukailagi 490, Navukailagi 505, Navukailagi 475, Navukailagi 356, Navukailagi 496, Navukailagi 575, Navukailagi 300. **Kadavu:** Mt. Washington 760, Moanakaka 60, Moanakaka 60, Vunisea 200, Namalata 100, Namalata 120, Namalata 50, Namalata 139, Daviqele 300, Namara 300. **Koro:** Mt. Kuitarua 500, Mt. Kuitarua 505, Mt. Kuitarua 485, Nasoqoloa 300, Mt. Kuitarua 380. **Lakeba:** Tubou 100 b, Tubou 100 c. **Moala:** Maloku 80, Naroi 75, Maloku 1, Mt. Korolevu 375, Mt. Korolevu 300. **Ovalau:** Draiba 300, Ovalau. **Taveuni:** Lavena 300, Lavena 235, Lavena 217, Lavena 219, Lavena 229, Soqulu Estate 140, Qacavulo Point 300, Somosomo 200. **Vanua Levu:** Wainibeqa 135, Wainibeqa 53, Kilaka 98, Mt. Kasi Gold Mine 300, Nakasa 300, Yasawa 300, Kasavu 300, Nakanakana 300, Vusasivo Village 190, Drawa 270, Vuya 300, Mt. Vatudiri 641, Mt. Vatudiri 570, Mt. Delaikoro 699, Vusasivo Village 400 b, Rokosalase 180, Rokosalase 150, Rokosalase 97, Rokosalase 118, Rokosalase 94, Lomaloma 587, Lomaloma 630, Mt. Delaikoro 391, Lagi 300, Dreketi 48, Wainunu, Suene, Lasema a. **Viti Levu:** Nabukavesi 40, Mt. Evans 800, Mt. Evans 800, Mt. Evans 800, Mt. Evans 700, Nakavu 300, Vatubalavu 300, Lautoka Port 5 b, Koronivia 10, Suva 10, Suva, Savione 750 a, Monasavu Dam 600, Nadakuni 300, Nadakuni 300 b, Korobaba 300, Lami 200, Nakobalevu 340, Colo-i-Suva 200, Waimoque 850, Colo-i-Suva 325, Lami 3, Colo-i-Suva 186 d, Volivoli 55, Volivoli 25, Mt. Evans 700, Naboutini 300, Nabukelevu 300, Navai 1020, Galoa 300, Nakavu 200, Vunisea 300, Nabukavesi 300, Mt. Rama 300, Naikorokoro 300, Veisari 300 (3.8 km N), Waivudawa 300, Veisari 300 (3.5 km N), Mt. Nakobalevu 200, Nadarivatu 750, Nausori, Waiyanitu.

Nylanderia vitiensis (Mann), REVISED STATUS
(Plate 50)

Prenolepis (*Nylanderia*) *vitiensis* Mann, 1921: 474, fig. 28; worker, male described. Type locality: FIJI, Kadavu, Vunisea. Syntypes: 2 workers, 1 male (MCZC type no. 8717, examined); 5 workers, 1 male (USNM, examined). Combined in *Paratrechina* (Emery, 1925: 221); in *Nylanderia* (LaPolla et al., 2010). Junior synonym of *vaga* (Dlussky, 1994).

Nylanderia vitiensis is a shiny yellow brown species with very small eyes and very sparse pubescence. The species can be separated from *N. vaga* by the smaller eyes, the paler and more yellow color, and the finer and less erect setae of the antennal scapes. *Nylanderia vitiensis* is most difficult to separate from *N. glabrior*, but see notes under the discussion of that species for diagnostic differences. Dlussky's (1994) synonymy of *N. vitiensis* with *N. vaga* is considered here to be in error, and the species status of the former is reinstated. *Nylanderia vitiensis* is widespread and frequently encountered in Fiji, both in the leaf litter and on vegetation. It is known to nest in rotting logs.

Material examined. **Gau:** Navukailagi 675, Navukailagi 408, Navukailagi 432, Navukailagi 490, Navukailagi 505, Navukailagi 475, Navukailagi 356, Navukailagi 575, Navukailagi 300. **Kadavu:** Mt. Washington 760, Moanakaka 60, Moanakaka 60, Mt. Washington 700, Namara 300, Vunisea. **Ovalau:** Levuka 550, Levuka 450, Draiba 300. **Taveuni:** Lavena 300, Lavena 217, Lavena 219, Lavena 229. **Vanua Levu:** Mt. Delaikoro 910, Kilaka 146, Nakasa 300, Yasawa 300, Kasavu 300, Nakanakana 300, Drawa 270, Vuya 300, Mt. Wainibeqa 152 c, Mt. Vatudiri 641, Mt. Vatudiri 570, Mt. Delaikoro 699, Vusasivo Village 400 b, Rokosalase 180, Rokosalase 150, Lomaloma 587, Mt. Delaikoro 391. **Viti Levu:** Nabukavesi 40, Mt. Evans 800, Mt. Evans 800, Mt. Evans 700, Mt. Tomanivi 700 b, Navai 700, Mt. Tomanivi 700, Nakavu 300, Vatubalavu 300, Nadala 300, Colo-i-Suva Forest Park 140, Naqaranabuluti 1000, Nadakuni 300, Nadakuni 300 b, Korobaba 300, Lami 200, Nakobalevu 340, Colo-i-Suva 200, Colo-i-Suva 460, Colo-i-Suva 372, Colo-i-Suva 186 d, Mt. Evans 700, Naboutini 300, Nabukelevu 300, Galoa 300, Nakavu 200, Vunisea 300, Nabukavesi 300, Mt. Rama 300, Naikorokoro 300, Waivudawa 300, Nadarivatu 750, Colo-i-Suva.

Nylanderia sp. FJ03

(Plate 51)

Nylanderia sp. FJ03 is a large, dark species with a densely pubescent mesopleuron. The species is most likely to be confused with *N. vaga*, from which it can usually be separated by the uniform distribution of abundant pubescence on the mesopleuron. However, there are some specimens of *N. vaga* that approach *Nylanderia* sp. FJ03 in the amount of pubescence, and these are very difficult to distinguish. The species we define here as *Nylanderia* sp. FJ03 is likely the same as much of the material referred to elsewhere in the Pacific as *N. bourbonica* (Forel). However, we have been informed that much of the material referred to as *N. bourbonica* does not actually belong to that species, and that the true *N. bourbonica* does not occur in Fiji (Shattuck, pers. comm.).

Material examined. **Taveuni:** Soqulu Estate 140. **Vanua Levu:** Mt. Delaikoro 910, Banikea 398. **Viti Levu:** Monasavu 800 a, Monasavu Dam 600, Nakobalevu 340.

Genus Paraparatrechina

Diagnosis of worker caste in Fiji. Antenna 12-segmented. Antennal club indistinct. Antennal scapes lacking erect hairs. Scape length less than 1.5x head length. Eyes medium to large (> 3 facets); do not break outline of head. Metanotum impressed. Metapleuron lacking a distinct gland orifice. Propodeum unarmed and with convex declivity. Waist 1-segmented but can be obscured by gaster. Petiolar node present and with distinct anterior and posterior faces. Gaster armed with acidopore. Head, mesosoma and gaster with pairs of long thick hairs.

Paraparatrechina was recently revived from synonymy with new status by LaPolla et al. (2010) to define a monophyletic group of mostly small paleotropical ants with compact mesosoma formerly placed in *Paratrechina*. The genus is characterized by a uniform erect setal pattern on the mesosoma. All species possess erect setae on the mesosoma that are distinctly paired, with two pairs on the pronotum, one pair on the mesonotum and one on the propodeum represents a morphological synapomorphy for the genus (LaPolla et al., 2010).

Fiji's single representative of the genus, *Paraparatrechina oceanica*, is most easily confused with species of *Nylanderia* and possibly *Paratrechina longicornis*. It can be separated from the former by the following characters: (1) antennal scapes and femora with erect hairs absent; (2) mesonotum with one pair of standing hairs; (3) propodeum with one pair of standing hairs; and (4) smaller size (HW < 0.45 mm). It is most easily separated from the latter, which also has erect hairs absent from the antennal scapes, by the following characters: (1) antennal scapes distinctly less than 1.5x length of head; (2) eyes do not break outline of head in full face view; and (3) smaller size (HW < 0.45 mm).

Paraparatrechina oceanica Mann, REVISED STATUS
(Plate 52)

Prenolepis (*Nylanderia*) *oceanica* Mann, 1921: 476, fig. 28; worker described. Type locality: FIJI, Nadarivatu (W. M. Mann). Syntypes: 2 workers, 1 queen (MCZC type no. 8718, examined); 3 workers (USNM, examined). Combination in *Paratrechina* (Emery, 1925: 221); in *Paraparatrechina* (LaPolla et al., 2010: 128). Junior synonym of *minutula* (Dlussky, 1994).

Paraparatrechina oceanica, as defined by Mann's type series from Nadarivatu, is a very small species with a dark brown to black head, mesosoma and gaster that all contrast with pale brown to pale yellow appendages. Like *P. minutula* (Forel), the standing setae on the head are sparse on the head and gaster. Mann describes the type series as possessing two pairs of standing hairs both on the pronotum and mesonotum and one pair on the propodeum. However, all the material examined matching the description matches the distribution as observed in *P. minutula* with two pairs of standing hairs on the pronotum, and one pair of standing hairs each on the mesonotum and the propodeum. Aside from the difference in color, Mann (1921) notes the similarity of his specimens to *P. minutula*, and distinguishes the former by its more robust form and broader subquadrate head. The two can also be separated by the scape length, which is relatively longer in *P. oceanica* (Steve Shattuck, pers. comm.).

There is a great range of size and color within what is treated here as *P. oceanica*. In Koronivia (Viti Levu), the specimens [e.g., CASENT0187436] most are a paler reddish color, and several workers from Kadavu [CASENT0181352, CASENT0181393] are of similar pale color and slightly smaller. In the Nadarivatu area, the specimens take on a much larger size. The larger size also holds for the syntypes from Nadarivatu.

In addition to being collected from malaise trapping and litter sifting, *P. oceanica* was found foraging on the ground, stones and vegetation, and also nesting in decaying wood and ant-plants.

Material examined. **Beqa:** Mt. Korovou 326, Malovo 182. **Gau:** Navukailagi 625, Navukailagi 632, Navukailagi 505, Navukailagi 356, Navukailagi 575, Navukailagi 300. **Kadavu:** Moanakaka 60, Moanakaka 60, Namara 300. **Koro:** Mt. Kuitarua 500, Mt. Kuitarua 505, Nasoqoloa 300. **Lakeba:** Tubou 100 b, Tubou 100 c. **Moala:** Naroi 75, Mt. Korolevu 375, Mt. Korolevu 300. **Ovalau:** Levuka 500, Draiba 300. **Taveuni:** Soqulu Estate 140. **Vanua Levu:** Wainibeqa 87, Kilaka 61, Mt. Kasi Gold Mine 300, Nakasa 300, Yasawa 300, Kasavu 300, Nakanakana 300, Vusasivo Village 190, Vusasivo 50, Mt.

Vatudiri 570, Vusasivo Village 400 b, Rokosalase 180, Mt. Delaikoro 391, Lagi 300, Labasa. **Viti Levu:** Nabukavesi 40, Mt. Evans 800, Mt. Evans 800, Mt. Evans 800, Mt. Tomanivi 700 b, Navai 700, Mt. Tomanivi 700, Koronivia 10, Mt. Batilamu 840 c, Savione 750 a, Monasavu 800, Monasavu Dam 600, Nadakuni 300 b, Korobaba 300, Lami 200, Nakobalevu 340, Colo-i-Suva 200, Waimoque 850, Colo-i-Suva 186 d, Naboutini 300, Nabukelevu 300, Galoa 300, Nakavu 200, Vunisea 300, Nabukavesi 300, Mt. Rama 300, Naikorokoro 300, Veisari 300 (3.8 km N), Waivudawa 300, Veisari 300 (3.5 km N), Nadarivatu 750, Nausori, Waiyanitu.

Genus Paratrechina

Diagnosis of worker caste in Fiji. Antenna 12-segmented. Antennal club indistinct. Antennal scapes with erect hairs. Scape length greater than 1.5x head length. Eyes medium to large (> 3 facets); break outline of head. Metanotum impressed. Metapleuron with a distinct gland orifice. Propodeum unarmed and with convex declivity. Waist 1-segmented but can be obscured by gaster. Petiolar node present and with distinct anterior and posterior faces. Gaster armed with acidopore. Head, mesosoma and gaster with pairs of long thick hairs.

Paratrechina was recently redefined as a monotypic genus (LaPolla et al., 2010). The single included species, *P. longicornis*, is a highly successful tramp ant that occurs across the tropics, including Fiji. It can be separated from the species of *Nylanderia*, which also bear short stout hairs arranged in pairs, by the following combination of characters: (1) antennal scapes approximately 1.5x length of head; (2) eyes strongly convex and breaking outline of head in full face view; (3) standing hairs white; (4) antennal scapes lacking standing setae.

Paratrechina longicornis (Latreille)

(Plate 53)

Formica longicornis Latreille, 1802: 113; worker described. Type locality: SENEGAL [not examined]. Senior synonym of *gracilescens* (Roger, 1863b: 10). Senior synonym of *currens* (Emery, 1892: 166). Senior synonym of *vagans* (Dalla Torre, 1893: 179). For complete synoptic history see Bolton, et al. (2006).

Paratrechina longicornis, commonly known as the Black Crazy Ant or the Longhorn Ant, is a dark colored species with extraordinarily long appendages, rows of long thick white hairs, and bulging eyes that break the outline of the head. In the field it can often be recognized by the rapid and erratic movement for which it is called a ‘crazy’ ant. It recruits strongly to baits and food resources with long foraging lines.

In a recent review of the species, Wetterer (2008) describes *P. longicornis* as a ubiquitous agricultural and household pest throughout the tropics and subtropics, and a pervasive indoor pest in the temperate areas. Wilson and Taylor (1967) offered the following discussion, “*P. longicornis* is one of the most widespread and abundant of all pantropical species. It is especially well adapted to dry habitats and abounds in the most urbanized portions of many tropical towns and cities. It probably originated within the Old World tropics, perhaps specifically in southeastern Asia or Melanesia.” Based on their phylogenetic analysis of the *Prenolepis* genus-group, Lapolla et al. (2010) suggest that the species is native to Southeast Asia.

Although it is ubiquitous in the cities and villages of Fiji, the species has not been observed penetrating into forested habitats, and is represented by very few specimens collected during the recent survey.

Material examined. **Naroi:** Nanunu-i-Ra Island. **Viti Levu:** Lautoka Port 5 b, Suva, Lami 3, Ellington Wharf 1, Nadarivatu 750.

Genus Plagiolepis

Diagnosis of worker caste in Fiji. Antenna 11-segmented. Antennal club indistinct. Scape length less than 1.5x head length. Head circular with length approximately equal to width. Eyes medium to large (> 3 facets) do not break outline of head. Metanotum not impressed. Metapleuron with a distinct gland orifice. Propodeum unarmed and with convex declivity. Waist 1-segmented but can be obscured by gaster. Petiolar node present and with distinct anterior and posterior faces. Gaster armed with acidopore; distinct constriction not visible between abdominal segments 3+4. Head, mesosoma and gaster lacking pairs of long thick hairs.

Plagiolepis alluaudi Emery
(Plate 54)

Plagiolepis alluaudi Emery, 1894b: 71; worker described. Type locality: SEYCHELLES IS. [not examined]. Senior synonym of *augusti*, *foreli*, *mactavishi* and *ornate* (Smith, 1958: 196). For complete synoptic history see Bolton et al. (2006).

Plagiolepis alluaudi is a minute, pale yellow species. Several minute to small, pale colored *Paraparatrechina* species from Fiji can be confused with *Pl. alluaudi*. *Plagiolepis alluaudi* can be distinguished by the lack of stiff standing hairs present on the mesosoma and face, and the less strongly impressed mesometanotum. *Camponotus* species often appear close to *Plagiolepis* in identification keys because neither genus typically has an impressed metanotum. However, no *Camponotus* species approaches the minute size of *Pl. alluaudi*. Furthermore, *Camponotus* species almost all have polymorphic worker castes and tend to have oval shaped heads that are longer than wide, while the head of *Pl. alluaudi* is circular.

Although an accomplished tramp species, *Plagiolepis alluaudi* is not known to cause significant harm to ecological or agricultural systems. Wilson and Taylor (1967) tentatively concluded it is native to Africa. It is more often found on vegetation in forested areas than in urban landscapes.

Material examined. **Kadavu:** Moanakaka 60. **Moala:** Naroi 75, Naroi. **Viti Levu:** Mt. Evans 800, Mt. Naqarababuluti 864, Volivoli 55.

Subfamily Myrmicinae

Genus Adelomyrmex

Diagnosis of worker caste in Fiji. Head shape ovoid to rectangular. Antenna 12-segmented. Antennal club 2-segmented. Mandibles triangular; armed with a distinct tooth on the basal margin in addition to those on the masticatory margin. Propodeum armed with spines or teeth. Waist 2-segmented.

Adelomyrmex is represented in Fiji by two species (*A. hirsutus* and *A. samoanus*). All of the material obtained during the recent survey was collected by sifting the leaf litter, and no nests of either species have been discovered in Fiji. The genus exhibits a peculiar disjunction between the Neotropical region and the western Pacific islands of New Guinea, New Caledonia, Fiji and Samoa. The presumed sister genus, *Baracidris*, is confined to the tropical regions of Africa. In his recent revision of *Adelomyrmex*, Fernández (2003) invoked the contraction of an ancient Gondwanian ancestral lineage of the sister genera to explain the disjunction. Neither the conservative (Brady et al., 2006) nor more liberal (Moreau et al., 2006) recent phylogenies of the Formicidae found evidence for so ancient an origin for the crown group of Myrmicinae, much less the most recent common ancestor of the two aforementioned genera. The alternative hypothesis, accounting for the disjunction of *Adelomyrmex* by long-distance dispersal from the Neotropics, faces its own challenges. Aside from the *Brachylophus* iguanas of Fiji and Tonga (Gibbons, 1981; Keogh et al., 2008; Pregill & Steadman, 2004; Zug, 1991), there are very few instances of Neotropical and western Pacific disjunctions. The discovery of morphologically homogeneous *A. hirsutus* populations across the Fijian archipelago, including smaller islands, such as Moala, suggests that the lineage maintains a relatively strong dispersal capacity. A phylogeny of the Pacific species with Neotropical outgroups would be provide an excellent tool for illuminating this peculiar biogeographical pattern.

Key to *Adelomyrmex* workers of Fiji.

1 Dorsum of head smooth and shiny with wide, deep, well-defined and evenly spaced foveae distributed across the entire surface except for a longitudinal median strip. Gaster with first tergite and sternite covered by same foveate pattern as found on head. Metanotal groove weakly impressed. Petiole more quadrate. Fiji endemic .... ***A. hirsutus***

– Dorsum of head with longitudinal rugae medially and rugoreticulate laterally. Gaster with first tergite and sternite weakly impressed basally by shallow overlapping foveae that give a shagreened appearance. Metanotal groove conspicuously impressed. Petiole more triangular. Pacific native. .... ***A. samoanus***

Adelomyrmex hirsutus Mann

(Plate 55)

Adelomyrmex (*Arctomyrmex*) *hirsutus* Mann, 1921: 458, fig. 21; worker described. Type locality: FIJI, Viti Levu, Nadarivatu (Mann). Holotype [single specimen]: 1 worker (USNM, examined). In generic revision (Fernández, 2003: 19).

Adelomyrmex hirsutus is a dark shiny species with wide well-defined foveae evenly distributed across the head and gaster. Although there is some minor variation in the sculpture across the archipelago, there is no morphologic evidence of highly structured populations. The specimens from New Caledonia that are reported as *A. hirsutus* (Wilson & Taylor, 1967) were apparently not examined by Fernández (2003), who did not include the island as part of the *Adelomyrmex* distribution, although they do lend circumstantial credence to the Gondwanian origin hypothesis. The two New Caledonian specimens examined for the present study match poorly against the type specimen and the additional Fijian material. Though near *A. hirsutus*, the New Caledonian specimens entirely lack the distinctive foveae of the Fijian specimens, and in profile the petiole is characterized by a significantly narrower node and longer peduncle.

The single specimen Mann collected was a worker of *A. hirsutus* found beneath a stone, and his conclusion that the species was hypogaeic is supported by recent collections. All of the specimens collected during the recent survey were taken from sifted litter.

Material examined. **Koro:** Tavua 220, Nasoqoloa 300. **Moala:** Naroi 75. **Vanua Levu:** Kasavu 300. **Viti Levu:** Mt. Evans 700, Mt. Batilamu 1125 b, Korobaba 300, Waivudawa 300.

Adelomyrmex samoanus Wilson & Taylor

(Plate 56)

Adelomyrmex (*Arctomyrmex*) *samoanus* Wilson & Taylor, 1967: 77, fig. 62; worker described. Type locality: SAMOA, Utumapu, Upolu. Holotype: 1 worker (MCZC type no. 31112, examined). Paratypes: 12 workers (MCZC type no. 31112, examined). In generic revision (Fernández, 2003: 27).

Adelomyrmex samoanus is a yellow brown species with a rugoreticulate face, a weakly sculptured mesosoma, and a subtriangular petiolar node. Aside from a slightly more foveolate gaster, the three specimens known from Fiji match well with the examined type material. The original description makes note of at least one morphologically aberrant specimen from Samoa, suggesting the discrepancies observed in the Fijian specimens may be within the scale of population level differences. The species was collected from a single litter sample on Koro.

Material examined. **Koro:** Nasau 465 a.

***Genus* Cardiocondyla**

Diagnosis of worker caste in Fiji. Head shape ovoid to rectangular. Antenna 12-segmented. Antennal club 3-segmented. Antennal scapes fail to reach posterior margin of head by at least the length of the first funicular segment. Antennal scrobes absent. Sides of head lacking carinate ridge extending below eye-level from mandibular insertions to posterolateral head margin. Mandibles triangular. Mesosoma lacking erect hairs. Propodeum armed with spines or teeth. Waist 2-segmented. Petiole pedunculate. Postpetiole swollen, in dorsal view wider than long and much broader than petiole.

Cardiocondyla is represented in Fiji by five known species, all of which occur broadly across the Pacific. In Fiji, *Cardiocondyla* is predominately restricted to disturbed habitats. Distinguishing among species is a difficult exercise, though one made easier by the taxonomic revision published by Seifert (2003). The characters used in the key presented here offer a less technical, and perhaps less reliable, alternative to the key published by Seifert, and the reader is referred to the latter for confirmation of ambiguous specimens.

Key to *Cardiocondyla* workers of Fiji.

1 Metanotal groove conspicuously impressed 2
– Metanotal groove weakly impressed.. 3
2 Postpetiolar node in dorsal view with rounded obtusely angled anterior corners and with anterior margin weakly indented. Propodeal spines shorter. Pacific native ... ***C. obscurior***
– Postpetiolar node in dorsal view with sharp acutely angled anterior corners and with anterior margin strongly indented. Propodeal spines longer. Introduced ....***C. emeryi***
3 Postpetiole bulging in profile view with sternite conspicuously convex. Maximum height of postpetiole subequal to that of petiole. Propodeal spines relatively short. Pacific native. .... ***C. nuda***
– Postpetiole not bulging in profile view, and sternite flat to very weakly convex. Maximum height of postpetiole less than that of petiole. Propodeal spines relatively short to moderate in length 4
4 Propodeal spines reduced to triangular angles. Pacific native .... ***C. kagutsuchi***
– Propodeal spines of moderate length. Pacific native... ***C. minutior***

Cardiocondyla emeryi Forel

(Plate 57)

Cardiocondyla emeryi Forel, 1881: 5; worker described. Type locality: VIRGIN IS. [not examined]. Senior synonym of *nereis* (Wilson & Taylor, 1967: 53); of *monilicornis* (Baroni Urbani, 1973: 200); of *mahdii, mauritia, rasalamae* (Bolton, 1982: 312). In generic revision (Seifert, 2003: 276). Current subspecies: nominal plus *fezzanensis*. For complete synoptic history see Bolton et al. (2006).

Cardiocondyla emeryi is believed to be native to Africa (Seifert, 2003), but is now one of many *Cardiocondyla* tramp species that has been spread across the globe by human activity. Together with *C. obscurior*, it can be separated from other congeners in Fiji by its conspicuously impressed metanotal groove. Unlike *C. obscurior*, the postpetiolar node in dorsal view lacks sharp acutely angled anterior corners, and the anterior margin is not strongly indented. *Cardiocondyla emeryi* also has longer propodeal spines than *C. obscurior. Cardiocondyla emeryi* has been observed in Fiji forming strong recruiting trails to food baits. It is distinctly bicolored in the field, and the species shows a preference for drier habitats.

Material examined. **Koro:** Mt. Kuitarua 500. **Moala:** Naroi 75. **Viti Levu:** McDonald's Resort 10 b, Suva 10.

Cardiocondyla kagutsuchi Terayama

(Plate 58)

Cardiocondyla kagutsuchi Terayama, 1999: 100, figs. 1-9; worker, queen, male, ergatoid male described. Type locality: JAPAN [not examined]. In generic revision (Seifert, 2003: 252).

Cardiocondyla kagutsuchi is known to occur widely across Asia and the Pacific Islands. Together with its purported sister species, *C. mauritanica*, this species has often been confused for *C. nuda*. For the most reliable means of separating these species, the reader is referred to the key in Seifert (2003). Clouse (2007a) offers a less technical means of differentiation. He observes that the postpetiolar sternite of *C. nuda* is more convex than that of *C. kagutsuchi*, making the postpetiolar height of the former species subequal to the petiolar height, whereas the postpetiolar height of the latter species is shorter than the petiolar height. The other species in Fiji with a weakly impressed metanotal groove is *C. minutior*, which can often be separated from *C. kagutsuchi* and *C. nuda* by the longer propodeal spines.

Material examined. **Viti Levu:** Lautoka Port 5 b.

Cardiocondyla minutior Forel
(Plate 59)

Cardiocondyla nuda var. *minutior* Forel, 1899: 120; worker described. Type locality: HAWAII [not examined]. Subspecies of *nuda* (Smith, 1944: 38). Junior synonym of *nuda* (Wilson & Taylor, 1967: 55). Revived from synonymy (Heinze, 1999: 251). Senior synonym of *tsukuyomi*; raised to species (Seifert, 2003: 283). For complete synoptic history see Bolton et al. (2006).

A rather nondescript species, *C. minutior* is another of the widespread *Cardiocondyla* known from Fiji. See the notes under *C. kagutsuchi* for how to distinguish it from other *Cardiocondyla* in Fiji. In the field, *C. minutior* is often found in disturbed habitats where a worker or two are foraging. Unlike *C. emeryi*, this species is a poor recruiter, and has been observed tandem running to bring nestmates to resources.

Material examined. **Gau:** Navukailagi 364. **Koro:** Mt. Kuitarua 500. **Viti Levu:** Mt. Tomanivi 1294, Galoa 300, Monasavu 800 a, Navai 863, McDonald's Resort 10 b, Koronivia 10, Suva, Ellington Wharf 1, Nuku 50, Nausori. **Yasawa:** Wayalailai Resort 55 b.

Cardiocondyla nuda (Mayr)
(Plate 60)

Leptothorax nudus Mayr, 1866: 508; worker described. Type locality: FIJI, Ovalau [not examined]. Combined in *Cardiocondyla* (Forel, 1881: 6). In generic revision (Seifert, 2003: 245). Current subspecies: nominal plus *sculptinodis*.

Cardiocondyla nuda was first described by from Ovalau. Since then its name has been erroneously applied to many species of similar appearance, including *C. kagutsuchi* and *C. minutior*, both of which occur in Fiji. For a discussion on how to separate *C. nuda* from other *Cardiocondyla* in Fiji, see the notes under *C. kagutsuchi*. Only several specimens of *C. nuda* were taken during the recent survey, including one that was nesting under a stone.

Material examined. **Viti Levu:** Lami 432, Nuku 50, Saiaro.

Cardiocondyla obscurior Wheeler

(Plate 61)

Cardiocondyla wroughtonii var. *obscurior* Wheeler, W. M. 1929: 44; worker, queen described. Type locality: TAIWAN [not examine]. Raised to species rank and senior synonym of *bicolor* (Seifert, 2003: 271).

Cardiocondyla obscurior can be distinguished from other *Cardiocondyla* in Fiji by the combination of the following characters: (1) the metanotal groove is strongly impressed, (2) the anterior corners of the postpetiolar node are sharply angled causing the postpetiolar anterior margin to appear strongly concave, and (3) the entire gaster is uniformly darker than the rest of the body. *Cardiocondyla obscurior* is one of the more commonly collected species of the genus in Fiji. It is known from many of the islands, and its presence in malaise traps suggests an arboreal habit.

Material examined. **Beqa:** Malovo 182. **Kadavu:** Moanakaka 60. **Macuata:** Vunitogoloa 10. **Moala:** Naroi 75. **Vanua Levu:** Vusasivo Village 400 b, Rokosalase 97, Lagi 300. **Viti Levu:** Mt. Evans 800, Mt. Evans 800, Navai 700, McDonald's Resort 10 b, Mt. Batilamu 840 c, Ellington Wharf 1, Volivoli 55.

Genus **Carebara**

Diagnosis of minor worker caste in Fiji. Head shape ovoid to rectangular. Antenna 9-segmented. Antennal club 2-segmented. Mandibles triangular. Propodeum armed with denticles. Waist 2-segmented.

Represented in Fiji by a single species, *Carebara* is most often mistaken for a very small *Solenopsis* species such as *S. papuana*, on account of its color, stature and two-segmented antennal club. *Carebara* can be distinguished by its nine-segmented antennae (the club of which outsizes the rest of the funiculus combined) and its conspicuously armed propodeum. Moreover, the single species that occurs in Fiji is dimorphic, and the major exhibits an oversized quadrate head with "horns" projecting from its posterolateral corners.

Carebara was recently proposed as the senior synonym for *Oligomyrmex* (Fernández, 2004), and the proposed change is adopted here. However, the Old World taxa, including the Pacific species, were not reviewed in any detail, and their status as true congeners of the New World *Carebara* is not conclusively demonstrated.

Carebara atoma (Emery)

(Plate 62)

Oligomyrmex atomus Emery, 1900: 328, pl. 8, fig. 30; major worker, minor worker described. Type locality: NEW GUINEA [not examined]. Combined in *Carebara* (Fernández, 2004: 235). Type specimens and Fijian material discussed (Taylor, 1991a: 606).

The workers of this minute species are among the smallest ants in the world, and can even be difficult to spot in a litter sample under the microscope. It was suggested first by Wilson and Taylor (1967) that the Fijian and Samoan populations, with their sparser mesosomal sculpture, might represent a distinct species from the New Guinea type series. However, given the lack of sympatry of the two morphotypes, they decided against describing a new species. Subsequently, Taylor (1991a) discovered specimens of both morphotypes from the same locality in New Guinea and offered a convincing argument based on sympatry that the name *C. atoma* Emery was being applied to two distinct species. Taylor's investigation of the putative type material, however, could not definitively assign the name to a single morphotype. Of the three pins examined, only one with a single soldier bears a type label. Of the two other pins, one has workers of the more sculptured species mounted, and the other has workers of the smoother species mounted, but neither bear any label indicating type status. Reconciliation of the discussed specimens must await a more focused investigation of the Pacific *Carebara* and the designated type material than the present study can afford. Until then, we keep with the convention of applying the name *C. atoma* Emery to the species present on Fiji.

Interestingly, *Carebara atoma* is one of the few species with a wide distribution across the archipelago which hasn't been recorded from Viti Levu. All of the specimens examined were taken from sifted litter.

Material examined. **Beqa:** Mt. Korovou 326. **Gau:** Navukailagi 490, Navukailagi 300. **Kadavu:** Moanakaka 60, Mt. Washington 700. **Koro:** Nasau 465 a, Nasoqoloa 300. **Moala:** Naroi 75. **Ovalau:** Draiba 300. **Vanua Levu:** Mt. Kasi Gold Mine 300, Nakasa 300, Vuya 300, Vusasivo Village 400 b, Rokosalase 180, Mt. Delaikoro 391.

Genus Eurhopalothrix

Diagnosis of worker caste in Fiji. Head shape triangular. Antenna 7-segmented. Antennal club 2-segmented or indistinct. Eyes located on upper margin of antennal scrobes. Mandibles triangular. Propodeum armed with spines or teeth. Waist 2-segmented; lacking spongiform tissue. Hairs flagellate or spatulate on at least some portion of head or body.

One of the more cryptic ant genera known from Fiji, *Eurhopalothrix* is represented by two described species, *E. emeryi* Forel and *E. insidiatrix* Taylor. A putative third species reported here as *Eurhopalothrix* sp. FJ52, is known from a single male specimen collected by a malaise trap during the recent survey. The genus is currently placed in the Bascicerotini, and occurs in the Indo-Australian and Neotropical regions. The wide-ranging *E. procera* (Emery), a putative close relative to the Fijian species, is the only Old World congener to occur east of Fiji.

Eurhopalothrix is difficult to confuse with any of the other genera present on Fiji. The triangular face, reduced number of antennal segments, and spatulate hairs can be mistaken for *Strumigenys*, or more likely *Pyramica* (on account of the triangular mandibles). However, *Eurhopalothrix* is several times the size and significantly more robust than the largest Fijian dacetine, the antenna is seven-segmented as opposed to six-segmented, the antennal scrobes are measurably more excavated, and the sculpture is conspicuously more furrowed.

In Fiji the *Eurhopalothrix* species are restricted to the litter of interior forests, but the alates are frequently captured by malaise traps. The distribution of the males suggests the actual range of both described species is greater than would be known by sole examination of the workers caste. It is presently impossible to associate the males collected by the malaise traps with workers, thus their identity cannot be determined.

Key to *Eurhopalothrix* workers of Fiji.

1 Posterior margin of head and anterior pronotal dorsum sculptured with thick intersecting rugae that form many enclosed piligerous pockets. Mesosoma in profile sculptured with thick rugae running in various directions and intersecting often ....***E. insidiatrix***

– Posterior margin of head and anterior pronotal dorsum sculptured with thick transverse rugae that do not intersect to form enclosed piligerous pockets. Mesosoma in profile sculptured with thick rugae running in parallel sublongitudinal lines but occasionally intersecting. ....***E. emeryi***

Eurhopalothrix emeryi (Forel)

(Plate 63)

Rhopalothrix emeryi Forel, 1912a: 58; queen described. Type locality: "AUSTRALIE" [dubious, see Brown & Kempf (1960: 230)]. Combined in *Eurhopalothrix* and senior synonym of *elegans* Mann, 1921: 467, fig. 25 (Type locality: FIJI, Viti Levu, Nadarivatu) (Brown & Kempf, 1960: 230). Lectotype: 1 worker (USNM, examined). Paralectotypes: 14 workers, 2 queens (USNM, examined); 4 workers (MCZC, examined). In revision of Fijian species (Taylor, 1980a: 235, figs. 7-10).

Eurhopalothrix emeryi workers can be distinguished from *E. insidiatrix* by the characters given in the key, namely the more groomed appearance of the cephalic and mesosomal rugae. Some confusion of geographic origin surrounds the holotype queen designated by Forel as *emeryi*. The only locality data associated with the specimen is "Austriale". The species has no other record of occurring in Australia, and no other ant species is known to be exclusively restricted to only Australia and Fiji. Therefore, the presumption of previous studies (Brown & Kempf, 1960; Taylor, 1980a) that the holotype is of Fijian origin but the locality was erroneously labeled, is maintained here.

In studying the material determined by Mann as *E. elegans*, Taylor (1980a) discovered that the name was applied to two distinct species. Mann clearly states the type series as being collected from Nadarivatu, hence Taylor designated a worker of that series as the lectotype, and assigned the material from Suene (Vanua Levu), Waiyanitu (Viti Levu) and Ovalau to *E. insidiatrix*.

Assuming one of the two more commonly collected males belongs to *E. emeryi*, then the species must be assumed to have a much wider range than known from collections of workers and queens alone. Mann writes that the species nests in small colonies beneath stones or logs deep in the woods, and that individuals are hard to discern on account of their neutral color and habit of remaining motionless when disturbed. All of the worker specimens collected during the recent survey were taken from sifted litter.

Material examined. **Vanua Levu:** Kilaka 61, Kilaka 98, Yasawa 300, Kasavu 300, Nakanakana 300, Drawa 270, Mt. Delaikoro 391. **Viti Levu:** Mt. Evans 800, Mt. Evans 700, Nadakuni 300 b, Korobaba 300, Nakobalevu 340, Nakavu 200, Vunisea 300, Waivudawa 300, Nadarivatu 750.

Eurhopalothrix insidiatrix Taylor
(Plate 64)

Eurhopalothrix insidiatrix Taylor, 1980a: 238, figs. 11-14; worker, queen described. Type locality: FIJI, Vanua Levu, Suene. Paratypes: 11 workers (MCZC, examined); 15 workers, 1 queen (USNM, examined).

Eurhopalothrix insidiatrix workers can be distinguished from *E. emeryi* by the characters given in the key, namely the more reticulated and random appearance of the cephalic and mesosomal rugae. Although some minor variation exists among populations of this species, it is clearly distinct from *E. emeryi*, with which it is sympatric at several localities in Viti Levu. Compared to the material examined, the type series is drawn from a population with relatively large workers. All of the worker specimens collected during the recent survey were taken from sifted litter.

Material examined. **Kadavu:** Mt. Washington 800. **Ovalau:** Ovalau. **Vanua Levu:** Suene. **Viti Levu:** Mt. Evans 800, Waimoque 850, Vunisea 300, Waiyanitu.

***Eurhopalothrix* sp. FJ52**
(Plate 65)

A single male specimen (CASENT0194603) from Waivudawa (Viti Levu) diverges significantly from the males of the more common two species. In addition to being yellow-red in color, the posterolateral corners of the head are acutely marginated, giving them a decidedly pinched appearance, and the propodeal spines are conspicuously more developed than those of the other species. Although there are no doubt additional species in Fiji represented only by males collected from malaise traps, this is the only case in which such a species can be unambiguously recognized.

Material examined. **Viti Levu:** Veisari 300 (3.5 km N).

***Genus* Lordomyrma**

Diagnosis of worker caste in Fiji. Head shape ovoid to rectangular. Antenna 12-segmented. Antennal club 3-segmented. Antennal scapes fail to reach posterior margin of head by at least the length of the first funicular segment. Antennal sockets not surrounded by a raised sharp-edged ridge. Anterior clypeal margin variously shaped, but never armed with three broad and blunt teeth. Frontal lobes relatively far apart so that the posteromedian portion of the clypeus, where it projects between the frontal lobes, is much broader than one of the lobes. Anterior margin of clypeus lacking a rectangular projection that extends over base of mandibles. Sides of head lacking carinate ridge extending below eye-level from mandibular insertions to posterolateral head margin. Mandibles triangular; lacking a distinct basal tooth. Mesosoma with depression distinctly separating promesonotum from propodeum; erect hairs present. Propodeum armed with spines or teeth. Propodeal lobes shorter than propodeal spines. Propodeal spines distinctly longer than diameter of propodeal spiracle. Waist 2-segmented. Petiole pedunculate; lacking large anteroventral subpetiolar process. Postpetiole not swollen; in dorsal view not distinctly broader than long or distinctly wider than petiole. Tip of sting tapered to a point and lacking a triangular to pendant-shaped extension.

The genus *Lordomyrma* is comprised of relatively uncommon and often elegantly sculptured ants occurring in the Australian and Oriental regions. Species occur in Japan, Borneo, New Guinea, eastern Australia, New Caledonia, the Solomon Islands and Fiji, in addition to the Philippines and perhaps other regions of the Pacific (Lucky & Sarnat, 2010; Taylor, 2009). The Fijian *Lordomyrma* are most likely to be confused for *Rogeria*, but differ in their possession of antennal scrobes, long propodeal spines, shorter petiolar peduncle.

Fiji has hosted a prolific radiation of these ants as reviewed by a recent revision of the genus across the archipelago (Lucky & Sarnat, 2008; Sarnat, 2006). There is strong support that all the Fijian *Lordomyrma* are derived from a single common ancestor that traces back to New Guinea (Lucky & Sarnat, 2010), and it is possible that Fiji was colonized via the ancient chain of islands, known as the Vitiaz Arc, during the Miocene.

Within the archipelago, the genus has been collected from nearly all the islands sampled during the recent survey, including Moala. Aside from the most common species of *Lordomyrma* (*L. tortuosa*), however, most of the other Fijian members of the genus are limited in their distribution, and many are known only from one island. Little is known about the biology of the Fijian *Lordomyrma* beyond their association with undisturbed mesic forests, maintenance of small inconspicuous colonies in soil and rotting logs, and their collection from the leaf litter and, to a lesser extent, the forest canopy. The following key and species accounts are modified from several recent studies of the Fijian taxa (Lucky & Sarnat, 2008; Sarnat, 2006), to which the reader is referred for more comprehensive treatments that include more diagnostic measurements and illustrations.

Key to *Lordomyrma* workers of Fiji.

1 In dorsal view, dorsal face of the propodeum gently sloping and continuously smooth, unbroken by a transverse carinate ridge or ridges between metanotal groove and insertion of propodeal spines. In larger workers promesonotum massive and spherical, produced well above the level of head and propodeum. Moderate sized to large workers (HW 0.73 mm–1.19 mm) 2

– In dorsal view, dorsal face of the propodeum with a transverse carinate ridge or ridges between metanotal groove and insertion of propodeal spines. Promesonotal shape, workers size and propodeal spines of variable length (HW 0.59 mm–0.89 mm) 6

2 Entire dorsum of head covered by thick crenulated rugae that become reticulated towards the posterior margin. Propodeal spines upturned. Very large species (HW 1.03 mm–1.19 mm) .... ***L. vanua***

– Dorsum of head with thick rugae absent, or only present laterad of the frontal carinae, but never down the median. Propodeal spines upturned to downcurved. Smaller species (HW 0.73 mm–0.90 mm) 3

3 Frontal carinae becoming confluent with well-developed, arcuate carinae posterior to the eye. Carinae present mesad of, and paralleling, frontal carinae. Propodeal spines straight. Large species (HW 0.82 mm–0.90 mm) .... ***L. tortuosa***

– Frontal carinae terminating before or just behind posterior level of eye. Carinae posterior to eye short and poorly-developed to absent. In full face view, carinae absent mesad of frontal carinae. Propodeal spines straight, downcurved or upcurved. Medium to large species 4

4 Propodeal spines robust and strongly upcurved. Reddish-brown species .... ***L. stoneri***

– Propodeal spines downcurved to straight but never upcurved. Reddish-brown to dark-brown 5

5 Propodeal spines weakly produced, straight to downcurved. Propodeal dorsum steeply sloped. Apical face of petiole sloping posteriorly with rounded apex. Weak sculpturing present behind eye. Smaller reddish-brown species (HW 0.73 mm–0.83 mm). .... ***L. desupra***

– Propodeal spines strongly produced, downcurved. Propodeal dorsum shallowly sloped. Apical face of petiole vertical with weakly peaked apex. Sculpturing absent behind eye. Larger dark-brown species (HW 0.87 mm–0.90 mm) .... ***L. vuda***

6 In profile, propodeal spines strongly produced, length greater than width of procoxae. Petiole more robust with a gently sloping posterior face and overlain by a coarse rugoreticulum. Antennal scrobes broad and weakly defined, or if well-defined, then also possessing upturned propodeal spines 7

– In profile, propodeal spines weakly produced, straight to downcurved, shorter than or equal to width of procoxae; petiole more slender, subtriangular with a steeply sloping posterior face and overlain by a fine rugoreticulum. Antennal scrobes narrow and more deeply impressed 10

7 Entire face densely rugose between frontal carinae, rugoreticulate laterally and posteriorly. Procoxae with well-defined transverse striae .... ***L. rugosa***

– Median portion of face smooth with scattered foveolae, if carinae are present they are restricted to the area immediately mesad of the frontal carinae. Procoxae lacking well-developed transverse striae 8

8 Posterior corners of the head overlain by rugoreticulum. Frontal carinae becoming confluent with rugoreticulum posterior to the eye forming a broad, well-defined antennal scrobe. Propodeal spines upcurved. Small species (HW < 0.63 mm) .... ***L. curvata***

– Posterior corners of the head either smooth or with a few weak carinae, but never overlain by rugoreticulum. Frontal carinae terminating before reaching any carinae posterior to eye and never forming a well-defined antennal scrobe. Propodeal spines straight. Larger species (HW > 0.63 mm) 9

9 Longest hairs on promesonotum, petiole, postpetiole and gaster exceeding length of eye. Series of strong carinae present between eyes and ventrolateral margin of face. Large species (HW 0.89 mm) .... ***L. levifrons***

– Hairs on promesonotum, petiole, postpetiole and gaster shorter than length of eye. Strong carinae absent between eyes and ventrolateral margin of face. Smaller species (HW 0.64 mm–0.76 mm) .... ***L. polita***

10 In full face view, median region of head between frontal carinae filled with fine and tightly packed longitudinal striae. Sides and dorsum of pronotum rugose to rugulose. .... ***L. striatella***

– In full face view, median region of head between frontal carinae smooth and shiny with scattered foveolae. Sides and dorsum of pronotum smooth, lacking striae or rugae. .... ***L. sukuna***

Lordomyrma curvata Sarnat

(Plate 66)

Lordomyrma curvata Sarnat, 2006: 15, figs. 2-3; worker described. Type locality: FIJI, Vanua Levu, Kasavu Village [16°42′S 179°39′E, 300m]. Holotype: 1 worker (FNIC). Paratypes:

worker (USNM).

Known only from Vanua Levu, *L. curvata* is a very small reddish-brown species with long upturned propodeal spines and the well-developed rugoreticulum present posterior to the eyes. While *L. rugosa* and *L. striatella* both possess rugoreticulate occipital corners, only in *L. curvata* are the antennal scrobes and area between the frontal carinae covered by a smooth and shiny surface. The only other species that possesses strongly upcurved spines is *L. stoneri*, from which *L. curvata* can be readily differentiated by the rugoreticulate posterior corners of the head and substantially smaller size.

Material examined. **Vanua Levu:** Nakanakana 300, Drawa 270.

Lordomyrma desupra Sarnat

(Plate 67)

Lordomyrma desupra Sarnat, 2006: 17, figs. 4-5; worker described. Type locality: FIJI, Viti Levu, Monasavu Rd., 1.75 km SE Waimoque Settlement, 17°40′13″S 177°59′38″E, 850m (E. M. Sarnat). Holotype: worker (FNIC). Paratypes: workers (ANIC, CASC, LACM, MCZC, USNM).

Lordomyrma desupra is a medium to large reddish-brown species with a large to massive promesonotum, straight to downcurved propodeal spines of modest length, and reduced cephalic and body sculpturing. Like its close relatives, *L. tortuosa* and *L. stoneri*, the species lacks a transverse carina between the propodeal dorsum posterior to the metanotal groove and possesses a robustly produced promesonotum that, in larger workers, bulges above the level of its head and propodeum. Of the three other species that share these characters, *L. desupra* can be distinguished from *L. tortuosa* by its weaker facial sculpture, from *L. stoneri* by its smaller, more slender appearance and straight propodeal spines, and *L. vanua* by its thick crenulated cephalic rugae. *Lordomyrma desupra* shows a wider variation in the size of workers than normally encountered within the Fijian *Lordomyrma*. In smaller workers, the size of the promesonotum and propodeal spines is markedly smaller.

The collection of the species from *Hydnophytum* (Rubiaceae) ant-plants and from canopy fogging, together with its absence from sifted litter collections, suggests *L. desupra* is a component of Fiji's arboreal ant fauna.

Material examined. **Viti Levu:** Mt. Tomanivi 700 b, Vaturu Dam 575 b, Vaturu Dam 620, Monasavu 800, Monasavu Dam 600, Waimoque 850, Mt. Nakobalevu 200.

Lordomyrma levifrons (Mann)

(Plate 68)

Rogeria tortuosa subsp. *levifrons* Mann, 1921: 453; worker described. Type locality: FIJI, Viti Levu, Nadarivatu (W. M. Mann). Syntypes: workers (USNM, examined). Combined in *Lordomyrma* (Kugler, 1994: 26). Raised to species status (Sarnat, 2006: 20, figs. 6-7).

This large, robust, long-spined ant is known only from Mann's collections and is the only species of *Lordomyrma* he described that was not collected during the recent survey. *Lordomyrma levifrons* is distinguished from its Fijian congeners by its long propodeal spines, fine-tipped hairs present on its face, mesosoma, petiole, and gaster.

Lordomyrma polita (Mann)

(Plate 69)

Rogeria tortuosa subsp. *polita* Mann, 1921: 453; workers, queen described. Type locality: FIJI, Viti Levu, Nadarivatu (W. M. Mann). Syntypes: workers (USNM, examined); workers (MCZC, examined). Combined in *Lordomyrma* (Kugler, 1994: 26). Raised to species status (Sarnat, 2006: 21, figs. 8-9).

Lordomyrma polita is a medium-sized reddish-brown species with long straight propodeal spines, short hair, shallowly impressed antennal scrobes, reduced facial sculpture and a robust petiole. *Lordomyrma polita* is readily discernible from the other long-spined Fijian species, *L. levifrons*, by its short hairs. There exists considerable variation with respect to sculpture of the promesonotum among specimens of *L. polita* (Sarnat, 2006), though strong support for monophyly is found for the two specimens included in phylogenetic analysis (Lucky & Sarnat, 2010).

Thus far, *L. polita* is known from the highlands surrounding Mt. Tomanivi, the drier western forest near Vaturu Dam, the interior of Vanua Levu, and the island of Koro. This species has been observed nesting in logs in the Navai area and Vanua Levu, and nesting between epiphyte roots and a tree trunk on the island of Koro.

Material examined. **Koro:** Mt. Nabukala 520, Nasau 420 b. **Vanua Levu:** Mt. Vatudiri 641. **Viti Levu:** Mt. Evans 800, Vaturu Dam 575 b, Monasavu 800.

Lordomyrma rugosa (Mann)

(Plate 70)

Rogeria rugosa Mann, 1921: 455; worker, queen described. Type locality: FIJI, Viti Levu, Nadarivatu. Syntypes: workers (MCZC, examined); workers (USNM, examined). Combined in *Lordomyrma* (Kugler, 1994: 26). In revision of Fijian species (Sarnat, 2006: 23, figs. 10-11).

Lordomyrma rugosa is one of the most distinctive species of *Lordomyrma* in Fiji. Like *L. levifrons*, *L. polita* and *L. curvata*, this species possesses long propodeal spines, well-developed, upturned propodeal lobes, and a robust petiole. It differs from the general appearance of the aforementioned species in its small eyes, darker coloration, and the heavy rugoreticulum covering all surfaces of its face. The only other Fijian congener with such strong facial sculpturing is *L. striatella*, from which *L. rugosa* can be distinguished by its larger size, coarser sculpture, rugoreticulate antennal scrobes, longer propodeal spines and lobes, and more robust petiole. Additionally, *L. rugosa* is the only known species of all Fijian *Lordomyrma* to bear strong striations on its mandibles and procoxae.

Lordomyrma rugosa is only known from the Nadarivatu, Mt. Tomanivi area. Mann (1921) noted that the colonies are small and live beneath stones or in the ground, and that the workers are slow moving. We observed a colony nesting directly in the soil, with a small inconspicuous hole flush with the ground serving as the entrance.

Material examined. **Viti Levu:** Navai 930, Mt. Tomanivi 950, Waimoque 850

Lordomyrma stoneri (Mann)

(Plate 71)

Rogeria tortuosa subsp. *stoneri* Mann, 1925: 5; worker described. Type locality: FIJI, Viti Levu, Tamavua. Syntypes: workers (USNM, examined). Combined in *Lordomyrma* (Kugler, 1994: 26). Raised to species status (Sarnat, 2006: 25, figs. 12-13).

One of the larger species of *Lordomyrma* in Fiji, *L. stoneri* has an attractive and shiny reddish-brown integument and strong, upturned spines. The species lacks a carinate anterior margin on the dorsal face of its propodeum and all examined specimens possess a robustly produced promesonotum that bulges above the level of its head and propodeum. Of the other three Fijian species that possess these characters, *L. stoneri* can be distinguished from *L. tortuosa* by its weaker cephalic sculpture, and from *L. desupra* and *L. vuda* by its strongly upturned propodeal spines. The only other species that has strongly upturned spines is *L. curvata*, which is almost half the size of *L. stoneri*, has a strongly carinate anterior margin of the dorsal face of its propodeum, and has a strong rugoreticulum on the posterolateral corners of its head.

Lordomyrma stoneri is constricted to a narrow range of mountains in south-eastern Viti Levu, close to Suva, where it is sympatric with its close relatives, *L. tortuosa* and *L. desupra*. Little is known about the biology of this species, expect that it has been collected by litter sifting.

Material examined. **Viti Levu:** Nakobalevu 340, Waivudawa 300.

Lordomyrma striatella (Mann)

(Plate 72)

Rogeria striatella Mann, 1921: 454; worker described. Type locality: FIJI, Kadavu, Vanua Ava (W. M. Mann). Syntypes: workers (USNM, examined). Combined in *Lordomyrma* (Kugler, 1994: 26). In revision of Fijian species (Sarnat, 2006: 27, figs. 14-15).

Lordomyrma striatella is a close relative of *L. sukuna*. Together, they are characterized by a narrow well-developed antennal scrobe, a slender subtriangular petiole, striations on the propodeal declivity between the insertion of the spines, relatively short propodeal spines, weakly-produced propodeal lobes, fine rugoreticulate sculpturing, long hairs on the dorsal surfaces, and dark coloration. *Lordomyrma striatella* can be easily separated from *L. sukuna* by the thin longitudinal striae running the length of its face within the bounds of the frontal carinae. While *L. rugosa* also has strong sculpturing between its frontal carinae, *L. striatella* can be distinguished by its more strongly developed and smooth antennal scrobe, more triangular petiole, smaller and more slender appearance, and weaker propodeal spines and lobes.

This species is recorded from collections scattered across Viti Levu, Ovalau, Beqa I., and Kadavu. Many of the collections have been made from the leaf litter, and Mann reports them as being abundant from Kadavu where he found them nesting beneath stones.

Material examined. **Beqa:** Mt. Korovou 326. **Kadavu:** Moanakaka 60. **Ovalau:** Draiba 300. **Taveuni:** Qacavulo Point 300. **Viti Levu:** Mt. Naqaranabuluti 1050, Mt. Tomanivi 950, Galoa 300, Vunisea 300.

Lordomyrma sukuna Sarnat

(Plate 73)

Lordomyrma sukuna Sarnat, 2006: 29, figs. 16-17; worker described. Type locality: FIJI, Viti Levu, Mt. Naqaranibuluti 1.3 km W Emperor Gold Mine Rest House, 17°34′10″S 177°58′20″E, 1050m (E. M. Sarnat). Holotype: worker (FNIC). Paratypes: workers (ANIC, CASC, LACM, MCZC, USNM).

Lordomyrma sukuna is a medium-sized black species with long hair, a slender petiole, short propodeal spines and reduced facial sculpture. See notes under *L. striatella* for discussion on how to separate the two species. *Lordomyrma sukuna* demonstrates considerable variation among the different island populations (Sarnat, 2006). The Viti Levu specimens from the Navai region were taken from logs and under stones while the Ovalau and Taveuni specimens were collected from sifted litter, suggesting these ants are components of the ground fauna. The type series is from a colony collection of 30 workers that was made from a nest in soil beneath a stone, identifiable by excavated earth adjacent to the entrance.

Material examined. **Beqa:** Mt. Korovou 326. **Kadavu:** Moanakaka 60. **Ovalau:** Draiba 300. **Taveuni:** Qacavulo Point 300. **Viti Levu:** Mt. Naqaranabuluti 1050, Mt. Tomanivi 950, Galoa 300, Vunisea 300.

Lordomyrma tortuosa (Mann)

(Plate 74)

Rogeria tortuosa Mann, 1921: 452, figs. 18a-b; worker described. Type locality: FIJI, Ovalau (W. M. Mann). Syntypes: workers (MCZC, examined); workers (USNM, examined). Combined in *Lordomyrma* (Kugler, 1994: 26). In revision of Fijian species (Sarnat, 2006: 33, figs. 18-19).

Lordomyrma tortuosa is a large-sized, shiny, reddish-brown species with a massive promesonotum, modestly-sized straight propodeal spines and strong arcuate carinae on face. *Lordomyrma tortuosa*, together with *L. desupra*, *L. stoneri* and *L. vuda* lacks a transverse carina on the dorsal face of its propodeum posterior to the metanotal groove and possesses a robust promesonotum that bulges above the level of its head and propodeum. It can be readily distinguished from these three by the frontal carinae that join with the arcuate carinae posterior of the eye, and the presence of longitudinal carinae that run immediately inward from the frontal carinae. Although the number and strength of these carinae vary, the variation does not appear to follow a distinguishable geographic pattern.

With many records from eight of the archipelago's islands, *L. tortuosa* is the most geographically widespread of *Lordomyrma* species occurring in Fiji. The species is often collected from leaf litter, and nests of small colonies have been found in logs and under stones. Additionally, *L. tortuosa* appears to be restricted to the lower elevations of the islands. Mann

(1921) noted that he often found workers of this species foraging on mossy stones in ravines, and we have also observed workers gleaning the surfaces of stones on the banks of rivers.

Material examined. **Beqa:** Mt. Korovou 326. **Gau:** Navukailagi 625, Navukailagi 387, Navukailagi 490, Navukailagi 475. **Kadavu:** Mt. Washington 700. **Koro:** Mt. Nabukala 500, Nasau 470 (3.7 km), Mt. Kuitarua 440 b, Nasau 420 b, Nasau 465 a. **Moala:** Mt. Korolevu 375. **Taveuni:** Devo Peak 1188, Lavena 217, Mt. Devo 892, Tavoro Falls 160, Tavoro Falls 100, Soqulu Estate 140. **Vanua Levu:** Kasavu 300, Nakanakana 300, Drawa 270, Vuya 300, Mt. Vatudiri 570, Lomaloma 630, Mt. Delaikoro 391. **Viti Levu:** Mt. Evans 800, Mt. Evans 800, Mt. Evans 700, Nadakuni 300 b, Korobaba 300, Nakobalevu 340, Vunisea 300, Naikorokoro 300.

Lordomyrma vanua Sarnat
(Plate 75)

Lordomyrma vanua Lucky & Sarnat, 2008: 42, figs. 4-5; worker described. Type locality: FIJI, Vanua Levu, Mt. Delaikoro, 3.7 km E Dogoru Village 16°34.515′S 179°18.983′E, 699m (E. P. Economo). Holotype: worker (FNIC). Paratype: worker (USNM).

Lordomyrma vanua is a large shiny black species with deep, widely spaced furrows running longitudinally on head and mesosoma. *Lordomyrma vanua*, with its heavily rugose head and mesosoma, is similar to *L. rugosa*, but can be distinguished by its smooth forecoxae, smooth propodeal declivity, smooth anterodorsal region of the promesonotum, broader and more widely-spaced rugae, and larger size. The other species that *L. vanua* might be confused with is *L. striatella*, from which it can be separated by its weaker antennal scrobes, broader and more widely spaced rugae, more well-developed propodeal spines, more robust petiole, and larger size. Despite its morphological resemblance to *L. rugosa*, molecular phylogenetic analysis places *L. vanua* as a closer relative to species such as *L. tortuosa* and its relatives, which also lack the transverse carinae across the dorsal propodeum. Thus far, *L. vanua* has only been collected twice, both times from the litter of Mt. Delaikoro on Vanua Levu.

Material examined. **Vanua Levu:** Mt. Delaikoro 699.

Lordomyrma vuda Sarnat
(Plate 76)

Lordomyrma vuda Sarnat, 2006: 34, figs. 20-21; worker described. Type locality: FIJI, Viti Levu, Koroyanitu National Park, Savione Falls, 2 km ESE Abaca Village, 17°40′33.6″S 177°33′00.5″E, 650m (E. M. Sarnat). Holotype: FNIC. Paratypes: workers (ANIC, MCZC, USNM).

Lordomyrma vuda is a large dark-brown species with sparse cephalic sculpture, long appendages, and strongly produced downcurved propodeal spines. Like *L. desupra*, *L. stoneri* and *L. tortuosa*, and *L. vanua*, it lacks a transverse carinate margin posterior to the metanotal groove on the dorsal face of its propodeum. Like *L. desupra* and *L. stoneri*, it lacks a developed facial sculpture and the presence of longitudinal carinae that run inward from, and parallel to, the frontal carinae. The downcurved spines of *L. vuda* distinguish the species from *L.*

stoneri. *Lordomyrma vuda* can be distinguished from *L. desupra* by the more vertical, peaked appearance of its petiole node, the more robust propodeal spines, the more shallowly sloped propodeal dorsum, its larger size and darker coloration.

Thus far, *L. vuda* has only been collected from two nearby localities in western Vuda Province (Viti Levu). Both collections were made from workers foraging on stones, with one locality being adjacent to a river. So far, *L. vuda* is the only species of the genus that appears to be restricted to the drier, leeward mountain ranges of western Viti Levu.

Material examined. **Viti Levu:** Mt. Batilamu 840 c, Savione 750 a.

Genus *Metapone*

Diagnosis of worker caste in Fiji. Head shape rectangular. Antenna 11-segmented. Antennal club 3-segmented. Antennal scrobes present. Anterior margin of clypeus with a rectangular projection that extends over base of mandibles. Mandibles triangular. Propodeum lacking spines or teeth. Waist 2-segmented. Petiole lacking peduncle; with large anteroventral subpetiolar process.

Metapone is an easily recognizable but rarely encountered part of the Fijian ant fauna. It is believed the genus is a specialist on termites, both taking them for prey and occupying their galleries in large stumps and logs (Hölldobler et al., 2002). In this context, the Fijian species, with their armored, compact, cylindrical bodies look well designed for hunting and living inside small tunnels. Their elongated heads are rectangular, with a projecting and lobed clypeus overhanging powerfully built mandibles. The antennae are dorsoventrally compressed, and can retract entirely into the deeply excavated antennal scrobes. The short legs are laterally compressed with swollen femora, the propodeum is unarmed, and the quadrate petiole is massively built.

The genus was unknown from Fiji until the alates were captured in malaise traps. No workers of *Metapone* have been collected, but the alates were found on many of the archipelago's islands, including Viti Levu, Vanua Levu, Taveuni, Koro and Gau. The species listed here is believed to be undescribed (R. W. Taylor, pers. comm.).

Metapone sp. FJ01

(Plate 77)

There are approximately 13 specimens of queens and males that were captured across the archipelago in malaise traps. Although there is slight variation in size and sculpture, the great majority of these specimens appear to represent a single widespread species. A single male specimen (CASENT0181675) from Gau, however, appears distinct from all other males in its larger size and conspicuously coarser sculpture. Whereas the posterior dorsum of the head, along with the dorsal surfaces of the pronotum and mesonotum are sculptured with fine rugulae and punctations, the same surfaces on the Gau specimen are more shiny without the micropunctation, and sculptured with robust and coarsely reticulate rugae.

Without a more thorough review of Pacific *Metapone*, it is difficult to say whether the differences noted above fall within intraspecific variation or whether they warrant distinction of a separate species. One clue is that the queen known from Gau is not remarkable in its morphology relative to those of other islands. For the purposes of this study the aberrant male is treated as belonging to the same species as all the other Metapone known from Fiji.

Notes on the description of the queen.
Head rectangular, approximately one third longer than wide. Eyes large, length subequal to antennal scape, situated behind midline of head. Clypeus narrowing anteriorly into a blunt squared shelf that overhangs mandibles. Mandibles powerfully built with masticatory margin armed with five teeth of approximately equal size. Antenna 10-segmented, scape broad and dorsoventrally flattened, length of funiculus approximately equal to length of scape, and terminal three segments enlarged into a flat club. Antennal scrobes deeply excavated and capable of receiving entire antennae. Mesosoma elongate. Propodeum with obtusely angled margins, but unarmed. Legs short and laterally flattened with swollen femora. Petiolar node robust and quadrate with subpetiolar process, in dorsal view trapezoidal with posterior corners projecting and diverging as acute angles causing the posterior margin to broadly concave. Postpetiolar node broadly attached to gaster.

Face very shiny with weak longitudinal striations mesad of the frontal carinae, and well-defined striations laterad of the frontal carinae. Head with venter smooth and shiny. Mesosoma with well-defined longitudinal striations on dorsal surfaces of pronotum and mesonotum, the rest of the surfaces very shiny with weaker striations. Dorsal surfaces of propodeum and petiole with scattered fine and weak punctures. Postpetiole shagreened. Gaster weakly shagreened. Pilosity of varying length fine and light.

Material examined. **Gau:** Navukailagi 387, Navukailagi 496. **Koro:** Mt. Kuitarua 500. **Taveuni:** Devo Peak 1188. **Vanua Levu:** Lomaloma 587. **Viti Levu:** Mt. Evans 800, Mt. Evans 800, Veisari 300 (3.5 km N).

Genus Monomorium

Diagnosis of worker caste in Fiji. Head shape ovoid to rectangular. Antenna 12-segmented. Antennal club 3-segmented. Sides of head lacking carinate ridge extending below eye-level from mandibular insertions to posterolateral head margin. Mandibles triangular. Mesosoma with erect hairs present. Propodeum either unarmed or with weak angle, but never with well-developed spines or teeth. Waist 2-segmented. Petiole pedunculate; lacking large anteroventral subpetiolar process. Postpetiole not swollen; in dorsal view not distinctly broader than long or distinctly wider than petiole.

The *Monomorium* of Fiji are small species of varying shapes, sizes and colors, but the propodeum of each is armed at most with a weak angle, and all have three-segmented antennal clubs. The genus is most often mistaken for *Solenopsis* (which has two-segmented antennal clubs), or *Cardiocondyla* (which has distinct propodeal spines or angles). *Monomorium* is represented in Fiji by six species of which several are exotic, several are native, and one (*M. vitiense*) is tentatively considered endemic to the archipelago.

Key to *Monomorium* workers of Fiji.

1 Eyes small and composed of 1-2 facets. Propodeum weakly to moderately angulate. Sculpture entirely absent or restricted to mesopleuron and metapleuron. Pilosity abundant on dorsal surfaces. Pale yellow to yellow brown. Monomorphic worker caste.............................. 2

– Eyes larger and composed of 10 or more facets; propodeum evenly rounded. Sculpture, pilosity and color variable. Monomorphic or polymorphic 3
2 Mesopleuron and metapleuron with punctate sculpture. Propodeal angles small but conspicuously emarginated into fine points. Pacific native....................... ***M. sechellense***
– Entire mesosoma glassy smooth and free of sculpture. Propodeum angulate but dorsolateral corners not emarginated into fine points. Fiji endemic .... ***M. vitiense***
3 Entire head and mesosoma finely punctate. Uniformly pale. Monomorphic worker caste. Introduced.. ***M. pharaonis***
– Head smooth and shiny, sculpture absent or restricted to transverse carinae on posterior margin. Mesosoma smooth and shiny or with punctate sculpture restricted to mesopleuron and metapleuron but never on pronotum. Color variable. Monomorphic or polymorphic worker caste... 4
4 Transverse carinae present on posterior margin of head and propodeal dorsum. Head and body not conspicuously elongate. Mesopleuron and metapleuron punctate. Dark gaster contrasting with paler head and mesosoma. Pilosity abundant at least on head and gaster. Polymorphic worker caste. Introduced.. ***M. destructor***
– All surfaces glassy smooth and shiny without punctation or striations. Head and body conspicuously elongate. Color variable but never with dark gaster contrasting with paler head and mesosoma. Pilosity sparse. Monomorphic worker caste 5
5 Head and gaster dark, typically contrasting with paler mesosoma. Petiole twice as long as tall. Eyes more ovoid. Introduced .... ***M. floricola***
– Posterior of head and dorsum of gaster infuscated, but mostly the same pale yellow color as mesosoma. Petiole not twice as long as tall. Eyes less ovoid. Distribution unknown. .... ***Monomorium* sp. FJ02**

Monomorium destructor (Jerdon)

(Plate 78)

Atta destructor Jerdon, 1851: 105; worker described. Type locality: INDIA [not examined]. Combined in *Monomorium*, senior synonym of *ominosa* (and its junior synonym *atomaria*) (Dalla Torre, 1893: 66). Senior synonym of *basalis* (Forel, 1894: 86); of *vexator* (Donisthorpe, 1932: 468); of *gracillima* (Bolton, 1987: 324). In revision of Malagasy species (Heterick, 2006: 96). For additional synoptic history see Bolton et al. (2006).

Monomorium destructor, commonly known as the Singapore ant, is considered to be a significant threat to native biological diversity and human health. The polymorphic workers, punctate sculpture on the propodeum and mesopleuron, and transverse ridges running across the posterior margin of the head makes *M. destructor* unique among its Fijian congeners. The dark gaster contrasting with the rest of the body can also be used as a relatively reliable character for species recognition both in the field and under the microscope. The species is believed to have originated in northern Africa, but its original range prior to human dispersal across the globe might also have included the Middle East and even parts of Asia (Wetterer, 2009a). In Fiji, *M. destructor* has been observed recruiting heavily to baits in Lautoka (Viti Levu).

Material examined. **Viti Levu:** Lautoka Port 5 b.

Monomorium floricola (Jerdon)

(Plate 79)

Atta floricola Jerdon, 1851: 107; worker described. Type locality: INDIA [not examined]. Combination in *Monomorium*; senior synonym of *specularis* (Mayr, 1879: 671). Senior synonym of *poecilum* (Emery, 1894b: 51); of *cinnabari* (Wheeler, 1913a: 486); of *floreanum* (Brown, 1966: 175); of *angusticlava, impressum* (Bolton, 1987: 390); of *rina, philippinensis* (Heterick, 2006: 122). Discussed and figured (Wilson & Taylor, 1967: 64, fig. 49). For additional synoptic history see Bolton et al. (2006).

Monomorium floricola is a minute, colorful, glassy-smooth species. In the field, it can be spotted by its elongated, shiny black body that is often lighter in the middle section. However, the mesosoma approaches the color of the head and gaster in a number of Fijian specimens. Closer inspection reveals a sparse scattering of erect white pilosity, an elongate head with multifaceted ovate eyes, and a petiole that is twice as long as tall. In Fiji, *M. floricola* is most similar to *Monomorium* sp. FJ02, which is equivalent in shape and glassy sculpture, but can be separated by its darker head and gaster, longer petiolar peduncle, and larger more ovate eyes. In the Pacific, the more uniformly dark specimens of this species are most commonly confused with the similarly smooth and elongate *M. liliuokalanii*, but the petiole length of the latter is conspicuously less than twice its height.

Monomorium floricola is one the world's most broadly distributed tramp ants, although little is known about the ecological impacts of *M. floricola* (Wetterer, 2010a). Colonies of *M. floricola* are capable of recruiting in high numbers to baits and natural food resources, and the species often invades human habitations. Wilson and Taylor (1967) described the species as predominately arboreal in habit, and Emery (1921) suggested it originated in tropical Asia. In Fiji the species was collected across the disturbance gradient, from urban port cities to interior forest. Although collected from malaise traps, it was not taken from sifted leaf litter.

Material examined. **Gau:** Navukailagi 564. **Lakeba:** Tubou 100 a, Tubou 100 c. **Macuata:** Vunitogoloa 36. **Moala:** Naroi 75. **Taveuni:** Lavena 217. **Vanua Levu:** Kilaka 146. **Viti Levu:** Nabukavesi 40, Mt. Tomanivi 700 b, Ocean Pacific 1, Ocean Pacific 2, Lautoka Port 5 b, Volivoli 55, Nadarivatu 750. **Yasawa:** Nabukeru 144, Nabukeru 120.

Monomorium pharaonis (Linnaeus)

(Plate 80)

Formica pharaonis Linnaeus, 1758: 580; worker described. Type locality: EGYPT [not examined]. Combined in *Monomorium* (Mayr, 1862: 752). Senior synonym of *antiguensis* and *domestica* (Roger, 1862: 294); of *contingua* and *fragilis* (Mayr, 1886: 359); of *minuta* (Emery, 1892: 165); of *vastator* (Donisthorpe, 1932: 449). For additional synoptic history see Bolton et al. (2006).

Monomorium pharaonis, commonly known as the Pharaoh ant, is a relatively small, yellow species with a dull appearance caused by the punctate sculpture covering most of its surfaces. The combination of large multifaceted eyes, an entirely punctate mesosoma, sparse pilosity, and uniform yellow color differentiate *M. pharaonis* from other Fijian *Monomorium*. This species is one of the most accomplished tramp ants in the world, and has demonstrated a remarkable ability to persist in human habitations, hothouses and food processing plants across the globe. Emery (1921) suggested it originated in tropical Asia.

Material examined: **Vanua Levu:** Savusavu. **Viti Levu:** Suva.

Monomorium sechellense Emery
(Plate 81)

Monomorium fossulatum subsp. *sechellense* Emery, 1894a: 69, text-fig.; worker, queen described. Type locality: SEYCHELLE IS. [not examined]. Synonym of *fossulatum* (Wilson & Taylor, 1967: 65, fig. 50) [*fossulatum* is given as senior name, but *sechellense* has priority (Bolton, 1995: 266)].

Monomorium sechellense is a small yellow brown species with a single eye facet and a pronotum armed with small but distinct angles. In Fiji, *M. sechellense* is easily confused with *M. vitiense* on account of the small eye and angulate propodeum. In addition to being larger in size, *M. sechellense* can be distinguished by the punctate sculpture on the mesopleuron and metapleuron and the clearly margined propodeal angles. In Fiji the species is known from many of the islands, and has often been collected from leaf litter in marginal and interior forest habitats. Colonies were found nesting under logs and stones. Workers were collected foraging on trees, but the species was not recorded from any of the malaise traps. Wilson and Taylor (1967) suggest the species may be native to tropical Asia.

Material examined. **Gau:** Navukailagi 415, Navukailagi 356, Navukailagi 300. **Kadavu:** Moanakaka 60. **Koro:** Kuitarua 480, Nasoqoloa 300. **Vanua Levu:** Mt. Kasi Gold Mine 300, Nakasa 300, Vusasivo Village 400 b, Mt. Delaikoro 391. **Viti Levu:** Mt. Evans 700, Colo-i-Suva 186 d, Naboutini 300, Galoa 300, Nabukavesi 300.

Monomorium vitiense Mann, REVISED STATUS
(Plate 82)

Monomorium (*Monomorium*) *vitiensis* Mann, 1921: 444; worker described. Type locality: FIJI, Viti Levu, Somosomo (W. M. Mann). Syntypes: 3 workers (MCZC type no. 8700, examined); 12 workers (USNM, examined). Synonymized with *talpa* Emery (Dlussky, 1994).

A pale yellow species with a minute eye composed of one to several facets, an angulate propodeum, and an entirely smooth and shiny mesosoma. The combination of these characters separates it from all other *Monomorium*, including the closely related *M. sechellense*. Although the name designated by Mann is used here, further study may reveal it to be synonymous with one of the other *fossulatum* group species. Mann reported that *M. vitiense* is close to *M. talpa* Emery from New Guinea, but has a distinctly longer head and the antennal scapes do not extend beyond its posterior margin. The figure of *M. talpa* given in Wilson and Taylor (1967), however, looks appreciably similar to *M. vitiense*, with its long head, short antennal scapes, and the low profile of the promesonotum. *Monomorium talpa* was listed as a junior synonym of *M. australicum* Forel by Heterick (2001). The synonymy was based on examination of positively identified material (including that of Wilson and Taylor) of both taxa, all of which were reported to have a strongly domed promesonotum. The low promesonotal profile of *M. talpa* figured in Wilson and Taylor, serves as a contradiction to the synonymy, and it is thus difficult to judge the validity of the synonymy without examination of the type material together with more Pacific material.

A cursory examination of *fossulatum* group specimens from the MCZC and PSWC reveals a variety of forms in which relative head length, relative scape length, promesonotal profile, and color vary to a wide degree. Several specimens, including workers from New Guinea and Moorea agree strongly with *M. vitiense*. Until these specimens can be checked against the type material of *M. australicum* and *M. talpa*, it is difficult to say what name is best applied to them. Dlussky's (1994) synonymy of *M. vitiense* with *M. talpa* may well stand, but a more careful review of the types of all the involved species and additional Pacific material must be conducted before proper judgment can passed. Until such a study is initiated, it is proposed here that the full species status of *M. vitiense* be reinstated.

In Fiji, *Monomorium vitiense* was found in sifted leaf litter and nesting in logs.

Material examined. **Koro:** Nasau 470 (3.7 km). **Taveuni:** Somosomo 200. **Vanua Levu:** Nakasa 300. **Viti Levu:** Colo-i-Suva 186 d, Naboutini 300, Nabukelevu 300.

Monomorium sp. FJ02

(Plate 83)

Monomorium sp. FJ02 is a small, yellow, glassy smooth species with multifaceted eyes and a rounded propodeum. The species is similar in form to *M. floricola*, but can be readily distinguished by the pale yellow head and gaster. See the notes under *M. floricola* for additional characters to separate the two. A cursory examination of material at the USNM reveals a similarity of size and shape to specimens of *M. intrudens* F. Smith from Japan. The Japanese specimens (some of which were intercepted in North Carolina, USA) differ in their darker gasters.

Known from the port cities of Lautoka and Suva, this species is likely not native to Fiji. Two specimens at the USNM collected by the botanist A.C. Smith in 1952 from Gau are similar, though one of these is darker. In Lautoka, a colony was observed recruiting to sugar solution bait. When other species arrived, individuals of *Monomorium* sp. FJ02 would come to a standstill and raise their gasters in the air. A droplet of venom would be extruded onto the tip of the sting, and the individual would wait in position until the other species ran into it, or the threat passed.

Material examined. **VITI LEVU:** McDonald's Resort (3); Lautoka Port; Lautoka Shirley Park; Suva Ratu Sukuna Park. **GAU:** Ngau.

Genus **Myrmecina**

Diagnosis of worker caste in Fiji. Head shape ovoid to rectangular. Antenna 12-segmented. Antennal club 3-segmented. Sides of head with carinate ridge extending below eye-level from mandibular insertions to posterolateral head margin. Mandibles triangular; lacking single tooth on basal margin. Propodeum armed with an anterior pair of shorter teeth and a posterior pair of longer spines. Petiole lacking peduncle; lacking large anteroventral subpetiolar process.

Myrmecina is easily recognized by the distinct subocular carina, the cylindrical apedunculate petiole, and the double pair of propodeal spines. The Fijian species are shiny black with deeply sulcate grooves on the head, mesosoma and petiole. It is difficult to confuse this genus with any of other present in Fiji, with the possible exception of *Lordomyrma*. There are 33 species and subspecies described, with few exceptions, from wet tropical regions across the globe. Fiji represents easternmost limit of the Old World *Myrmecina* range. The genus is mostly restricted to the leaf litter, soil and rotting logs. The majority of the Fijian records are known from males, which were captured by malaise trapping on nearly all the major islands. No females were collected by malaise trapping, and workers are known from only two localities.

Myrmecina cacabau Mann was known from only a single worker collected in Nadarivatu until it was rediscovered sixty years later at the same locality (Taylor, 1980b). The only other workers of *Myrmecina* were collected on Mt. Washington (Kadavu) during the recent survey. The Kadavu series shows considerable difference in the sculpture and propodeal spines from the Viti Levu series. These differences alone might be attributed to the somewhat notorious population variation observed in *Myrmecina*. However, there is also a distinct and consistent variation in the morphology between the males of Viti Levu and those of all other islands, including Kadavu. The Kadavu workers, together with the males of all localities besides Viti Levu, are tentatively treated here as a species distinct from *M. cacabau*.

Key to *Myrmecina* workers and males of Fiji.

1 *Workers*: Head with thin, uniform parallel carinae. Dorsum of mesosoma with carinae distinctly terminating before attaining anterior margin of pronotum. Posterior pair of propodeal spines shorter, upright and subparallel to anterior pair. *Males*: Mesonotum with notauli inconspicuous and reduced to a few widely separated lightly impressed punctures .... ***M. cacabau***

– *Workers*: Head with thick wavy rugae; dorsum of mesosoma with carinae distinctly attaining the anterior margin of pronotum. Posterior pair of propodeal spines longer, pointing posteriorly and clearly diverging from anterior pair. *Males*: Mesonotum with notauli developed as series of punctures that converge posteriorly to form a convex median triangle on the dorsal surface.. ***Myrmecina* sp. FJ01**

Myrmecina cacabau (Mann)

(Plate 84)

Archaeomyrmex cacabau Mann, 1921: 449, fig. 17; worker described. Type locality: FIJI, Viti Levu, Nadarivatu (W. M. Mann). Holotype [single specimen]: worker (USNM, examined). Combined in *Myrmecina* (Brown, 1971a: 1). Rediscovered and imaged (Taylor, 1980b: 122, figs. 1-3).

Myrmecina cacabau workers are known only from the type locality at Nadarivatu (Viti Levu). The characters listed in the key, namely the more uniform, fine and straight cephalic sculpture, the weaker pronotal carinae and the shorter, more upright posterior pair of propodeal spines separate this species from *Myrmecina* sp. FJ01. The type specimen was searched for by Brown (1971a) extensively, but in vain, in the two collections into which Mann's Fiji material was deposited (MCZC, USNM). The rediscovery of the holotype is reported here, after it was found in a unit tray at the USNM labeled as *M. submarta* Mann.

Mann collected a single worker, "from the trunk of a tree, in some portion of which it probably nests." If, in fact, *Myrmecina cacabau* nests arboreally, it would serve as further evidence of the species status of *Myrmecina* sp. FJ01, which was found to nest directly in the soil. However, the two workers collected by G. Kuschel were from a litter sample, which suggests the species at least forages on the ground. Despite extensive collections in the Nadarivatu area, workers of the species were not encountered during the recent survey. Two male specimens believed to belong to this species were collected in malaise traps from Viti Levu.

Material examined. **Viti Levu:** Mt. Evans 800, Vaturu Dam 530.

Myrmecina sp. FJ01

(Plate 85)

This species can be separated from *M. cacabau* by the characters given in the key, namely the coarser and wavier cephalic sculpture, stronger pronotal carinae, and less upright propodeal spines. Approximately a dozen workers of this species were collected from midway up Mt. Washington in Kadavu. The workers recruited to cookie crumbs placed on the forest floor, which they took back to their nest. The nest was tunneled into bare soil with an unadorned hole of small diameter serving as the entrance.

The males associated here with this species are distinct from those of Viti Levu by the median triangle formed on the mesonotum by the two distinct and converging notauli. The presumed males of *Myrmecina* sp. FJ01, which share the conspicuous notauli, were collected from Vanua Levu, Taveuni and Kadavu.

Material examined. **Kadavu:** Mt. Washington 700. **Taveuni:** Devo Peak 1188, Devo Peak 1187 b, Mt. Devo 734, Mt. Devo 892, Mt. Devo 1064. **Vanua Levu:** Kilaka 61, Wainibeqa 150.

Genus Pheidole

Diagnosis of worker caste in Fiji. Head shape ovoid to rectangular. Antenna 12-segmented. Antennal club 3-segmented. Antennal scapes reach or exceed posterior margin of head. Antennal sockets not surrounded by a raised sharp-edged ridge. Anterior clypeal margin variously shaped, but never armed with three broad and blunt teeth. Frontal lobes relatively close together so that the posteromedian portion of the clypeus, where it projects between the frontal lobes, is at most only slightly broader than one of the lobes. Sides of head lacking carinate ridge extending below eye-level from mandibular insertions to posterolateral head margin. Mandibles triangular; lacking a distinct basal tooth. Mesosoma with depression distinctly separating promesonotum from propodeum; erect hairs present. Propodeum armed with spines or teeth. Propodeal lobes shorter than propodeal spines. Waist 2-segmented. Petiole pedunculate; lacking large anteroventral subpetiolar process. Postpetiole not swollen; in dorsal view not distinctly broader than long or distinctly wider than petiole. Tip of sting tapered to a point and lacking a triangular to pendant-shaped extension.

Pheidole is the most diverse ant genus in the world (Wilson, 2003), and is represented in Fiji by nearly 20 species. The most striking radiation of the *Pheidole* in Fiji is that of the *roosevelti* group (Sarnat, 2008b; Sarnat & Moreau, 2011), which has evolved from a small

Fijian ancestor with simple propodeal spines into a variety of large species with ornately modified propodeal spines and mesonotal projections. Another group that has diversified in Fiji is a cluster of species related to *P. knowlesi* and *P. vatu*, but the taxonomic boundaries of the group are currently unclear. In addition to these endemic species, there are also *Pheidole* that are widespread throughout the Pacific, including *P. oceanica*, *P. fervens* and *P. umbonata*. The only *Pheidole* in Fiji introduced into the Pacific is *P. megacephala*.

Key to major and minor workers of *Pheidole* in Fiji.

1 *Major and minor worker*: Propodeal spines short to moderate, distance from propodeal spiracle to tip of propodeal spine distinctly less than two thirds length of petiole. Propodeal spines always simple and unmodified 2

– *Major and minor worker*: Propodeal spines long, distance from propodeal spiracle to tip of propodeal spine subequal to or longer than two thirds length of petiole. Distal portion of propodeal spines simple, angulate or bifurcate 10

2 *Major and minor worker*: Postpetiole in profile appearing swollen relative to petiole. *Major worker*: Hypostomal bridge lacking well-defined median or lateral teeth. Antennal scrobe absent. Frontal carinae inconspicuous. Posterolateral lobes smooth and shiny. Head subcordate in face view. Large (HW > 1.0 mm). *Minor worker*: Postpetiole approximately as long as petiole. Sculpture restricted to mesopleuron and metapleuron. Introduced .... ***P. megacephala***

– *Major and minor worker*: Postpetiole in profile not appearing swollen relative to petiole. *Major worker*: Hypostomal bridge with one median tooth and two lateral teeth. Antennal scrobe, frontal carinae, posterolateral lobes, size variable. Head subquadrate in face view. *Minor worker*: Postpetiole distinctly shorter than petiole. Sculpture variable. Size variable 3

3 *Major worker*: Antennal scrobe absent. Frontal carinae inconspicuous. Pronotum smooth and shiny. Posterolateral lobes mostly smooth and shiny. Head not impressed posterior to vertex in profile, pronotal humeri developed as obtuse angles. Small (HW < 1.0 mm). *Minor worker*: Head entirely smooth and shiny. Mesosoma mostly smooth and shiny with sculpture restricted to mesopleuron and metapleuron. Antennal scapes relatively short and surpassing posterior margin of head by a distance less than the combined length of terminal two antennal segments. Pacific native .... ***P. umbonata***

– *Major worker*: Antennal scrobe present (sometimes weakly). Frontal carinae distinct; pronotum and posterolateral lobes variable. Head variably impressed in profile, pronotal humeri variably developed. Variable size. *Minor worker*: If head entirely smooth and shiny then antennal scapes surpass posterior margin of head by a distance approximate to the combined length of terminal two antennal segments. Mesosoma variably sculptured. Antennal scapes of variable length 4

4 *Major worker*: Posterolateral lobes strongly rugoreticulate. Ventrolateral portion of head strongly sculptured. Pronotal humeri weakly projecting. Large species (HW > 1.2 mm). *Minor worker*: Head entirely smooth and shiny. Mesosoma mostly smooth and shiny with sculpture restricted to mesopleuron and metapleuron. Antennal scapes relatively long and surpassing posterior margin of head by a distance greater than the combined length of terminal two antennal segments 5

– *Major worker*: Posterolateral lobes, ventrolateral portion of head strongly sculptured. Mesonotal declivity gently sloping. Pronotal humeri moderately to strongly projecting. Smaller species (HW < 1.2 mm). *Minor worker*: Head with some combination of striation and punctation, but never entirely smooth and shiny. Mesosoma variably sculptured. Antennal scapes moderate to relatively short and surpassing posterior margin of head by a distance distinctly less than the combined length of terminal two antennal segments 6

5 *Major worker*: Carinae between eye and mandible branching and reticulated. *Minor worker*: Length of propodeal spine greater than diameter of propodeal spiracle. Pacific native. .... ***P. oceanica***

– *Major worker*: Carinae between eye and mandible unbranching and not reticulated. *Minor worker*: Length of propodeal spine equal to or less than diameter of propodeal spiracle. Pacific native .... ***P. fervens***

6 *Major worker*: Head posterior to vertex smooth and shiny or with weak carinulae but never strongly rugoreticulate. Antennal scrobe smooth to weakly punctate. Head in profile weakly impressed posterior to vertex. *Minor worker*: Either all surfaces of head and mesosoma densely punctate *or* pronotum and head ventrad of eye smooth and shiny... 7

– *Major worker*: Head posterior to vertex strongly rugoreticulate. Antennal scrobe densely punctate. Head in profile strongly impressed posterior to vertex. *Minor worker*: Either all surfaces of head completely punctate then mesosoma mostly smooth and with promesonotum strongly domed and mesonotal declivity nearly vertical *or* head entirely punctate except for distinct smooth patch ventrad of eyes and mesosoma entirely punctate 9

7 *Major worker*: Posterolateral lobes either entirely smooth and shiny or with a few carinulae, but never punctate. Carinae between eyes and antennal insertions unbranching and lacking reticulation. Antennal scrobe narrow. *Minor worker*: Head dorsum with at least median portion shiny and with occasional striae, but never punctate. Pronotum mostly smooth with transverse striae, but never punctate. Fiji endemic .... ***P. knowlesi***

– *Major worker*: Posterolateral lobes with scattered carinulae and moderately to strongly punctate, never entirely smooth and shiny. Carinae between eyes and antennal insertions branching and reticulate. Antennal scrobe broad. *Minor worker*: Entire dorsum of head densely punctate 8

8 *Major worker*: Antennal scrobe posteriorly bordered by transverse and arcuate carinulae. Area posterior to antennal scrobe with many branching and reticulated carinulae. Metapleuron mostly smooth and shiny with striae, but not punctate. *Minor worker*: Pronotum and ventral surface of head smooth and shiny. Fiji endemic.......... ***P. wilsoni***

– *Major worker*: Antennal scrobe not posteriorly bordered by transverse and arcuate carinulae. Area posterior to antennal scrobe with mostly longitudinal carinulae that rarely branch or become reticulate. Metapleuron mostly punctate with striae. *Minor worker*: All surfaces of head and mesosoma densely punctate. Fiji endemic .... ***Pheidole* sp. FJ09**

9 *Major worker*: Gaster with basal third of first tergite striate and punctate. Anterior margin of clypeus notched. *Minor worker*: All surfaces of head completely punctate then mesosoma mostly smooth and with promesonotum strongly domed and mesonotal declivity nearly vertical. Fiji endemic.......... ***P. vatu***

– *Major worker*: Gaster entirely smooth. Anterior margin of clypeus flat and lacking notch. *Minor worker*: Head entirely punctate except for distinct smooth patch ventrad of eyes. Mesosoma entirely punctate. Promesonotum sloping more gently towards propodeum with vertical mesonotal declivity. Fiji endemic .... ***Pheidole* sp. FJ05**

10 *Major and minor worker*: Pronotum armed with long distinct anterolateral projecting spines. Mesonotum armed with short posterior projecting spines. Pacific native. .... ***P. sexspinosa***

– *Major and minor worker*: Pronotum lacking long distinct anterolateral projecting spines.

Mesonotum variable, but never armed with spines 11

11 *Major and minor worker*: Mesonotal declivity distinctly convex. Propodeal spines always simple and not angulate or bifurcate distally. Smaller species (*Major worker* HW < 1.4 mm. *Minor worker*: HW < 0.6 mm) 12

– *Major and minor worker*: Mesonotal declivity either vertical or concave. Propodeal spines either simple or angulate or bifurcate distally. Larger species (*Major worker* HW > 1.6 mm. *Minor worker*: HW > 0.6 mm). *P. roosevelti* group .... 13[7]

12 *Major worker*: Gaster with basal quarter of first tergite striate and punctate. Propodeal spines thick basally and downcurved. Larger (HW > 1.0 mm). *Minor worker*: Propodeal spines thick basally and downcurved. Postpetiole bulbous. All surfaces except for gaster densely punctate. Larger (HW > 0.5 mm). Fiji endemic .... ***P. caldwelli***

– *Major worker*: Gaster entirely smooth and shiny. Propodeal spines slender and upcurved. Smaller (HW < 1.0 mm). *Minor worker*: Propodeal spines slender and upcurved. Postpetiole not bulbous. Face striate-punctate. Projecting pronotal humeri. Raised angles on the mesonotum. Larger (HW < 0.5 mm). Fiji endemic .... ***P. onifera***

13 *Major worker*: Mesonotal process, in profile, truncated into a blunt angle lacking lamellate or acute posterior margin. Propodeal spines simple, evenly tapering to a single straight acuminate point without becoming bifurcate or angulate apically. posterolateral lobes, in full face view, with distinct transverse rugae extending from median cleft to posterolateral corners. Scapes short. *Minor worker*: Propodeal spines simple, evenly tapering to a single straight acuminate point without becoming bifurcate or angulate apically. Mesonotal process truncated into a blunt process lacking lamellate or distinct posterior margin. Head as broad as long. Fiji endemic .... ***P. simplispinosa***

– *Major worker*: Mesonotal process, in profile, with acute posterior angle or lamella. Propodeal spines usually modified apically with bifurcate or angulate tip. Posterolateral lobes, in full face view, variably sculptured but never with distinct transverse rugae extending from median cleft to posterolateral corners. *Minor worker*: Propodeal spines modified apically with bifurcate or angulate tip, but never evenly tapering to a single straight acuminate point. Mesonotal process with lamellate or acute posterior margin. Head longer than broad 14

14 *Major worker*: Posterolateral lobes either smooth and shiny or with parallel carinae, but never rugoreticulate. *Minor worker*: Head, in full face view, smooth and shiny above level of eyes. Promesonotum, in dorsal view, smooth and shiny 15

– *Major worker*: Posterolateral lobes, in full face view, rugoreticulate, such that longitudinal rugae are intersected by transverse rugae. *Minor worker*: Head, in full face view, rugose to rugoreticulate above level of eyes. Promesonotum, in dorsal view, transversely rugose to rugoreticulate 17

15 *Major worker*: Posterolateral lobes, in full face view, smooth and shiny lacking rugae or carinae. Median ocellus present and well developed. Intercarinular spaces on head smooth and shiny. Postpetiole with anterior face and dorsum smooth and shiny lacking rugulae. Gaster with basal portion of first tergite smooth and shiny. *Minor worker*: Propodeal spines with dorsal edge distinctly shorter than anterior edge. Posterior of head weakly pinched dorsoventrally. Color of petiole, postpetiole and gaster same as mesosoma and head. Fiji endemic .... ***P. colaensis***

– *Major worker*: Posterolateral lobes, in full face view, sculptured with rugae or carinae. Median ocellus present or absent. Intercarinular spaces on head smooth and shiny to

[7] See Sarnat (2008b) for more comprehensive key and species descriptions.

foveolate. Postpetiole with anterior face and dorsum smooth and shiny to rugulose-foveolate. Gaster with basal portion of first tergite smooth and shiny to densely sculptured. *Minor worker*: Propodeal spines with dorsal edge subequal to or longer than anterior edge. Posterior of head variable. Color of petiole, postpetiole and gaster lighter than mesosoma and head. Scapes longer, metafemur longer.. 16

16 *Major worker:* Propodeal spines, in profile, with dorsal edge as long as or longer than anterior edge. Mesonotal process, in dorsal view, broad basally. Petiole with posterior face smooth and shiny. Head wider, metafemur longer, scapes longer (HW 2.20 mm–2.35 mm). *Minor worker*: Propodeal spines with dorsal edge distinctly longer than anterior edge. Posterior of head strongly pinched dorsoventrally, appearing flattened in profile and shield-like in full face view. Fiji endemic.. ***P. pegasus***

– *Major worker*: Propodeal spines, in profile, with dorsal edge distinctly shorter than anterior edge. Mesonotal process, in dorsal view, strongly attenuated basally. Petiole with posterior face rugoreticulate. Head narrower, metafemur shorter, scapes shorter (HW 2.05 mm–2.12 mm). *Minor worker*: Propodeal spines with dorsal edge approximately as long as anterior edge. Head venter, in profile, with genal carinae modified into elevated flanges. Head, in full face view, oval shaped lacking posterolateral corners forming obtuse angles. Scapes and metafemur shorter... ***P. uncagena***

17 *Major worker*: Posterolateral lobes, in full face view, with rugoreticulum terminating before attaining posterior margin. In dorsal view, length of median basigastral sculpturing immediately posterior to postpetiole attachment longer than length of postpetiole. Head shorter (HL 1.95 mm–2.04 mm). *Minor worker*: head venter smooth and shiny. In profile, genal carinae inconspicuous. Fiji endemic.. ***P. furcata***

– *Major worker*: Posterolateral lobes, in full face view, with rugoreticulum attaining posterior margin. In dorsal view, length of median basigastral sculpturing immediately posterior to postpetiole attachment shorter than length of postpetiole. Head longer (HL 2.06 mm–2.38 mm). *Minor worker*: Head venter sculptured. Strongly produced genal carinae present.. 18

18 *Major worker*: Head, in full face view, with intercarinular spaces densely and distinctly foveolate. Postpetiolar dorsum, in dorsal view, rugulose with foveolate interspaces. *Minor worker*: Head, in full face view, with strongly branching network of longitudinal and transverse rugae. Spaces between head rugoreticulum strongly foveolate. Pronotum, in dorsal view, rugoreticulate. Mesonotal process, in dorsal view, broadly lamellate and with a medially excised posterior margin. Fiji endemic ....***P. roosevelti***

– *Major worker*: Head, in full face view, with intercarinular spaces smooth and shiny to weakly impressed, but never densely nor distinctly foveolate. Postpetiolar dorsum, in dorsal view, smooth and shiny. *Minor worker*: Head, in full face view, with discontinuous longitudinal rugae that branch occasionally, but become rugoreticulate only on posterolateral corners of head. Spaces between head rugae smooth and shiny. Pronotum, in dorsal view, shiny with transverse rugae. Mesonotal process, in dorsal view, narrowly lamellate with flat to weakly concave posterior margin.. ***P. bula***

Pheidole caldwelli Mann

(Plate 86)

Pheidole caldwelli Mann, 1921: 434, fig. 14; major worker, minor worker, queen described. Type locality: FIJI, Viti Levu, Nadarivatu (W. M. Mann). Syntypes: 24 minor workers, 4 major workers (MCZC type no. 8697, examined); 11 minor workers, 10 major workers, 3 queens (USNM, examined).

Pheidole caldwelli is a large, heavily sculptured species with long downcurved propodeal spines. The cephalic sculpture of the major worker is most similar to *P. vatu* and its close relatives, but *P. caldwelli* is a much larger species with very long propodeal spines, and the postpetiolar node is not armed with laterally projecting spines. Interestingly, there is also similarity between *P. caldwelli* and *P. simplispinosa* of the *roosevelti* group, especially with regard to the larger size, long propodeal spines and steep mesonotal declivity.

Mann reported *P. caldwelli* as being very common in the Nadarivatu area, where he found it beneath stones and logs. Although the recent survey did not recover the species from the Nadarivatu area, it was collected from the nearby Monasavu Dam area and the upland forests of southeastern Viti Levu. The recent survey also collected a single specimen from a litter sample on Moala. All collections were taken from sifted leaf litter. The species was named after Mann's travelling companion while in Fiji, Mr. Charles Caldwell of Suva.

Material examined. **Moala:** Mt. Korolevu 300. **Viti Levu:** Monasavu Dam 800, Korobaba 300, Nabukavesi 300, Waivudawa 300, Nadarivatu 750.

Pheidole fervens Smith, F.
(Plate 87)

Pheidole fervens Smith, F. 1858: 176; major worker described. Type locality: SINGAPORE [not examined]. Senior synonym of *cavannae*, *javana*, *nigriscapa*, and material of the unavailable name *tahitiana* (Wilson & Taylor, 1967: 45, fig. 33); of *desucta* (Eguchi, 2001: 53); of *amia*, *azumai*, *dharmsalana*, *dolenda*, *pungens*, *soror* (Eguchi, 2004: 197). Current subspecies: nominal plus *pectinata*. For additional synoptic history see Bolton et al. (2006).

Pheidole fervens is a common species in the Pacific. The minor workers are small, pale to medium yellow with long appendages, very small propodeal spines, and sculpturing that is restricted to the middle and posterior portions of the mesosoma. The major workers are larger, darker and have heavily sculptured heads. The majors also have distinct antennal scrobes and longer propodeal spines than the minors.

Pheidole oceanica is easily confused with *Pheidole fervens*. The major workers can be distinguished most reliably by the sculpture located between the eye and the antennal insertions. Whereas the carinae of *P. fervens* are branched, those of *P. oceanica* are unbranched. Also, the antennal scrobes of *P. fervens* are more densely punctate than those of *P. oceanica*. The minor workers of *P. oceanica* are especially difficult to differentiate from *P. fervens*. Whereas the propodeal spiracles of *P. oceanica* are smaller than the propodeal spines, the propodeal spiracles of *P. fervens* are equal to or larger than their propodeal spines. *Pheidole megacephala* can be separated from *P. fervens* by more obvious means. The head of the *P. megacephala* major worker lacks antennal scrobes and are almost entirely absent of sculpture. The minor workers have shorter antennal scapes and a more swollen appearing postpetiole.

Pheidole fervens recruits strongly to baits and food resources and forms long and busy foraging trails. Minor workers are much more abundant than major workers, both outside of and within the nest. Specimens were collected from malaise traps, sifted litter and foraging lines on the ground and vegetation.

Material examined. **Gau:** Navukailagi 364. **Koro:** Mt. Kuitarua 500, Mt. Kuitarua 505, Mt. Kuitarua 485. **Ovalau:** Draiba 300. **Viti Levu:** Mt. Evans 800, Mt. Evans 700, Abaca 525, Colo-i-Suva Forest Park 220, Ocean Pacific 1, Korobaba 300, Mt. Evans 700, Nabukelevu 300.

Pheidole knowlesi Mann

(Plate 88)

Pheidole knowlesi Mann, 1921: 436, fig. 13a; major worker, minor worker, queen described. Type locality: FIJI, Vanua Levu, Suene (W. M. Mann). Syntypes: 2 minor workers, 3 major workers, 1 male (MCZC type no. 8698, examined).

Pheidole knowlesi subsp. *extensa* Mann, 1921: 438; major worker, minor worker described. Type locality: FIJI, Viti Levu, Nadarivatu (W. M. Mann). Syntypes: 3 minor workers, 4 major workers, 3 queens (MCZC type no. 8699, examined); 9 major workers, 3 queens (USNM, examined). NEW SYNONYMY

Note: Specimens from the non-type locality of Vanua Ava housed at the MCZC bear a red cotype label, but are not true syntypes.

Pheidole knowlesi is a light yellow brown to dark red brown species of moderate size and relatively reduced sculpturing that is commonly encountered in Fiji. It is most similar to *P. wilsoni* and *Pheidole* sp. FJ09, but can be reliably separated from them by the more shiny appearance and lack of sculpturing on the posterior portion of the head (although some individuals will have carinae extending to the posterior margin). *Pheidole knowlesi*, together with the aforementioned species, can be separated from *P. vatu* and its close relatives by the less impressed region of the head between the vertex and posterior margin when viewed in profile, and by the lack of strong rugoreticulum on the posterior margin of the head.

Pheidole knowlesi, as it is defined here, may prove upon further study to be a complex of several distinct species. At present, however, it is difficult to find consistent morphological characters to separate what are treated here as different populations. One promising character that deserves further investigation is the length of cephalic carinae found on the dorsal surfaces of the head of the major worker. Mann (1921) used this character to separate *P. knowlesi* from *P. extensa*. A corresponding character, as expressed in the minor workers, is the amount of sculpturing on the pronotum, especially surrounding the pronotal humeri.

While there appear to be relatively discrete breaks in the aforementioned characters among populations, sympatry of the two conditions is difficult to find. In fact, the distinction between the two general morphotypes is weakest in the Monasavu Dam area (Viti Levu) where they achieve the closest geographical proximity to each other. There, some specimens of a series of majors lack carinae on their posterolateral lobes but their associated minors are identical to those of a nearby nest series in which the long carinae are present on the majors.

Head color is another character that tends to show considerable variation. In general, head color tends to be lighter in the more sculptured specimens, but this is not uniformly true. The Gau specimens, for example, which have reduced sculpturing, tend to be as dark as most of those with more extensive sculpturing. Although the dark patches that appear around the frontal lobes are often more pronounced in specimens with reduced sculpturing, the same patterns can appear in specimens with extensive sculpturing.

The two morphotypes generally replace each other across the Fijian archipelago, with the less sculptured populations being more widespread and occupying generally lower elevation sites while the more sculptured populations are restricted to the high elevations of Fiji's central mountain range in the Nadarivatu and Monasavu areas. Aside from the one site where the two forms approach each in the Monasavu area, the two forms are not known to occur at the same locality.

Several series are notable for their more prominent aberration from the typical *P. knowlesi*. For example, the Gau specimens tend to be smaller and darker, and their majors have a more impressed and sculptured antennal scrobe. The major and minors (CASENT0184104, CASENT0184008) from a Taveuni series are quite large and with exceptionally little sculpture on the major and minor workers, especially on the dorsum of the mesosoma.

Pheidole knowlesi is perhaps the most frequently encountered of Fiji's native *Pheidole*. It was collected from sifted litter, malaise traps, canopy fogging and by hand. Nests were found in dead branches, logs, in litter deposits on tree trunks, under stones, and in live and fallen ant-plants. Mann mentions that *P. extensa* is a harvesting species, and named *P. knowlesi* after Mr. C. W. Knowles, Fiji's Superintendent of Agriculture.

Material examined. **Gau:** Navukailagi 717, Navukailagi 675, Navukailagi 387, Navukailagi 415, Navukailagi 535, Navukailagi 432, Navukailagi 490, Navukailagi 505, Navukailagi 480, Navukailagi 522, Navukailagi 475, Navukailagi 356, Navukailagi 496, Navukailagi 564, Navukailagi 575. **Kadavu:** Mt. Washington 760, Mt. Washington 800, Moanakaka 60, Mt. Washington 700, Vanua Ava b. **Koro:** Mt. Kuitarua 500, Mt. Kuitarua 440 b, Nasau 465 a. **Ovalau:** Levuka 500. **Taveuni:** Devo Peak 1187 b, Tavuki 734, Mt. Devo 1064, Mt. Devo 775 a, Tavoro Falls 100. **Vanua Levu:** Kilaka 146, Kilaka 61, Kilaka 98, Mt. Vatudiri 641, Mt. Vatudiri 570, Mt. Delaikoro 699, Mt. Delaikoro 734, Vusasivo Village 342 b, Rokosalase 118, Suene. **Viti Levu:** Mt. Tomanivi 1300, Mt. Evans 800, Mt. Tomanivi 700 b, Navai 700, Mt. Tomanivi 700, Monasavu 800 a, Vaturu Dam 575 b, Navai 930, Colo-i-Suva Forest Park 220, Naqaranabuluti 860, Monasavu Dam 800, Mt. Batilamu 840 c, Savione 750 a, Mt. Naqaranabuluti 1050, Mt. Tomanivi 950, Monasavu Dam 1000, Monasavu 800, Nasoqo 800 a, Nasoqo 800 b, Nasoqo 800 d, Monasavu Dam 600, Korobaba 300, Nakobalevu 340, Colo-i-Suva 200, Waimoque 850, Colo-i-Suva 460, Colo-i-Suva 325, Colo-i-Suva 372, Navai 1020, Nuku 50, Nabukavesi 300, Mt. Rama 300, Naikorokoro 300, Waivudawa 300, Nadarivatu 750.

Pheidole megacephala (Fabricius)

(Plate 89)

Formica megacephala Fabricius, 1793: 361; major worker described. Type locality: not given [not examined]. Combined in *Pheidole*; senior synonym of *trinodis* (Roger, 1863b: 30). Senior synonym of *edax* (Dalla Torre, 1892: 90); of *perniciosa* (Emery, 1915: 235); of *pusilla* (and its junior synonyms *janus*, *laevigata* Smith, *laevigata* Mayr) (Wheeler, 1922: 812); of *suspiciosa* (Donisthorpe, 1932: 455); of *testacea* (Brown, 1981: 530). Current subspecies: nominal plus *costauriensis*, *duplex*, *ilgi*, *impressifrons*, *melancholica*, *nkomoana*, *rotundata*, *scabrior*, *speculifrons*, *talpa*. Discussed (Eguchi, 2001: 77; Wilson, 2003: 549; Wilson & Taylor, 1967: 46, fig. 14). For additional synoptic history see Bolton et al. (2006).

Pheidole megacephala is a common tramp species across the tropical regions of the globe. The minor workers are small, yellow to brown with small propodeal spines, reduced sculpture, and a swollen postpetiole. Sculpture on the minor worker is restricted to the middle and posterior portions of the mesosoma. The major workers lack antennal scrobes and do not have any sculpture posterior to the eyes. *Pheidole oceanica* and *P. fervens* are the two ants in Fiji most often confused with *P. megacephala*. Majors of *P. oceanica* and *P. fervens* have strongly sculptured posterolateral lobes and antennal scrobes, while majors of *P. megacephala* have unsculptured posterolateral lobes and lack antennal scrobes. The minor workers of *P. megacephala* are best separated from the aforementioned species by the swollen appearance of the postpetiole and the shorter antennal scapes.

Pheidole megacephala is known to cause significant damage to native biological diversity, including vertebrates, and also significant damage to agricultural systems. The species recruits strongly to baits and food resources and forms long and busy foraging trails. In Fiji it was collected heavily in the most disturbed habitats, such as cities and villages, but also manages to penetrate into interior forests.

Material examined. **Gau:** Navukailagi 625, Navukailagi 408, Navukailagi 432, Navukailagi 356, Navukailagi 300. **Kadavu:** Namalata 100, Namalata 50, Namalata 139, Daviqele 300. **Moala:** Naroi 75. **Taveuni:** Lavena 235. **Vanua Levu:** Vusasivo Village 400 b, Dreketi 48. **Viti Levu:** Mt. Tomanivi 700 b, McDonald's Resort 10 b, Lautoka Port 5 b, Koronivia 10, Korobaba 300, Nakobalevu 340, Volivoli 50, Volivoli 55, Volivoli 25, Vunisea 300, Nadarivatu 750.

Pheidole oceanica Mayr

(Plate 90)

Pheidole oceanica Mayr, 1866: 510; major worker, minor worker, queen, male described. Type locality: FIJI, Ovalau [not examined]. Senior synonym of *boraborensis*, *pattesoni*, *upoluana* (Wilson & Taylor, 1967: 48, fig. 35).

Pheidole oceanica is a common species in the Pacific with minor workers that are pale to dark with long appendages and very small propodeal spines. Sculpture on the minor worker is restricted to the middle and posterior portions of the mesosoma. The major workers are larger, darker and are have oversized, heavily sculptured heads. The majors also have distinct antennal scrobes and longer propodeal spines than the minors. *Pheidole fervens* and, to a lesser extent, *P. megacephala*, can be confused with *P. oceanica*. See the notes under those species for characters that can be used to separate the two.

Pheidole oceanica was described by Wilson and Taylor (1967) as being native to the Pacific region. It should also be noted that this species demonstrates a considerable amount of morphological variation with respect to size, color and sculpture, but these characters do not correlate with any geographic patterns (Wilson & Taylor, 1967). The species is fairly ubiquitous in Fiji, and was collected from malaise traps, sifted leaf litter, and from foraging lines on the ground and low vegetation. Nests were found in dead logs, under logs and under stones. A nest was also found excavated in bare soil marked by a turret entrance.

Material examined. **Beqa:** Mt. Korovou 326, Malovo 182. **Gau:** Navukailagi 387, Navukailagi 408, Navukailagi 356, Navukailagi 564, Navukailagi 300. **Kadavu:** Moanakaka 60, Vunisea 200, Namalata 100, Namalata 120, Namalata 50, Namalata 139, Daviqele 300, Namara 300, Vunisea, Vanua Ava b. **Koro:** Mt. Kuitarua 485, Nasoqoloa 300. **Lakeba:** Tubou 100 a, Tubou 100 b, Tubou 100 c. **Moala:** Mt. Korolevu 375, Mt. Korolevu 300. **Ovalau:** Levuka 500, Levuka 400, Levuka 450, Draiba 300, Ovalau. **Taveuni:** Devo Peak 1187 b, Lavena 300, Lavena 235, Lavena 217, Lavena 229, Tavoro Falls 160, Tavoro Falls 100, Soqulu Estate 140. **Vanua Levu:** Kilaka 146, Wainibeqa 53, Wainibeqa 150, Yasawa 300,

Kasavu 300, Nakanakana 300, Drawa 270, Vusasivo 50, Vusasivo Village 400 b, Vusasivo Village 342 b, Vusasivo Village 400 b, Rokosalase 180, Rokosalase 150, Rokosalase 97, Rokosalase 118, Rokosalase 94, Mt. Delaikoro 391, Lagi 300, Labasa. **Viti Levu:** Nabukavesi 40, Mt. Evans 800, Mt. Evans 800, Mt. Evans 700, Nakavu 300, Vaturu Dam 575 b, Vaturu Dam 550, Savione 750 a, Nadakuni 300, Nadakuni 300 b, Korobaba 300, Nakobalevu 340, Colo-i-Suva 200, Colo-i-Suva 325, Lami 171, Colo-i-Suva 105 b, Colo-i-Suva 186 d, Volivoli 55, Mt. Evans 700, Naboutini 300, Nabukelevu 300, Navai 1020, Galoa 300, Nuku 50, Vunisea 300, Nabukavesi 300, Naikorokoro 300, Veisari 300 (3.8 km N), Waivudawa 300, Colo-i-Suva 400, Nausori. **Yasawa:** Nabukeru 144.

Pheidole onifera Mann

(Plate 91)

Pheidole onifera Mann, 1921: 427, fig. 12; major worker, minor worker described. Type locality: FIJI, Viti Levu, Nadarivatu (W. M. Mann). Syntypes: 10 minor workers, 2 major workers, 1 queen (MCZC type no. 8694, examined); 6 minor workers, 5 major workers (USNM, examined).

Note: Specimens from the non-type locality of Ovalau housed at the MCZC bear a red cotype label, but are not true syntypes.

Pheidole onifera is a yellow brown species with strong sculpture, prominent pronotal humeri, weakly projecting angles on the posterolateral margins of the mesonotum, a steep mesonotal declivity, and long upcurved propodeal spines. The major workers have a head with cephalic sculpture similar to *P. vatu* and *P. caldwelli*, but it can be separated from those species by the lighter yellow color, the raised angles on the mesonotal posterolateral margins, and the upcurved propodeal spines. The minor worker has a striate-punctate face, projecting pronotal humeri, raised angles on the mesonotum, and long upcurved spines. Mann reported finding small colonies situated beneath stones. Specimens were collected during the recent survey from several malaise traps and many sifted litter samples. Nests were also found in ant-plants, trees and logs.

Material examined. **Beqa:** Mt. Korovou 326. **Gau:** Navukailagi 356. **Kadavu:** Daviqele 300. **Moala:** Mt. Korolevu 375. **Ovalau:** Levuka 500, Levuka 450. **Vanua Levu:** Kasavu 300, Drawa 270, Lagi 300. **Viti Levu:** Mt. Evans 800, Korobaba 300, Nakobalevu 340, Colo-i-Suva 200, Colo-i-Suva 325, Colo-i-Suva 372, Vunisea 300, Naikorokoro 300, Waivudawa 300, Nadarivatu 750.

Pheidole sexspinosa Mayr

(Plate 92)

Pheidole sexspinosa Mayr, 1870: 977; major worker, minor worker described. Type locality: TUVALU [not examined]. Combined in *P.* (*Pheidolacanthinus*) (Mann, 1919: 307). Senior synonym of *adamsoni* (Wilson & Taylor, 1967: 52, fig. 37). Current subspecies: nominal plus *biroi* Emery.

Pheidole sexspinosa is immediately recognizable by the long robust pronotal spines projecting anteriorly, short mesonotal spines projecting posteriorly, and long propodeal spines. The species is known in Fiji from a single minor worker collected in a malaise trap set up in a mangrove forest on the southern coast Viti Levu. Despite this first record for Fiji, the species is widespread in the Pacific, ranging from New Guinea to Vanuatu and into Micronesia (Clouse, 2007a; Wilson & Taylor, 1967).

Material examined. **Viti Levu:** Ocean Pacific 1.

Pheidole umbonata Mayr
(Plate 93)

Pheidole umbonata Mayr, 1870: 978; major worker, minor worker described. Type locality: TONGA [not examined]. In Mann (1921: 431). Senior synonym of *zimmermani* (Wilson & Taylor 1967: 50, fig. 36). Current subspecies: nominal plus *fusciventris* Emery.

Pheidole umbonata is a variably shaded yellow brown species with a generally shiny appearance and reduced sculpture and reduced propodeal spines. On most major workers, the carinae extend to the posterior margin of the head, but on some specimens the carinae terminate at the vertex. The dorsal surfaces of the pronotum, waist segments and gaster are smooth and shiny, and that of the propodeum is punctate. Sculpture of the minor workers is reduced to the mesopleuron, metapleuron and propodeum. Color is another variable character. Although most of the Fijian specimens are orange, there are a few with substantially darker coloration from Vanua Levu.

Wilson and Taylor (1967) noted of the variation this species demonstrates across the Pacific, and cite *P. umbonata* as "the only example of true geographic variation within these [central and eastern Pacific] populations," of any species they included in their study. They discuss the species in some detail, and interested readers are referred to their work for a more thorough review than is provided here. Suffice it to say, *P. umbonata* varies with respect to color and size across its Pacific range. *Pheidole umbonata* is quite common in Fiji, and has been collected from even the more remote islands included in the recent survey. The species appears to prefer lower elevations, however, and was found nesting in logs, under moss and under stones.

Material examined. **Beqa:** Mt. Korovou 326. **Kadavu:** Moanakaka 60, Moanakaka 60. **Koro:** Nabuna 115, Nasoqoloa 300. **Lakeba:** Tubou 100 a, Tubou 100 b, Tubou 100 c. **Moala:** Maloku 80, Naroi 75, Mt. Korolevu 300. **Ovalau:** Draiba 300. **Taveuni:** Lavena 229, Tavoro Falls 100, Qacavulo Point 300, Somosomo 200. Vanua Balavu: Lomaloma. **Vanua Levu:** Mt. Kasi Gold Mine 300, Nakasa 300, Yasawa 300, Nakanakana 300, Vusasivo Village 190, Vusasivo Village 400 b, Lagi 300, Wainunu, Suene, Lasema a. **Viti Levu:** Colo-i-Suva Forest Park 220, Lami 432, Colo-i-Suva 186 d, Nabukelevu 300, Galoa 300, Nuku 50, Nabukavesi 300. **Yasawa:** Nabukeru 120.

Pheidole vatu Mann
(Plate 94)

Pheidole vatu Mann, 1921: 431, fig. 13b; major worker, minor worker described. Type locality: FIJI, Viti Levu, Nadarivatu (W. M. Mann). Syntypes: 6 minor workers, 6 major workers (MCZC type no. 8695, examined).

Both the majors and minors of *Pheidole vatu* can be recognized by the high profile of the promesonotum and the near vertical slope of the mesonotal declivity. The head of the major workers have a punctate ground sculpture overlain by carinae on the vertex that become rugoreticulate on the posterolateral lobes and laterad of the antennal scrobes. The head, in profile, is impressed between the vertex and posterior margin, and the pronotal humeri are bluntly projecting. The promesonotum is massive and is transversely striate in dorsal view. The mesopleuron is smooth and shiny, and the gaster is striate on the basal fourth of the first gastral tergite and sternite. The minor worker has an entirely punctate head, and a mostly smooth and shiny mesosoma with a few punctate patches on anterior pronotum, katepisternum and metapleuron. The petiole and postpetiole are punctate laterally and smooth and shiny dorsally. The only other Fijian *Pheidole* with so steep a mesonotal declivity are those of the *roosevelti* group. The *roosevelti* group species are much larger, with long and often highly modified propodeal spines.

There is appreciable variation within *P. vatu* as defined here, especially with regard to size and color. For example, the integument of the single Vanua Levu specimen (CASENT0183693) is generally small and of a yellow red color. Another two specimens from Viti Levu (CASENT0185636, CASENT0185597) are entirely dark like the types, but with contrasting yellow red heads. A single major worker (CASENT0183561) differs from the typical *P. vatu* by its lack of pronotal sculpture, reduced gastral sculpture, and much smaller size. The specimen is treated here as an aberrant form of *P. vatu*, but it will be interesting to note if future collections turn up more specimens with similar or intermediate morphologies.

Pheidole vatu was primarily collected during the recent survey from sifted litter samples taken from forest interior habitats. Nests were found under stones and under a fallen branch.

Material examined. **Gau:** Navukailagi 415, Navukailagi 432, Navukailagi 356, Navukailagi 300. **Kadavu:** Mt. Washington 700. **Koro:** Nasau 420 b, Nasau 465 a. **Moala:** Mt. Korolevu 375, Mt. Korolevu 300. **Ovalau:** Draiba 300. **Taveuni:** Lavena 300. **Vanua Levu:** Kilaka 146, Yasawa 300, Kasavu 300, Nakanakana 300. **Viti Levu:** Navai 700, Nakavu 300, Vatubalavu 300, Navai 863, Navai 930, Monasavu Dam 800, Nasoqo 800 d, Korobaba 300, Lami 200, Nakobalevu 340, Waimoque 850, Colo-i-Suva 186 d, Naboutini 300, Nabukelevu 300, Galoa 300, Nabukavesi 300, Mt. Rama 300, Naikorokoro 300, Waivudawa 300, Nadarivatu 750.

Pheidole wilsoni Mann

(Plate 95)

Pheidole wilsoni Mann, 1921: 433; major worker, minor worker described. Type locality: FIJI, Kadavu, Vanua Ava (W. M. Mann). Syntypes: 3 minor workers, 3 major workers (USNM, examined); (MCZC 8696, not examined).

Pheidole wilsoni is a species of moderate size most notably recognized by the weakly reticulated carinulae on the posterolateral lobes of the major and the dorsally punctate and ventrally smooth head of the minor. It can be separated from the sympatrically occurring *P. knowlesi* and *Pheidole* sp. FJ09 by the following characters of the major: weak reticulated carinulae and punctation occurring on the posterolateral lobes; posterior portion of antennal scrobe broadly impressed; carinulae surrounding posterior portion of antennal scrobe often weakly arcuate. Although there is not significant geographic variation in this species, as defined here, there does appear to be a nearly bimodal distribution of size, with the posterolateral lobes of the larger major workers being more robustly developed than those of the smaller major workers. Whether this size variation is representative of intraspecific variation, highly structured populations, or incipient species must be explored more thoroughly in future studies.

Pheidole wilsoni is another of Fiji's frequently encountered *Pheidole*. Specimens were taken primarily from sifted leaf litter and hand collection in forest interior habitats. Nests were found in dead branches, logs of various moisture contents, under moss on trees, under stones, and in ant-plants. Foragers were also attracted to cookie baits.

Material examined. **Gau:** Navukailagi 625, Navukailagi 415, Navukailagi 408, Navukailagi 432, Navukailagi 505, Navukailagi 480, Navukailagi 475, Navukailagi 356, Navukailagi 496, Navukailagi 575. **Kadavu:** Mt. Washington 760, Mt. Washington 700, Vunisea 200, Daviqele 300, Namara 300. **Lakeba:** Tubou 100 c. **Moala:** Mt. Korolevu 300. **Ovalau:** Levuka 450, Draiba 300. **Taveuni:** Lavena 300, Tavoro Falls 160. **Vanua Levu:** Mt. Delaikoro 910, Nakasa 300, Yasawa 300, Kasavu 300, Nakanakana 300, Drawa 270, Vuya 300, Mt. Vatudiri 641, Mt. Vatudiri 570, Mt. Delaikoro 699, Vusasivo Village 400 b, Vusasivo Village 400 b, Mt. Delaikoro 391, Lagi 300, Labasa. **Viti Levu:** Mt. Evans 800, Vatubalavu 300, Nadala 300, Vaturu Dam 575 b, Navai 930, Mt. Naqarababuluti 912, Naqaranabuluti 860, Mt. Batilamu 840 c, Mt. Naqaranabuluti 1050, Naqaranabuluti 1000, Mt. Tomanivi 950, Nadakuni 300, Nadakuni 300 b, Korobaba 300, Lami 200, Nakobalevu 340, Colo-i-Suva 200, Waimoque 850, Lami 171, Colo-i-Suva 186 d, Mt. Evans 700, Naboutini 300, Nabukelevu 300, Galoa 300, Nakavu 200, Vunisea 300, Nabukavesi 300, Mt. Rama 300, Naikorokoro 300, Waivudawa 300, Nausori.

Pheidole sp. FJ05

(Plate 96)

The majors of this species are robustly built with a heavily sculptured head that is posteriorly impressed in profile view, coarsely rugoreticulate posterolateral lobes, broad antennal scrobes, obtusely projecting humeri, and a glassy smooth gaster. Minors are noted by their low promesonotal profile and punctate sculpture that covers the cephalic dorsum, the entire mesosoma, and the lateral portions of the waist segments. The Koro specimens of major workers deviate slightly from those of Viti Levu, and Gau to a lesser extent, by a more densely punctate ground sculpture, slightly deeper impression between the vertex and posterior margin of head, a postpetiolar node that is broader and more smooth and shiny in dorsal view, and a patch of sculpturing at the base of the gaster. This species should be compared to *P. nindi* Mann (Solomon Is.), *P. philemon* Forel (Australia), and *P. recondita* Clouse (Micronesia).

Pheidole sp. FJ05 was primarily collected from sifted leaf litter and hand sampling in interior forests. Nests were found in dead branches, under stones and in ant plants.

Notes on the description of the major and minor workers.

Major worker. Head brown, same color as rest of body; distinctly concave between vertex and posterior margin; carinate on vertex becoming rugoreticulate on posterolateral lobes. Anterior clypeal margin flat to weakly impressed medially. Antennal scrobe shallow and broad. Pronotal humeri bluntly projecting. Promesonotum massive; punctuate with a few weak striae in dorsal view. Mesonotal declivity nearly vertical. Petiole and postpetiole punctate laterally, smooth to weakly striate dorsally. Postpetiole with moderately projecting lateral teeth. Gaster either entirely smooth or with a small patch of sculpture basally.

Minor worker. Head densely punctate dorsally, smooth and shiny ventrally. Mesonotal declivity gently sloped. Entire mesosoma punctate except for a smooth patch on the side of the pronotum. Pronotal humeri weakly developed. Petiole and postpetiole punctate laterally, smooth and shiny dorsally. Body and head brown with antennae and tarsi yellow.

Material examined. **Gau:** Navukailagi 625, Navukailagi 675, Navukailagi 408, Navukailagi 490, Navukailagi 505, Navukailagi 475, Navukailagi 575. **Koro:** Nasau 470 (4.4 km), Mt. Kuitarua 440 b, Nasau 465 a. **Viti Levu:** Vaturu Dam 575 b, Waimoque 850, Naikorokoro 300.

Pheidole sp. FJ09

(Plate 97)

Pheidole sp. FJ09 is a species of similar appearance to *P. knowlesi* and *P. wilsoni*, and the discussions of those species offer characters for distinguishing them. The major workers have a mostly smooth and shiny ground sculpture with patches of punctation, thin regular longitudinal carinae that begin on frontal lobes, extend past the vertex, and usually terminate before reaching the posterior margin of head. The sculpture between the eye and mandible is unbranching. Some specimens possess dark splotches on their frontal lobes. The minor workers are very distinctive, with the head and mesosoma entirely covered in dense punctation. The Viti Levu specimens from Colo-i-Suva are slightly aberrant in their glassy smooth ground sculpture. Interestingly, the males of this species are ergatoid.

Material examined. **Beqa:** Mt. Korovou 326, Malovo 182. **Gau:** Navukailagi 597, Navukailagi 415, Navukailagi 535, Navukailagi 408, Navukailagi 432, Navukailagi 490, Navukailagi 505, Navukailagi 480, Navukailagi 475, Navukailagi 496, Navukailagi 564, Navukailagi 575, Navukailagi 300. **Kadavu:** Moanakaka 60, Moanakaka 60, Vunisea 200, Namalata 120, Namalata 50, Daviqele 300, Namara 300. **Koro:** Mt. Nabukala 520, Nasau 470 (3.7 km), Mt. Kuitarua 440 b, Nasau 465 a, Nasoqoloa 300, Mt. Kuitarua 380. **Lakeba:** Tubou 100 b, Tubou 100 c. **Moala:** Mt. Korolevu 375, Mt. Korolevu 300. **Ovalau:** Levuka 400, Levuka 450, Draiba 300. **Taveuni:** Devo Peak 1187 b, Lavena 300, Lavena 235, Lavena 234, Lavena 217, Lavena 229, Mt. Devo 892, Tavoro Falls 160, Soqulu Estate 140, Lavena 235, Qacavulo Point 300. **Vanua Levu:** Nakasa 300, Yasawa 300, Kasavu 300, Nakanakana 300, Drawa 270, Vuya 300, Mt. Wainibeqa 152 c, Mt. Vatudiri 641, Mt. Vatudiri 570, Mt. Delaikoro 699, Vusasivo Village 400 b, Rokosalase 143, Rokosalase 180. **Viti Levu:** Mt. Evans 700, Nakavu 300, Colo-i-Suva Forest Park 220, Mt. Batilamu 1125 b, Savione 750 a, Korobaba 300, Lami 200, Nakobalevu 340, Colo-i-Suva 200, Colo-i-Suva 186 d, Naboutini 300, Nabukelevu 300, Navai 1020, Galoa 300, Nakavu 200, Vunisea 300, Nabukavesi 300, Naikorokoro 300, Waivudawa 300, Nausori.

roosevelti group

The *Pheidole roosevelti* group is composed of seven species endemic to the montane forests of the Fiji Islands. The group was revised by Sarnat (2008b), and was subjected to phylogenetic analysis along with congeners from Fiji and across the Pacific (Sarnat & Moreau, 2011). The group is diagnosable from all other *Pheidole* by the posteriorly projecting mesonotal process, concave mesonotal declivity, and lack of pronotal spines or processes, and long propodeal spines. All the species except for one (*P. simplispinosa*) have highly modified propodeal spines that are either distally angulate or bifurcated.

In addition to their unique appearance, species of the *roosevelti* group are characterized by a distinctive natural history. All species prefer the cooler, undisturbed wet forests of higher elevation mountains. Although their ranges are often small, with some species known only from the moss forests of single mountain summits, they can be locally quite abundant. All species appear to make nests in bare soil, the entrances to which are marked by conspicuous vertical turrets which stand 2–5 cm.

Phylogenetic analysis of this group and its relatives from Fiji and other Pacific regions has revealed that the *roosevelti* group is monophyletic. *Pheidole simplispinosa* is sister to the rest of the species, and the entire clade appears to have descended from a Fijian stock related to *P. knowlesi*, *P. vatu* and their close relatives. The group is only distantly related to the species groups (such as the *cervicornis* group, *quadrispinosa* group, and *quadricuspis* group) that compose the subgenus *Pheidolacanthinus*, demonstrating the peculiar spine modifications are a remarkable example of convergence.

Pheidole bula Sarnat

(Plate 98)

Pheidole bula Sarnat, 2008b: 12, figs. 32–34, 53–55, 74–76; minor worker, major worker, queen described. Type locality: FIJI, Viti Levu, Mt. Tomanivi, 3.4 km E Navai Village, 1.ii.2005, 1320m, -17.61481°,178.01825°, exposed mountain summit, nesting under stone (E. M. Sarnat, EMS#1789). Holotype: major worker (FNIC, CASENT0171113). Paratypes: workers, soldiers, queens (FNIC, USNM, ANIC).

Pheidole bula is one of the smaller members of the *roosevelti* group, with modestly projecting spines and strong sculpturing. The species is most readily distinguished from its close relatives by the smooth and shiny spaces between its facial rugae. While *P. roosevelti* and *P. furcata* both have facial rugoreticulum (majors) or rugae (minors) similar to *P. bula*, the interspaces between their rugae are filled with densely packed foveolae, giving them a duller appearance. The minor worker can be separated from all other minors of *roosevelti* group by the strong sculpturing of the ventral surface of its head.

Pheidole bula was encountered only at the summit of Mt. Tomanivi, Fiji's tallest mountain. The population of *P. bula* may therefore be in a precarious situation. With perhaps its closest extant relative occupying the lower elevations, and with no higher elevation to retreat to, it is possible that the current trends in climate change will consign *P. bula* to extinction in the near future.

Material examined: **Viti Levu:** Mt. Tomanivi 1294, Mt. Tomanivi 1300, Mt. Tomanivi 700 b, Navai 863.

Pheidole colaensis Mann

(Plate 99)

Pheidole (*Electropheidole*) *colaensis* Mann, 1921: 441; major worker, minor worker described. Type locality: FIJI, Viti Levu, Nadarivatu (W. M. Mann). Syntypes: minor workers, major workers (USNM, examined). In *roosevelti* group revision (Sarnat, 2008b: 17, figs. 35–37, 56–58, 77–79).

Pheidole colaensis is most readily distinguished from other *roosevelti* group species by its shiny integument and reduced sculpture. *Pheidole colaensis* is the only species in which the posterolateral lobes of the major caste are entirely free of sculpture. While the minor workers of the Vanua Levu species *P. pegasus* and *P. uncagena* also lack facial sculpturing above eye level, the propodeal spines of *P. colaensis* bear a distinctly shorter dorsal edge.

Pheidole colaensis appears to be restricted to the few high elevation ranges of Viti Levu, but the species is locally abundant where it occurs. *Pheidole colaensis* is widely sympatric with *P. roosevelti*, with the former tending to occupy the higher elevations and the latter preferring a slightly lower range. Although *P. colaensis* majors are scarce and timid, the minors can be observed foraging about the leaf litter some distance from their nests. The nest entrance of *P. colaensis* typically consists of a single turret built of small soil pellets that rises 3–5 cm above the ground, and leads to chambers over one meter deep that contain many hundreds or thousands of workers. The multiple dealate queens recovered from nest excavations suggest that the species might be polygynous. The queens of *P. colaensis*, like those of *P. bula* and *P. furcata*, are small with a reduced mesosoma.

Material examined. **Viti Levu:** Mt. Tomanivi 1294, Mt. Evans 800, Navai 700, Mt. Tomanivi 700, Monasavu Dam 800, Mt. Batilamu 1125 b, Mt. Naqaranabuluti 1050, Mt. Tomanivi 950, Navai 1023, Mt. Tomanivi 1105, Nasoqo 800 a, Nasoqo 800 b, Nasoqo 800 c, Nasoqo 800 d, Waimoque 850, Colo-i-Suva 460, Colo-i-Suva 325, Colo-i-Suva 372, Navai 1020.

Pheidole furcata Sarnat

(Plate 100)

Pheidole furcata Sarnat, 2008b: 19, figs. 38–40, 59–61, 80–82; soldier, worker, queen described. Type locality: FIJI: Kadavu, Mt. Washington, 1.4 km SSW Lomaji Village, 5.ix.2006, 760m, -19.11806°, 177.98750°, high elevation moss forest, turret nest in bare soil (E. M. Sarnat, EMS#2407). Holotype: major worker (FNIC, CASENT0171111). Paratypes: major workers, minor workers, queens (FNIC, USNM, ANIC, MCZC, LACM).

Pheidole furcata, owing to its strong facial sculpture, is most similar to *P. bula* and *P. roosevelti*. Whereas the facial sculpture of majors of *P. furcata* terminates before reaching the posterior margin, the facial sculpture of *P. bula* and *P. roosevelti* extends the full distance to the posterior margin. The minor of *P. furcata* is separated from these other two species by the completely smooth and shiny ventral surface of its head.

Known only from Mt. Washington, *P. furcata* is the only species of the *roosevelti* group that occurs on Kadavu. Whereas many of the *roosevelti* species are locally abundant where they occur, no foragers of *P. furcata* were observed at the type locality during the afternoon spent on the mountain. The collection was made by locating the signature earthen turret entrance of a nest rising several centimeters above the surrounding bare soil. Like *P. bula* and *P. colaensis*, the queens of this species have a strongly reduced mesosoma.

Material examined. **Kadavu:** Mt. Washington 760, Lomaji 580.

Pheidole pegasus Sarnat

(Plate 101)

Pheidole pegasus Sarnat, 2008: 21, figs. 41–43, 62–64, 83–85; major worker, minor worker, queen described. Type locality: FIJI, Vanua Levu, Mt. Delaikoro, 4.3 km SE Dogoru

Village, 31.vii.2006, 910m, -16.59028°, 179.31580°, high elevation moss forest, from turret nest in bare soil (E. M. Sarnat, EMS#2370). Holotype: major worker (FNIC, CASENT0171108). Paratypes: major workers, minor workers, queen (FNIC, USNM, ANIC, LACM).

Pheidole pegasus, on account of its large size, long limbs, glassy integument, and extraordinarily long propodeal spines, is arguably the most distinctive species of the *roosevelti* group. The only species that it can be confused with is its sister species, *P. uncagena*, with which it is sympatric. The major of *P. pegasus* can be distinguished from that of *P. uncagena* by the long dorsal edge of the propodeal spine and a broadly attached mesonotal process. The most distinctive differences between the minors of the two species, besides the longer spines and limbs of *P. pegasus*, are both found on the head. Whereas *P. pegasus* has a strongly ventrodorsally flattened subquadrate head and inconspicuous genal carinae, the head of *P. uncagena* is subovate, less flattened, and bears genal carinae that are produced conspicuously as elevated flanges.

Although *P. pegasus* is known only from the summit of Mt. Delaikoro, the species may occur on other unexplored high peaks of Vanua Levu. Where it does occur, it is locally abundant. The single turret of the nest belonging to the type series had a 5 mm diameter entrance hole, and was also insulated by a tidy ring of vegetation debris apparently placed there by workers. The queens of the species are large, with a strongly developed mesosoma.

Material examined. **Vanua Levu:** Mt. Delaikoro 910.

Pheidole roosevelti Mann

(Plate 102)

Pheidole roosevelti Mann, 1921: 438, Fig. 15; major worker, minor worker, queen described. Type locality: FIJI, Viti Levu, Nadarivatu. Syntypes: workers (MCZC no. 23173, examined); workers (USNM, examined). In revision of *roosevelti* group (Sarnat, 2008: 22, figs. 44–46, 65–67, 86–88.)

Pheidole roosevelti is a large species, most recognizable by the heavy sculpturing present on its face and promesonotum. The two other species with rugoreticulate faces are *P. furcata* from Kadavu, and *P. bula* from Viti Levu. In addition to the differences elaborated within the discussions of these other species, the majors and minors of *P. roosevelti* can be separated by the strongly produced facial rugoreticulum overlying a densely foveolate ground sculpture and thickly rugoreticulate mesosoma. Unlike *P. furcata* and *P. bula*, in which the queen caste is characterized by its smaller size and much reduced mesosoma, the queens of *P. roosevelti* bear closer resemblance to their northern relatives (*P. pegasus*, *P. simplispinosa*).

With the possible exception of *P. simplispinosa*, this species exhibits the most intraspecific variation of any in the *roosevelti* group. The observed variation in shape and sculpture may, in part, be due to the wide range occupied by the species. Unlike many of its close relatives, *P. roosevelti* does not appear to be restricted to the upper elevation limits of Fiji's mountain ranges, thus allowing its population to span significantly more suitable habitat.

Pheidole roosevelti is quite abundant where it occurs, and foragers can often be observed foraging on the ground and on vegetation. They nest in chambers deep underground in the soil, and the entrance to the nest is a turret approximately 2 cm tall and 0.5 cm wide that is composed of soil pellets. One nest found on the top of a mountain in Ovalau had three such turrets leading to the chambers below.

Material examined. **Gau:** Navukailagi 625, Navukailagi 717, Navukailagi 675, Navukailagi 415, Navukailagi 481, Navukailagi 432, Navukailagi 475, Navukailagi 575. **Ovalau:** Levuka 550, Levuka 400, Levuka 450, Draiba 300. **Viti Levu:** Mt. Naqarababuluti 912, Monasavu Dam 800, Mt. Batilamu 840 c, Mt. Batilamu 1125 b, Savione 750 a, Korobaba 300, Lami 304, Nakobalevu 340, Mt. Evans 700, Naikorokoro 300.

Pheidole simplispinosa Sarnat

(Plate 103)

Pheidole simplispinosa Sarnat, 2008: 26, figs. 47–49, 68–70, 89–91; major worker, minor worker, queen described. Type locality: FIJI, Koro, Mt. Kuitarua, 3.7 km NW Nasau Village, 20.vi.2005, 470m, -17.29083°, 179.40183°, primary rainforest, nesting in soil (E. M. Sarnat, EMS#2084). Holotype: major worker (FNIC, CASENT0171106). Paratypes: major workers, minor workers, queen (FNIC, USNM, ANIC).

Pheidole simplispinosa is the most distinctive of all *roosevelti* group species. It is the only member of the group in which the spines are simple and evenly straight without becoming modified into distal angles or bifurcations. The mesonotal process so prominent in other all other *roosevelti* group species is truncated into a blunt process such that the angle between the dorsal face of the mesonotum and the mesonotal declivity is obtuse. Beyond the simplified spine and mesonotal process, *P. simplispinosa* is also the smallest species of this group and has the shortest limbs relative to its size. The queens of *P. simplispinosa*, like those of *P. roosevelti* and *P. pegasus*, are characterized by the well-developed mesosoma. A significant variation in sculpture is associated with the geography of *P. simplispinosa*, with the Koro populations being the most sculptured, the Taveuni populations are the least sculptured, and the Vanua Levu populations are intermediate between the two.

Pheidole simplispinosa has a range within the Fiji archipelago rivaled only by *P. roosevelti*. Like *P. roosevelti*, this species tolerance of lower elevation habitat may serve as some explanation for its wide range. The two species, however, are entirely allopatric. Whereas *P. roosevelti* occupies the more southern islands of Viti Levu and Ovalau, *P. simplispinosa* occurs in the northern islands of Vanua Levu, Taveuni and Koro. Although single turret nests were observed, the species is also capable of constructing nests with multiple entrances. One such nest, from Mt. Delaikoro on Vanua Levu, was composed of irregular mounds of excavated soil.

Material examined. **Koro:** Mt. Nabukala 520, Nasau 470 (3.7 km), Mt. Kuitarua 440 b, Mt. Kuitarua 380. **Taveuni:** Devo Peak 1188, Tavuki 734, Mt. Devo 892, Mt. Devo 1064, Mt. Devo 775 a, Soqulu Estate 140. **Vanua Levu:** Kilaka 61, Kilaka 98, Kasavu 300, Nakanakana 300, Mt. Vatudiri 570, Mt. Delaikoro 699.

Pheidole uncagena Sarnat

(Plate 104)

Pheidole uncagena Sarnat, 2008: 28, figs. 50–52, 71–73; major worker, minor worker described.

Type locality: FIJI, Vanua Levu, Mt. Delaikoro, 4.3 km SE Dogoru Village, 31.vii.2006, 910 m, -16.59028°, 179.31580°, high elevation moss forest, turret nest in bare soil (E. M. Sarnat, EMS#2372). Holotype: major worker (FNIC, CASENT0171110). Paratypes: workers, soldiers (FNIC, USNM, ANIC).

Pheidole uncagena is most easily confused with its sister species, *P. pegasus*. Both are sympatric on Vanua Levu, and are characterized by a smooth and shiny integument, paler coloration, and long propodeal spines. The features that best separates *P. uncagena*, not only from *P. pegasus* but from all other *roosevelti* group species, are the modified genal carina that appear almost hook-like in oblique lateral view, and the strongly attenuated mesonotal process, which is best seen in dorsal view.

Some variation exists between the Vanua Levu type series and the minor workers collected in malaise traps from Taveuni. The propodeal spines of the type series are bifurcate, with a distinct anterior point in addition to the posterior point, whereas the anterior point of the Taveuni specimens are reduced to blunt angle, and the posterior points are longer than those exhibited by the Vanua Levu specimens. Additionally, the genal carinae of the type series come to a more definite point, whereas those of the Taveuni specimens are blunter.

Although no queen of *P. uncagena* has been collected, the similarities it shares with *P. pegasus* predict that it will be a large queen with a well-developed mesonotum. The type series was taken from a nest in bare soil with multiple turret entrances. The recovery of this species from malaise traps suggest that the workers at least, occasionally, forage in the arboreal stratum.

There is one collection from Kadavu arising from the recent malaise survey. Although we have no *a priori* reason to suspect this record, we find it to be dubious and possibly due to a labeling error. This is for two reasons, a) it would represent an unusual disjunction for this group, and b) we have visited this locality and only observed the Kadavu endemic *Pheidole furcata*. Species of the *roosevelti* group are usually locally common and conspicuous, where present at all.

Material examined. **Kadavu:** Mt. Washington 800. **Taveuni:** Devo Peak 1188, Devo Peak 1187 b, Mt. Devo 1064. **Vanua Levu:** Mt. Delaikoro 910, Kilaka 146, Kilaka 61, Kilaka 113.

Genus Poecilomyrma

Diagnosis of worker caste in Fiji. Head shape ovoid to rectangular. Antenna 12-segmented. Antennal club 3-segmented. Antennal scrobes absent. Anterior clypeal margin variously shaped, but never armed with three broad and blunt teeth. Sides of head lacking carinate ridge extending below eye-level from mandibular insertions to posterolateral head margin. Mandibles triangular; lacking a distinct basal tooth. Mesosoma evenly convex; lacking depression separating promesonotum from propodeum; erect hairs present. Propodeum armed with spines or teeth. Propodeal lobes longer than propodeal spines. Propodeal spines distinctly longer than diameter of propodeal spiracle. Waist 2-segmented. Petiole pedunculate; lacking large anteroventral subpetiolar process.

Poecilomyrma is the only ant genus that is strictly endemic to the Fijian archipelago. It is also one of the most beautiful ants of Fiji, with a deeply grooved and very shiny integument, reduced propodeal spines, extended and spinose propodeal lobes, a long slender petiole and often striking red and black coloration. The genus was erected by Mann (1921), who considered it to be very close to *Podomyrma*, perhaps on account of the weakly armed propodeum and the arboreal habitus. Recent molecular work, however, does not find strong support for a sister group relationship between the two taxa (P. S. Ward, pers. comm.).

Mann named *P. senirewae* and its subspecies *P. myrmecodiae* both from the Nadarivatu area. The type series of *P. senirewae* is actually a mix of both species, and Mann incorrectly believed that the black-headed specimens were minor workers and the red-headed specimens were major workers. He explains that the series was taken from, "a small colony nesting in a hollow twig of a recently felled *kauri* tree, and a couple of individuals found on leaves." The type series of *P. senirewae* was taken from a *Myrmecodia* ant plant. The confusion Mann experienced with these ants is quite understandable, as additional collections have only contributed to the vexing question of where species boundaries lie. Mann's prediction that these ants might be "widely distributed though locally hard to find," has been borne out during the recent survey. Both the geographical and morphological range of this genus was significantly expanded. *Poecilomyrma* is now known from all seven of the largest islands (only males known from Koro), but records remain quite rare.

The morphological variation encountered in *Poecilomyrma* is very difficult to organize into any geographical or phylogenetic pattern, and even after an earnest study it is to some degree arbitrary as to whether the genus is split into seven species or lumped into one. To begin with, there are three distinct color morphs. Morphotype #1 (as exemplified by *P. senirewae*) is all red except for the gaster. Morphotype #2 (as exemplified by *Poecilomyrma* sp. FJ05) has a black gaster and head, a red mesosoma, and infuscated waist segments. Morphotype #3 is entirely black. The sculpture of the mesosoma varies substantially from uniform parallel carinae to fully reticulated rugae. The cephalic sculpture, postpetiole sculpture and length of propodeal spines and lobes all vary, as well. Size also varies conspicuously. However, the aforementioned characters do not appear to be correlated with each other in any meaningful way. For example, particular sculpture patterns do not tend to correlate with color morphotype or with size or with geography.

Despite the great variance among the series of specimens collected, the variance within any particular series is quite low. That is to say, variation among local populations is quite high while variation within local populations is quite low. Similarly highly structured populations are observed in the Fijian *Cerapachys*. Both of these genera, in Fiji, have wingless ergatoid queens, which may be a factor in the limited gene flow. The challenges of species delineation would be better met by molecular tools capable of penetrating deeper than this cursory study of surface sculpture and color. The males are occasionally collected by malaise trapping, and they exhibit a number of variable characters as well, especially with regard to overall size, relative eye size, and color.

Key to workers of *Poecilomyrma*.

1 Head color red 2

– Head color black 4

2 Cephalic hairs are subequal to eye length. Cephalic and body sculpture with sharp, carinate edges. Pronotal humeri not projecting; in dorsal view promesonotum irregularly sculptured with crenulated longitudinal and transverse carinae. Smaller size (HW 0.97 mm, n = 1). Head narrow (CI 0.85, n =1) .... ***Poecilomyrma* sp. FJ03**

– Cephalic hairs distinctly shorter than eye length. Cephalic and body sculpture with blunt, rounded edges. Pronotal humeri variably produced. In dorsal view promesonotum with variable sculpture. Size variable (HW 0.93 mm–1.19 mm, n = 10). Head shape variable (CI 0.82–0.92, n = 10) 3

3 Pronotal humeri produced as strongly projecting angles. In dorsal view promesonotum irregularly sculptured with crenulated rugoreticulum. Size large (HW 1.09 mm–1.19 mm, n = 8). Head broad (CI 0.88–0.92, n = 8).. ***P. senirewae***

– Pronotal humeri absent to weakly produced, but never forming strongly projecting angles. In dorsal view promesonotum regularly sculptured with approximately uniform longitudinal rugae. Size small (HW 0.93 mm–0.96 mm, n = 2). Head narrow (CI 0.82–0.83, n = 2). .... ***Poecilomyrma* sp. FJ02**

4 Mesosoma black.. ***Poecilomyrma* sp. FJ08**

– Mesosoma red 5

5 Pronotal humeri produced as strongly projecting angles. Cephalic sculpture more weakly produced with finer rugae... 6

– Pronotal humeri absent to weak, but never produced as strongly projecting angles. cephalic sculpture strongly produced with robust rugae .... ***Poecilomyrma* sp. FJ07**

6 Smaller species (HW 0.82 mm–0.96 mm, n = 9) with relatively short limbs (SI 0.76–0.84, FI 0.95–1.07, n = 9). In dorsal view promesonotum variably sculptured...7

– Larger species (HW 1.02 mm–1.13 mm, n = 13) with relatively long limbs (SI 0.83–0.93, FI 1.14–1.24, n = 13). In dorsal view promesonotum irregularly sculptured with crenulated rugoreticulum .... ***P. myrmecodiae***

7 Larger species (HW 0.89 mm–0.96 mm, n = 6). In dorsal view promesonotum regularly sculptured with approximately uniform longitudinal rugae separated by deep furrows. Legs and coxae dark and contrasting with mesosoma .... ***Poecilomyrma* sp. FJ06**

– Smaller species (HW 0.81 mm–0.85 mm, n = 3). In dorsal view promesonotum with sculpture regular or irregular, but never separated by deep furrows. Legs and coxae similar in color to mesosoma.. ***Poecilomyrma* sp. FJ05**

Poecilomyrma myrmecodiae Mann, NEW STATUS

(Plate 105)

Poecilomyrma senirewae subsp. *myrmecodiae* Mann, 1921: 448; worker described. Type locality: FIJI, Viti Levu, Mt. Victoria [Mt. Tomanivi] (W. M. Mann). Syntypes: 24 workers (USNM, examined); 6 workers (MCZC type no. 2100, examined).

Poecilomyrma myrmecodiae is a large species with a black head infuscated posteriorly, a red mesosoma and petiole, an infuscated postpetiole, a black gaster and dark appendages. The species is similar in size to *P. senirewae*, with which it is sympatric, but can be differentiated by the dark head. The larger size, relatively longer limbs and more deeply furrowed sculpture sets it apart from *Poecilomyrma* sp. FJ05, with which it is also sympatric. Characters in the key separate it from other similarly colored taxa. An ergatoid queen (CASENT0194604) was collected in a small nest from a *Hydnophytum* (Rubiaceae) ant plant, and is characterized by similar long flexous pilosity and more margined carinae and rugae as described for the putative ergatoid *P. senirewae* as discussed under that species. The type series consists of 21 workers collected from a *Hydnophytum* bulb by Mann. The species, as defined here, appears restricted to the central and southeastern mountain ranges of Viti Levu.

Material examined. **Viti Levu:** Monasavu Dam 800, Monasavu Dam 1000, Monasavu 800, Waimoque 850, Colo-i-Suva 372, Navai 1020, Waivudawa 300, Mt. Tomanivi.

Poecilomyrma senirewae Mann

(Plate 106)

Poecilomyrma senirewae Mann, 1921: 446, fig. 16; worker described. Type locality: FIJI, Viti Levu, Nadarivatu (W. M. Mann). USNM worker here designated Lectotype. Paralectotypes: 2 workers (MCZC, type no. 20999, designated); 4 workers (USNM, designated).

Note: The cotype series originally described by Mann consists of two species: a larger species with a head the same red color as the mesosoma (described as the major worker), and a smaller species with a darker red black head contrasting with the red mesosoma (described as the minor worker, and treated here as *Poecilomyrma* sp. FJ05). The description of the larger specimens appeared first in the publication on page 446, followed by the smaller specimens on page 448. Therefore, one specimen of the larger red-headed species is hereby designated as the lectotype, and the four workers of the same species from the same locality are designated as paralectotypes. The discussion of *P. senirewae* provided here is restricted only to the larger species with the red head described by Mann as the major worker. The smaller species with the darker head described by Mann as the minor worker is treated here as *Poecilomyrma* sp. FJ05.

Poecilomyrma senirewae is a large reddish species with darker appendages and a black gaster. The mesosomal sculpture is strongly reticulated, and the pronotal humeri project as acute teeth. Specimens matching the type series are known from the Nadarivatu area, Mt. Tomanivi and Koroyanitu (all Viti Levu). It is sympatric with *P. myrmecodiae* at Mt. Tomanivi and with *Poecilomyrma* sp. FJ05 at Nadarivatu and Koroyanitu. *Poecilomyrma* sp. FJ03 (Gau) has a similar color pattern, but differs in its uniformly longitudinal and non-reticulated mesosomal sculpture and reduced pronotal humeri. There is a specimen collected by E. O. Wilson from Nadala that may represent the ergatoid queen of this species. The pilosity is longer and more flexous, and the carinae and rugae are more strongly margined.

Material examined. **Viti Levu:** Navai 770, Navai 930, Mt. Naqarababuluti 912, Naqaranabuluti 860, Mt. Batilamu 840 c, Mt. Tomanivi 950, Nadarivatu 750.

***Poecilomyrma* sp. FJ03**

(Plate 107)

Poecilomyrma sp. FJ03 is known only from a single locality on Gau. It can be separated from *P. senirewae* and *Poecilomyrma* sp. FJ02 by its shorter hairs, more uniform mesosomal sculpture and reduced pronotal humeri.

Material examined. **Gau:** Navukailagi 432.

***Poecilomyrma* sp. FJ05**

(Plate 108)

Poecilomyrma sp. FJ05 includes the 'minor workers' described under *P. senirewae* by Mann

(1921). Although a more thorough examination of that material is required, the single worker (CASENT0181642) from Koroyanitu (Viti Levu) and the two workers collected by James Wetterer from Waisoi (Viti Levu) differ markedly in their sculpture. The mesosoma of the former has parallel rugae while that of the latter has weaker and less uniform carinulae. The description given by Mann fits the latter specimens more closely than the former. Characters given in the key can separate *Poecilomyrma* sp. FJ05 from other *Poecilomyrma* with black heads. A colony collected by J. K. Wetterer consists of 31 workers, three males and one ergatoid queen, also taken from a *Hydnophytum* (Rubiaceae).

Material examined. **Viti Levu:** Savione 750 a, Nadarivatu 750.

Poecilomyrma sp. FJ06

(Plate 109)

Known from several localities in Kadavu, *Poecilomyrma* sp. FJ06 is distinguished from other black headed *Poecilomyrma* by the relatively short dark appendages and deeply furrowed parallel grooves of the mesosoma. A small colony was found nesting in a dead branch.

Material examined. **Kadavu:** Mt. Washington 760, Moanakaka 60, Mt. Washington 700.

Poecilomyrma sp. FJ07

(Plate 110)

Known from Vanua Levu and Ovalau, *Poecilomyrma* sp. FJ07 is unique among the dark-headed species in its combination of reduced pronotal humeri and robustly developed rugae.

Material examined. **VANUA LEVU:** 3km NW Waisali Vlg. (570m); 3.6km SE Dogoru Vlg.; 4km SE Lomaloma Vlg. (630m). **OVALAU:** 2.4km W Levuka; 1.6km WSW Levuka.

Poecilomyrma sp. FJ08

(Plate 111)

Poecilomyrma sp. FJ08 is known from several localities on Gau and one on Taveuni. It is the only *Poecilomyrma* with a black mesosoma, and also has reduced pronotal humeri. The mesonotal dorsum of the Taveuni specimen is more strongly crenulated than those of the Gau specimens. *Poecilomyrma* sp. FJ08 is also sympatric with the red-headed *Poecilomyrma* sp. FJ03 on Gau.

Material examined. **Ovalau:** Levuka 550, Levuka 400. **Vanua Levu:** Mt. Vatudiri 570, Mt. Delaikoro 699, Lomaloma 630.

Genus **Pristomyrmex**

Diagnosis of worker caste in Fiji. Head shape circular to ovoid. Antenna 11-segmented.

Antennal club 3-segmented. Antennal sockets surrounded by a raised sharp-edged ridge. Anterior clypeal margin armed with three broad and blunt teeth. Sides of head lacking carinate ridge extending below eye-level from mandibular insertions to posterolateral head margin. Mandibles triangular; with four large teeth on the masticatory margin and one on the basal margin. Mesosoma with erect hairs present. Propodeum either unarmed or armed with spines distinctly longer than diameter of propodeal spiracle. Waist 2-segmented. Petiole pedunculate; lacking a large anteroventral subpetiolar process. Postpetiole not swollen; in dorsal view not distinctly broader than long or distinctly wider than petiole.

Pristomyrmex is restricted to the Old World and has its center of diversity in the Oriental region (Wang, 2003). Most species inhabit the rainforest and forage as predators or scavengers, and tend to nest in soil, leaf litter or rotten wood. The Fijian *Pristomyrmex* are most likely confused with *Tetramorium* in Fiji on account of the rims surrounding the antennal insertions, but can be separated by the 11-segmented antenna, more circular faces, mandibular teeth, clypeal apron and lack of antennal scrobes.

The two species in Fiji are endemic to the archipelago. Both belong to the *laevigatus* group, which is Asian in origin and extends through New Guinea into the Solomon Islands. They are most often encountered in the leaf litter, and their small colonies can be found nesting in rotting logs and under stones. A global revision of the genus is available, though only one of the Fijian species was recognized at the time of publication (Wang, 2003).

Key to *Pristomyrmex* workers of Fiji.

1 Propodeal spines present and robust .... ***P. mandibularis***
– Propodeal spines absent or developed as small angles .... ***Pristomyrmex* sp. FJ02**

Pristomyrmex mandibularis Mann

(Plate 112)

Pristomyrmex mandibularis Mann, 1921: 444; worker described. Type locality: FIJI, Viti Levu, Nadarivatu (W. M. Mann). Syntypes: 9 workers (MCZC type no. 8702, examined); 11 workers (USNM, examined). In generic revision (Wang, 2003: 505, figs. 225-228).

Note: Specimens from the non-type localities of Waiyanitu, Nasoqo and Ovalau housed at the USNM bear red cotype labels but are not true syntypes.

Pristomyrmex mandibularis is a very shiny robustly built species that can be separated from its only Fijian congener, *Pristomyrmex* sp. FJ02, by the unmistakable presence of propodeal spines. Color, shape and sculpture vary considerably in the species. Color ranges from reddish brown to reddish black. The shape of the promesonotum shows a weak trend with color, such that the pronotal humeri and mesonotal dorsolateral margins tend to be more angulate in the reddish specimens and more obtuse in the darker specimens. Similarly, the reddish specimens tend towards a strongly foveate face while darker specimens tend towards a smoother face nearly free of foveae. Wang (2003) mentions most of the variation in his monograph, but does not discuss any correlations he noticed among the character states. Furthermore, there are some variations, such as the lateral longitudinal carinae being fully present to fully absent (varies within nest series), that are not discussed.

Although the extremes of the spectrum occur sympatrically on Koro, intermediates are found in Viti Levu, suggesting that gene flow may be more common between some populations than others. In Koro, there are a number of workers (e.g., CASENT0171142, CASENT0181718, CASENT0194559) collected from the leaf litter that strongly display the darker, less angulate, less foveate syndrome. Also on Koro, and even at the exact site of CASENT0194559 (Wailolo Crk., 2.1 km SW Nabuna Village), are specimens (e.g., CASENT0171044, CASENT017100, CASENT0181831) that strongly display the redder, more angulate, more foveate syndrome.

If no other collections were known but those from Koro, the clear sympatry of the two morphotypes without intermediate specimens would suggest a lack of gene flow and would serve as cause for dividing them into two species. However, intermediates are found elsewhere in Fiji. The workers from the western mountain ranges (e.g., CASENT0181773) tend strongly towards the more foveate syndrome, and those from southeastern Viti Levu (e.g., CASENT0181648) tend strongly towards the less foveate syndrome. Somewhat predictably, specimens (e.g., CASENT0181759) from the southern central ranges are intermediate between the two, suggesting gene flow is occurring across the Viti Levu populations. There are no known examples of the two syndromes occurring sympatrically on Viti Levu.

The few collections from Kadavu match the less foveate syndrome. In Vanua Levu, there is one specimen (CASENT0181756) that matches the foveate syndrome, one that is somewhat intermediate (CASENT0181579), and one specimen (CASENT0181637) that is quite aberrant in its morphology, especially with regard to petiole shape, and is difficult to classify. None of the Vanua Levu specimens are sympatric. The two specimens examined from Taveuni tend towards the more foveate syndrome. The specimens from Gau all trend strongly towards the less foveate syndrome.

The species occurs widely across Fiji, and alates were collected effectively by malaise traps on the more remote islands, such as Moala and Lakeba. Workers, however, were collected primarily by litter sifting and from nests taken from beneath stones.

Material examined. **Gau:** Navukailagi 625, Navukailagi 387, Navukailagi 408, Navukailagi 432, Navukailagi 490, Navukailagi 475, Navukailagi 496, Navukailagi 564, Navukailagi 300. **Kadavu:** Mt. Washington 760, Mt. Washington 800, Moanakaka 60, Moanakaka 60, Mt. Washington 700, Lomaji 580. **Koro:** Nabuna 115, Tavua 220, Mt. Kuitarua 500, Mt. Kuitarua 505, Nasau 420 b, Nasau 465 a, Nasoqoloa 300, Mt. Kuitarua 380. **Lakeba:** Tubou 100 b, Tubou 100 c. **Moala:** Mt. Korolevu 300. **Ovalau:** Ovalau. **Taveuni:** Lavena 234, Tavuki 734, Mt. Devo 775 a, Soqulu Estate 140. **Vanua Levu:** Nakanakana 300, Mt. Vatudiri 641, Mt. Vatudiri 570, Banikea 398, Lomaloma 630, Lomaloma 630, Lasema a. **Viti Levu:** Nabukavesi 40, Mt. Evans 800, Mt. Evans 800, Mt. Evans 800, Mt. Evans 700, Vaturu Dam 530, Monasavu Dam 600, Korobaba 300, Nakobalevu 340, Colo-i-Suva 372, Mt. Evans 700, Naboutini 300, Galoa 300, Nabukavesi 300, Mt. Rama 300, Naikorokoro 300, Waivudawa 300, Nausori Highlands 400, Nasoqo, Waiyanitu.

***Pristomyrmex* sp. FJ02**

(Plate 113)

Pristomyrmex sp. FJ02 is generally similar in appearance to *P. mandibularis*, but can be easily distinguished by the propodeal spines which vary from short denticles to absent. Similar to *P. mandibularis*, the amount of foveae on the face varies from numerous to nearly absent. The mandibular structure of the species places it in the *laevigatus* group. *Pristomyrmex inermis* Wang from New Guinea, also a member of the *laevigatus* group, is the only other member of the genus known to possess an unarmed propodeum. However, *Pristomyrmex* sp. FJ02, can be distinguished from *P. inermis* by the petiole, which is strongly nodiform in the former and wedge-shaped in the latter. The petiole node of *Pristomyrmex* sp. FJ02, although taller, is much more similar to that of *P. mandibularis*. It would be interesting to know whether the species is more closely related to *P. inermis* than it is to *P. mandibularis*, as that would support a double invasion scenario of Fiji by more western Melanesian *Pristomyrmex* lineages.

The species is widely distributed across the archipelago, and broadly sympatric with *P. mandibularis*.

Material examined. **Gau:** Navukailagi 387, Navukailagi 432, Navukailagi 490, Navukailagi 496, Navukailagi 564, Navukailagi 575. **Kadavu:** Mt. Washington 800, Mt. Washington 700, Lomaji 580. **Koro:** Mt. Kuitarua 505, Mt. Kuitarua 485, Mt. Nabukala 500, Mt. Kuitarua 440 b, Nasau 465 a. **Lakeba:** Tubou 100 c. **Ovalau:** Draiba 300. **Taveuni:** Devo Peak 1188, Devo Peak 1187 b, Mt. Devo 892, Mt. Devo 1064, Soqulu Estate 140. **Vanua Levu:** Kilaka 61, Yasawa 300, Vusasivo Village 400 b, Lomaloma 630. **Viti Levu:** Mt. Evans 800, Mt. Evans 800, Mt. Evans 800, Navai 700, Mt. Tomanivi 700, Korobaba 300, Lami 200, Lami 260, Lami 400, Colo-i-Suva 460, Colo-i-Suva 325, Colo-i-Suva 372, Naboutini 300, Nabukelevu 300, Galoa 300, Vunisea 300, Nabukavesi 300, Naikorokoro 300, Veisari 300 (3.8 km N), Waivudawa 300, Veisari 300 (3.5 km N).

Genus **Pyramica**

Diagnosis of worker caste in Fiji. Head shape triangular. Antenna 6-segmented. Antennal club 2-segmented. Eyes located on lower margin of antennal scrobes. Mandibles triangular. Propodeum armed with spines or teeth. Waist 2-segmented. Spongiform attached to at least some portion of waist. Hairs flagellate or spatulate on at least some portion of head or body.

Pyramica is a dacetine genus closely related to *Strumigenys*, from which it can be separated in Fiji by the shape of the mandibles, which are triangular in the former and linear in the latter. Together, these two genera can be separated from all others known from Fiji by the combination of their triangular heads and wide array of specialized squamate hairs, spongiform tissue and lamellae. *Eurhopalothrix* is related to these dacetines, but can be separated by its significantly larger size and more deeply excavated sculpture. The Fijian *Pyramica* are not particularly diverse or abundant. They are most often encountered in the leaf litter. There is one species introduced into the Pacific, and several species that are apparently endemic to the archipelago.

Key to *Pyramica* workers of Fiji.

1 Dorsal surface of head not covered by short appressed spatulate hairs. Pronotal humeral hairs absent. Dorsal surface of mesosoma completely polished. Propodeal spines entirely covered by spongiform such that the tips are hidden from view. Dorsum of petiolar node polished smooth. Introduced .... ***P. membranifera***

– Dorsal surface of head covered by short appressed spatulate hairs. Pronotal humeral hairs present. Dorsal surface of mesosoma densely punctate. Propodeal spines with spongiform absent, or with a thin strip present, but never completely obscured. Dorsum of petiolar node punctate.. 2

2 Pronotum entirely punctate and without striae. Basigastral sculpture with thick costae that are broadly separated from each other by a distance greater than twice their width. Dark brown. Larger species (HW > 0.45 mm). Fiji endemic...................................... ***P. trauma***

– Pronotum punctate with short weak striae laterally. Basigastral sculpture with thin striae that are narrowly separated from each other by a distance distinctly less than twice their width. Reddish brown. Smaller species (HW < 0.45 mm). Fiji endemic. .... ***Pyramica* sp. FJ02**

Pyramica membranifera (Emery)

(Plate 114)

Strumigenys (*Trichoscapa*) *membranifera* Emery, 1869: 24, fig. 11; worker described. Type locality: ITALY [not examined]. Senior synonym of *foochowensis*, *marioni*, *santschii*, *silvestriana*, *simillima*, *vitiensis* (Syntypes: 2 workers [MCZC type no. 8727, examined]), *williamsi* (Brown, 1948: 114). In Fiji (Mann, 1921: 461, fig. 22c); in Polynesia (Wilson & Taylor, 1967: 35, fig. 24). In generic revision (Bolton, 2000: 322).

Pyramica membranifera is a small smooth species with a strongly convex cephalic dorsum that bears a single pair of standing hairs. It can be separated from Fiji's native *Pyramica* by the lack of pronotal humeral hairs, and the smooth dorsal surfaces of its mesosoma and petiole. Wilson and Taylor (1967) offered the following discussion of *P. membranifera*, "Brown (1949) states that *T. membranifera* [= *P. membranifera*] is probably of African origin. It has been spread by human commerce through a large part of the tropics and warm temperate zones, including such diverse areas as the Fiji Is., eastern China, West Indies and southeastern United States. The species has an ecological amplitude unusual for a dacetine, nesting in major habitats from dense woodland to dry, open cultivated fields." Wilson (1953) described the feeding behavior of the workers as being predaceous on a wide variety of small, soft-bodied arthropods.

In Fiji, the species is relatively rare, and in the recent survey was represented by a single collection of specimens from a litter sample taken at Rokosalase (Vanua Levu), although several older collections were examined in museums.

Material examined. **Lakeba:** Lakeba. **Munia:** Munia. **Vanua Levu:** Rokosalase 180, Suene, Lasema a. **Viti Levu:** Saiaro.

Pyramica trauma Bolton

(Plate 115)

Pyramica trauma Bolton, 2000: 408; worker described. Type locality: FIJI, Kadavu, Mt. Korogatule, nr. Matasawalevu, 18°59′S, 178°28′E, 4.vii.1987, 300 m, rainforest sieved litter, QM Berlesate No. 773 (G. Monteith). Holotype: worker (ANIC, not examined). Paratype: worker (BMNH, not examined).

Pyramica trauma is a dark brown species with two pairs of standing cephalic hairs, long stout pronotal humeral hairs, punctate dorsal surfaces of the mesosoma and petiole, and a very broad postpetiolar disc. The species is closely related to *Pyramica* sp. FJ02, but can be separated by its larger size, lack of pronotal striations, and coarser basigastral sculpture. Both species belong to the *capitata* group, and are presumed to be closely related to *P. charybdis* (Indonesia), *P. epipola* (Samoa), *P. phasma* (New Guinea), *P. tethys* (New Guinea) and *P. themis* (New Guinea).

Bolton (2000) lists the species as occurring on Viti Levu, Vanua Levu and the type locality of Kadavu. The recent survey recovered only three collections of *P. trauma*, two from Nadarivatu (Viti Levu), and one from Mt. Korobaba in southeastern Viti Levu. The Korobaba worker is lighter than the Nadarivatu workers.

Material examined. **Viti Levu:** Korobaba 300.

Pyramica sp. FJ02

(Plate 116)

Pyramica sp. FJ02 is nearly identical to *P. trauma*, and shares the same pilosity patterns and similar body shape. However, consistent differences are found between the two, namely those discussed under the notes of the latter species. Although the recent survey did not find the two in sympatric localities, it is possible the two do occur sympatrically based on the localities of *P. trauma* listed by Bolton (2000). The size and sculpturing are close to those listed for *P. epipola* which is known from Samoa, and which Bolton suggests could be more widely distributed. However, the arrangement of the cephalic hairs and the width of the postpetiolar disc are more in agreement with *P. trauma*.

Material examined. **Beqa:** Mt. Korovou 326. **Gau:** Navukailagi 415, Navukailagi 408, Navukailagi 490, Navukailagi 300. **Kadavu:** Moanakaka 60, Mt. Washington 700. **Koro:** Nasau 465 a, Nasoqoloa 300. **Moala:** Mt. Korolevu 300. **Vanua Levu:** Nakasa 300, Yasawa 300, Kasavu 300, Nakanakana 300, Drawa 270, Mt. Delaikoro 699, Rokosalase 180. **Viti Levu:** Vatubalavu 300, Mt. Naqaranabuluti 1050, Waimoque 850, Naboutini 300, Nabukelevu 300, Galoa 300, Nakavu 200, Nabukavesi 300, Naikorokoro 300.

Genus Rogeria

Diagnosis of worker caste in Fiji. Head shape ovoid to rectangular. Antenna 12-segmented. Antennal club 3-segmented. Antennal scapes fail to reach posterior margin of head by at least the length of the first funicular segment. Antennal scrobes absent. Antennal sockets not surrounded by a raised sharp-edged ridge. Anterior clypeal margin never armed with three broad and blunt teeth. Frontal lobes relatively far apart so that the posteromedian portion of the clypeus, where it projects between the frontal lobes, is much broader than one of the lobes. Sides of head lacking carinate ridge extending below eye-level from mandibular insertions to posterolateral head margin. Mandibles triangular; lacking a distinct basal tooth. Mesosoma evenly convex; lacking depression separating promesonotum from propodeum; erect hairs present. Propodeal spines short, approximately equal to diameter of propodeal spiracle. Waist 2-segmented. Petiole pedunculate; lacking large anteroventral subpetiolar process. Postpetiole not swollen; in dorsal view not distinctly broader than long or distinctly wider than petiole. Tip of sting tapered to a point; lacking a triangular to pendant-shaped extension.

Rogeria is most diverse in the Neotropics, and the presence of three species in the Pacific is an interesting biogeographical disjunction. Two Pacific species (*R. megastigmatica* Kugler and *R. stigmatica* Emery) belong to the *stigmatica* group, and the third (*R. exsulans* Wilson and Taylor) is proposed to be more closely related to species of the *creightoni* group (Kugler, 1994).

Rogeria stigmatica Emery
(Plate 117)

Rogeria stigmatica Emery, 1897: 589; worker described. Type locality: NEW GUINEA [not examined]. Description of queen and male, in Solomon Is. (Mann, 1919: 342); in Fiji (Mann, 1921: 451); in Polynesia (Wilson & Taylor, 1967: 74). Senior synonym of *manni*, *sublevinodis* (Kugler, 1994: 33).

Note: Specimens of *R. sublevinodis* Emery from Fiji (Waiyanitu and Vanua Ava) collected by Mann and housed at the MCZC bear red cotype labels (MCZC type no. 21058), but are not true syntypes. The type locality of *R. sublevinodis* is New Caledonia.

Rogeria stigmatica is a distinctive brown species with very short propodeal spines, a long petiolar peduncle and high petiolar node, abundant long pilosity, a shiny swollen gaster and a dense coverage of strongly reticulating carinulae on the head, mesosoma and waist segments. It is most similar to *Lordomyrma*, but can be separated by the shorter propodeal spines, lack of antennal scrobes, more gradually sloping mesosoma and longer petiolar peduncle. There is no noticeable variation within Fiji, and Wilson and Taylor (1967) report no variation among the populations of Fiji, Samoa and Tahiti.

The species is widely distributed not only across Fiji but throughout most of the western Pacific. *Rogeria stigmatica* displays one of the more interesting behaviors known of Pacific ants. The workers are capable of secreting long white chains of bubbles from their gaster that persist in the environment for a relatively long time. Mann (1921) offered the following account, "When the formicary is opened, the disturbed ants behave in a curious manner, secreting from their anal glands viscid matter in elongate threads that closely resemble worms. These threads twist in a life-like manner and the first time I saw them I actually took them to be small worms. The ants themselves, motionless and of the same color as the earth, were at first not visible and the twisting, apparently crawling 'worms' most conspicuous." This behavior was also observed during the recent survey. *Rogeria stigmatica* was found nesting in rotting wood, and workers were often collected from sifted litter.

Material examined. **Beqa:** Mt. Korovou 326. **Gau:** Navukailagi 415, Navukailagi 408, Navukailagi 300. **Kadavu:** Daviqele 300, Namara 300, Vanua Ava b. **Koro:** Tavua 220, Nasau 470 (3.7 km), Nasau 465 a, Nasoqoloa 300, Mt. Kuitarua 380. **Moala:** Naroi 75, Mt. Korolevu 300. **Ovalau:** Levuka 400, Levuka 450, Draiba 300. **Taveuni:** Qacavulo Point 300. **Vanua Levu:** Mt. Kasi Gold Mine 300, Nakasa 300, Yasawa 300, Kasavu 300, Nakanakana 300, Vusasivo Village 400 b, Vusasivo Village 400 b, Rokosalase 180, Mt. Delaikoro 391. **Viti Levu:** Mt. Evans 700, Nakavu 300, Nadakuni 300, Nadakuni 300 b, Korobaba 300, Nakobalevu 340, Colo-i-Suva 200, Colo-i-Suva 186 d, Mt. Evans 700, Naboutini 300, Nabukelevu 300, Galoa 300, Vunisea 300, Nabukavesi 300, Mt. Rama 300, Naikorokoro 300, Waivudawa 300, Sigatoka, Waiyanitu.

Genus Romblonella

Diagnosis of worker caste in Fiji. Head shape ovoid to rectangular. Antenna 12-segmented. Antennal club 3-segmented. Anterior margin of clypeus lacking a rectangular projection that extends over base of mandibles. Mandibles triangular. Propodeum armed with spines. Waist 2-segmented. Petiole lacking peduncle; lacking large anteroventral subpetiolar process.

Romblonella is a small genus comprised of eight species that range from the Philippines through New Guinea and the Torres Straight region of Australia, north into Micronesia and south into the Solomons and Fiji (Taylor, 1991b). The single species from Fiji is quite rare, as it is one of the few that was not collected by Mann or the recent survey.

Romblonella liogaster (Santschi), NEW STATUS
(Plate 118)

Tetramorium scrobiferum var. *liogaster* Santschi, 1928a: 69; worker described. Type locality: FIJI, Lau, Vanua Balavu, 20.ix.1924 (E. H. Bryan). Holotype: worker (BPBM type no. 360, examined). Combined in *Romblonella* (Bolton, 1995: 382).
Romblonella vitiensis Smith, M.R. 1953: 79; worker described. Type locality: FIJI, Wakaya, 17.x.1924, (E. H. Bryan). Holotype: worker (BPBM type no. 2238, examined). NEW SYNONYMY.

Romblonella liogaster is a robustly built medium-sized ant with a subquadrate, apedunculate petiole, a strongly arching mesosoma, deeply excavated antennal scrobes, and evenly distributed stout white hairs that are shorter than the length of the eye. In Fiji, the species is most easily confused with dark species of *Tetramorium* such as *T. pacificum* and *T. manni*, especially on account of the similar rugoreticulate sculpture on the head, mesosoma and waist. *Romblonella liogaster* can separated from these species by its lack of a petiolar peduncle, the lack of raised ridges surrounding the antennal insertions, and the simple sting.

The first specimens of this species were collected by E. H. Bryan during his 1924 expedition to the archipelago. He collected one specimen from Vanua Balevu (Lau Group), and another from Wakaya (near Ovalau), and both specimens were deposited in the Bishop Museum (Hawaii). Santschi (1928) described the Vanua Balevu specimen as a subspecies of *R. scrobiferum* Emery, noting that the color and shape differed from the type of *R. scrobiferum* and *R. elysii* Mann in being slightly more convex in the dorsal profile and the gaster being absolutely smooth, rather than punctate. Other differences include the sculpture of *R. liogaster*, which is entirely reticulated on the dorsal mesosoma, rather than predominately longitudinal, as in *R. scrobiferum*.

These species, however, were all combined in *Tetramorium* until recently (Bolton, 1995). Perhaps the original combination obscured the type specimen of *liogaster* from M. R. Smith, who discovered the second collection of Bryan in the Bishop and described it as *vitiensis* under the newly established genus, *Romblonella*. Although the gaster is missing from Smith's type, his description clearly stated it as being shiny without the strong sculpture of *scrobiferum*. The specimens otherwise look identical. The closest relatives of *R. liogaster* appear to be *R. scrobiferum* Mann (New Guinea, New Britain), *R. heatwolei* Taylor (New Guinea, Queensland) and *R. elysii* Mann (Solomons).

Romblonella liogaster is one of the very few species that was not collected by Mann or during the recent survey. It is interesting that Bryan, who was by no means an ant specialist, was able collect two specimens from different islands. A third specimen (from Ovalau) was collected by Krauss and is deposited at the ANIC.

Genus Solenopsis

Diagnosis of worker caste in Fiji. Head shape ovoid to rectangular. Antenna 10-segmented. Antennal club 2-segmented. Mandibles triangular; lacking tooth on basal margin. Propodeum lacking spines or teeth. Waist 2-segmented.

Solenopsis can sometimes be mistaken for *Monomorium* because both genera lack propodeal spines. However, the Fijian *Monomorium* all have 12-segmented antenna with three-segmented antennal clubs, and the two species of Fijian *Solenopsis* both have 10-segmented antenna with two-segmented antennal clubs.

Key to *Solenopsis* workers of Fiji.

1 Eyes large and multifaceted. Medium to large species (HW > 0.25 mm). Polymorphic worker caste. Introduced.......... ***S. geminata***

– Eyes vestigial and composed of single facet. Minute species (HW < 0.25 mm). Monomorphic worker caste. Pacific native.......... ***S. papuana***

Solenopsis geminata (Fabricius)

(Plate 119)

Atta geminata Fabricius, 1804: 423; queen described. Type locality: CENTRAL AMERICA [not examined]. In Polynesia (Wilson & Taylor, 1967: 14, fig. 43). Current subspecies: nominal plus *micans*. For additional synoptic history see Bolton et al. (2006).

Solenopsis geminata is commonly referred to as the Tropical Red Fire Ant. It is a medium-sized reddish species with abundant thin and erect pilosity and a polymorphic worker caste. The largest workers have disproportionately large and square-shaped heads. *Solenopsis geminata*, is easily distinguishable from *S. papuana* (the only congener known from Fiji) by its larger size, large multifaceted eyes and polymorphic worker caste.

Solenopsis invicta, the Red Imported Fire Ant, is not currently known from Fiji, but it has recently spread to other parts of the Pacific, including Australia, China and Taiwan. Several incursions have been made into New Zealand, but were eradicated before large populations established. It is recommended that new collections of large *Solenopsis* in Fiji be carefully examined to make sure potential incursions are identified quickly.

Solenopsis geminata is easily confused with *S. invicta*. The workers of both species overlap strongly in color, size and shape, and are impossible to differentiate in the field. The most reliable character for separating the two is the absence or presence of a middle tooth on the anterior clypeal margin. The middle tooth tends to be absent in *S. geminata* and present in *S. invicta*.

Solenopsis geminata is also an aggressive species with a painful sting and is known to cause damage to ecological and agricultural systems. Thus far, records are known only from the drier regions of Viti Levu. Local inhabitants from the village of Navai, near Mt. Tomanivi, recall the species arriving after World War II, and believe that it was brought on American war vessels.

Material examined. **Viti Levu:** Navai 770, Abaca 525, Navai 863, Lautoka Port 5 b, Savione 750 a, Volivoli 55.

Solenopsis papuana Emery

(Plate 120)

Solenopsis papuana Emery, 1900: 330; worker, queen described. Type locality: NEW GUINEA [not examined]. In Fiji (Mann, 1921: 444). Senior synonym of *cleptis*, *dahlii*, *vitiensis*: Wilson & Taylor, 1967: 60.

Solenopsis papuana is a minute pale yellow to brown yellow species with no antennal scrobes, minute eyes, no propodeal spines, polished sculpture, abundant thin and erect pilosity, and a monomorphic worker caste. It can be separated from *S. geminata* by characters listed in the discussion of the latter. Several small and pale *Monomorium* species can be mistaken for *S. papuana* because both genera lack propodeal spines and many of the *Monomorium* are also quite small, and several have minute eyes composed of only several facets. However, the *Monomorium* all have 12-segmented antennae with three-segmented antennal clubs, and *S. papuana* has 10-segmented antennae with two-segmented antennal clubs.

There are two specimens examined from the recent survey that vary in subtle but consistent characters from the vast majority of the material treated here as *S. papuana*. One individual from Viti Levu (CASENT0182173) and another from Koro (CASENT0182542) are distinct from all other examined Fiji collections. They differ from the more common form in their larger size, smaller eyes and more abundant short pilosity on the promesonotum and gaster. Although it is quite possible these individuals represent a different species, they are treated here as *S. papuana* until a more thorough is undertaken.

Mann described *S. cleptis* (1919) from the Solomons, and its variety *S. vitiensis* (1921) from Fiji. Both were synonymized by Wilson and Taylor (1967) after reviewing the material and concluding that Mann had compared his specimens to those of *S. maxillosa* Emery rather than *S. papuana*. A closer comparison of the *S. papuana* and *S. vitiensis* type material is required to determine which of the aforementioned Fijian morphotypes most closely approximates it. Regardless, the paucity of morphological characters available for this group is likely to belie a greater diversity than is accounted for by the available names.

Solenopsis papuana is a small species, but nests will recruit in vigorous force to baits and food resources. They can defend the bait from competing ants with use of their stings.

Material examined. **Beqa:** Mt. Korovou 326. **Gau:** Navukailagi 625, Navukailagi 632, Navukailagi 415, Navukailagi 408, Navukailagi 432, Navukailagi 505, Navukailagi 356, Navukailagi 575. **Kadavu:** Moanakaka 60, Mt. Washington 700. **Koro:** Kuitarua 480, Nasoqoloa 300. **Moala:** Naroi 75, Mt. Korolevu 375, Mt. Korolevu 300. **Ovalau:** Levuka 400, Levuka 450, Draiba 300. **Taveuni:** Lavena 300, Lavena 235. **Vanua Levu:** Mt. Kasi Gold Mine 300, Mt. Vatudiri 641, Mt. Delaikoro 699, Vusasivo Village 400 b, Mt. Delaikoro 391, Lagi 300. **Viti Levu:** Mt. Evans 700, Nakavu 300, Vaturu Dam 575 b, Colo-i-Suva Forest Park 220, Monasavu Dam 800, Nadakuni 300, Nadakuni 300 b, Korobaba 300, Lami 200, Nakobalevu 340, Colo-i-Suva 200, Waimoque 850, Colo-i-Suva 186 d, Naboutini 300, Galoa 300, Nakavu 200, Vunisea 300, Nabukavesi 300, Mt. Rama 300, Naikorokoro 300, Waivudawa 300.

Genus Strumigenys

Diagnosis of worker caste in Fiji. Head shape triangular. Antenna 6-segmented. Antennal club

2-segmented. Eyes located on lower margin of antennal scrobes. Mandibles linear. Propodeum armed with spines or teeth. Waist 2-segmented. Spongiform attached to at least some portion of waist. Hairs appearing flagellate or spatulate on at least some portion of head or body.

Strumigenys is one of the most diverse genera in both Fiji and tropical regions around the world. It is immediately recognizable in Fiji by the combination of the triangular head and the linear mandibles. *Pyramica* is a close relative that also occurs on the archipelago, but it can be separated by the triangular shape of its mandibles. The genus is represented in Fiji by several independent endemic radiations from the *godeffroyi* group, including six species from the *smythiesii* complex and four species from the *signeae* complex. Endemic representatives of several other species groups are present in Fiji, but these are perhaps more recent arrivals that have not radiated into the diversity observed in the aforementioned groups. *Strumigenys* tend to be specialized predators that hunt small arthropods with their trap-like apically forked jaws. They are most often encountered in the leaf litter, but individuals can also be found patrolling above the surface on logs and tree trunks.

Key to *Strumigenys* workers of Fiji.[8]

1 Preapical dentition of each mandible either with two preapical teeth or with one tooth and a denticle. Ventrolateral margin of head immediately in front of the eye with an abrupt and very conspicuous preocular notch or indentation. Small species (HW < 0.5 mm). Introduced .... ***S. rogeri***

– Preapical dentition of each mandible either absent or with a single tooth or denticle. Ventrolateral margin of head immediately in front of the eye either lacking an abrupt and very conspicuous preocular notch or indentation, or if present, than a large species (HW > 0.7 mm) 2

2 Dorsal (outer) surface of hind basitarsus with one or more freely projecting filiform or flagellate hairs that are very long and suberect to erect; this specialized pilosity may also be present on the middle basitarsus and the middle and hind tibiae 3

– Dorsal (outer) surface of hind basitarsus lacking freely projecting long filiform or flagellate hairs; any pilosity present is simple to spatulate and usually decumbent to appressed; projecting long fine simple hairs absent from middle basitarsus and from middle and hind tibiae 12

3 First gastral tergite entirely covered with fine dense longitudinal sulcate sculpture; no other form of sculpture present on sclerite, differentiated basigastral costulae absent. Fiji endemic .... ***S. panaulax***

– First gastral tergite not entirely covered with longitudinal sulcate sculpture; tergite usually with basigastral costulae that may extend up to half the length of the sclerite, which is usually unsculptured posteriorly; occasionally tergite entirely smooth or with another form of sculpture distal of the basigastral costulae 4

4 With mesosoma in profile propodeal declivity equipped with a broad and conspicuous lamella; propodeal tooth may be replaced by the lamella or completely buried in the lamella, or lamella may subtend ventral margin of the tooth for most or all of its length; posterior (free) margin of lamella may be convex, straight or irregular but it is not narrowly concave, nor is it close to and parallel with the edge of the declivity 5

– With mesosoma in profile propodeal declivity equipped with simple carina or at most a narrow cuticular flange; carina or narrow flange does not subtend ventral margin of tooth for most or all of its length; posterior (free) margin of carina or narrow flange concave,

[8] Adapted from Bolton (2000).

close to and parallel with the edge of the declivity 11

5 With head in full-face view upper scrobe margin with two or more flagellate or filiform hairs that freely project laterally; at least with one in apicoscrobal position and another anterior to this (dorsolateral margin of posterolateral lobe to apex of scrobe may have additional laterally projecting hairs) 6

– With head in full-face view upper scrobe margin usually with a single hair that freely projects laterally, in apicoscrobal position; this hair may be flagellate, filiform, or short and stiff; sometimes lacking a hair in this position (dorsolateral margin of posterolateral lobe to apex of scrobe may have laterally projecting hairs)... 7

6 Freely projecting filamentous hairs present along entire lateral margin of head from posterolateral lobe to antennal insertions. Cephalic and pronotal surfaces smooth with large and deep irregular pits. First gastral segment lacking any sculpturing the entire length of the tergite. Fiji endemic .... ***Strumigenys* sp. FJ18**

– Freely projecting filamentous hairs lacking between apicoscrobal hair and eye level. Cephalic and pronotal surfaces punctate-reticulate, lacking large irregular pits. First gastral segment with basigastral costulae distinctly longer than length of the postpetiolar disc. Fiji endemic ....***S. chernovi***

7 With mesosoma in profile dorsum of pronotum lacking additional hairs equal to length of the humeral hair.. 10

– With mesosoma in profile dorsum of pronotum usually with one additional pair of hairs equal to length of humeral hair; rarely with more than one pair................................... 8

8 Eyes composed of single facet. First gastral segment with basigastral costulae distinctly shorter than length of the postpetiolar disc. Posterior margin of head in full face view shallowly excised such that the depth of the concavity is less than length of the apicoscrobal hair. Fiji endemic .... ***Strumigenys* sp. FJ14**

– Eyes composed of four or more facets. Basigastral costulae and posterior margin of head variable 9

9 First gastral segment with basigastral costulae distinctly longer than length of postpetiolar disc. Posterior margin of head in full face view deeply excised such that the depth of the concavity is greater than length of the apicoscrobal hair. Fiji endemic.
.... ***Strumigenys* sp. FJ17**

– First gastral segment with basigastral costulae distinctly shorter than the length of postpetiolar disc. Posterior margin of head in full face view shallowly excised such that depth of concavity is less than length of apicoscrobal hair. Fiji endemic......... ***S. jepsoni***

10 Apicoscrobal hair present and long, filiform or flagellate; this hair very different in form and length from any other on the margin both anterior and posterior to it. In profile dorsum of promesonotum covered with a thick pelt of short curved ground pilosity, appearing weakly fury. Pacific native .... ***S. godeffroyi***

– Apicoscrobal hair absent; entire margin with a dense row of uniformly shaped small curved hairs. In profile dorsum of promesonotum lacking a thick pelt of short curved ground pilosity, not appearing weakly fury. Fiji endemic... ***S. scelesta***

11 With head in full-face view the upper scrobe margin with two or more flagellate or filiform hairs that freely project laterally; at least with one in apicoscrobal position and another anterior to this (dorsolateral margin of posterolateral lobe to apex of scrobe may have additional laterally projecting hairs). Fiji endemic ....***S. daithma***

– With head in full-face view the upper scrobe margin usually with a single hair that freely projects laterally, in apicoscrobal position; this hair may be flagellate, filiform, or short and stiff; sometimes lacking a hair in this position (dorsolateral margin of posterolateral lobe to apex of scrobe may have laterally projecting hairs). Fiji endemic. ......***S. ekasura***

12 Fully closed mandible in full-face view very broad proximally and strikingly tapered distally, not linear or curvilinear. Outer margin of mandible flared outwards or strongly convex prebasally, mandible not straight, not evenly convex, not evenly bowed outwards. In ventral view outer margin of mandible distal of level of anterior clypeal margin abruptly changing direction 13

– Fully closed mandible in full-face view usually obviously linear or curvilinear, sometimes slightly increasing in width towards the base. Outer margin of mandible not flared outwards or strongly convex prebasally, mandible straight, evenly convex, or evenly bowed outwards. In ventral view outer margin of mandible forms a more or less continuous line that does not abruptly change direction distal of level of anterior clypeal margin 14

13 In full face view distance of mandible from clypeal border to preapical tooth subequal to length of preapical tooth. Posteromedian margin of head with single deep circular puncture. Dense tuft of white filamentous hairs arising from the lateral promesonotal border above the procoxae. With mesosoma in profile the propodeal declivity equipped with a broad and conspicuous lamella. Dorsum of mesosoma with abundant long white erect hairs. Fiji endemic .... ***Strumigenys* sp. FJ19**

– In full face view distance of mandible from clypeal border to preapical tooth distinctly greater than length of preapical tooth. Posteromedian margin of head lacking single deep circular puncture. Dense tuft of white filamentous hairs lacking from the lateral promesonotal border above the procoxae. With mesosoma in profile the propodeal declivity equipped with a simple carina or at most a narrow cuticular flange. Dorsum of mesosoma lacking abundant long white erect hairs. Fiji endemic..................***S. basiliska***

14 Pronotal humeral hair absent 15

– Pronotal humeral hair present.. 16

15 Dorsal mesosoma coarsely longitudinally sulcate, with a ploughed appearance. Disc of postpetiole longitudinally sulcate, sulci narrower and finer than on mesosoma. Cephalic dorsum coarsely longitudinally rugose. Ventral spongiform curtain of petiole narrow, at maximum only a fraction the depth of the peduncle. Fiji endemic..................***S. sulcata***

– Dorsal mesosoma densely reticulate-punctate. Disc of postpetiole lacking sulcate sculpture. Cephalic dorsum sharply and densely reticulate-punctate. Ventral spongiform curtain of petiole deep, at maximum at least equal to the depth of the peduncle. Fiji endemic. ....***S. praefecta***

16 Ventrolateral margin of head immediately in front of the eye with an abrupt and very conspicuous preocular notch or indentation.. 17

– Ventrolateral margin of head immediately in front of the eye lacking an abrupt and very conspicuous preocular notch or indentation.. 18

17 First gastral segment sculptured basally and strongly polished and shiny the remainder of its length. Fiji endemic..***S. nidifex***

– First gastral segment strongly punctate-striate the entire length. Fiji endemic. .... ***Strumigenys* sp. FJ01**

18 Preapical dentition of mandible a stout tooth that is distinctly shorter than the width of the mandible at the point where it arises; preapical tooth never as long as the maximum width of the mandible. Mesopleuron and procoxae punctate-reticulate. Fiji endemic. ....***S. tumida***

– Preapical dentition of mandible a slender spiniform tooth that is always longer than the width of the mandible at the point where it arises; preapical tooth usually at least as long as the maximum width of the mandible. Mesopleuron and procoxae smooth and shiny. 19

19 Cephalic dorsum lacking a pair of erect hairs in front of highest point of vertex. Pronotal dorsum lacking erect hairs in addition to the humeral pair; humeral hair long fine and flagellate. Dorsum of head and pronotum finely and densely but chaotically punctate-rugulose. Disc of postpetiole sculptured. Fiji endemic .... ***S. frivola***

– Cephalic dorsum with a pair of erect hairs in front of the highest point of vertex. Pronotal dorsum with erect hairs in addition to the humeral pair; humeral hair short stiff and simple. Dorsum of head and pronotum finely and densely reticulate-punctate. Disc of postpetiole smooth 20

20 Pronotal humeral hairs flagellate .... ***Strumigenys* sp. FJ13**

– Pronotal humeral hairs simple. Pacific native .... ***S. mailei***

Strumigenys basiliska Bolton

(Plate 121)

Strumigenys basiliska Bolton, 2000: 750; worker described. Type locality: FIJI, Viti Levu, Nadarivatu Reserve, 11.vii.1987, QM Berlesate no. 775, 17.34°S, 177.57°E, 800 m, rainforest, sieved litter (G. Monteith). Holotype: worker (ANIC, not examined). Paratypes: workers (ANIC, not examined); workers (BMNH, examined).

Strumigenys basiliska is a distinctive species that can be recognized by the highly polished dark reddish brown color, reduced pilosity, very short arcuate mandibles, and lack of propodeal lamellae. Although the overall appearance of this species is relatively uniform across the archipelago, the number of erect setae on the mesosoma varies among specimens. It is difficult to ascertain, however, how much of the observed variation reflects true taxonomic differences rather than the result of specimen damage. There are some relatively large series in which the number and placement of erect setae are consistent among all specimens. For example, all the specimens known from Gau have a pair of erect setae at the posterior margin of the promesonotum but lack a pair of erect setae on the humeri. The same condition is found on most of a nest collection from Mt. Tomanivi (Viti Levu). However, there are specimens from Vanua Levu and Koro with erect setae on both the posterior margin of the promesonotum and the humeri, and there are also specimens from many islands on which no erect hairs can be seen.

The type series from Nadarivatu (Viti Levu, near Mt. Tomanivi) is described by Bolton as lacking pronotal humeral hairs. The only other material examined by Bolton was from nearby Monasavu Dam. Although a closer study may demonstrate these different populations to represent different species, they are treated here as one. Bolton tentatively places *S. basiliska* in the *biroi* group, but admits that the Fijian species is so aberrant from the condition of the other species in the group that the resemblance might be the result of convergence.

Strumigenys basiliska is widespread throughout the inner islands of the archipelago, though no records from Taveuni or Kadavu are known.

Material examined. **Gau:** Navukailagi 625, Navukailagi 408, Navukailagi 432, Navukailagi 490, Navukailagi 505, Navukailagi 475, Navukailagi 575. **Koro:** Nasau 465 a, Nasoqoloa 300. **Moala:** Mt. Korolevu 375, Mt. Korolevu 300. **Ovalau:** Levuka 400. **Vanua Levu:** Nakasa 300, Yasawa 300, Kasavu 300, Nakanakana 300, Mt. Vatudiri 570, Rokosalase 180. **Viti Levu:** Vatubalavu 300, Mt. Tomanivi 950,

Nasoqo 800 c, Nadakuni 300 b, Korobaba 300, Galoa 300, Nakavu 200, Vunisea 300, Nabukavesi 300, Mt. Rama 300, Naikorokoro 300, Waivudawa 300.

Strumigenys chernovi Dlussky

(Plate 122)

Strumigenys chernovi Dlussky, 1993: 57, figs. 2, 3; worker, queen described. Type locality: FIJI, Viti Levu, Suvy [Suva?], ii.1977 (Y. Chernov) [not examined]. Redescription of worker, in global revision (Bolton, 2000: 805).

Bolton (2000) described *Strumigenys chernovi* as being characterized by the two projecting hairs on the upper scrobe margin, flagellate pronotal humeral hair, a densely sculptured promesonotum that contrasts with the polished propodeal dorsum, a densely sculptured postpetiolar disc with a large membranous ventral lobe, and very long basigastral costulae.

Strumigenys chernovi is part of the *smythiesii* complex, which also includes the Fijian species *S. ekasura*, *S. jepsoni*, *S. panaulax*, and *S. scelesta*. All of these species have short scapes (SI 68–83), and an entirely sculptured postpetiolar disc. *Strumigenys panaulax* can be separated by its entirely sulcate first gastral tergite. *Strumigenys ekasura* is separated by the presence of only a single apicoscrobal hair and the lack of a lamella on the propodeal declivity. *Strumigenys chernovi* is separated from *S. ekasura* by the two freely laterally projecting fine hairs on each upper scrobe margin, one at the eye level and one apicoscrobal, and the broad propodeal lamella with a shallowly convex posterior margin. *Strumigenys frivola* might belong to this complex, as well, but can be separated from the other species by its lack of erect flagellate hairs on the dorsal surface of the hind basitarsus. *Strumigenys jepsoni* is unique among the Fijian species of *smythiesii* complex in that the pronotal humeral hairs are stiff and simple rather than flagellate. The Pacific tramp species, *Strumigenys godeffroyi*, is apparently a more distant relative to these other species in the group, and it can be separated by the many flagellate hairs present on all dorsal surfaces, including the gaster.

Strumigenys chernovi is widespread, recorded from most of the major islands with the exceptions of Kadavu and Taveuni.

Material examined. **Gau:** Navukailagi 675, Navukailagi 415, Navukailagi 408, Navukailagi 490, Navukailagi 475, Navukailagi 575, Navukailagi 300. **Koro:** Tavua 220, Nasoqoloa 300. **Ovalau:** Draiba 300. **Vanua Levu:** Yasawa 300, Kasavu 300, Rokosalase 180, Mt. Delaikoro 391. **Viti Levu:** Monasavu Dam 800, Nasoqo 800 c, Korobaba 300, Lami 200, Nakobalevu 340, Naboutini 300, Galoa 300, Nakavu 200, Nabukavesi 300, Naikorokoro 300.

Strumigenys daithma Bolton

(Plate 123)

Strumigenys daithma Bolton, 2000: 756; worker described. Type locality: FIJI, Viti Levu, Road E of Monasavu Dam, 26.vii.1987, 17°43′S, 178°03′E, QM Berlesate No. 788, 1000 m, rainforest, sieved litter, ANIC Ants vial 45.195 (G. Monteith). Holotype: worker (ANIC, not examined).

Strumigenys daithma is the only known Fijian representative of the *caniophanes* group, which

Bolton describes as ants with fine and soft pilosity, and sculpturing that is denser and coarser than usual. No specimens of *S. daithma* are known besides the holotype. This Fijian endemic is broadly disjunct with respect its presumed closest relatives in the *caniophanes* group, which are otherwise restricted to the following countries: China, Taiwan, Thailand, Malaysia, Indonesia (Sumatra), Philippines (Luzon).

Strumigenys ekasura Bolton
(Plate 124)

Strumigenys ekasura Bolton, 2000: 807; worker described. Type locality: FIJI, Vanua Levu, Kontiki, 19 km. E of Savusavu, 18.vii.1987, QM Berlesate No. 782, 16°48′S, 179°26′E, 20 m, secondary rainforest, sieved litter (G. Monteith). Holotype: worker (ANIC, not examined).

Strumigenys ekasura is part of the *smythiesii* complex, and is characterized by abundant free-stranding flagellate pilosity, a sculptured postpetiolar disc, and conspicuous basigastral pilosity. The species is most similar to *S. chernovi*, from which it can be separated by the single laterally projecting fine hairs on each upper scrobe margin and lack of propodeal lamellae. See the notes under *S. chernovi* for additional characters to separate it from its Fijian relatives. The species is widespread and is recorded from most of the major islands with the exception of Taveuni and Ovalau.

Material examined. **Gau:** Navukailagi 432, Navukailagi 575, Navukailagi 300. **Kadavu:** Moanakaka 60, Namara 300. **Koro:** Tavua 220, Nasau 420 b, Nasoqoloa 300. **Moala:** Mt. Korolevu 375. **Vanua Levu:** Nakasa 300, Yasawa 300, Kasavu 300, Nakanakana 300, Vuya 300, Mt. Delaikoro 699, Vusasivo Village 400 b, Mt. Delaikoro 391, Lagi 300. **Viti Levu:** Monasavu Dam 1000, Korobaba 300, Colo-i-Suva 200, Naboutini 300, Nakavu 200, Naikorokoro 300.

Strumigenys frivola Bolton
(Plate 125)

Strumigenys frivola Bolton, 2000: 817; worker described. Type locality: FIJI, Viti Levu, Nadarivatu Reserve, 850 m, 11.vii.1987, pyrethrum/trees and logs (G. & S. Monteith). Holotype: worker (ANIC, not examined). Paratype: worker (BMNH, examined).

Strumigenys frivola is an opaque yellow brown species with punctate sculpture and short recumbent pilosity covering its head, most of the mesosoma, and both waist segments. The basigastral costulae are conspicuous and long, and the propodeum lacks lamellae. Bolton suggests that the species might belong to the *smythiesii* complex, but places it for convenience in the *rofocala* complex because it lacks erect flagellate hairs on the dorsal surface of the hind basitarsus. The species is apparently quite rare, as it is known only from the type locality at Nadarivatu (Viti Levu), a nest collection from nearby Mt. Tomanivi, and a fogging sample at the nearby Monasavu Dam area.

Material examined. **Viti Levu:** Mt. Tomanivi 950, Monasavu Dam 1000.

Strumigenys godeffroyi Mayr

(Plate 126)

Strumigenys godeffroyi Mayr, 1866: 516; worker described. Type locality: SAMOA [not examined]. Senior synonym of *butteli*, *indica* (Brown, 1949: 17); of *geococci* (Bolton, 2000: 791). For additional synoptic history see Bolton et al. (2006).

Strumigenys godeffroyi is immediately recognizable among the Fijian dacetines by the plethora of long fine flagellate hairs that occur on all dorsal surfaces, including those of the gaster. The closest relatives of *S. godeffroyi* are discussed under the notes for *S. chernovi*. *Strumigenys godeffroyi* is a tramp species that has spread across the Pacific and has been recorded from most of Fiji's major islands.

Material examined. **Gau:** Navukailagi 415, Navukailagi 408, Navukailagi 356, Navukailagi 300. **Kadavu:** Moanakaka 60, Vunisea. **Koro:** Kuitarua 480, Nasoqoloa 300, Mt. Kuitarua 380. **Moala:** Mt. Korolevu 375, Mt. Korolevu 300. **Taveuni:** Qacavulo Point 300, Somosomo 200. **Vanua Levu:** Nakasa 300, Yasawa 300, Kasavu 300, Nakanakana 300, Drawa 270, Vuya 300, Mt. Wainibeqa 152 c, Vusasivo Village 400 b, Vusasivo Village 400 b, Rokosalase 180, Rokosalase 94, Mt. Delaikoro 391, Lagi 300, Wainunu, Suene. **Viti Levu:** Nakavu 300, Abaca 525, Colo-i-Suva Forest Park 220, Colo-i-Suva 200, Lami 171, Colo-i-Suva 186 d, Naboutini 300, Nabukelevu 300, Galoa 300, Nakavu 200, Vunisea 300, Nabukavesi 300, Mt. Rama 300, Naikorokoro 300, Lomolaki, Sigatoka, Nausori.

Strumigenys jepsoni Mann

(Plate 127)

Strumigenys jepsoni Mann, 1921: 462, fig. 22a; worker described. Type locality: FIJI, Vanua Levu, Suene (W. M. Mann). Syntypes: 2 workers (MCZC type no. 21099, examined); 4 workers (USNM, examined). In global revision (Bolton, 2000: 809).

Strumigenys jepsoni can be distinguished from the other Fijian members of the *smythiesii*-complex (*S. chernovi*, *S. ekasura*, *S. godeffroyi*, *S. jepsoni*, *S. panaulax*, and *S. scelesta*) by its stiff and simple (as opposed to flagellate) humeral hair. A more detailed description is given in Bolton (2000). *Strumigenys jepsoni* was named after Mr. F. P. Jespon, Entomologist of Fiji. The species was not collected during the recent survey, and the only other record of the species, besides the type series, is from a collection by G. Kuschel in Ndreketi [Dreketi] (Vanua Levu).

Material examined. **Koro:** Nasoqoloa 300. **Vanua Levu:** Suene.

Strumigenys mailei Wilson & Taylor

(Plate 128)

Strumigenys mailei Wilson & Taylor, 1967: 38, fig. 28; worker described. Type locality: SAMOA, Afiamalu, 800 m, 15.iii.1962, rain forest, berlesate of moss on tree, 10-13 m from ground, no. 581 (R. W. Taylor). Holotype: 1 worker (MCZC type no. 31114, examined). Paratype: 1 worker (MCZC type no. 31114, examined) [Paratype locality: FIJI, Viti Levu, Navai Mill, nr. Nadarivatu, 800 m, 17.ix.1938 (E. C. Zimmerman)]. In global revision (Bolton, 2000: 823).

Strumigenys mailei is a light yellow brown to reddish brown species with a narrow head that is shallowly impressed posteriorly, long mandibles, and no hairs on the pronotal humeri. See the notes below for characters used to separate it from its close Fijian relatives. The *signeae* complex has four Fijian species (*S. mailei*, *S. praefecta*, *S. sulcata*, *S. tumida*). All of these species lack flagellate hairs, and can thus be separated from nearly all the other species within the *godeffroyi* group. The Fijian species of the *signeae* group are characterized by having the antennal scrobe absent behind the eye, the propodeal declivity with a narrow carina (no lamella), and the petiole with a long and low node that in profile shows the anterior face to be much shorter than the dorsum.

Of the Fijian *signeae* complex species, *S. sulcata* is easy to separate by the coarse sulcate sculpture of its head and mesosoma. *Strumigenys tumida* is easy to separate by its short preapical tooth on the mandible, the densely punctate-reticulate and swollen postpetiolar disc, and the uniformly punctate-reticulate metapleuron on sides of the propodeum. *Strumigenys mailei* and *S. praefecta* lack the striking features of the two aforementioned species. *S. mailei* can be separated from *S. praefecta* by the presence of humeral hair, traces of oblique costulae posteriorly on the side of the pronotum, a shorter, broader petiolar node, and a more slender apical antennomeres.

Material examined. **Taveuni:** Lavena 300, Tavuki 734, Mt. Devo 775 a. **Vanua Levu:** Mt. Delaikoro 699. **Viti Levu:** Monasavu Dam 800, Nasoqo 800 b, Nasoqo 800 c, Nasoqo 800 d, Navai.

Strumigenys nidifex Mann
(Plate 129)

Strumigenys nidifex Mann, 1921: 464, fig. 23; worker, queen, male described. Type locality: FIJI, Viti Levu, Nadarivatu (W. M. Mann). Syntypes: 7 workers, 2 queens (MCZC type no. 8710, examined); 1 worker, 4 queens, 1 male (USNM, examined). In revision of *szalayi* group (Brown, 1971b: 81); in global revision (Bolton, 2000: 905). For additional synoptic history see Bolton et al. (2006).

Note: Specimens from the non-type locality of Buka Levu (USNM), Waiyanitu (MCZC), and Veisari (MCZC) bear red cotype labels but are not true syntypes.

Strumigenys nidifex is one of the most conspicuous dacetines of the Fijian fauna. It is the largest of the species, with a distinctive broad head that is impressed posterior to the vertex and bears a preocular impression on the ventrolateral margin of the head. All surfaces of the head, mesosoma, waist and coxae are punctate-reticulate. The nearest relative to *S. nidifex* is *Strumigenys* sp. FJ01. The gaster of the former is sculptured basally, but is otherwise strongly polished and shiny. The gaster of the latter is strongly punctate-striate the entire length of the first segment. The males of the two differ, with those of *S. nidifex* being larger with a distinctly elongated head, while the male of *Strumigenys* sp. FJ01 is smaller with a much rounder head. The queen of *Strumigenys* sp. FJ01 is also distinctly smaller, especially with regard to the mesonotum. However, the series of *S. nidifex* from Koro are all smaller, and the queens of that series are more similar to those of *Strumigenys* sp. FJ01 than to those of its conspecifics on other islands, with the caveat that its gaster is polished and shiny. Both species occur sympatrically in the Nadarivatu area. The only other species in Fiji with a preocular notch is *S. rogeri*, which is easily separated by its significantly smaller size, yellow brown color, and mandible with two pre-apical teeth. The other large *Strumigenys* in Fiji all lack the preocular impression.

Strumigenys nidifex is a member of the *szalayi* group, which was revised by Brown (1971b), and for which an excellent account of a laboratory-raised colony is described. E. O. Wilson collected the colony from Nadala (Viti Levu) during his expedition to Fiji, and sent it back to his laboratory in the United States. The colony was kept for two years, during which time it produced queens and hunted a variety of small arthropods. The oviposition and hunting behavior is explained in some detail. *Strumigenys nidifex* is thus one of the very few native Fijian ants for which any detailed natural history is known. Other species of the *szalayi* group occur in the Philippines, the lower elevations of New Guinea, the Solomons, Micronesia, Samoa, northern Queensland and Vanuatu. The distribution of this group makes it an excellent candidate for testing the importance of the Vitiaz Arc with regard to the evolution of the Melanesian ant fauna.

Material examined. **Gau:** Navukailagi 625, Navukailagi 432, Navukailagi 356. **Kadavu:** Mt. Washington 760, Mt. Washington 700, Buka Levu. **Koro:** Kuitarua 480, Nasau 470 (3.7 km), Mt. Kuitarua 440 b, Nasau 420 b, Nasau 465 a. **Taveuni:** Tavuki 734, Mt. Devo 775 a. **Vanua Levu:** Mt. Delaikoro. **Viti Levu:** Nadala 300, Navai 930, Mt. Tomanivi 950, Nasoqo 800 d, Monasavu Dam 600, Waimoque 850, Naikorokoro 300, Nadarivatu 750, Nausori, Navai, Waiyanitu, Vesari.

Strumigenys panaulax Bolton
(Plate 130)

Strumigenys panaulax Bolton, 2000: 811; worker described. Type locality: FIJI, Vanua Levu, Ndelaikoro, 800 m , 27.x.1977, litter, No. 77/130 (G. Kuschel). Holotype: (ANIC, not examined). Paratypes: (ANIC, BMNH, not examined).

Strumigenys panaulax is a small reddish brown species covered with thick decumbent hairs and with striations that extend across the entire length of the first gastral segment. The latter character separates it from all other members of the *smythiesii* complex, which is discussed in more detail under the notes for *S. chernovi*. The species was represented in the recent survey by a single leaf litter collection on Vanua Levu.

Material examined. **Vanua Levu:** Rokosalase 180.

Strumigenys praefecta Bolton
(Plate 131)

Strumigenys praefecta Bolton, 2000: 826; worker, queen described. Type locality: FIJI, Viti Levu, Namosi/Queens Rd Divide, 23.vii.1987, QM Berlesae No. 787, 18°05′S, 178°10′E, 500 m , rainforest, litter and moss (G. Monteith). Holotype: worker (ANIC, not examined). Paratypes: (ANIC, not examined); worker (BMNH, examined).

Strumigenys praefecta is a mid-sized dark reddish brown species with very short antennal scrobes, no humeral hairs, a punctate-reticulate head and mesosoma with a mostly polished smooth mesopleuron and metapleuron, a mostly polished postpetiolar dorsum, and basigastral costae approximately the length of the postpetiole. The species belongs to the *signeae* group, which is discussed in more detail under the notes for *S. mailei*. There are several good nest series from the Nadarivatu area (Viti Levu) in which the workers are consistently larger and redder. The smaller morphotype, which agrees with the type series, occurs very close by in the Monasavu Dam area, in addition to the islands of Koro and Gau. It is interesting that the small morphotype is maintained across a large geographic scale, but that all the close-by Nadarivatu specimens show a consistently larger size and more red coloration. Bolton (2000) discussed a variety of material that differed from the type series, including specimens from Nadarivatu, but decided to treat them all as *S. praefecta*. The same conservative approach is maintained here.

The species may be arboreal, as much of the examined material, including the larger morph, was collected from under moss on trees, under bark on trees, foraging on tress, and from canopy fogging.

Material examined. **Gau:** Navukailagi 625. **Koro:** Nasau 470 (3.7 km), Nasau 420 b, Nasau 465 a. **Viti Levu:** Navai 863, Mt. Naqaranabuluti 1050, Monasavu Dam 1000, Monasavu Dam 600, Waimoque 850, Nakavu 200, Naikorokoro 300.

Strumigenys rogeri Emery
(Plate 132)

Strumigenys rogeri Emery, 1890: 68, pl. 7, fig. 6; worker described. Type locality: ANTILLES IS. [not examined]. Senior synonym of *incise* (Donisthorpe, 1915: 341); of *sulfurea* (Brown, 1954: 20). In global revision (Bolton, 2000: 604). For additional synoptic history see Bolton et al. (2006).

Strumigenys rogeri is a light yellow brown species that can be immediately identified by its preocular notch on the ventrolateral margin of the head, two preapical teeth on the mandibles, and small size. The only other dacetines in Fiji with preocular notches are *S. nidifex* and its close relation, *Strumigenys* sp. FJ01, both of which are very large dark heavily punctate species with a single preapical tooth on the mandible. *Strumigenys rogeri* is a tramp species believed to be native to Africa and is widespread across Fiji.

Material examined. **Koro:** Tavua 220, Kuitarua 480, Nasoqoloa 300. **Moala:** Naroi 75, Mt. Korolevu 375. **Taveuni:** Qacavulo Point 300. **Vanua Levu:** Mt. Kasi Gold Mine 300, Yasawa 300, Kasavu 300, Vuya 300, Mt. Delaikoro 699, Vusasivo Village 400 b, Rokosalase 180, Mt. Delaikoro 391. **Viti Levu:** Nakavu 300, Nadakuni 300, Lami 200, Colo-i-Suva 200, Colo-i-Suva 186 d, Naboutini 300, Nabukelevu 300, Galoa 300, Nakavu 200, Nabukavesi 300, Naikorokoro 300.

Strumigenys scelesta Mann
(Plate 133)

Strumigenys scelestus Mann, 1921: 463, fig. 22b; worker described. Type locality: FIJI, Taveuni, in mountains near lake [Lake Tagimaucia] (W. M. Mann). Holotype [single specimen]: worker (USNM, examined). In global revision (Bolton, 2000: 812).

Strumigenys scelesta is known only from the single holotype which is in rather poor condition. It is therefore difficult to comment on the specialized hairs, which could either be naturally absent or worn off by damage. The recent survey recovered several specimens that bear close resemblance to *S. scelesta*. They are considered here, with some reservation, to belong to that species.

The shape and size of the recently collected specimens match the holotype precisely. In addition to the short curved mandibles, the head has the distinctive shape of *S. scelesta* in full face view, with linear lateral margins and a very shallowly concave posterior margin. The eyes, however, appear to be several facets larger than those of the holotype, and the color has more red than yellow. Whereas the holotype has distinctive short curved hairs restricted mainly to the head and mesosoma, the recently collected specimens have the distinctive hairs evenly distributed across all dorsal surfaces, including the gaster. All the specimens have the short curved hairs on the legs, and they all appear to lack apicoscrobal hairs and humeral hairs. As Bolton predicted, the fresh specimens have basitarsal hairs present, as well, which have apparently rubbed off from the holotype. The species is clearly part of the *smythiesii* group, however, which is discussed in more detail under the notes for *S. chernovi*.

The specimens considered here to be *S. scelesta* were collected from the lower elevations of Vanua Levu and Viti Levu.

Material examined. **Vanua Levu:** Mt. Wainibeqa 152 c. **Viti Levu:** Galoa 300.

Strumigenys sulcata Bolton

(Plate 134)

Strumigenys sulcata Bolton, 2000: 828; worker, queen described. Type locality: FIJI, Vanua Levu, Ndreketi [= Dreketi], 25.x.1977, litter, wood, No. 77/127 (G. Kuschel). Holotype: worker (ANIC, not examined). Paratypes: (ANIC, not examined); worker (BMNH, examined).

Strumigenys sulcata is a relatively large dark reddish brown species with the dorsal surfaces of the mesosoma and postpetiole deeply sulcate. The rugae of the dorsum of head and sides of mesosoma are also primarily longitudinal, but less regular and more reticulated. The petiole lacks a distinct node. Instead, the peduncle arches gradually into the posterior margin. Like other members of the *signeae* complex, *S. sulcata* lacks lamella lining the propodeal spines or any flagellate hairs. The complex is discussed in more detail under the notes for *S. mailei*. *Strumigenys sulcata* is known from most of the archipelago's major islands, and nests have been found in wet, decaying logs.

Material examined. **Koro:** Mt. Kuitarua 440 b, Nasau 420 b, Mt. Kuitarua 380. **Taveuni:** Tavuki 734. **Vanua Levu:** Mt. Delaikoro 910, Kasavu 300, Mt. Vatudiri 570, Mt. Delaikoro 699, Mt. Delaikoro. **Viti Levu:** Navai 870, Monasavu Dam 800, Mt. Batilamu 840 c, Naqaranabuluti 1000, Mt. Tomanivi 950, Korobaba 300, Nakobalevu 340, Waimoque 850, Naikorokoro 300, Nausori Highlands 400.

Strumigenys tumida Bolton

(Plate 135)

Strumigenys tumida Bolton, 2000: 830; worker, queen described. Replacement name for *Strumigenys wheeleri* Mann, 1921: 466, fig. 24. Type locality: FIJI, Viti Levu, Nadarivatu (W. M. Mann). Syntypes: 2 workers, 1 queen (MCZC type no. 23319, examined). Junior secondary homonym of *Epitritus wheeleri* Donisthorpe, 1916: 121 (now in *Strumigenys*).

Strumigenys tumida belongs to the *signeae* complex, and can be distinguished by other species in that group by the extremely swollen postpetiole, which is conspicuously twice as broad as the petiole in dorsal view. See the notes under *S. mailei* for additional discussion of the Fijian members of the *signeae* complex. Further description is available in Mann (1921) and Bolton (2000).

Material examined: **Gau:** Navukailagi 432, Navukailagi 356, Navukailagi 300. **Koro:** Nasoqoloa 300. **Taveuni:** Lavena 300, Mt. Devo 775 a, Qacavulo Point 300. **Vanua Levu:** Yasawa 300, Kasavu 300. **Viti Levu:** Monasavu Dam 800, Nasoqo 800 d, Korobaba 300, Naikorokoro 300.

Strumigenys sp. FJ01

(Plate 136)

Strumigenys sp. FJ01 is a close relative to *S. nidifex*, and a discussion of the two is given in under the notes of the latter. The species is known from a single collection from the Nadarivatu area (Viti Levu), which is the type locality of *S. nidifex*. The two are sympatric, which, together with the clear morphological differences, validates the distinction of two separate species.

Material examined. **Viti Levu:** Mt. Naqaranabuluti 1050.

Strumigenys sp. FJ13

(Plate 137)

Strumigenys sp. FJ13 is a very small species (HW < 0.35 mm; HL < 0.45 mm) with short and simple apicoscrobal hairs, flagellate humeral hairs, and no basitarsal hairs. The propodeal spines are subtended by weak lamellae that closely parallel the curve of the declivity. The species is punctate-reticulate on the head and mesosoma, except for the polished mesopleuron and metapleuron. The basigastral sculpturing is conspicuously longer than the length of the postpetiolar disc. Known only from two specimens (Ovalau and Moala), it is possible that *Strumigenys* sp. FJ13 is a more widespread species in the Pacific. In Fiji, it is most similar to the species related to *S. mailei*, but can be easily separated by the smaller size and the flagellate humeral hairs.

Material examined. **Moala:** Mt. Korolevu 375. **Ovalau:** Levuka 450.

Strumigenys sp. FJ14

(Plate 138)

Strumigenys sp. FJ14 is a small pale species with small eyes of one to several facets, abundant long flexous hairs on all dorsal surfaces, and shorter subdecumbent hairs on the promesonotum. In one of the Gau specimens (CASENT0184969), it is clear that the long hairs were broken, and now appear short and simple. The species is a relative of *S. jepsoni* and is very similar to *Strumigenys* sp. FJ17, but differs in the smaller eye.

Material examined. **Gau:** Navukailagi 432, Navukailagi 505.

Strumigenys sp. FJ17

(Plate 139)

Strumigenys sp. FJ17 is a relatively small pale yellow brown species with long flexous apicoscrobal hair, long flexous humeral hairs, a pair of long flexous mesonotal hairs, and a single basitarsal hair. The waist segments and gaster have abundant flagellate hairs. Shorter subdecumbent hairs are present on the promesonotum. The lamellae attending the propodeal spines vary from thin and parallel to the declivity, to wider and not paralleling the declivity. The dorsal surfaces of the head, mesosoma and waist segments are punctate-reticulate, and the mesopleuron and much of the metapleuron is polished. The mesonotum has one to several pairs of short erect hairs, but no flagellate hairs other than those on the pronotal humeri. *Strumigenys* sp. FJ17 is widespread, occurring even on the smaller islands in the archipelago.

Material examined. **Beqa:** Mt. Korovou 326. **Koro:** Nasoqoloa 300. **Moala:** Mt. Korolevu 375, Mt. Korolevu 300. **Vanua Levu:** Vusasivo Village 400 b. **Viti Levu:** Mt. Batilamu 1125 b, Colo-i-Suva 200, Colo-i-Suva 186 d, Nabukelevu 300, Nadarivatu 750.

Strumigenys sp. FJ18

(Plate 140)

Strumigenys sp. FJ18 is a relatively large and highly distinctive rich reddish brown species with abundant flagellate hairs on its dorsal surfaces and a unique sculpture characterized by a smooth integument interrupted by deep irregular pits. The antennal scrobe, pronotal dorsum, mesopleuron, metapleuron, sides of the propodeum, postpetiolar dorsum, and gaster are all highly polished. The propodeal spines are subtended by broad and convex lamellae. Several small intercalary teeth are visible between the apical fork.
The species is represented by three specimens, two from Gau and one from Viti Levu, all between 400–475 meters and collected from extracted leaf litter.

Material examined. **Gau:** Navukailagi 475, Navukailagi 300. **Viti Levu:** Vunisea 300.

Strumigenys sp. FJ19

(Plate 141)

Strumigenys sp. FJ19 is one of the most peculiar representatives of the genus in Fiji. It is a very small light reddish brown and highly polished species with bowed mandibles that are shorter than the length of the terminal antennal segment, a head that becomes very wide posteriorly, fairly abundant long stiff filiform hairs, and a dense tuft of white filamentous hairs arising from the lateral promesonotal border above the procoxae. *Strumigenys* sp. FJ19 is most similar to *S. basilica*, and can be distinguished by the shorter mandibles with longer teeth, a broader head posteriorly, a deep puncture on the posteromedial margin, and the abundant long erect hairs over all dorsal surface of mesosoma, waist and gaster. The only collection of this species is from Nabukavesi (Viti Levu).

Material examined. **Viti Levu:** Nabukavesi 300.

Genus Tetramorium

Diagnosis of worker caste in Fiji. Head shape ovoid to rectangular. Antenna 12-segmented. Antennal club 3-segmented. Antennal scapes fail to reach posterior margin of head by at least the length of the first funicular segment. Antennal scrobes present. Antennal sockets surrounded by a raised sharp-edged ridge. Anterior clypeal margin variously shaped, but never armed with three broad and blunt teeth. Frontal lobes relatively far apart so that the posteromedian portion of the clypeus, where it projects between the frontal lobes, is much broader than one of the lobes. Anterior margin of clypeus without a rectangular projection that extends over base of mandibles. Mandibles triangular; lacking a distinct basal tooth. Mesosoma evenly convex; lacking depression separating promesonotum from propodeum. Mesosoma with erect hairs. Propodeum armed with spines or teeth. Propodeal lobes shorter than propodeal spines. Propodeal spines distinctly longer than diameter of propodeal spiracle. Waist 2-segmented. Petiole pedunculate. Postpetiole not swollen; in dorsal view not distinctly broader than long or distinctly wider than petiole. Tip of sting with a triangular to pendant-shaped extension.

Tetramorium is among the most diverse myrmicine genera in the Old World and its species tend to be generalist foragers in many different habitats. They are often good recruiters, nest in logs, under stones and in twigs, and are large enough to find in the field without the use of specialized collecting techniques. In addition to the many endemic species found in the region, a large number of invasive and tramp *Tetramorium* occur in the Pacific.

Key to *Tetramorium* workers in Fiji.

1 Pilosity extremely dense and abundant with bifid and occasionally trifid hairs present at least on propodeum. Strongly rugoreticulate sculpture on head, mesosoma and waist segments; propodeal spines long. Usually dark in color 2

– Pilosity extremely sparse to abundant, but never with bifid or trifid hairs present. Sculpture, propodeal spines and color variable 3

2 Pilosity on mesosoma conspicuously longer than maximum length of eye. Clypeus sculpture striate without becoming reticulate. Postpetiole shorter than petiole. Introduced. .... ***T. lanuginosum***

– Pilosity on mesosoma equal to or shorter than maximum length of eye. Clypeus sculpture rugoreticulate. Postpetiole height subequal to that of petiole. Fiji endemic ....***T. manni***

3 Pilosity of mesosomal dorsum sparse, thick, blunt and shorter than length of eye. Propodeal spines short. Petiolar node squared dorsally. Yellow brown. Smaller (HW < 0.55 mm). 4

– Pilosity of mesosomal dorsum abundant, fine, flexuous, and equal to or longer than length of eye. Propodeal spines moderate to long. Petiolar node, color variable. Larger (HW > 0.55 mm) 5

4 Palp formula 4:3. Posterior margin of head broader, flatter and with more squared corners.

Petiolar node and peduncle thicker in profile. Frontal carinae more conspicuous. Antennal scrobes more conspicuous. Cephalic sculpture stronger. Introduced........... ***T. simillimum***

– Palp formula 3:2. Posterior margin of head narrower, more impressed medially, and with more rounded corners. Petiolar node and peduncle narrower in profile. Frontal carinae less distinct. Antennal scrobes less distinct. Cephalic sculpture weaker. Introduced. .... ***T. caldarium***

5 Petiolar peduncle relatively long and arched. Petiolar node dome-shaped. Both waist segments smooth and shiny lacking rugoreticulate sculpture. Anterior margin of clypeus lacking median notch. Propodeal spines of moderate length. Uniform yellow brown. Smaller (HW < 0.65). Pacific native .... ***T. tonganum***

– Petiolar peduncle shorter and less arched. Petiolar node subquadrate or wave-shaped, but not dome-shaped. Waist segments with rugoreticulate sculpture. Anterior margin of clypeus with median notch present. Propodeal spines long. Variously colored. Larger (HW > 0.65).. 6

6 Petiolar node subquadrate with anterodorsal and posterodorsal corners squared and level. Color reddish brown with darker gaster. Introduced................................. ***T. bicarinatum***

– Petiolar node wave-shaped with anterodorsal corner rounded and lower, and posterodorsal corner squared and higher. Color either uniformly reddish brown or uniformly dark brown; never with gaster darker than head and mesosoma.. 7

7 Color uniformly dark brown to brown black with brown legs. Pacific native.... ***T. pacificum***

– Color uniformly reddish brown with yellowish legs. Pacific native...................... ***T. insolens***

Tetramorium bicarinatum (Nylander)

(Plate 142)

Myrmica bicarinata Nylander, 1846: 1061; worker, queen described. Type locality: U.S.A. Combined in *Tetramorium* (Mayr, 1862: 740). Senior synonym of *Formica guineensis* (now in *Pheidole*) (Fabricius, 1793: 357). Revived from synonymy, combined in *Tetramorium*, and senior synonym of *cariniceps* (and its junior synonym *kollari*), *modesta* Smith and *reticulate* (Bolton, 1977: 94). In Fiji (Mann, 1921: 459); In Polynesia (Wilson & Taylor, 1967: 71, fig. 56). For additional synoptic history see Bolton et al. (2006).

Tetramorium bicarinatum is a medium-sized ant with a reddish head, mesosoma and waist contrasting with a dark gaster. This species can be distinguished from its congeners in Fiji by the combination of the following characters: (1) square-shaped petiolar node, (2) long propodeal spines, (3) abundant long thin pilosity, and (4) a reddish head, mesosoma and waist contrasting with a dark gaster. *Tetramorium bicarinatum* has often been referred to under the name *T. guineensis* in older literature.

Tetramorium bicarinatum was once believed to be native to Africa, but more recent examination of its distribution (Bolton, 1977; Wetterer, 2009b) suggests an Indo-Pacific or SE Asian origin. Today, *T. bicarinatum* is widespread throughout much of the tropics and subtropics, except for continental Africa and West Asia, where it is largely absent. Wetterer (2009b) reports that in Oceania, *T. bicarinatum* is known from every tropical country except Niue and Nauru, and cites Haupt (1893) for the first record in Fiji. The species is also found in temperate areas inside greenhouses and heated buildings. *Tetramorium bicarinatum* can achieve dense populations in Fiji's disturbed habitats and is likely to adversely affect native biodiversity.

Material examined. **Kadavu:** Moanakaka 60, Mt. Washington 700. **Koro:** Mt. Kuitarua 500, Mt. Kuitarua 505, Nasau 476, Mt. Kuitarua 440 b, Nasau 465 a. **Lakeba:** Tubou 100 b, Tubou 100 c. **Ovalau:** Levuka 450. **Taveuni:** Devo Peak 1188, Lavena 235, Lavena 219, Mt. Devo 892, Tavoro Falls 100. **Vanua Levu:** Kilaka 146, Kilaka 98, Banikea 398, Rokosalase 180, Rokosalase 94, Rokosalase 150, Rokosalase 97, Rokosalase 118, Rokosalase 94. **Viti Levu:** Nabukavesi 40, Mt. Evans 800, Mt. Evans 800, Mt. Evans 800, Mt. Tomanivi 700 b, Navai 700, Monasavu Dam 830, Monasavu 800 a, Ocean Pacific 1, Ocean Pacific 2, Lautoka Port 5 b, Colo-i-Suva 372, Volivoli 55, Volivoli 25, Navai 1020, Nuku 50. **Yasawa:** Wayalailai Resort 55 b, Tamusua 118, Nabukeru 144.

Tetramorium caldarium (Roger)

(Plate 143)

Tetrogmus caldarius Roger, 1857: 12; worker, queen described. Type locality: GERMANY. Junior synonym of *simillimum* (Roger, 1862: 297). Revived from synonymy, combined in *Tetramorium*, and senior synonym of *hemisi*, *minutum* (Bolton, 1979: 169). Senior synonym of *transformans* (Bolton, 1980: 310).

Tetramorium caldarium is a small reddish ant with a square petiolar node, short propodeal spines, and short sparse pilosity. The species is nearly indistinguishable from *T. simillimum*, and accurate separation of these two species may require a taxonomic specialist or a side by side comparison with previously determined specimens. Bolton (pers. comm.) has observed that *T. caldarium* has a 3:2 palp formula, while that of *T. simillimum* is 4:3. Unfortunately, it is very difficult to view the palps of these small species in pinned specimens. Other characters discussed by Bolton (1979) for separating *T. caldarium* from *T. simillimum* include the former's (1) less strongly developed frontal carinae, (2) more feeble antennal scrobes, (3) weaker cephalic ground sculpture, and (4) more ovate head shape with narrower posterior margin.

Additional differences observed of determined PSWC specimens during the course of the present study include the narrower petiolar peduncle, narrower petiolar node, and the more medially impressed posterior margin of the head of *T. caldarium*. All of the described differences are difficult to determine unless the species are seen side by side. *Wasmannia auropunctata* is easily confused with both *T. simillimum* and *T. caldarium*. Although the highly destructive species is not known from Fiji as of this study, it is recommended any determinations of *T. simillimum* or *T. caldarium* be checked against *W. auropunctata*. The latter can be separated by its 11-segmented antennae, two-segmented antennal clubs, and long propodeal spines.

Tetramorium caldarium is believed to be native to Africa and is now widely distributed across the Pacific and other tropical regions. The effect this species has on native biodiversity is not well known. The only *T. caldarium* specimens determined by this study were collected from the port city of Lautoka (Viti Levu).

Material examined. **Viti Levu:** Lautoka Shirley Park 5.

Tetramorium insolens (Smith, F.)

(Plate 144)

Myrmica insolens Smith, F. 1861: 47; queen described. Type locality: SULAWESI [not examined]. Worker described; senior synonym of *macra, pallidiventre, wilsoni* (Syntypes: 8 workers [USNM, examined], 3 workers [MCZC type no. 8708, examined]) (Bolton, 1977: 99). Combined in *Tetramorium* (Emery, 1901: 567). In Fiji (as *wilsoni*) (Mann, 1921: 459). For additional synoptic history see Bolton et al. (2006).

Tetramorium insolens is a medium-sized, uniformly reddish ant. This species can be distinguished from its Fijian congeners by the combination of the following characters: (1) wave-shaped petiolar node, (2) long propodeal spines, (3) abundant long thin pilosity, and (4) uniformly reddish color. *Tetramorium insolens* has established populations outside of its native range, but it is not believed to cause significant damage to ecological or agricultural systems. The species is most often encountered foraging on vegetation in disturbed or edge forest habitat. Like its close relative, *T. pacificum*, it often nests in hollow twigs.

Material examined. **Beqa:** Malovo 182. **Gau:** Navukailagi 356. **Kadavu:** Moanakaka 60, Moanakaka 60. **Koro:** Nasoqoloa 300. **Lakeba:** Tubou 100 a, Tubou 100 b, Tubou 100 c. **Moala:** Mt. Korolevu 300. **Ovalau:** Levuka 400, Draiba 300. **Taveuni:** Lavena 229. **Vanua Levu:** Vusasivo Village 190, Rokosalase 97, Rokosalase 94. **Viti Levu:** Nabukavesi 40, Mt. Evans 800, Mt. Evans 800, Ocean Pacific 1, Korobaba 300, Nakobalevu 340, Colo-i-Suva 372, Colo-i-Suva 186 d, Volivoli 55, Naboutini 300, Nabukelevu 300, Nakavu 200, Nabukavesi 300, Waiyanitu.

Tetramorium lanuginosum Mayr

(Plate 145)

Tetramorium lanuginosum Mayr, 1870: 976; worker described. Type locality: JAVA. Combined in *Triglyphothrix* (Emery, 1891: 4); in *Tetramorium* (Bolton, 1985: 247). Senior synonym of *australis, ceramensis, felix, flavescens, laevidens, mauricei, orissana, striatidens, tricolor* (Bolton, 1976: 350). In Polynesia (Wilson & Taylor, 1967: 70, fig. 55). For additional synoptic history see Bolton et al. (2006).

Tetramorium lanuginosum is a small ant with a reddish head, mesosoma and waist contrasting with a dark gaster, and a dense pelt of bifurcated long white pilosity. This species can be distinguished from most of its Fijian congeners by the combination of the following characters: (1) a rounded petiolar node, (2) long propodeal spines, (3) extremely dense long thin pilosity, and (4) a reddish head, mesosoma and waist contrasting with a dark gaster. It is most similar to *T. manni*, which also has dense bifurcate pilosity, a rounded petiolar node, and rugoreticulate sculpture. *Tetramorium lanuginosum* can be separated from *T. manni* by the longer pilosity, striate clypeus, shorter postpetiole, and more upturned propodeal spines.

Tetramorium lanuginosum (often referred to as *Triglyphothrix striatidens* in older literature) is widely distributed across the Pacific and the globe. It is most often encountered in forest leaf litter. The species is not known to cause significant damage to ecological or agricultural systems. *Tetramorium lanuginosum* is believed to be native to Asia (Bolton, 1985; Wilson & Taylor, 1967). Wetterer (2010b), in addition to presenting arguments for an expanded native range that also includes parts of Oceania, suggests that the species is unlikely to become a serious pest except, perhaps, on small islands. The species in likely introduced in Fiji, as the first published records of its occurrence there are from 2004 (Ward & Wetterer, 2006).

Material examined. **Koro:** Nasoqoloa 300. **Viti Levu:** Ratu Sukuna Park 5.

Tetramorium manni Bolton
(Plate 146)

Tetramorium manni Bolton, 1985: 247. Type locality: FIJI, Viti Levu, Nadarivatu (W. M. Mann). Syntypes: 7 workers, 1 queen (MCZC type no. 8709, examined); 16 workers, 2 queens (USNM, examined). Replacement name for *Triglyphothrix pacifica* Mann, 1921: 460; worker described. Junior secondary homonym of *Tetramorium pacificum* Mayr, 1870: 976.

Tetramorium manni is the only species of the genus endemic to Fiji. It can be identified by the dense pelt of short bifurcate pilosity, a petiole with a relatively long peduncle and rounded node, dark color, and thickly rugoreticulate sculpture. It is most similar to *T. lanuginosum* but can be separated by characters given in the key and listed in the discussion of that species.

Tetramorium manni, as defined here, includes a measurable amount of morphological variation that appears to follow some geographical and elevational trends. Specimens collected from higher elevations on Viti Levu, Vanua Levu and Taveuni tend to be larger with more robust propodeal spines and propodeal angles and generally fewer bifid hairs on the gaster and other surfaces. Specimens collected from lower elevation sites on Viti Levu, Ovalau, Kadavu, Vanua Levu and Taveuni tend to be smaller with dense pelts of bifid hairs on the gastral tergites. However, there are no examples of fine-scale sympatry of these two general morphotypes, and the frequency of specimens bearing intermediate states of these characters is high enough to cast some doubt on claims that multiple species are involved. The species is most similar to *T. antennata* Mann and *T. vombis* Bolton, both of which are endemic to the Solomons. While Bolton uses the smooth postpetiolar dorsum to separate *T. manni* from these species, the character shows variability within the Fijian specimens, though it is always reduced.

Mann collected one colony nesting beneath a stone at Nadarivatu (Viti Levu), and the species was collected most frequently in leaf litter during the recent survey. Individual foragers were observed on occasion foraging on low vegetation.

Material examined. **Gau:** Navukailagi 717, Navukailagi 675, Navukailagi 432, Navukailagi 475. **Kadavu:** Moanakaka 60, Moanakaka 60. **Ovalau:** Draiba 300. **Taveuni:** Devo Peak 1188, Lavena 300, Mt. Devo 1064, Qacavulo Point 300. **Vanua Levu:** Kilaka 98, Nakasa 300, Yasawa 300, Kasavu 300, Nakanakana 300, Drawa 270, Vuya 300. **Viti Levu:** Mt. Evans 800, Mt. Tomanivi 700, Nadala 300, Vaturu Dam 575 b, Navai 930, Mt. Batilamu 840 c, Mt. Tomanivi 950, Monasavu 800, Korobaba 300, Nakobalevu 340, Naikorokoro 300, Waivudawa 300, Nadarivatu 750, Nausori Highlands 400.

Tetramorium pacificum Mayr
(Plate 147)

Tetramorium pacificum Mayr, 1870: 976; worker, queen described. Type locality: TONGA [not examined]. Senior synonym of *subscabrum* (Bolton, 1977: 102). In Polynesia (Wilson & Taylor, 1967: 14, fig. 57). Synonymies discussed (Schlick-Steiner et al., 2006: 182).

Tetramorium pacificum is a medium-sized dark colored ant that is most often encountered on vegetation. The species can be distinguished from its Fijian congeners by the combination of the following characters: (1) wave-shaped petiolar node, (2) long propodeal spines, (3) abundant long thin pilosity, and (4) uniformly dark color. In Fiji, *T. pacificum* is most similar to *T. insolens*, but can be consistently separated by the darker brown to black color and less arching petiolar node.

Tetramorium pacificum has established populations outside of its native range, but it is not believed to cause significant damage to ecological or agricultural systems. The species is most often encountered on vegetation in disturbed or edge forest habitat, and tends to nest in hollow twigs. *Tetramorium pacificum* is among the most widespread and commonly encountered ant species in Fiji. Nests of *T. pacificum* tend to be smaller than those of *T. bicarinatum* and the workers do not recruit as strongly as those of the latter.

Material examined. **Beqa:** Mt. Korovou 326, Malovo 182. **Gau:** Navukailagi 387, Navukailagi 408, Navukailagi 490, Navukailagi 505, Navukailagi 475, Navukailagi 356, Navukailagi 496, Navukailagi 564, Navukailagi 300. **Kadavu:** Moanakaka 60, Mt. Washington 700, Lomaji 580, Namalata 100, Namalata 139. **Koro:** Mt. Kuitarua 500, Mt. Kuitarua 505, Nasau 476, Kuitarua 480, Mt. Kuitarua 485, Nasau 470 (3.7 km), Nasau 465 a, Nasoqoloa 300. **Lakeba:** Tubou 100 a, Tubou 100 b, Tubou 100 c. **Moala:** Maloku 80, Maloku 120, Maloku 1, Mt. Korolevu 375, Mt. Korolevu 300. **Ovalau:** Levuka 450, Draiba 300. **Taveuni:** Devo Peak 1188, Devo Peak 1187 b, Lavena 300, Lavena 235, Lavena 234, Lavena 217, Lavena 219, Lavena 229, Tavuki 734, Mt. Devo 892, Mt. Devo 1064, Tavoro Falls 100, Soqulu Estate 140, Qacavulo Point 300. **Vanua Levu:** Kilaka 146, Wainibeqa 87, Kilaka 61, Wainibeqa 135, Wainibeqa 53, Wainibeqa 150, Kilaka 98, Yasawa 300, Kasavu 300, Nakanakana 300, Vusasivo Village 190, Drawa 270, Vusasivo 50, Vuya 300, Mt. Vatudiri 641, Mt. Vatudiri 570, Vusasivo Village 400 b, Rokosalase 143, Rokosalase 180, Rokosalase 150, Rokosalase 97, Rokosalase 118, Rokosalase 94, Lomaloma 587, Lomaloma 630, Lomaloma 630, Mt. Delaikoro 391, Lagi 300. **Viti Levu:** Nabukavesi 40, Mt. Evans 800, Mt. Evans 800, Mt. Evans 800, Mt. Evans 700, Mt. Tomanivi 700 b, Navai 700, Abaca 525, Vaturu Dam 575 b, Colo-i-Suva Forest Park 220, Ocean Pacific 1, Ocean Pacific 2, Mt. Naqarababuluti 912, Naqaranabuluti 860, Sigatoka 30 a, Vaturu Dam 700, Vaturu Dam 620, Savione 750 a, Monasavu 800, Monasavu Dam 600, Lami 200, Nakobalevu 340, Colo-i-Suva 460, Colo-i-Suva 325, Colo-i-Suva 372, Volivoli 55, Mt. Evans 700, Nabukelevu 300, Navai 1020, Galoa 300, Nabukavesi 300, Naikorokoro 300, Veisari 300 (3.8 km N), Waivudawa 300, Veisari 300 (3.5 km N), Nausori. **Yasawa:** Tamusua 118.

Tetramorium simillimum (Smith, F.)

(Plate 148)

Myrmica simillima Smith, F. 1851: 118; worker described. Type locality: GREAT BRITAIN [not examined]. Combined in *Tetramorium* (Mayr, 1861: 61). In Fiji (Mann, 1921: 459); in Polynesia (Wilson & Taylor, 1967: 73, fig. 59). For additional synoptic history see Bolton et al. (2006).

Tetramorium simillimum is a small yellow-brown ant with short sparse pilosity and a square petiolar node. It is most similar to *T. caldarium*, but can be separated by the characters in the key and those discussed under that species. It is difficult to accurately distinguish between the two, however, unless specimens of both species can be compared side by side. *Tetramorium simillimum* is believed to be native to Africa and is now widely distributed across the Pacific and other tropical regions. The species achieve dense populations in Fiji's disturbed habitats and is likely to adversely affect native biodiversity.

Material examined. **Koro:** Nabuna 115, Nasoqoloa 300. **Lakeba:** Tubou 100 a, Tubou 100 b. **Moala:** Naroi 75, Maloku 1, Mt. Korolevu 300. **Ovalau:** Levuka 400. **Taveuni:** Lavena 229. **Vanua Levu:** Vusasivo Village 400 b. **Viti Levu:** Mt. Evans 800, Ratu Sukuna Park 5, Koronivia 10, Mt. Batilamu 1125 b, Volivoli 55.

Tetramorium tonganum Mayr

(Plate 149)

Tetramorium tonganum Mayr, 1870: 976; worker described. Type locality: TONGA [not examined]. Senior synonym of *magitae* (Bolton, 1977: 129). In Fiji (Mann, 1921: 459). In Polynesia (Wilson & Taylor, 1967).

Tetramorium tonganum is a small yellow brown ant with long flexous pilosity, a long petiolar peduncle, and a strongly rounded petiolar node. The species can be distinguished from its Fijian congeners by the combination of the following characters: (1) a rounded petiolar node and a long thin petiolar peduncle, (2) moderate length propodeal spines, (3) sparse long thin pilosity, and (4) uniformly yellow brown color. *Tetramorium tonganum* has established populations outside of its native range, but it is not believed to cause significant damage to ecological or agricultural systems. The species is most often encountered on vegetation in disturbed or edge forest habitat.

Material examined. **Gau:** Navukailagi 625, Navukailagi 415, Navukailagi 475, Navukailagi 300. **Kadavu:** Moanakaka 60, Mt. Washington 700. **Koro:** Mt. Kuitarua 500, Nasau 476, Nasau 470 (3.7 km). **Lakeba:** Tubou 100 b, Tubou 100 c. **Moala:** Mt. Korolevu 300. **Ovalau:** Levuka 400. **Taveuni:** Lavena 300, Tavoro Falls 160, Lavena 235. **Vanua Levu:** Kilaka 146, Yasawa 300, Vusasivo Village 190, Rokosalase 143, Rokosalase 180, Rokosalase 150, Rokosalase 97, Mt. Delaikoro 391, Lagi 300. **Viti Levu:** Nabukavesi 40, Mt. Evans 800, Mt. Evans 800, Mt. Evans 800, Mt. Evans 700, Lami 200, Nakobalevu 340, Colo-i-Suva 200, Colo-i-Suva 186 d, Volivoli 55, Naboutini 300, Galoa 300, Nakavu 200, Mt. Rama 300, Naikorokoro 300.

Genus Vollenhovia

Diagnosis of worker caste in Fiji. Head shape ovoid to rectangular. Antenna 12-segmented. Antennal club 3-segmented. Antennal scapes fail to reach posterior margin of head by at least the length of the first funicular segment. Frontal lobes relatively far apart so that the posteromedian portion of the clypeus, where it projects between the frontal lobes, is much broader than one of the lobes. Anterior margin of clypeus lacking a rectangular projection that extends over base of mandibles. Sides of head lacking a carinate ridge extending below eye-level from mandibular insertions to posterolateral head margin. Mandibles triangular; not armed with single tooth on basal margin. Mesosoma evenly convex; lacking a depression separating promesonotum from propodeum. Propodeum armed at most with denticles, but never with well-developed spines or teeth. Waist 2-segmented. Petiole lacking peduncle; with large anteroventral subpetiolar process.

The *Vollenhovia* of Fiji are small slightly cylindrical species. Although one species has the propodeum armed with a pair of small denticles, those of the other species are rounded and unarmed. *Vollenhovia* in Fiji can be confused with *Cardiocondyla*, but their petioles have much shorter peduncles and bear distinct teeth or keels on the ventral surface. Little is known about the Fijian species, except that the genus is more diverse on the archipelago than previously thought. Although several new species were recently described from Micronesia (Clouse, 2007b), a revision of the Pacific taxa is required in order to determine whether most of the Fijian species are endemic to the archipelago or more widespread.

Key to *Vollenhovia* species of Fiji.

1 Face and mesosoma carinate or striate and lacking foveae or punctations. .... ***Vollenhovia* sp. FJ03**
– Face and mesosoma foveate or punctate, not carinate or striate 2
2 Petiolar node strongly nodiform and distinctly taller than long. Eyes smaller (< 12 facets). .... ***Vollenhovia* sp. FJ01**
– Petiolar node triangular to subquadrate. Eyes larger (> 15 facets) 3
3 Petiole node triangular. Propodeum armed with denticles .... ***V. denticulata***
– Petiolar node subquadrate to subtriangular. Propodeum unarmed 4
4 Petiolar and postpetiole densely sculptured with reticulating foveae. Dorsal surface of mesosoma foveate-striate. First gastral segment striate .... ***Vollenhovia* sp. FJ04**
– Petiole and postpetiole mostly smooth and shiny. Dorsal surface of mesosoma lightly foveate and lacking striations. First gastral segment smooth and shiny without striations. .... ***Vollenhovia* sp. FJ05**

Vollenhovia denticulata Emery
(Plate 150)

Vollenhovia denticulata Emery, 1914: 405, pl. 13, fig. 4; worker described. Type locality: NEW CALEDONIA [not examined]. In Fiji (Wilson & Taylor, 1967: 58).

Vollenhovia denticulata is recognized in Fiji by the propodeum, which is armed with small angulate denticles. The species was first reported from Fiji by Wilson & Taylor (1967), who discuss examination of a series collected by Mann from Lasema (Vanua Levu) in addition to noting its occurrence on Vanuatu. A second report of this species in Fiji is from Dlussky (1994) based on a specimen collected in 1977 by Chernov (Ward & Wetterer, 2006). Mann's collections, together with those of the recent survey, are nearly indistinguishable from *V. dentata* Mann described from the Solomon Is. and *V. samoaensis* Mayr described from Samoa. Although Wilson and Taylor write that *V. dentata* differs by the entirely unarmed propodeum, the type series does, in fact, have conspicuous propodeal denticles matching those of both *V. denticulata* and *V. samoaensis*. The Fiji material is treated here as *V. denticulata*, but a more thorough examination of the genus across the Pacific is required before its relationship to populations on other islands can be understood. Single specimens have been collected from Viti Levu, Beqa, Kadavu and Gau.

Material examined. **Beqa:** Malovo 182. **Gau:** Navukailagi 300. **Kadavu:** Moanakaka 60. **Vanua Levu:** Lasema a. **Viti Levu:** Lami 171.

Vollenhovia sp. FJ01
(Plate 151)

Vollenhovia sp. FJ01 is a dark species with strongly foveate to punctate sculpture on the head and mesosoma, a smooth and shiny propodeal dorsum, and a tall petiole. There is some variation observed in the accumulated material, with the largest break occurring between the more striate and sculptured Nadarivatu area specimens and the smoother specimens from lower elevation Viti Levu and other islands.

Taxonomic notes for worker.
Face and mesosoma uniformly covered in reticulating foveae with longitudinal striae medially. Posterolateral portion of head behind eyes with a smooth to weakly striate patch. Propodeum entirely unarmed, the dorsal surface smooth and shiny. Petiole strongly nodiform, distinctly taller than long with anterior face weakly concave. Subpetiolar process developed as a broad lamellate keel. Dorsal surfaces of waist and gaster mostly smooth with sparse and widely spaced punctations.

Specimens from southeastern Viti Levu, Kadavu and Gau with eyes smaller (4-6 facets), face with longitudinal striations, posterolateral portion of head behind eyes smooth, waist segments more polished, petiole node taller. Specimens from Nadarivatu area (CASENT0182596, CASENT0182151, CASENT0182639, CASENT0171049) with larger eyes (8-10 facets), face lacking striations, posterolateral portion of head behind eyes weakly striate, waist segments more punctate, and petiole node shorter.

Material examined. **Gau:** Navukailagi 415, Navukailagi 408. **Kadavu:** Mt. Washington 700. **Viti Levu:** Mt. Naqaranabuluti 1050, Naqaranabuluti 1000, Mt. Tomanivi 950, Korobaba 300, Colo-i-Suva 200, Naboutini 300, Nabukavesi 300, Naikorokoro 300, Waivudawa 300.

Vollenhovia sp. FJ03
(Plate 152)

Vollenhovia sp. FJ03 is a distinct member of the Fijian *Vollenhovia*, the typical foveae and punctations are replaced here by long elegant longitudinal striations that cover the face, mesosoma, waist and even gaster. The only two specimens known for this species are both from southeastern Viti Levu.

Taxonomic notes for worker.
Face and mesosoma, waist and first gastral segment uniformly covered with well-defined longitudinal carinae. Posterolateral portion of head behind eyes striate. Eyes large (> 20 facets). Propodeum unarmed. Petiole node subquadrate. Subpetiolar process produced as a thin lamellate tooth. Red brown. Long flexuous pilosity.

Material examined. **Viti Levu:** Waivudawa 300.

Vollenhovia sp. FJ04
(Plate 153)

Vollenhovia sp. FJ04 is a dark brown species covered by dense punctate-striate sculpture, including the gaster. It is known from a single specimen from Gau collected in a malaise trap.

Taxonomic notes for worker.
Face and mesosoma uniformly covered in reticulating foveae, weakly striate medially. Posterolateral portion of head behind eyes striate. Eyes large (> 20 facets). Promesonotum with striae and rugoreticulate. Propodeum unarmed. Petiole evenly rounded. Subpetiolar process developed as a broad lamellate keel. Petiole and postpetiole rugoreticulate. First gastral segment striate.

Material examined. **Gau:** Navukailagi 387.

Vollenhovia sp. FJ05

(Plate 154)

Vollenhovia sp. FJ05 is a pale yellow brown species with a subquadrate petiolar node. It can be distinguished from its Fijian congeners by the characters given in the key.

Taxonomic notes for worker.
Face and mesosoma uniformly covered in reticulating foveae. Face striate medially. Mesosoma foveate. Propodeum unarmed. Posterolateral portion of head behind eyes smooth to weakly striate. Petiole subquadrate and taller apically than posteriorly.

Material examined: **Viti Levu:** Nasoqo 800 d.

Subfamily Ponerinae

Genus Anochetus

Diagnosis of worker caste in Fiji. Antenna 12-segmented. Eyes large (> 20 facets); distinctly anterior to head midline. Clypeus with anterior margin flat to convex, but never forming a distinct triangle that projects anteriorly beyond the base of the mandibles. Posterior margin of head uninterrupted by median longitudinal groove. Mandibles linear; inserted towards the middle of the anterior head margin; armed with apical fork. Waist 1-segmented. Petiole narrowly attached to gaster. Petiolar node evenly rounded; lacking apical spine. Gaster armed with sting; tip pointing posteriorly or straight down, but never anteriorly.

Anochetus graeffei Mayr

(Plate 155)

Anochetus graeffei Mayr, 1870: 961; worker described. Type locality: SAMOA [not examined].
Senior synonym of *amati*, *minutus*, *oceanicus*, *punctiventris* (Wilson, 1959b: 507); of

rudis, *ruginotus*, *taylori* (Brown, 1978: 557). In Fiji (Mann, 1921: 426).

Anochetus graeffei is the only species of this genus known from Fiji. The single waist segment and linear trap-jaws group it together with Fiji's two *Odontomachus* species, from which it can be separated by its smaller size, scale-shaped petiolar node lacking an apical spine, and nuchal carinae that do not converge in a V at the midline of the posterodorsal margin of head. The species is reported to be highly variable across the Pacific (Brown, 1978; Wilson, 1959b), especially with regards to the sculpturing of the first gastral tergite and color. These differences are observed in Fiji, and are particularly prominent in the queen caste. While the vast majority of the Fiji material examined is dark and smooth, there is one nest series (e.g., CASENT0186295, CASENT0186295) collected on Taveuni in which the specimens are a distinctly lighter yellow brown and the sculpturing is heavier. The sculpture of the queen is considerably more punctate on the first gastral segment than is that of the more typical Fijian morphotype, including one specimen also collected from Taveuni (CASENT0194580).

Wilson (1959b), in his remarks on the odontomachines being represented by only *A. graeffei* and *O. simillimus* on New Caledonia, suggested the species were introduced there by humans. The same may well hold true for Fiji. The species is widespread across the archipelago and is recorded from all the major islands with the exceptions of Ovalau and Gau.

Material examined. **Kadavu:** Moanakaka 60, Vunisea 200, Daviqele 300. **Koro:** Nabuna 115, Tavua 220, Mt. Kuitarua 440 b, Nasau 420 b, Nasau 465 a, Nasoqoloa 300. **Moala:** Naroi 75. **Taveuni:** Lavena 235, Lavena 217, Tavoro Falls 160, Tavoro Falls 100, Qacavulo Point 300. **Vanua Levu:** Nakasa 300, Yasawa 300, Kasavu 300, Nakanakana 300, Drawa 270, Vuya 300, Mt. Wainibeqa 152 c, Vusasivo Village 400 b, Rokosalase 180, Mt. Delaikoro 391. **Viti Levu:** Mt. Evans 700, Nadakuni 300 b, Korobaba 300, Lami 200, Volivoli 55, Mt. Evans 700, Naboutini 300, Nabukelevu 300, Galoa 300, Nakavu 200, Nabukavesi 300, Naikorokoro 300.

Genus Hypoponera

Diagnosis of worker caste in Fiji. Antenna 12-segmented. Eyes minute to large; located distinctly anterior to head midline. Clypeus with anterior margin flat to convex, but never forming a distinct triangle that projects anteriorly beyond the base of the mandibles. Frontal lobes narrow; separated by the posterior extension of the clypeus, and not by a longitudinal median groove. Mandibles triangular; with more than 5 teeth. Petiole narrowly attached to gaster. Hind legs lacking pectinate tarsal claws. Hind tibia with pectinate spur, but lacking simple spur. Waist 1-segmented. Petiolar node wedge-shaped in profile; distinctly taller than long; with posterior face convex to flat. Subpetiolar process lacking fenestra anteriorly; never with a pair of teeth or sharp angle posteriorly. Gaster armed with sting; distinct constriction between abdominal segments 3+4; tip pointing posteriorly or straight down, but never anteriorly.

Hypoponera and *Ponera* in Fiji have been reviewed by Mann (1921), Wilson (1958a) and Taylor (1967). In addition to providing a key to species, Wilson proposed close relationships between the endemic Fijian species and relatives from western Melanesia. Both Wilson and Brown note, in their respective studies of the Melanesian fauna, that Fiji's endemic *Hypoponera* and *Ponera* tend towards gigantism to a degree not observed in their more western relatives. Wilson offers that the Fijian species have radiated into ecological niches more often filled by a number of *Pachycondyla* lineages not known from the archipelago.

Key to *Hypoponera* workers of Fiji.

1 In full face view, posterolateral corners of head covered by crowded small diameter overlapping punctures giving them an opaque dull appearance. Eyes small (< 7 facets). Petiolar node relatively broad with anterior and posterior faces weakly narrowing apically at equal angles 2

– In full face view, posterolateral corners of head with very weakly impressed more widely spaced punctures giving them a polished and shiny appearance. Eyes and petiolar node variable 5

2 Anteroventral surface of head smooth and shiny. Antennal scapes attaining or surpassing posterior margin of head. Introduced. .... ***H. opaciceps***

– Anteroventral surface of head as punctate as dorsal surface. Antennal scapes not attaining posterior margin of head........ 3

3 Darker reddish brown species with lighter yellow brown legs. Fiji endemic........ ***H. eutrepta***

– Lighter yellow brown species with lighter legs 4

4 Large species (HW > 0.7 mm). Posterior margin of head deeply concave. Head broader. Fiji endemic .... ***H. turaga***

– Small species (HW < 0.6 mm). Posterior margin of head weakly concave. Head narrower. Distribution unknown........ ***Hypoponera* sp. FJ16**

5 Antennal scapes conspicuously surpassing posterior margin of head. Petiolar node with anterior face straight to concave and posterior face convex such that the node curves weakly anteriorly. Dark brown to reddish brown species 6

– Antennal scapes may attain posterior margin of head, but they do not conspicuously surpass it. Petiolar node either subquadrate or subtriangular, but both faces narrowing at equal angles such that the node does not appear to curve anteriorly. Color variable, but often yellow brown........ 8

6 Eyes larger with more than six facets. Pacific native........ ***H. pruinosa***

– Eyes small and reduced to fewer than six facets........ 7

7 Petiolar node in profile compressed with posterior face conspicuously more convex than anterior face. In dorsal view petiolar node subequal in width as pronotum. Metanotal groove weakly impressed. Dark brownish-red species. Fiji endemic .... ***H. monticola***

– Petiolar node in profile subtriangular with posterior face and anterior face equally convex. In dorsal view petiolar node conspicuously narrower than pronotum. Metanotal groove strongly impressed. Pale brownish-yellow species. Fiji endemic........ ***H. vitiensis***

8 Antennal scapes just reach posterior margin of head. Petiolar node subtriangular and strongly narrowing apically. Uniformly shiny yellow brown. Pacific native........ ***H. confinis***

– Antennal scapes fail to attain posterior margin of head. Petiolar node subquadrate and weakly narrowing apically. Variously colored, but shade of appendages noticeably lighter than mesosoma. Introduced........ ***H. punctatissima***

Hypoponera confinis (Roger)

(Plate 156)

Ponera confinis Roger, 1860: 284; worker described. Type locality: SRI LANKA [not examined]. Combination in *Hypoponera*, and senior synonym of *nautarum* (Wilson & Taylor, 1967: 26).

Hypoponera confinis is a small shiny yellow to yellow-brown species with a single facet eye and a tall subtriangular petiolar node. It is most likely to be confused in Fiji with *H. punctatissima*, but can be distinguished by the small eye, paler color and more triangular petiolar node. Specimens of the Fiji material matched favorably with that at the MCZC determined by Bill Brown and E. O. Wilson, some of which were compared to original type specimens. Taylor (1967) wrote that *H. confinis* is a fairly widespread tramp species ranging at least from India and Ceylon through Melanesia and Polynesia to the Society Islands, and that it seems to have a capacity for survival in disturbed situations. He also postulated that in Samoa, *H. confinis* is responsible for circumscribing the range of its native relative, *H. woodwardi* Taylor, to the Upolu's higher elevations by means of competitive exclusion. *Hypoponera confinis*, together with *Tetramorium bicarinatum* and *Linepithema humile*, are the only ants that have managed to establish on the remote Juan Fernández Archipelago (Chile) (Ingram et al., 2006).

Material examined. **Gau:** Navukailagi 300. **Vanua Levu:** Mt. Wainibeqa 152 c, Vusasivo Village 400 b.

Hypoponera eutrepta (Wilson), REVISED STATUS
(Plate 157)

Ponera eutrepta Wilson, 1958a: 344. Replacement name for *Ponera biroi* subsp. *rugosa* Mann, 1921: 415; worker described. Type locality: FIJI, Viti Levu, Nadarivatu (W. M. Mann). Syntypes: 11 workers, 1 queen (MCZC type no. 8687, examined). Junior primary homonym of *Ponera rugosa* Le Guillou, 1842: 318 (now in *Diacamma*). Combination in *Hypoponera* (Bolton, 1995: 214). Junior synonym of *opaciceps* (Mayr) (Dlussky, 1994).

Hypoponera eutrepta is a robust, relatively large species with a very small eye, a densely punctate sculpture, a weakly striate katepisternum, and a petiolar node that in profile is thick, broad and weakly tapering apically. The color varies from yellow brown to dark brown, and size varies considerably. This is a broadly variant species, and the brief examination afforded by the present study reveals variation in sculpture, petiole shape and number of eye facets. Wilson (1958a) made a similar observation, "Of special interest is the extraordinary variability shown in *eutrepta* in several characters that are only weakly variable in the western Melanesian members of the *biroi* group, namely in total size, in scape index, and in petiole form." A more careful study of these specimens may reveal interesting patterns of population structure. *Hypoponera eutrepta* is likely a close relative of *H. turaga*, and is distinguished by its distinctly smaller size and darker coloration.

This Fijian endemic was considered by Dlussky (1994) to be a junior synonym of the widespread tramp *H. opaciceps*. However, the two are clearly different. For example, the ventral portion of the head is smooth and shiny in *H. opaciceps*, and strongly punctate in *H. eutrepta*. The species is widespread across the archipelago and produces abundant workers that regularly inhabit the leaf litter and alate queens readily captured by malaise trapping.

Material examined. **Beqa:** Mt. Korovou 326. **Gau:** Navukailagi 675, Navukailagi 387, Navukailagi 415, Navukailagi 408, Navukailagi 432, Navukailagi 490, Navukailagi 505, Navukailagi 356, Navukailagi 496, Navukailagi 557, Navukailagi 564, Navukailagi 575, Navukailagi 300. **Kadavu:** Mt. Washington 760, Mt. Washington 800, Moanakaka 60, Mt. Washington 700, Lomaji 580, Daviqele 300, Namara 300, Vunisea. **Koro:** Mt. Kuitarua 500, Mt. Kuitarua 505, Kuitarua 480, Mt. Nabukala 520, Mt. Nabukala 500,

Nasau 470 (3.7 km), Nasau 420 b, Nasau 465 a. **Moala:** Mt. Korolevu 375. **Ovalau:** Levuka 400, Levuka 450, Draiba 300. **Taveuni:** Devo Peak 1188, Devo Peak 1187 b, Devo Peak 1187 c, Devo Peak 1187, Lavena 234, Tavuki 734, Mt. Devo 892, Mt. Devo 1064, Mt. Devo 775 a. **Vanua Levu:** Wainibeqa 53, Yasawa 300, Mt. Wainibeqa 152 c, Mt. Delaikoro 699, Vusasivo Village 400 b, Vusasivo Village 400 b, Lomaloma 630, Mt. Delaikoro 391, Labasa. **Viti Levu:** Mt. Evans 800, Mt. Evans 700, Vatubalavu 300, Nadala 300, Vaturu Dam 575 b, Navai 870, Navai 930, Colo-i-Suva Forest Park 220, Narokorokoyawa 700, Monasavu Dam 800, Mt. Batilamu 840 c, Mt. Batilamu 1125 b, Mt. Naqaranabuluti 1050, Mt. Tomanivi 950, Monasavu Dam 1000, Monasavu 800, Nasoqo 800 a, Nasoqo 800 b, Nasoqo 800 c, Nasoqo 800 d, Nadakuni 300 b, Korobaba 300, Lami 200, Lami 304, Nakobalevu 340, Colo-i-Suva 200, Waimoque 850, Colo-i-Suva 372, Colo-i-Suva 186 d, Naboutini 300, Nabukelevu 300, Navai 1020, Galoa 300, Nakavu 200, Vunisea 300, Nabukavesi 300, Mt. Rama 300, Naikorokoro 300, Veisari 300 (3.8 km N), Waivudawa 300, Veisari 300 (3.5 km N), Nadarivatu 750, Sigatoka, Nausori, Navai Forestry Camp, Waivaka.

Hypoponera monticola (Mann)

(Plate 158)

Ponera monticola Mann, 1921: 418, fig. 5; worker described. Type locality: FIJI, Viti Levu, Nadarivatu (W. M. Mann). Syntypes: 2 workers, 4 queens (MCZC type no. 8689, examined); 17 workers (USNM, examined). Combination in *Hypoponera* (Taylor, 1967: 12).

Hypoponera monticola is a very shiny yellow brown to dark red brown species with a small eye (< 6 facets), a thin petiolar node that has a flat to concave anterior face and a weakly convex posterior face. The species is most similar to *H. pruinosa*, from which it can be easily separated by the smaller eye. *Hypoponera monticola* is pervasive and prone to local variation, like its relatives *H. pruinosa* (a pacific native) and *H. vitiensis* (a rare endemic). Mann reported the species as nesting in small colonies beneath rotting wood and stones.

Material examined. **Gau:** Navukailagi 597, Navukailagi 625, Navukailagi 675, Navukailagi 387, Navukailagi 415, Navukailagi 408, Navukailagi 432, Navukailagi 490, Navukailagi 475, Navukailagi 356, Navukailagi 564, Navukailagi 575, Navukailagi 300. **Kadavu:** Mt. Washington 760, Mt. Washington 700. **Koro:** Tavua 220, Nasau 465 a, Nasoqoloa 300. **Moala:** Mt. Korolevu 375, Mt. Korolevu 300. **Ovalau:** Levuka 450. **Taveuni:** Mt. Devo 775 a, Qacavulo Point 300. **Vanua Levu:** Mt. Delaikoro 910, Nakasa 300, Yasawa 300, Kasavu 300, Nakanakana 300, Drawa 270, Mt. Vatudiri 570, Vusasivo Village 400 b, Rokosalase 180, Lomaloma 587, Lomaloma 630, Lagi 300. **Viti Levu:** Mt. Evans 800, Mt. Evans 700, Monasavu Dam 800, Mt. Naqaranabuluti 1050, Mt. Tomanivi 950, Nasoqo 800 a, Nasoqo 800 b, Nasoqo 800 c, Nasoqo 800 d, Nadakuni 300, Nadakuni 300 b, Korobaba 300, Lami 200, Nakobalevu 340, Colo-i-Suva 200, Waimoque 850, Mt. Evans 700, Nabukelevu 300, Nakavu 200, Vunisea 300, Nabukavesi 300, Mt. Rama 300, Naikorokoro 300, Waivudawa 300, Nadarivatu 750, Navai Forestry Camp.

Hypoponera opaciceps (Mayr)

(Plate 159)

Ponera opaciceps Mayr, 1887: 536; worker, queen described. Type locality: BRAZIL [not examined]. Combination in *Hypoponera* (Taylor, 1967: 11). Senior synonym of *perkinsi* (and its junior synonym *andrei*) (Wilson & Taylor, 1967: 28, fig. 15). Current subspecies: nominal plus *cubana, gaigei, gibbinota, jamaicensis, pampana, postangustata*. For additional synoptic history see Bolton et al. (2006).

Hypoponera opaciceps is a strongly punctate species that can be separated from its Fijian congeners by the combination of the smooth and shiny ventral surface of the head, and antennal scapes that attain the posterior margin of the head. It is most likely confused with *H. eutrepta*, but can be separated by the smooth ventral surface of the head. *Hypoponera punctatissima* has shorter antennal scapes that do not reach the posterior margin of the head. It is not known whether the material Dlussky (1994) reported from Fiji was truly that of *H. opaciceps*, or whether it was misidentified material of *H. eutrepta*.

Hypoponera opaciceps occurs widely across the Pacific Island region, but is mostly limited to forested habitat where it nests and forages in and around rotting logs, soil and leaf litter. The species is almost entirely blind, and is not often encountered foraging out in the open, nor is it known to recruit to food baits. Although *Hypoponera opaciceps* is considered an introduced species, it is not commonly regarded as a pest species, and little is known about the effects the species has on native biological diversity.

Material examined. **Gau:** Navukailagi 564. **Koro:** Mt. Kuitarua 500, Mt. Kuitarua 505, Mt. Kuitarua 485, Nasoqoloa 300. **Taveuni:** Lavena 235. **Viti Levu:** Volivoli 55.

Hypoponera pruinosa (Emery)

(Plate 160)

Ponera pruinosa Emery, 1900: 319, pl. 8, figs. 13, 14; worker described. Type locality: NEW GUINEA [not examined]. Combination in *Hypoponera* (Imai et al., 1984: 67). Senior synonym of *mocsaryi* (Wilson, 1958a: 335). In Solomons (Mann, 1919: 294).

Hypoponera pruinosa is a very shiny yellow brown to dark red brown species with relatively large eyes, erect hairs on the dorsum of the mesosoma, and a thin petiolar node that has a flat to concave anterior face and a weakly convex posterior face. This species bears strong resemblance to the *H. monticola*, but is distinguished by the larger eyes composed of seven or more facets. The Fiji material treated here as *H. pruinosa* matched specimens determined by Bill Brown and E. O. Wilson at the MCZC. None of the specimens examined, however, were compared with the type material, so the identifications remain tentative. Mann (1919) wrote that in the Solomons, *H. pruinosa* was the most commonly encountered ant in the genus, which at the time included both *Ponera* and *Hypoponera*.

Material examined. **Gau:** Navukailagi 415. **Kadavu:** Daviqele 300. **Taveuni:** Mt. Devo 775 a. **Vanua Levu:** Yasawa 300, Vuya 300. **Viti Levu:** Monasavu Dam 800, Nasoqo 800 b, Korobaba 300, Naboutini 300.

Hypoponera punctatissima (Roger)

(Plate 161)

Ponera punctatissima Roger, 1859: 246, pl. 7, fig. 7; worker, queen described. Type locality: GERMANY [not examined]. Combination in *Hypoponera* (Taylor, 1967: 12). Full synoptic history (Bolton & Fisher, 2011).

Hypoponera punctatissima is a relatively small and shiny species with a short subquadrate petiolar node and short antennal scapes that do not attain the posterior margin of head. In Fiji, it might be most readily confused with *H. vitiensis*, but can be differentiated by its multifaceted eyes, and shorter and more quadrate petiolar node. The other congener not native to Fiji is *H. opaciceps*, from which it can be differentiated by its relatively shorter scapes and shinier integument.

According to Wilson and Taylor (1967), *H. punctatissima* is a pantropical tramp that has been carried across the world by humans. Bolton and Fisher (2011) stated, "*H. punctatissima* is without doubt the world's most accomplished ponerine tramp-species. Its range incorporates all tropical and subtropical zoogeographical regions, including most oceanic islands, and it also penetrates well into the temperate zones of both hemispheres where it is frequently synanthropic." A brief synopsis of the world distribution is outlined in Delabie and Blard (2002).

There has been some confusion surrounding the taxonomy of *H. punctatissima*, especially as it relates to *H. ragusai* (Emery) (= *H. gleadowi*) (Bolton & Fisher, 2011). The two species are believed to be close relatives that originated in the Old World (Wilson & Taylor, 1967). Of these two, however, only *H. punctatissima* is definitively reported from the Pacific Island region. Both species produce ergatoid males, which have been implicated as helping colonies establish outside their native range (Taylor, 1967). Wilson and Taylor also commented that while *H. punctatissima* is the most widespread ponerine in Polynesia, it is not especially abundant. They report it being most frequently encountered in rotting logs at forest fringes or in disturbed but shaded situations. These observations match the situation in Fiji, where it occurs across the archipelago but is collected less frequently than many of its congeners.

Material examined. **Gau:** Navukailagi 496. **Kadavu:** Moanakaka 60. **Koro:** Mt. Kuitarua 500. **Lakeba:** Tubou 100 c. **Vanua Levu:** Mt. Delaikoro. **Viti Levu:** Mt. Evans 800, Volivoli 55.

Hypoponera turaga (Mann)

(Plate 162)

Ponera turaga Mann, 1921: 416; worker, queen described. Type locality: FIJI, Viti Levu, Nadarivatu (W. M. Mann). Syntypes: 4 workers (USNM, examined). Combination in *Hypoponera* (Bolton, 1995: 216).

Note: Specimens at the MCZC from the non-type localities of Waiyanitu and Nagasau bear red cotype labels but are not true syntypes.

Turaga, in Fijian, is roughly translated as king, or chief, and Mann no doubt applied the name to this species on account of its extremely large and robust size. *Hypoponera turaga* is most similar to *H. eutrepta* and *Hypoponera* sp. FJ16. It can be distinguished from the former by its larger size, paler yellow brown color, and more deeply concave posterior head margin. It is nearly twice the size of *Hypoponera* sp. FJ16, has a relatively shorter head, and a relatively narrower petiolar node. Mann's observation that the species is widespread in distribution but not common is borne out by the recent survey, from which a small number of specimens were collected in malaise traps and winkler extractions.

Material examined. **Ovalau:** Levuka 550. **Taveuni:** Mt. Devo 775 a, Nagasau. **Viti Levu:** Mt. Tomanivi 700 b, Monasavu Dam 800, Navai 1020, Waivudawa 300, Waiyanitu.

Hypoponera vitiensis (Mann), REVISED STATUS

(Plate 163)

Ponera vitiensis Mann, 1921: 414, fig. 4; worker, queen described. Type locality: FIJI, Viti Levu, Nadarivatu (W. M. Mann). Syntypes: 8 workers, 1 queen (MCZC type no. 8686, examined); 15 workers (USNM, examined). Combination in *Hypoponera* (Bolton, 1995: 216). Junior synonym of *H. confinis* (Dlussky, 1994: 53).

Hypoponera vitiensis is a medium-sized shiny yellow brown species with a single facet eye, antennal scapes that extend beyond the posterior margin of the head, and a tall subtriangular petiolar node. It is most likely to be confused in Fiji with *H. pruinosa*, but can be distinguished by the small eye. The species was originally described from Nadarivatu (Viti Levu), but has not been collected since. Dlussky (1994) synonymized *H. vitiensis* with the widespread *H. confinis* in his zoogeographic analysis of southwestern Pacific ants. After examining the type material of *H. vitiensis*, it is evident that the species is distinct from *H. confinis*, as demonstrated by the relatively longer antennal scapes of the former.

Hypoponera sp. FJ16

(Plate 164)

Hypoponera sp. FJ16 is a small strongly punctate yellow brown ant with a single eye facet and a short subquadrate petiolar node. The species looks in many ways like a miniature version of *H. turaga*, and can be distinguished from that species by the characters listed under its discussion. The species might also be confused with *H. eutrepta*, but the smaller size, paler coloration, more quadrate petiolar node can separate the former from the latter. This species appears to be fairly widely distributed, though rarely collected.

Material examined. **Beqa:** Mt. Korovou 326. **Kadavu:** Moanakaka 60. **Koro:** Nasau 470 (3.7 km), Nasoqoloa 300. **Taveuni:** Mt. Devo 892, Mt. Devo 1064. **Vanua Levu:** Mt. Vatudiri 570. **Viti Levu:** Navai 700, Navai 863, Nadakuni 300, Nabukavesi 300, Waivudawa 300, Navai Forestry Camp. **Yasawa:** Tamusua 118.

Genus Leptogenys

Diagnosis of worker caste in Fiji. Antenna 12-segmented. Eyes medium to large (> 10 facets); distinctly anterior to head midline. Clypeus distinctly triangular and projects anteriorly well beyond the base of the mandibles. Mandibles linear to triangular; inserted towards lateral corners of the of the anterior head margin; edentate and lacking an apical fork. Hind legs with pectinate tarsal claws. Waist 1-segmented. Petiole narrowly attached to gaster. Gaster armed with sting; distinct constriction between abdominal segments 3+4; tip pointing posteriorly or straight down, but never anteriorly.

Mann assigned the Fijian *Leptogenys* species to the subgenus *Lobopelta*, suggesting they belong to a group that also includes *L. conigera* Mayr and *L. adlerzi* Forel from Australia, *L. acutangula* Emery from New Caledonia, and *L. chinensis* from China. Wilson (1958b) referred to the Fijian taxa as belonging to the *L. chinensis* Emery group, and suggested they are a closely related species complex derived from a single ancestor. He proposed *L. hebrideana* Wilson (Vanuatu), *L. bituberculata* Emery (New Guinea), and *L. sagaris* Wilson (New Caledonia), as close relatives. Wilson further suggested that *L. navua* Mann and *L. humiliata* Mann from Fiji have, like other members of the *L. chinensis* group, undergone little morphological change, while the remaining Fijian species are morphologically quite divergent. He proposed that the Fijian taxa have diversified so radically because the weak "competition from the depauperate endemic ponerine-myrmicine fauna" allowed them the evolutionary opportunity to radiate.

With the possible exception of *L. letilae* Mann, most of the Fijian *Leptogenys* are quite rare and known from only one or two localities. There are a plethora of male specimens collected from malaise trapping, and a more thorough investigation of this group, especially with the aid of molecular techniques to associate males with workers, might allow for a more clear understanding of species limits and distributions.

Key to *Leptogenys* workers of Fiji.

1 Head with region between eyes, including the median, entirely covered by overlapping foveae or striae. Ventrolateral portion of head with irregular strong striae. Dorsal surfaces of mesosoma and petiole with weakly impressed foveae.. 2

– Head with region between eyes variable, but always with the median either smooth or with small, non-overlapping foveae; never entirely covered by overlapping foveae or striae. Ventrolateral portion of head occasionally with ovate punctures, but never strongly striate. Dorsal surfaces of mesosoma and petiole lacking foveae.................................. 3

2 Median portion of head with overlapping, irregularly shaped foveae, but conspicuous striae absent. Petiole approximately as long as tall. Larger species (HW > 1.0 mm). ....***L. foveopunctata***

– Median portion of head with overlapping irregularly shaped foveae that are bordered by conspicuous short striae. Petiole longer than tall. Smaller species (HW < 1.0 mm). .... ***Leptogenys* sp. FJ01**

3 Head broad, with width (measured across distance of anterolateral corners) greater than length (measured from anterolateral corner to level of posterior margin). Petiole approximately as long as tall. Large species (HW > 1.0 mm) ....***L. letilae***

– Head narrow, with width (defined above) conspicuously less than length (defined above). Petiole shape variable. Smaller species (HW < 1.0 mm)... 4

4 Eyes very large, their length conspicuously at least twice that of the mesosomal pilosity. Petiole distinctly longer than tall, with anterior face gently sloping towards vertex and posterior face conspicuously convex. Median of clypeus tipped anteriorly with three short and flattened hairs. Large elongate species (HW > 0.9 mm)....................***L. vitiensis***

– Eyes moderate to small, but their length conspicuously less than twice that of the mesosomal pilosity. Petiole shape variable, but anterior face forming a strong angle as it approaches dorsum and posterior face concave to weakly convex. Median of clypeus tipped anteriorly with one short and flattened hairs. Smaller species (HW < 0.9 mm).............................. 5

5 Eyes small, their length distinctly less than that of the mesosomal pilosity. Metapleuron entirely covered by a rough irregular sculpture. Smaller (HW < 0.6 mm). Head, mesosoma and gaster reddish brown...***L. humiliata.***

– Eyes moderate, their length distinctly subequal to mesosomal pilosity. Metapleuron not entirely covered by a rough irregular sculpture. Larger (HW > 0.6 mm). Head and mesosoma black, gaster variable 6

6 Head very narrow, approximately twice as long as wide. Antennal scapes not surpassing the posterior margin of the head by a length equal to the terminal two funicular segments combined. Head and mesosoma black, petiole and gaster reddish brown ....***L. navua***

– Head broader, length conspicuously less than twice width. Antennal scapes surpassing the posterior margin of the head by a length equal to or greater than terminal two funicular segments combined. Uniformly black.. ***L. fugax***

Leptogenys foveopunctata Mann
(Plate 165)

Leptogenys (*Lobopelta*) *foveopunctata* Mann, 1921: 421; worker described. Type locality: FIJI, Vanua Levu, Suene (W. M. Mann). Syntypes: 2 workers (USNM, examined).

Leptogenys foveopunctata is a large species with a relatively broad head. The foveae on the dorsal surface of the head are circular and well-defined; those on the lateral and ventral surfaces of the head, and on the dorsal surfaces of the mesosoma and petiole, are more ovate. Irregular sculpture is also present on the mesopleuron, metapleuron and posterolateral surface of the petiole. The petiolar node is large and robust, with the height of the posterior face being approximately equal to the length of the entire node. A clear majority of the mesosomal hairs are shorter than the length of the eye. *Leptogenys foveopunctata* is most similar to the other Fijian congener with a relatively broad head, *L. letilae*, but differs by the sculptured dorsal surfaces of the mesosoma and petiolar node. It differs from *Leptogenys fugax* by a relatively broader head, a less elongate petiolar node, and more sculptured dorsal surfaces of the mesosoma and petiole.

In addition to the two specimens collected from Vanua Levu by Mann, a single worker from Vanua Levu was also collected by G. Kuschel. The species was not recovered during the recent survey.

Material examined. **Vanua Levu:** Mt. Delaikoro, Suene.

Leptogenys fugax Mann
(Plate 166)

Leptogenys (*Lobopelta*) *fugax* Mann, 1921: 422, fig. 7; worker described. Type locality: FIJI, Viti Levu, Waiyanitu (W. M. Mann). Syntypes: 4 workers (MCZC type no. 8691, examined); 4 workers (USNM, examined).

Leptogenys fugax is a medium sized species with a relatively narrow head, an angulate anterior clypeal margin and a relatively elongate petiolar node. The dorsal surface of the head is marked by small diameter punctures, but the dorsal surfaces of the mesosoma and petiole are polished and lacking sculpture. The species is closely related to *Leptogenys* sp. FJ01 and *L. letilae*. It can be separated from the former by the distinctly smoother face, and from the latter by the smaller more gracile size, more angulate anterior margin of the clypeus, the relatively narrower face, and the more elongate petiole.

The material examined from five localities on Viti Levu agrees strongly with the type series from Waiyanitu (Viti Levu). The two worker specimens examined from the northern islands of Vanua Levu and Taveuni are both smaller and more gracile with weaker cephalic sculpture and even more elongate petiolar nodes that tend to be more angular in dorsal view. The fact that these more gracile and less sculptured forms are geographically closest to *L. foveopunctata* suggests that gene flow is not occurring between them, but rather represent allopatric populations of *L. fugax*.

Material examined. **Taveuni:** Tavuki 734. **Vanua Levu:** Nakanakana 300. **Viti Levu:** Mt. Tomanivi 950, Nasoqo 800 a, Nasoqo 800 d, Lami 200, Naikorokoro 300, Nadarivatu 750, Nasoqo, Waiyanitu.

Leptogenys humiliata Mann

(Plate 167)

Leptogenys (*Lobopelta*) *humiliata* Mann, 1921: 421; worker described. Type locality: FIJI, Viti Levu, Nadarivatu (W. M. Mann). Syntypes: 1 worker [badly damaged] (MCZC type no. 20499, examined); 1 worker (USNM, examined).

Leptogenys humiliata is the smallest of the Fijian *Leptogenys*. It is identifiable by the small eye which is shorter in length than most of the mesosomal pilosity. The head is narrow and foveate, the dorsolateral portions of the mesosoma are foveate, and the metapleuron is covered by a strong irregular sculpture. The species is known only from the two type series workers collected by Mann, which he described as dark reddish brown with lighter shades at the tips of the appendages and gaster.

Leptogenys letilae Mann

(Plate 168)

Leptogenys (*Lobopelta*) *letilae* Mann, 1921: 419, fig. 6; worker, queen, male described. Type locality: FIJI, Viti Levu, Nadarivatu W. M. Mann). Syntypes: 7 workers, 1 queen, 1 male (USNM, examined).

Leptogenys letilae is a large robust species. The petiole is tall in relation to its length, and the foveopunctate facial sculpture is distributed mainly towards the anterior portion of the face. The queen can be identified by the compressed shape of the petiolar node. The anterior margin of the clypeus is gently rounded compared to that of *L. fugax*, the head is broader and the petiolar node tends to be relatively higher. *Leptogenys letilae* is similar in size and shape to *L. foveopunctata*, but can be separated by the smooth polished dorsal surfaces of the mesosoma and the petiolar node. *Leptogenys letilae* was the most commonly collected species during the recent survey, where good nest series were collected from higher elevations of Viti Levu and a single worker was captured from Ovalau. The species is sympatric with the more slender *L. fugax*, but neither is known to be sympatric with *L. foveopunctata*.

Material examined. **Ovalau:** Levuka 550. **Viti Levu:** Navai 863, Mt. Naqarababuluti 912, Savione 750 a, Mt. Naqaranabuluti 1050, Navai 1023, Nasoqo 800 a, Nabukavesi 300, Nausori, Nausori Highlands 400.

Leptogenys navua Mann
(Plate 169)

Leptogenys (*Lobopelta*) *navua* Mann, 1921: 423, figs. 8, 9; worker, male described. Type locality: FIJI, Viti Levu, Waiyanitu (W. M. Mann). Syntypes: 5 workers (MCZC type no. 8692, examined); 1 worker, 1 male (USNM, examined).

Leptogenys navua is a small species with a very narrow head. The face is strongly foveate, as are the dorsolateral portions of the mesosoma. The petiolar node is conspicuously longer than tall, and together with the gaster is a more reddish shade of brown than the head and the mesosoma. The species is similar in size, shape and sculpture to *L. humiliata*, but can be distinguished by the larger eyes, narrower head, and darker head and mesosoma. Specimens of *L. navua* are known only from the type series and a collection by J. Wetterer from Navai (Viti Levu).

Material examined. **Naroi:** Nanunu-i-Ra Island. **Viti Levu:** Monasavu 800, Navai Forestry Camp, Waiyanitu.

Leptogenys vitiensis Mann
(Plate 170)

Leptogenys (*Lobopelta*) *vitiensis* Mann, 1921: 424, fig. 10; worker described. Type locality: FIJI, Viti Levu, Nadarivatu (W. M. Mann). Syntypes: 3 workers (MCZC type no. 8693, examined); 12 workers (USNM, examined).

Leptogenys vitiensis is one of the most distinct species of its genus in Fiji, recognizable by the oversized eyes which are approximately twice the length of the mesosomal hairs, and by the petiolar node which has a very long gentle sloping anterior face and a short weakly sloping posterior face. Mann suggested that *L. vitiensis* is more closely related to *L. acutangula* Emery from New Caledonia, than it is to its Fijian congeners. The morphology is strikingly different, but without phylogenetic tests it is difficult to verify Mann's hypothesis. The species appears to be arboreal, as evidenced by the considerably larger eyes and the fact that one worker was captured in a malaise trap. *Leptogenys vitiensis* has only been collected around the type locality of Nadarivatu (Viti Levu).

Material examined. **Viti Levu:** Mt. Tomanivi 1300, Mt. Tomanivi 700 b, Nadarivatu 750.

Leptogenys sp. FJ01
(Plate 171)

Leptogenys sp. FJ01 is medium sized species with a relatively broad head and an extremely sculptured face that includes short striae. It is apparently a close relative to *L. letilae* and *L. fugax*, but is distinct in the more heavily sculptured face (especially posteriorly), the flattened (rather than carinate) edge of the median clypeal carinae, the more attenuated posterior margin of the head, and heavy striations on the propleuron. It can be separated from *L. foveopunctata*, which also occurs on Vanua Levu, by the cephalic striae on the median portion of head, the smaller more gracile size and more elongate petiolar node. The species is known from a single specimen collected in a leaf litter sample.

Material examined. **Vanua Levu:** Mt. Vatudiri 570.

Genus Odontomachus

Diagnosis of worker caste in Fiji. Antenna 12-segmented. Eyes medium to large (>10 facets); located distinctly anterior to head midline. Posterior margin of head interrupted by median groove. Mandibles linear; inserted towards the middle of the anterior head margin; armed with apical fork. Petiole narrowly attached to gaster. Clypeus with anterior margin flat to convex, but never forming a distinct triangle that projects anteriorly beyond the base of the mandibles. Waist 1-segmented. Petiolar node armed with apical spine. Gaster armed with sting; tip pointing posteriorly or straight down, but never anteriorly.

Odontomachus is one of the most easily recognizable genera in Fiji on account of its large size, long linear mandibles, apically spined petiole, and striate sculpture. Of the two species present on Fiji, *O. angulatus* is endemic to the archipelago and *O. simillimus* is distributed throughout the Pacific and beyond. Both species are most often encountered in the leaf litter and tend to make nests in small mounds of debris and at the bases of tree trunks. They are active hunters of arthropods and forage individually. The Melanesian species were revised by Wilson (1959b).

Key to *Odontomachus* workers of Fiji.

1 Pronotum entirely covered by parallel arcuate striae; posterolateral lobes of head covered by parallel striae. Mandibular fork armed with short blunt teeth. Petiolar node apically tipped by a long thin spine. Black of gaster contrasting with reddish brown of mesosoma and head. Smaller and more compact. Pacific native .... ***O. simillimus***

– Pronotum with parallel arcuate striae weakly developed and discontinuous to absent. Posterolateral lobes of head polished and lacking striae. Mandibular fork armed with long teeth. Petiolar node attenuating apically but only becoming spinose at very apical portion. Gaster same reddish brown as mesosoma and head. Larger and more gracile. Fiji endemic .... ***O. angulatus***

Odontomachus angulatus Mayr
(Plate 172)

Odontomachus angulatus Mayr, 1866: 500, pl., fig. 10; worker described. Type locality: FIJI, Ovalau [not examined]. Queen described (Mann, 1921: 427). Senior synonym of *politus* (Brown, 1976: 102).

Odontomachus angulatus is Fiji's only endemic member of the genus. It can be distinguished from *O. simillimus* by the generally longer more gracile shape, the polished unsculptured lobes of the head laterad of the posteromedian carinae, the weakly striate to smooth pronotum, and the less spinose petiolar node. The species is far less common than *O. simillimus*, and tends to be restricted to higher elevation forests. The species is also encountered more frequently on vegetation. In addition to the highlands of Viti Levu, *O. angulatus* was also taken from Vanua Levu, Taveuni and Kadavu during the recent survey.

Material examined. **Kadavu:** Mt. Washington 700. **Taveuni:** Devo Peak 1188. **Vanua Levu:** Kilaka 61, Wainibeqa 150, Mt. Vatudiri 641, Mt. Delaikoro 699. **Viti Levu:** Mt. Evans 800, Mt. Evans 800, Mt. Evans 700, Abaca 525, Vaturu Dam 575 b, Vaturu Dam 620, Vaturu Dam 550, Vaturu Dam 530, Mt. Batilamu 840 c, Monasavu Dam 600, Nadakuni 300, Nadakuni 300 b, Korobaba 300, Lami 200, Nakobalevu 340, Waimoque 850, Colo-i-Suva 460, Colo-i-Suva 325, Colo-i-Suva 372, Mt. Evans 700, Nabukavesi 300, Naikorokoro 300, Waivudawa 300, Nadarivatu 750, Nausori, Vunidawa.

Odontomachus simillimus Smith, F.

(Plate 173)

Odontomachus simillimus Smith, F. 1858: 80, pl. 5, figs. 8, 9; queen described. Type locality: FIJI, Ovalau [not examined]. Junior synonym of *haematodus* (Roger, 1861: 24). Revived from synonymy; senior synonym of *fuscipennis* (Wilson, 1959b: 499). Senior synonym of *breviceps, pallidicornis* (Brown, 1976: 106). In Melanesia (Wilson, 1959b).

Odontomachus simillimus is a reddish brown ant with long apically forked linear mandibles and an apically spinose petiole. At over 6 mm long, it is easily spotted in the field. The only other Fijian congener is *O. angulatus*, from which *O. simillimus* can be separated by the strongly striate sculpturing of the pronotum and posterior portions of the head, in addition to the smaller and more compact size. The species is highly variable with respect to size. Mann and subsequent authors considered the species to be *O. haematodus* (Linnaeus), native to the Neotropics, and the mistake was not realized until Wilson's revision of the Melanesian Odontomachini (1959b).

Odontomachus simillimus is common, and is a ubiquitous resident of Fiji's ground fauna. Fijians refer to this to this species as *kanji* in their native tongue, and are respectful of its powerful sting. Nests are often located in the litter that accumulates at the base of trees.

Material examined. **Beqa:** Mt. Korovou 326, Malovo 182. **Gau:** Navukailagi 597, Navukailagi 336, Navukailagi 387, Navukailagi 535, Navukailagi 408, Navukailagi 432, Navukailagi 490, Navukailagi 505, Navukailagi 475, Navukailagi 496, Navukailagi 564, Navukailagi 575, Navukailagi 300. **Kabara:** Kabara. **Kadavu:** Moanakaka 60, Moanakaka 60, Mt. Washington 700, Lomaji 580, Vunisea 200, Namalata 100, Namalata 120, Namalata 50, Namalata 75, Namalata 139, Daviqele 300, Namara 300. **Koro:** Mt. Kuitarua 500, Mt. Kuitarua 505, Mt. Kuitarua 485, Mt. Nabukala 520, Nasau 470 (3.7 km), Nasau 465 a. **Lakeba:** Tubou 100 b, Tubou 100 c. **Macuata:** Vunitogoloa 10, Vunitogoloa 36. **Moala:** Maloku 80, Maloku 120, Mt. Korolevu 375, Mt. Korolevu 300. **Ovalau:** Draiba 300. **Taveuni:** Lavena 235, Lavena 234, Lavena 217, Lavena 219, Lavena 229, Tavoro Falls 100, Soqulu Estate 140, Qacavulo Point 300, Somosomo 200, Nagasau. **Vanua Levu:** Kilaka 61, Wainibeqa 53, Wainibeqa 150, Kilaka 98, Mt. Kasi Gold Mine 300, Nakasa 300, Yasawa 300, Kasavu 300, Nakanakana 300, Vusasivo Village 190, Drawa 270, Vusasivo 50, Vuya 300, Mt. Wainibeqa 152 c, Vusasivo Village 400 b, Rokosalase 180, Rokosalase 150, Rokosalase 97, Rokosalase 118, Rokosalase 94, Lomaloma 587, Lomaloma 630, Lomaloma 630, Mt. Delaikoro 391, Lagi 300. **Viti Levu:** Nabukavesi 40, Mt. Evans 800, Mt. Evans 800, Mt. Evans 800, Mt. Evans 700, Mt. Tomanivi 700 b, Nakavu 300, Vaturu Dam 575 b, Navai 863, Colo-i-Suva Forest Park 220, Colo-i-Suva Forest Park 140, Ocean Pacific 1, Ocean Pacific 2, Sigatoka 30 a, Koronivia 10, Suva, Vaturu Dam 620, Vaturu Dam 550, Vaturu Dam 530, Savione 750 a, Nadakuni 300, Nadakuni 300 b, Korobaba 300, Lami 432, Lami 200, Lami 260, Lami 400, Nakobalevu 340, Colo-i-Suva

200, Colo-i-Suva 460, Colo-i-Suva 325, Colo-i-Suva 372, Colo-i-Suva 186 d, Volivoli 50, Volivoli 55, Volivoli 25, Mt. Evans 700, Naboutini 300, Nabukelevu 300, Galoa 300, Nuku 50, Nakavu 200, Vunisea 300, Nabukavesi 300, Naikorokoro 300, Veisari 300 (3.8 km N), Waivudawa 300, Veisari 300 (3.5 km N), Nadarivatu 750, Nausori, Nasoqo, Waiyanitu, Korovau. **Yasawa:** Wayalailai Resort 55, Tamusua 118, Nabukeru 144, Nabukeru 120.

Genus Pachycondyla

Diagnosis of worker caste in Fiji. Antenna 12-segmented. Eyes minute to absent (≤3 facets); located distinctly anterior to head midline. Clypeus with anterior margin flat to convex, but never forming a distinct triangle that projects anteriorly beyond the base of the mandibles. Frontal lobes broad; separated by the posterior extension of the clypeus, and not by a longitudinal median groove. Mandibles triangular; with 5 teeth. Hind legs lacking pectinate tarsal claws. Hind tibia with both pectinate spur and simple spur. Petiole narrowly attached to gaster. Waist 1-segmented. Petiolar node wedge-shaped in profile; distinctly taller than long; posterior face convex to flat. Gaster armed with sting; distinct constriction between abdominal segments 3+4; tip pointing posteriorly or straight down, but never anteriorly.

Pachycondyla stigma (Fabricius)
(Plate 174)

Formica stigma Fabricius 1804: 400; queen described. Type locality: SOUTH AMERICA [not examined]. For additional synoptic history see Bolton et al. (2006).

Pachycondyla stigma is a relatively large, robust ponerine with strong mandibles, a thick petiole and a minute pair of eyes, and it is the only representative of its genus in Fiji. *Pachycondyla stigma* can be confused with *Hypoponera* and *Ponera* because all have approximately similar head shapes and body shapes, but it can be separated from these genera in Fiji by the larger size, smaller number and larger size of mandibular teeth, and broader frontal lobes. The presence of a simple tibial spine in addition to the pectinate tibial spine is the most reliable character used to separate *Pachycondyla* from the other two genera, but it can be difficult to recognize under the microscope.

Pachycondyla stigma was described by Wilson and Taylor (1967) as one of the most widespread ponerines in the world. A global synopsis of the species with respect to distribution, ecology and invasion success is provided by Wetterer (2011). *Pachycondyla stigma* occurs widely across the Pacific Island region, but is mostly limited to forested habitat where it nests and forages in and around rotting logs, soil and leaf litter. The species is almost entirely blind, and is not often encountered foraging out in the open, nor is it known to recruit to food baits. Although *Pachycondyla stigma* is considered an introduced species, it is not commonly regarded as a pest species, and little is known about the effects the species has on native biological diversity.

Material examined. **Taveuni:** Lavena 235, Lavena 234, Lavena 217, Lavena 219, Lavena 229, Tavoro Falls 160, Soqulu Estate 140. **Vanua Levu:** Wainibeqa 53, Wainibeqa 150, Kilaka 98, Drawa 270, Mt. Wainibeqa 152 c, Vusasivo Village 342 b, Rokosalase 180, Rokosalase 150, Rokosalase 97, Rokosalase 118, Rokosalase 94, Lomaloma 587, Lomaloma 630, Lagi 300. **Viti Levu:** Nabukavesi 40, Ocean Pacific 2, Nadakuni 300 b, Colo-i-Suva 200, Colo-i-Suva 372, Veisari 300 (3.5 km N).

Genus Platythyrea

Diagnosis of worker caste in Fiji. Antenna 12-segmented. Eyes medium to large (>3 facets); located distinctly anterior to head midline. Clypeus with anterior margin flat to convex, but never forming a distinct triangle that projects anteriorly beyond the base of the mandibles. Frontal lobes separated by a longitudinal median groove, not by the posterior extension of the clypeus. Mandibles triangular; with at least five distinct teeth or denticles. Hind legs lacking pectinate tarsal claws. Hind tibia with two pectinate spurs. Waist 1-segmented. Petiole narrowly attached to gaster. Petiolar node rectangular (approximately as tall as long) in profile; posterior face distinctly concave. Gaster armed with sting; distinct constriction between abdominal segments 3+4, tip pointing posteriorly or straight down, but never anteriorly.

Platythyrea parallela (F. Smith)

(Plate 175)

Ponera parallela Smith, F. 1859: 143; worker described. Type locality: INDONESIA, Aru I. [not examined]. Combination in *Platythyrea* (Donisthorpe, 1932: 454). Senior synonym of *aruana, coxalis, pusilla* (Wilson, 1958b: 151); of *pacifica* (Wilson & Taylor, 1967: 20); of *annamita, australis, cephalotes, ceylonensis, cylindrica, egena, inconspicua, javana, parva, philippinensis, pulchella, sechellensis, subtilis, tritschleri, victoriae, wroughtonii* (Brown, 1975: 8).

As the only member of its genus present on Fiji, *P. parallela* is recognizable by the concave propodeal declivity with submarginate posterolateral angles, the broad subquadrate petiolar node, and nearly complete lack of standing hairs. The first published record of this species appears in Ward and Wetterer (2006). Wilson and Taylor (1967) noted its absence from Fiji and New Caledonia. Collection records from the recent survey are restricted to a small handful of alates caught in malaise traps on the southern coast of Viti Levu.

Material examined. **Viti Levu:** Nabukavesi 40, Volivoli 55, Volivoli 25.

Genus Ponera

Diagnosis of worker caste in Fiji. Antenna 12-segmented. Eyes absent to large; located distinctly anterior to head midline. Clypeus with anterior margin flat to convex, but never forming a distinct triangle that projects anteriorly beyond the base of the mandibles. Frontal lobes narrow; separated by the posterior extension of the clypeus, and not by a longitudinal median groove. Mandibles triangular; with more than 5 teeth. Hind legs lacking pectinate tarsal claws. Hind tibia with pectinate spur; lacking simple spur. Petiole narrowly attached to gaster. Petiolar node wedge-shaped in profile; distinctly taller than long; posterior face convex to flat. Subpetiolar process with fenestra anteriorly; often with a pair of teeth or sharp angles posteriorly. Gaster armed with sting; distinct constriction between abdominal segments 3+4; tip pointing posteriorly or straight down, but never anteriorly.

The Fijian *Ponera* are mostly small species with long heads, small eyes and triangular mandibles armed with many small teeth and denticles. The Fijian species are very similar in shape and size to the *Hypoponera* species, but can be reliably differentiated by the distinct fenestra present on the subpetiolar process. They can be differentiated from *Pachycondyla* by the narrower frontal lobes, smaller size, smaller and more numerous mandibular teeth and the absence of a simple spine of their hind tibia. There are several species endemic to Fiji that were treated in Taylor's (1967) global revision that are mostly likely derived from Asian origins. The species are rare, and most commonly encountered in the leaf litter and in rotting logs.

Key to *Ponera* workers of Fiji.

1 Large species (HW > 0.55 mm). Mesometanotum conspicuously impressed in dorsal view. Antennal club indistinct. All surfaces opaque light yellow brown. Fiji endemic... ***P. manni***
– Smaller species (HW < 0.55 mm). Mesometanotum with suture present or absent, but not conspicuously impressed in dorsal view. Antennal club variable. Color variable, but if yellow brown then very small species (HW < 0.29–0.33 mm).. 2
2 Very small species (HW < 0.29–0.33 mm). Head very narrow (CI 72–78). Antennal scapes reaching approximately 2/3 distance to posterior margin of head. Antennal club five-segmented. All surfaces opaque light yellow brown. Pacific native .... ***P. swezeyi***
– Larger species (HW > 0.35 mm). Head broader (CI > 78). Antennal scapes reaching at least 3/4 distance to posterior margin of head. Antennal club variable. Body dark yellow brown to reddish brown and contrasting with distinctly paler yellow appendages..................... 3
3 Antennal club indistinct to 5-segmented. Mesometanotum suture completely absent in dorsal view. Petiolar node very broad in dorsal view. Fiji endemic................ ***P. colaensis***
– Antennal club distinctly 4-segmented. Mesometanotum suture weak, but visible in dorsal view. Petiolar node narrower in dorsal view. Distribution unknown ...... ***Ponera* sp. FJ02**

Ponera colaensis Mann
(Plate 176)

Ponera colaensis Mann, 1921: 417; worker described. Type locality: FIJI, Viti Levu, Waiyanitu (W. M. Mann). Holotype [single specimen]: 1 worker (USNM, examined). Queen described (Santschi, 1928a: 68). In generic revision (Taylor, 1967: 58).

Ponera colaensis is a relatively large species for the genus. It is dark reddish brown with yellow brown appendages, possesses an inconspicuous dorsal metanotal impression and a petiole that in dorsal view has a flat posterior margin and a strongly convex anterior margin. A full description of the species is available in Mann (1921) and Taylor (1967). Taylor placed this species in the *P. taipingensis* group, which consists of four other species ranging from peninsular Malaysia, eastern New Guinea, and Samoa. In addition to a somewhat aberrant record from the Lau group (Santschi, 1928a), the species is otherwise known only from the Nadarivatu area (Viti Levu) and Mt. Navatadoi (Vanua Levu).

Material examined. **Vanua Levu:** Vusasivo Village 342 b. **Viti Levu:** Navai 930, Mt. Naqaranabuluti 1050, Naqaranabuluti 1000, Mt. Tomanivi 950, Nadarivatu 750, Waiyanitu.

Ponera manni Taylor
(Plate 177)

Ponera manni Taylor, 1967: 86, figs. 76, 77; worker described. Type locality: FIJI, Viti Levu, Mt. Lomolaki [= Mt. Lomalagi] (in Naqaranibuluti Reserve), near Nadarivatu, 17.ii.1962 (R. W. Taylor, acc. 22). Holotype: 1 worker [damaged] (MCZC type no. 30924, examined).

Ponera manni is a relatively easy member of the genus in Fiji to identify on account of its large size, yellow brown coloration, broad head, and distinctly impressed mesometanotal suture, all of which separate it from *P. colaensis*. Taylor (1967) somewhat tenuously places *P. manni* in the *japonica* species group, rather than the *taipingensis* group to which *P. colaensis* belongs. The species was previously known from a single worker collected near Nadarivatu. Here we add records of presumed alates that were collected from a number of malaise traps across the archipelago in addition to another worker collected from lowland Viti Levu. Aside from the type locality, all the other collections are from relatively disturbed or lowland habitats.

Material examined. **Gau:** Navukailagi 387, Navukailagi 564. **Koro:** Mt. Kuitarua 500, Mt. Kuitarua 505. **Lakeba:** Tubou 100 a. **Vanua Levu:** Lomaloma 587. **Viti Levu:** Mt. Evans 800, Volivoli 55, Nuku 50, Nadarivatu 750. **Yasawa:** Tamusua 118.

Ponera swezeyi (Wheeler)
(Plate 178)

Pseudocryptopone swezeyi Wheeler, W. M. 1933: 16, fig. 6; worker, queen described. Type locality: HAWAII, vicinity of Honolulu. Syntypes: workers, queen (MCZC, examined). Combination in *Ponera* (Wilson, 1957: 370). In generic revision (Taylor, 1967: 85).

Ponera swezeyi is a small yellow brown species with a narrow face, an indistinct five-segmented antennal club, and a distinct mesometanotal suture. It is the smallest species of *Ponera* known from Fiji, and can be separated from its congeners by the characters in the key. In addition to Hawaii, the species is also known from Samoa. Several collections of *P. swezeyi* were made on Koro from litter sifting, and it was collected in a wet log from Taveuni. An alate queen was captured from a malaise trap in Sigatoka (Viti Levu).

Material examined. **Koro:** Tavua 220, Nasau 465 a, Nasoqoloa 300. **Taveuni:** Lavena 235. **Viti Levu:** Volivoli 55.

Ponera sp. FJ02
(Plate 179)

Ponera sp. FJ02 is a small brownish species (HW 0.42 mm) with a distinct four-segmented antennal club and antennal scapes that do not attain the posterior margin of the head. The species is likely to belong to the *taipingensis* group. Of the specimens examined of this group, the Fiji material compares most closely to the MCZC paratypes of *P. loi* Taylor described from Samoa. Both sets of specimens possess a median clypeal denticle, a similar antennal arrangement, and an inconspicuous mesometanotal suture. The Fijian material differs mostly in its smaller size and proportionally broader petiolar node in dorsal view. These series may well belong to the same species, but a more thorough study is required given the subtle differences used to distinguish among species in this group. Taylor (1967) presumed *P. loi* to be a rare species known only from forested areas near Afiamalu, Samoa, but he questions whether it is endemic to the island or is represented elsewhere in the Pacific. The Fijian specimens are clearly different from *P. incerta* (Wheeler), which is widespread across the Pacific, but has not been collected in Fiji. *Ponera* sp. FJ02 differs from *P. incerta* in its larger size, presence of a median clypeal denticles, and inconspicuous mesometanotal suture.

Although only a small handful of specimens were collected during the recent survey, *Ponera* sp. FJ02 appears to be relatively widespread, and is represented on Viti Levu, Vanua Levu, Taveuni, and Koro.

Material examined. **Koro:** Mt. Nabukala 520. **Macuata:** Vunitogoloa 4. **Taveuni:** Soqulu Estate 140. **Vanua Levu:** Yasawa 300, Vusasivo Village 400 b. **Viti Levu:** Mt. Tomanivi 950, Nabukavesi 300, Veisari 300 (3.5 km N).

Subfamily Proceratiinae

Genus Discothyrea

Diagnosis of worker caste in Fiji. Antenna 8- to 9-segmented; last antennal segment large and bulbous, longer than remaining funicular segments combined. Eyes minute to absent (≤3 facets). Anterior margin of clypeus not denticulate. Clypeus strongly projecting over mandibles in profile view. Waist 1-segmented. Petiole narrowly attached to gaster. Gaster armed with sting; distinct constriction between abdominal segments 3+4; gaster tucked beneath itself with the tip pointing anteriorly. Head and body lacking long flexous hairs.

Discothyrea is a distinctive genus recognizable by the small size, large terminal antennal segment, reduced frontal lobes, short antennal scapes, and strongly arched second gastral segment that causes the gaster to curve beneath itself. The genus may be confused with the closely related *Proceratium*, but can be distinguished by the single tooth at the tip of the mandibles, the overhanging anterior margin of the clypeus, and the configuration of the antennae. There are three undescribed species from Fiji, but no nests have been found to date. The evidence of this genus in Fiji comes from winged alates caught in malaise traps and several workers captured by litter sifting.

Key to *Discothyrea* workers of Fiji.[9]

1 Frontal carinae large, extending posteriorly to the level of the compound eyes and forming scrobes for the reception of the scape. In profile, posterior margin of head angled medially. Fiji endemic .... ***Discothyrea* sp. FJ02**

– Frontal carinae small, ending well forward of the compound eyes and not forming scrobes. In profile, posterior margin of head uniformly curved.. 2

2 Body sculpture more distinct, the foveae on the mesosoma and first gastral tergite deeper and

[9] Key adapted from Dr. Steve Shattuck's preliminary analysis of Indo-Australian *Discothyrea*.

more strongly defined, their edges distinct. Body dark yellow-brown, strongly contrasting with the yellowish legs. Fiji endemic .... ***Discothyrea* sp. FJ04**

– Body sculpture less well defined, the foveae shallow with ill-defined edges, those on the first tergite of the gaster only slightly larger than the hairs they contain, and in some cases the tergite nearly smooth. Body yellowish and at most only slightly darker than the legs. Pacific native.. ***Discothyrea* sp. FJ01**

***Discothyrea* sp. FJ01**

(Plate 180)

Discothyrea sp. FJ01 is a minute yellowish species with minute eyes (< 3 facets), short frontal carinae that do not reach eye level, and shallow and poorly defined foveae. It can be separated from *Discothyrea* sp. FJ04 by the body lighter color that does not contrast strongly with the leg color, and more poorly defined foveae of the first gastral segment. The species is reported to occur elsewhere in the Pacific (Shattuck, pers. comm.).

Material examined. **Viti Levu:** Mt. Tomanivi 950, Colo-i-Suva 186 d.

***Discothyrea* sp. FJ02**

(Plate 181)

Discothyrea sp. FJ02 is a small, reddish brown species with medium sized eyes (> 3 facets), frontal carinae that extend to eye level and form antennal scrobes, a broad petiole in profile, and a thick deeply foveate cuticle.

Material examined. **Gau:** Navukailagi 387. **Kadavu:** Lomaji 580. **Taveuni:** Mt. Devo 892. **Viti Levu:** Mt. Evans 800, Waimoque 850. **Vanua Levu:** Wainibeqa 150.

***Discothyrea* sp. FJ04**

(Plate 182)

Discothyrea sp. FJ04 is most similar to *Discothyrea* sp. FJ01, but can be distinguished by the characters listed under that species and by those in the key. A preliminary analysis of the genus across the region (Shattuck, pers. comm.) suggests the species is endemic to Fiji. The species was not recovered during the recent survey, but a specimen (ANIC32-040152) collected by S. & J. Peck from Mt. Tomanivi is deposited at the ANIC.

Material examined. **Viti Levu:** Navai 1000 (image examined).

***Genus* Proceratium**

Diagnosis of worker caste in Fiji. Antenna 12-segmented, last antennal segment not large and bulbous, distinctly shorter than remaining funicular segments combined. Eyes minute to absent (≤3 facets). Anterior margin of clypeus not denticulate. Clypeus not strongly projecting over mandibles in profile view. Waist 1-segmented. Petiole narrowly attached to gaster. Gaster armed with sting; distinct constriction between abdominal segments 3+4; tucked beneath itself with the tip pointing anteriorly. Head and body covered by abundant long and flexous hairs.

Proceratium is another of Fiji's more obscure genera. It is similar in form to *Discothyrea*, but can be separated by the characters listed under that genus. At least two of the three species known from Fiji are endemic to the archipelago, and are distinctive among all other species of the widespread *silaceum* group (Baroni Urbani & De Andrade, 2003). The genus is rarely collected in Fiji, but the specimens taken from the recent malaise trapping suggest they are more widespread than previously thought. Species of *Proceratium* are known to be specialist predators of arthropod eggs, but very little is known of the Fijian taxa.

Key to *Proceratium* workers of Fiji.

1 Petiolar node relatively short with dorsum appearing blunt in profile. Subpetiolar process obtusely angled. Dorsal surfaces of head and mesosoma conspicuously punctate. Small species (HW < 0.60 mm). Yellow brown .... ***Proceratium* sp. FJ01**
– Petiolar node strongly squamiform, relatively tall with dorsum appearing submarginate in profile. Subpetiolar process acutely angled to tooth-like. Dorsal surfaces of head and mesosoma polished without punctures. Larger species (HW > 0.70 mm). Reddish brown 2
2 Propodeal declivity submargined dorsolaterally with weak angles. Subpetiolar process developed as a distinct tooth. Larger species (HW > 0.80 mm)........................ ***P. relictum***
– Propodeal declivity rounded dorsolaterally and lacking weak angles. Subpetiolar process developed as an acute angle but not projecting as a distinct tooth. Smaller species (HW < 0.80 mm) .... ***P. oceanicum***

Proceratium oceanicum De Andrade
(Plate 183)

Proceratium oceanicum De Andrade, in Baroni Urbani & De Andrade, 2003: 310, fig. 127; worker described. Type locality: FIJI, Viti Levu, Nadarivatu, 16.ii.1962, R. W. Taylor, acc 25) [ANIC, not examined].

Proceratium oceanicum is a medium-sized dark reddish black species with a strongly squamiform petiolar node, a rounded unmargined propodeum, and an acutely angled subpetiolar process. *Proceratium oceanicum* is most similar to *P. relictum*, with which it is sympatric, but can be distinguished by the size, rounded propodeum and lack of a projecting tooth on the subpetiolar process.

Workers collected during the recent survey were taken from Mt. Devo (Taveuni) and Koroyanitu (Viti Levu) in addition to the type locality. The worker from Koroyanitu (CASENT0194739) is yellow brown, but it is was collected from a winkler sample and the different color is inferred to be the result of its younger age. The presumed males of *P. oceanicum* were collected in malaise traps from the two aforementioned islands in addition to Vanua Levu, and are included in the following material.

Material examined. **Taveuni:** Mt. Devo 892, Mt. Devo 775 a. **Vanua Levu:** Kilaka 98. **Viti Levu:** Mt.

Evans 700, Mt. Tomanivi 700 b, Navai 700, Naqaranabuluti 1000, Colo-i-Suva 372, Navai 1020, Veisari 300 (3.5 km N), Nausori.

Proceratium relictum Mann
(Plate 184)

Proceratium relictum Mann, 1921: 413, fig. 3; worker, queen described. Type locality: FIJI, Taveuni, Somosomo (W. M. Mann). Syntypes: 1 worker, 2 queens (MCZC type no. 20423, examined); 1 worker, 4 queens (USNM, examined). In generic revision (Baroni Urbani & De Andrade, 2003).

Proceratium relictum is the largest of the three *Proceratium* species in Fiji, and can be separated from the closely related *P. oceanicum* by its larger size, submarginate propodeal angels, and tooth-like subpetiolar process. Mann collected several females and a solitary worker under stones on Taveuni. The recent survey recovered workers from both Taveuni and Vanua Levu. Several dozen males were captured in malaise traps, all from either Taveuni or Vanua Levu, and it would appear that this species is restricted to those two northern islands.

Material examined. **Taveuni:** Devo Peak 1188, Devo Peak 1187 b, Tavuki 734, Mt. Devo 1064, Somosomo 200. **Vanua Levu:** Kilaka 61, Kilaka 154, Yasawa 300, Vusasivo Village 190, Mt. Delaikoro 699, Vusasivo Village 400 b, Rokosalase 180, Lomaloma 587, Lomaloma 630, Lomaloma 630.

***Proceratium* sp. FJ01**
(Plate 185)

Proceratium sp. FJ01 is a very small yellow brown species with a weakly punctate sculpture, a weakly squamate petiolar node, and an obtusely angled subpetiolar process. It can be separated from its congeners by the aforementioned characters. The species is close to *P. papuanum* Emery, but different in the narrower and more angulate frontal lobes, the lighter sculpture and the less angulate subpetiolar process. The Fijian taxa may be more similar to the specimens from the Solomon Islands discussed in Baroni Urbani & De Andrade (2003). Aside from the single worker collected at Navai (Viti Levu), males were captured in malaise traps from southeastern Viti Levu, Taveuni, Vanua Levu and Kadavu.

Material examined. **Taveuni:** Devo Peak 1187 b, Lavena 219, Mt. Devo 734, Mt. Devo 892, Mt. Devo 1064. **Vanua Levu:** Kilaka 98, Lomaloma 630. **Viti Levu:** Mt. Tomanivi 950, Colo-i-Suva 372, Veisari 300 (3.5 km N).

Omitted Taxa

The following species native to Rotuma are omitted from the present study. Although Rotuma is politically considered a Fijian dependency, it is geographical isolated from the archipelago by a considerable stretch of deep and open ocean relative to any of the other islands included here.

Polyrhachis (*Chariomyrma*) *rotumana* Wilson & Taylor, 1967: 99, fig. 83; worker described.

Type locality: ROTUMA [not examined].

Camponotus (*Myrmamblys*) *rotumanus* Wilson & Taylor, 1967: 98, fig. 82; worker described. Type locality: ROTUMA [not examined].

Dubious Records

Camponotus rufifrons (Smith, F.)

Formica rufifrons Smith, F. 1860: 95; worker, queen described. Type locality: INDONESIA, Batjan [not examined]. Combination in *Camponotus* (Mayr, 1862: 691); in *C.* (*Colobopsis*) (Dalla Torre, 1893: 250). Current subspecies: nominal plus *leucopus*.

The single record of this species from Fiji was the account written by Mayr (1866) of a specimen collected on Ovalau. While it is possible that a population *C. rufifrons* existed, and may even persist on the island, it has never been collected since. Moreover, the accuracy of the identification is also put in doubt by the Indonesian type locality of *C. rufifrons*. The species is not known to occur elsewhere in the Pacific, and it is an unlikely member of the Fijian ant fauna. The status of *C. rufifrons* must remain provisionary until the material Mayr collected is examined.

Linepithema humile (Mayr)

Hypoclinea humilis Mayr, 1868: 164; worker described. Type locality: ARGENTINA [not examined]. Full synoptic history (Wild, 2007).

Three specimens of this notoriously invasive species are labeled as being collected from Fiji in malaise traps, two from different localities on Taveuni and one from Kadavu. Several indirect lines of evidence suggest that these specimens did not, however, originate in Fiji. The first suspicions are raised in the paucity of specimens collected in the malaise traps. *Linepithema humile*, commonly known as the Argentine ant, is an aggressive recruiter that establishes high density populations. If populations truly exist in the labeled localities, one would expect more than a single specimen to be recovered. Furthermore, the species tends to be dominant where it is introduced, and despite rigorous hand collecting in all three localities, no individuals or nests were encountered. Lastly, all three samples can be traced back to a sorting facility in Santa Ynez, California, which is well populated by *Linepithema humile*. The manager of the sorting facility, Dr. Evert Schlinger, verifies that the ants have been present inside the facility, and have even been seen atop the lids of petri dishes covering malaise samples from Fiji. The only other record of this species from Fiji is reported as being intercepted at a port from an international yacht, and its Fijian origin is also considered dubious (Ward & Wetterer, 2006).

Tetramorium tenuicrine (Emery)

Xiphomyrmex tenuicrinis Emery, 1914: 416; worker described. Type locality: NEW CALEDONIA [not examined]. Combination in *Tetramorium* (Bolton, 1977: 98).

A pin with two workers identified as *T. tenuicrine* (Emery) is located at the MCZC with the following label data, "FIJI Viti Levu, 5 mi w. Koro vau [Korovou], 3.xii.1954, E. O. Wilson #33, lowland rainforest." The species is otherwise known only from New Caledonia, and has never before or since been collected in Fiji. Bolton (1977) reports the species from Fiji (based on the aforementioned specimen) but makes no mention of the unusual disjunction. It is presumed here that the specimen was collected by Wilson in New Caledonia during his Pacific expedition, and was inadvertently paired with a label designed for specimens collected during his Fijian leg of the trip.

Odontomachus haematodus (Linnaeus)

Formica haematoda Linnaeus, 1758: 582, queen described. Type locality: "America meridionali" [not examined]. For additional synoptic history see Bolton et al. (2006).

It is presumed that Mann's (1920) report of *O. haematodus* from Fiji was actually in reference to specimens of *O. simillimus* F. Smith. Mann reported the occurrence subsequent to the synonymy of the latter species with *O. haematodus*, and prior to its revived status published by Wilson (1959b).

Hypoponera ragusai (Emery)

Ponera ragusai Emery: 28; worker described. Type locality: ITALY, Sicily [not examined]. Junior synonym of *H. gleadowi* (Bolton & Fisher, 2011). For additional synoptic history see Bolton & Fisher (2011: 94).

There is only one published record of *Hypoponera ragusai* (Forel) from Fiji (Ward & Wetterer, 2006; Wilson, 1958a). Wilson and Taylor (1967) report that the Polynesian and Melanesian material treated in Wilson (1958a) as *Ponera gleadowi* (Forel) actually belonged to *Hypoponera punctatissima* (Roger). None of the material examined from the recent survey matches the description of *Hypoponera ragusai.*

Nomina Nuda

Strumigenys ursulus Dlussky, 1994: 54. NOMEN NUDUM. *Strumigenys ursulus* Dlussky appears in a checklist of ants from the southwestern Pacific and is stated to be endemic to Fiji. It is possible that the intended name was *Strumigenys chernovi* Dlussky, which was described just prior to the publication in question, and which does not appear in the checklist.

SPECIES PLATES

The following section provides full-page illustrated plates for 185 of Fiji's 187 ant species. Plates are composed of specimen images, a habitat-elevation chart, and a geographic distribution map. Plates for species in which only minor workers occur include images of an exemplar specimen in profile, full-face and dorsal view. Plates for species in which minor and major workers occur include images (where available) of exemplar specimens of majors and minors in profile and full-face view. Images of additional specimens, castes and views are available on Antweb.

Distribution maps for each figured species were generated from all examined material for which georeferenced localities were available. Localities at which the species were collected are marked by black circles. A reference map that includes island names, elevation and scale is presented in Figure 137. The habitat-elevation charts are intended to give a general overview of where the collection records fall in ecological space. Our five coarse habitat classifications capture the dominant ecological gradient in Fiji.

1. *Port cities.* Major urban areas and ports of entry for Fiji (Suva, Lautoka, Nadi, Savusavu).
2. *Human-dominated landscapes.* Villages, roadsides, agricultural areas, and pastures.
3. *Forest edge.* Localities at the border of native forest with agricultural areas, roads, etc.
4. *Disturbed forest.* Habitats retaining a general closed-canopy native forest structure but with some evidence of disturbance, such as selective logging, planted exotic trees (e.g., mahogany, fruit), or evidence of later-stage secondary status.
5. *Primary forest.* Closed canopy forests without observable evidence of recent human or natural disturbance.

An example chart is presented in Figure 138. Black dots represent unique localities at which the illustrated species was observed. To prevent overlap of collection records taken from the same locality or different localities with similar elevation/habitat types, we added a small random horizontal displacement to each point creating a vertical band for each habitat type. Gray dots represent the overall intensity of the sampling across different habitat-elevations, where dot density is proportional to the number of records (species-locality pairs) in that region of ecological space. Note in some cases there may be a discrepancy between the number of records on the geographic distribution map and the habitat-elevation chart. This is because some collections were associated with locality but not habitat data. We also included data from other reliable sources even if we had not examined the material ourselves. Thus, there may be modest differences between the figures and the material examined sections presented with each species. Blank charts are presented for species where no ecological data was available.

Although for most species these raw data provide the general picture of their geographic and ecological distribution, we caution against over-interpreting the apparent patterns. One pattern in particular deserves explanation. For many species collection frequency appears sparser in lowland primary habitats than lowland disturbed and mid-elevation primary habitats. This is most likely the result of the few primary forests remaining between 0–200m and correspondingly fewer samples. Those samples that do exist are concentrated in a small number of intact fragments, while lowland disturbed forest is more commonly encountered in Fiji. In general, if a species has been collected many times in lowland disturbed and mid-elevation primary forests, a lack of records from lowland primary forests should not be interpreted as evidence that those habitats are unsuitable to the species.

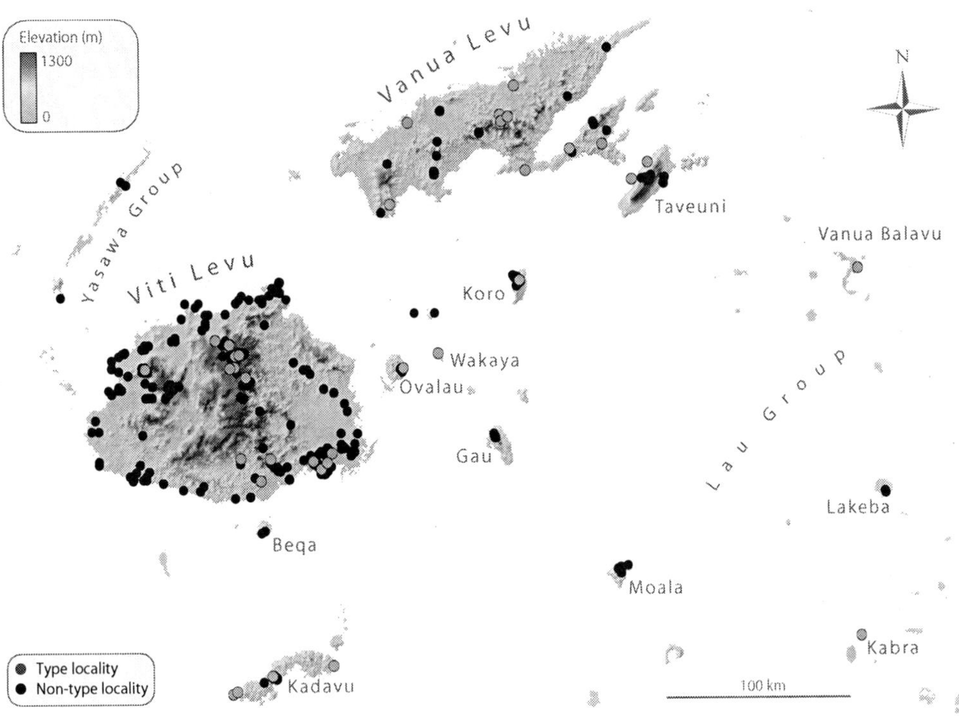

Figure 137. Collection localities for Fijian ant specimens. Gray dots represent localities from which type specimens are described, and black dots represent localities from which only non-type ant specimens are reported.

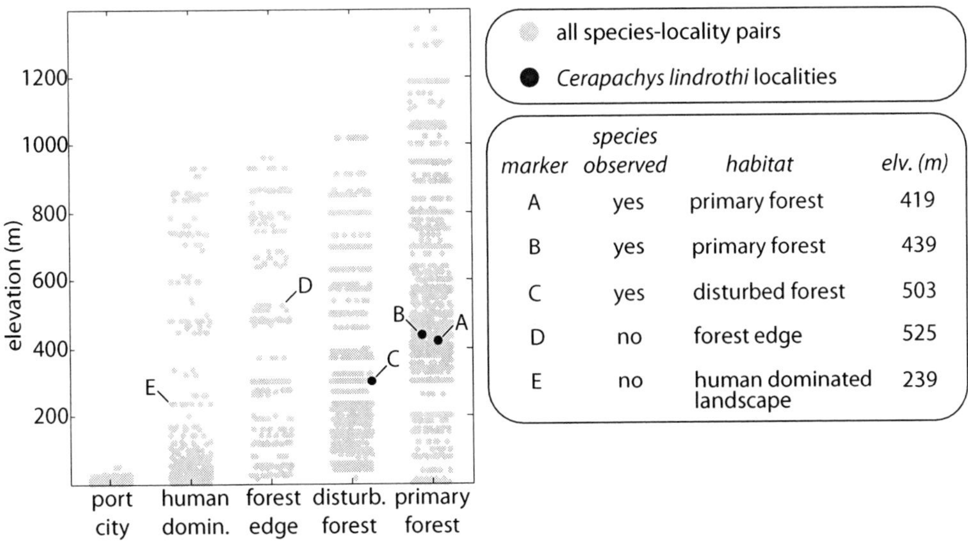

Figure 138. Example habitat-elevation chart of *Cerapachys lindrothi* with labels for five samples. *Cerapachys lindrothi* was observed in three samples (A–C). Markers D and E represent two of the hundreds of samples where *C. lindrothi* was not observed.

Plate 1. *Amblyopone zwaluwenburgi* worker, CASENT0187702.

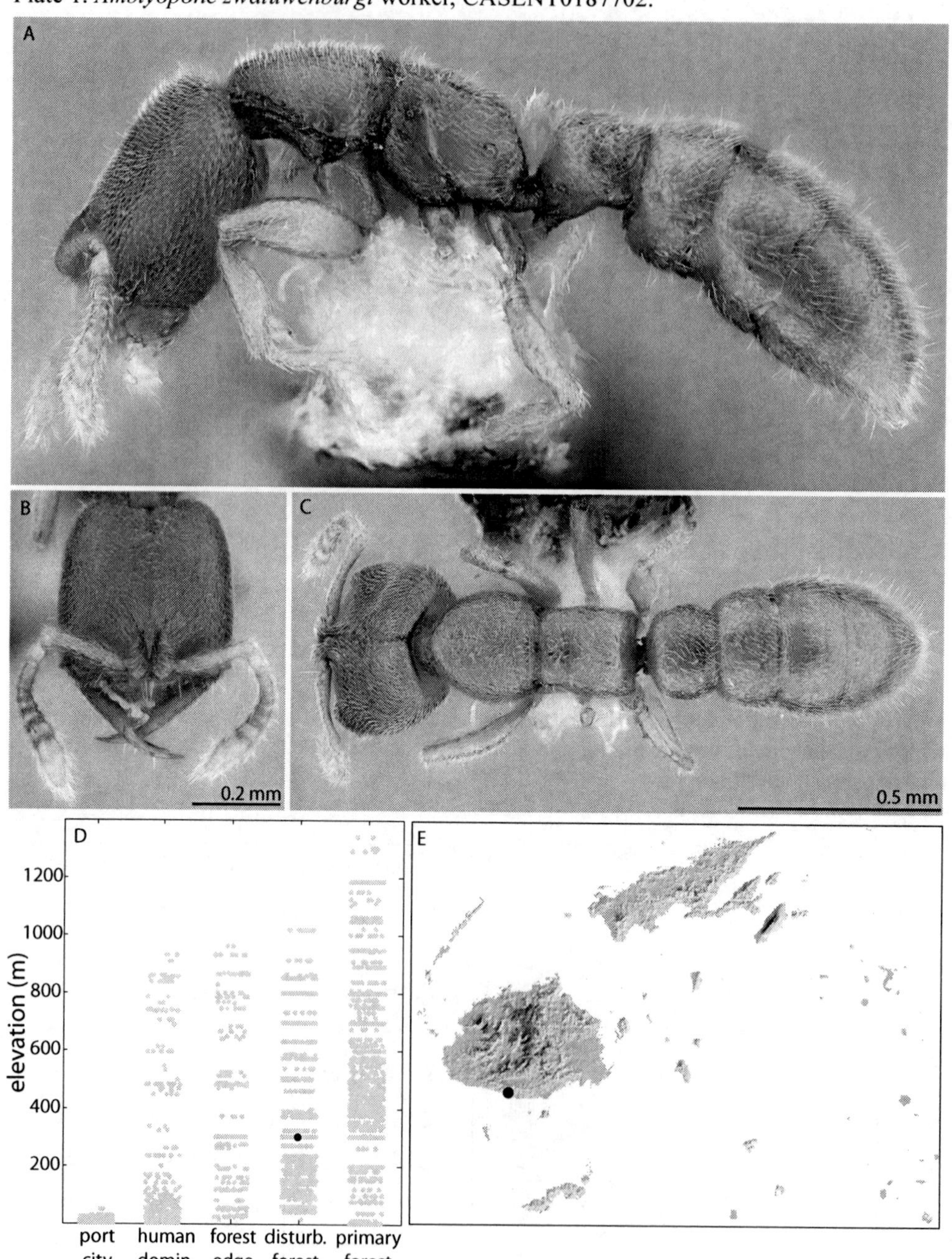

Plate 2. *Prionopelta kraepelini* worker, CASENT0219559.

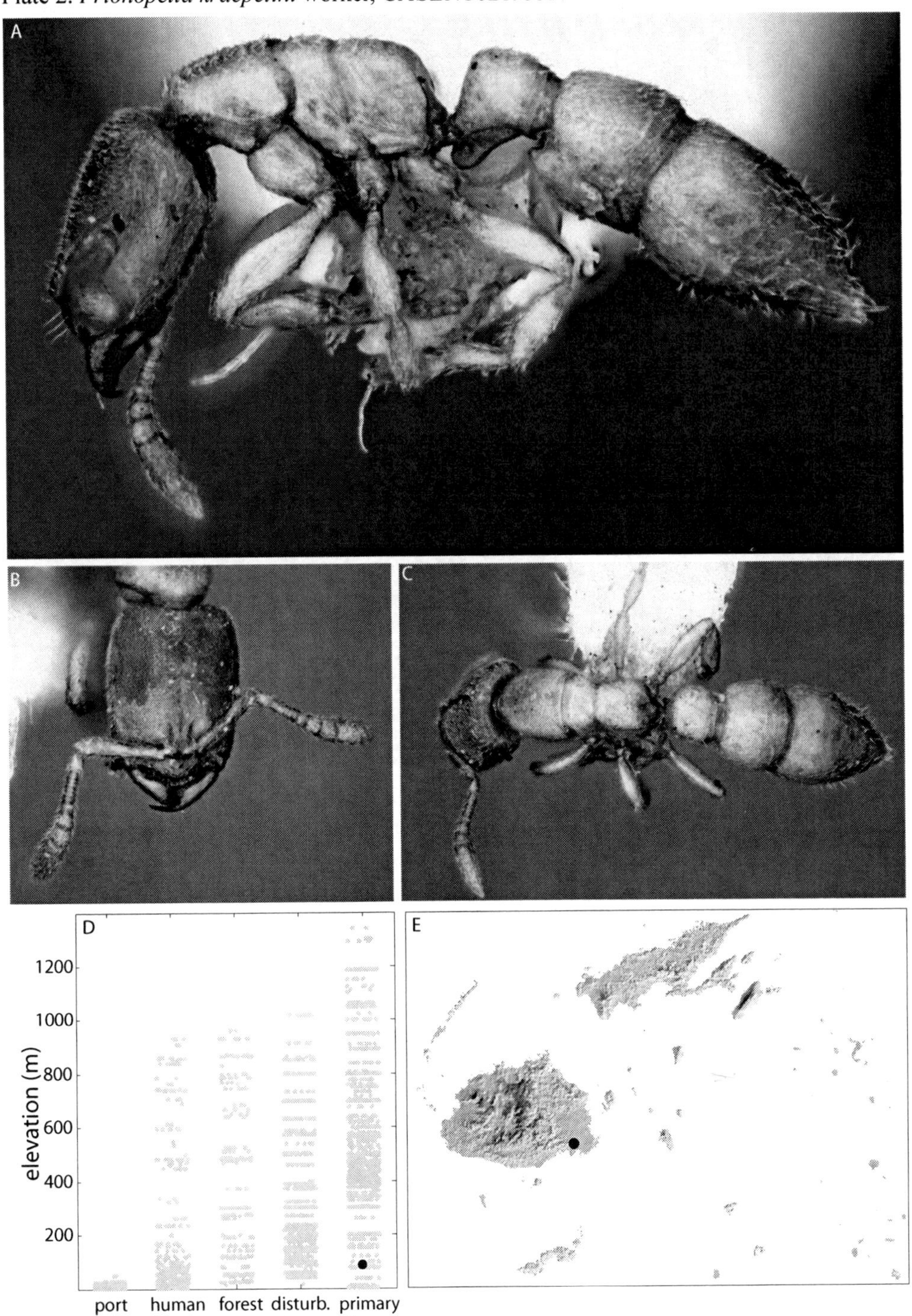

Plate 3. *Cerapachys cryptus* syntype worker, USNMENT00529135.

Plate 4. *Cerapachys fuscior* worker, CASENT0171152.

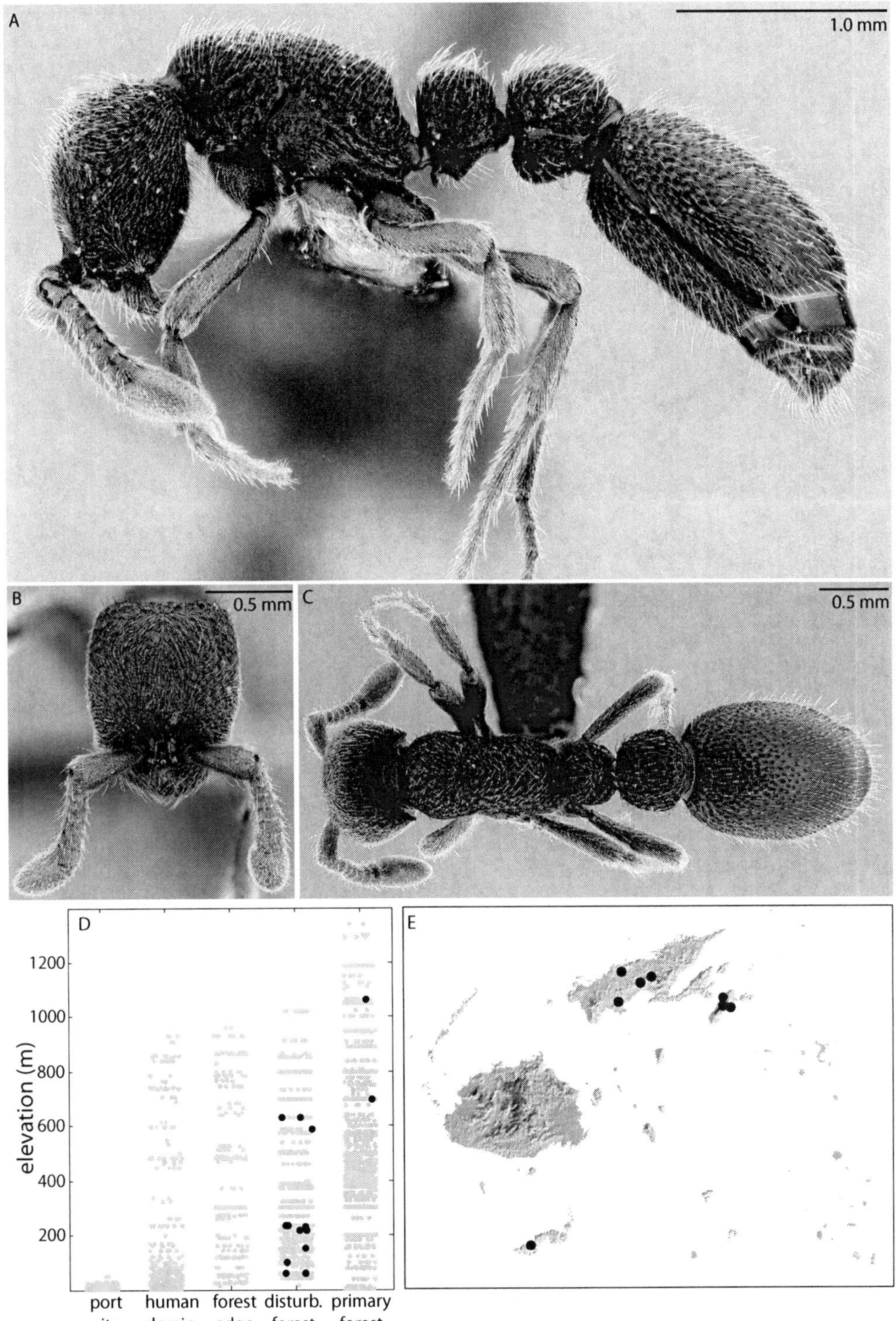

Plate 5. *Cerapachys* sp. FJ06 worker, CASENT0171145.

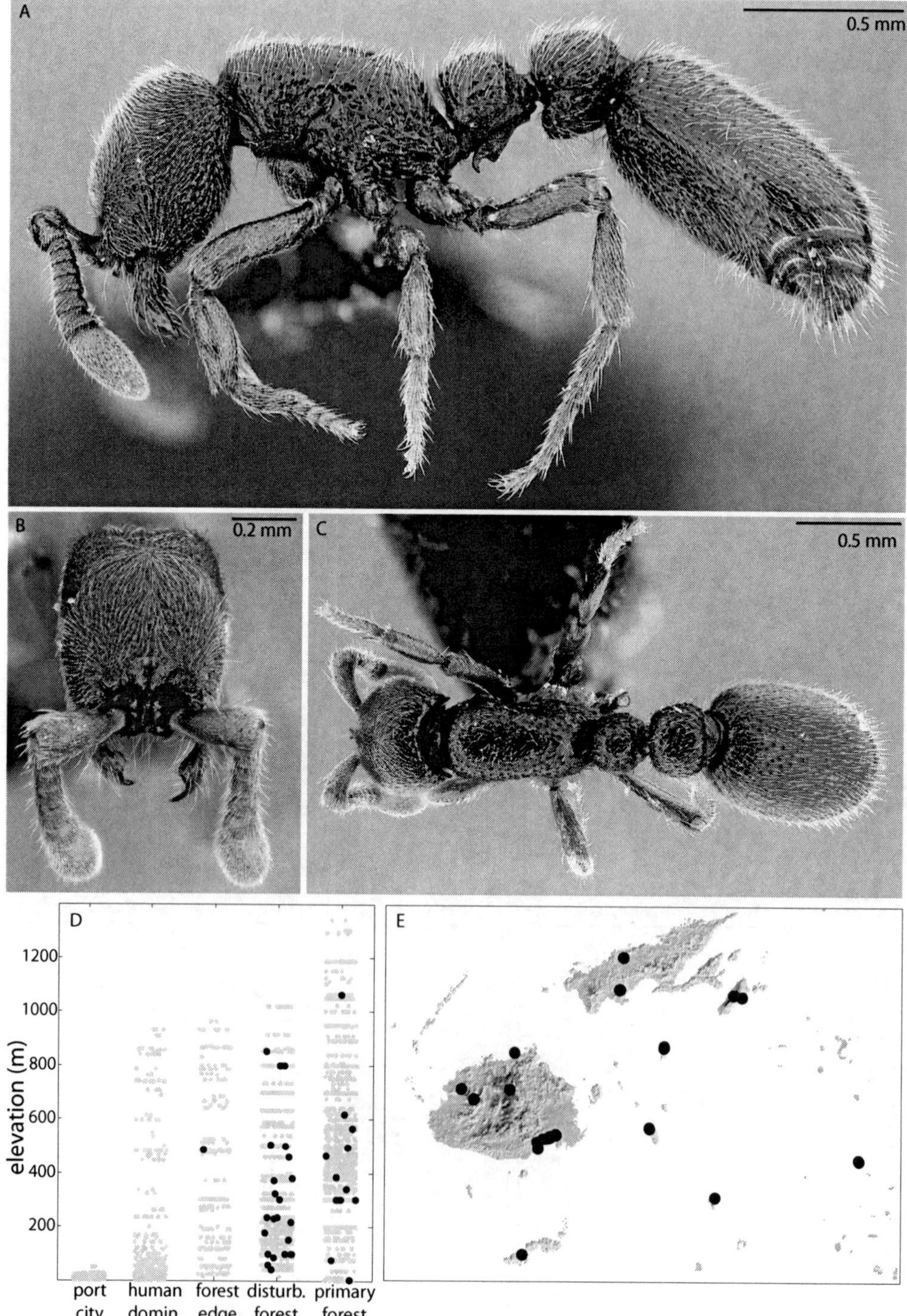

Plate 6. *Cerapachys lindrothi* worker, CASENT0171147.

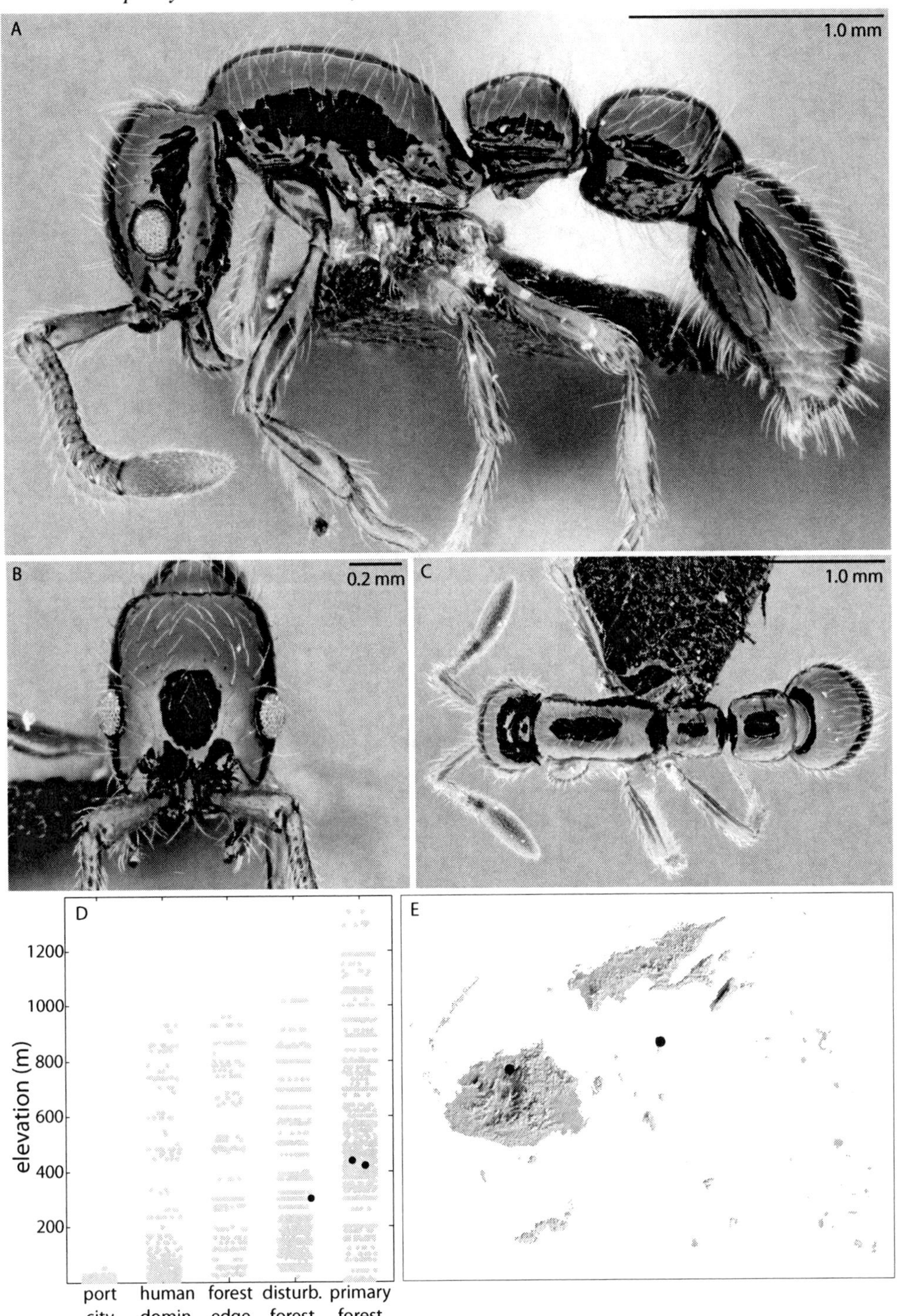

Plate 7. *Cerapachys zimmermani* worker, CASENT0175759.

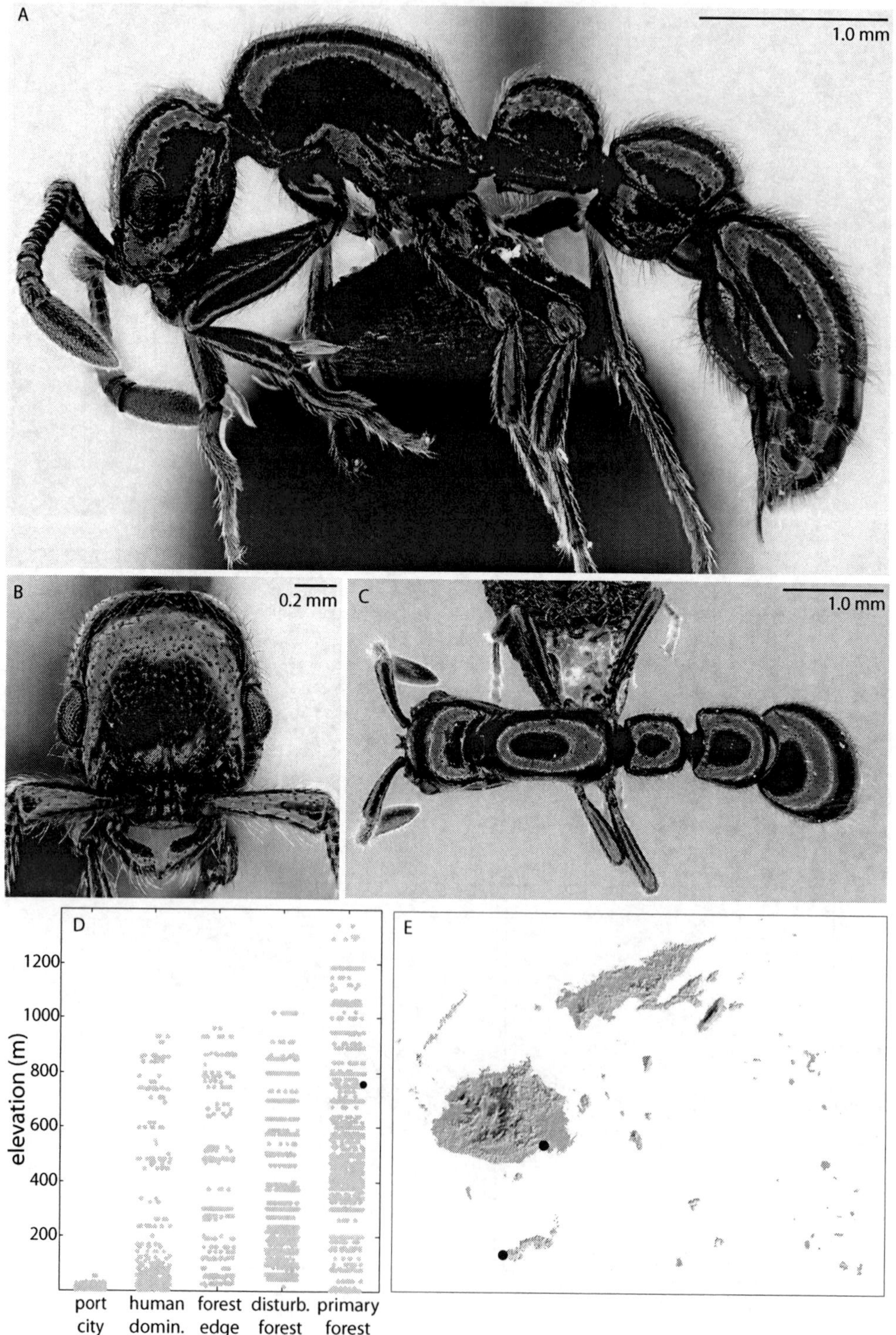

Plate 8. *Cerapachys* sp. FJ01 worker, CASENT0175817.

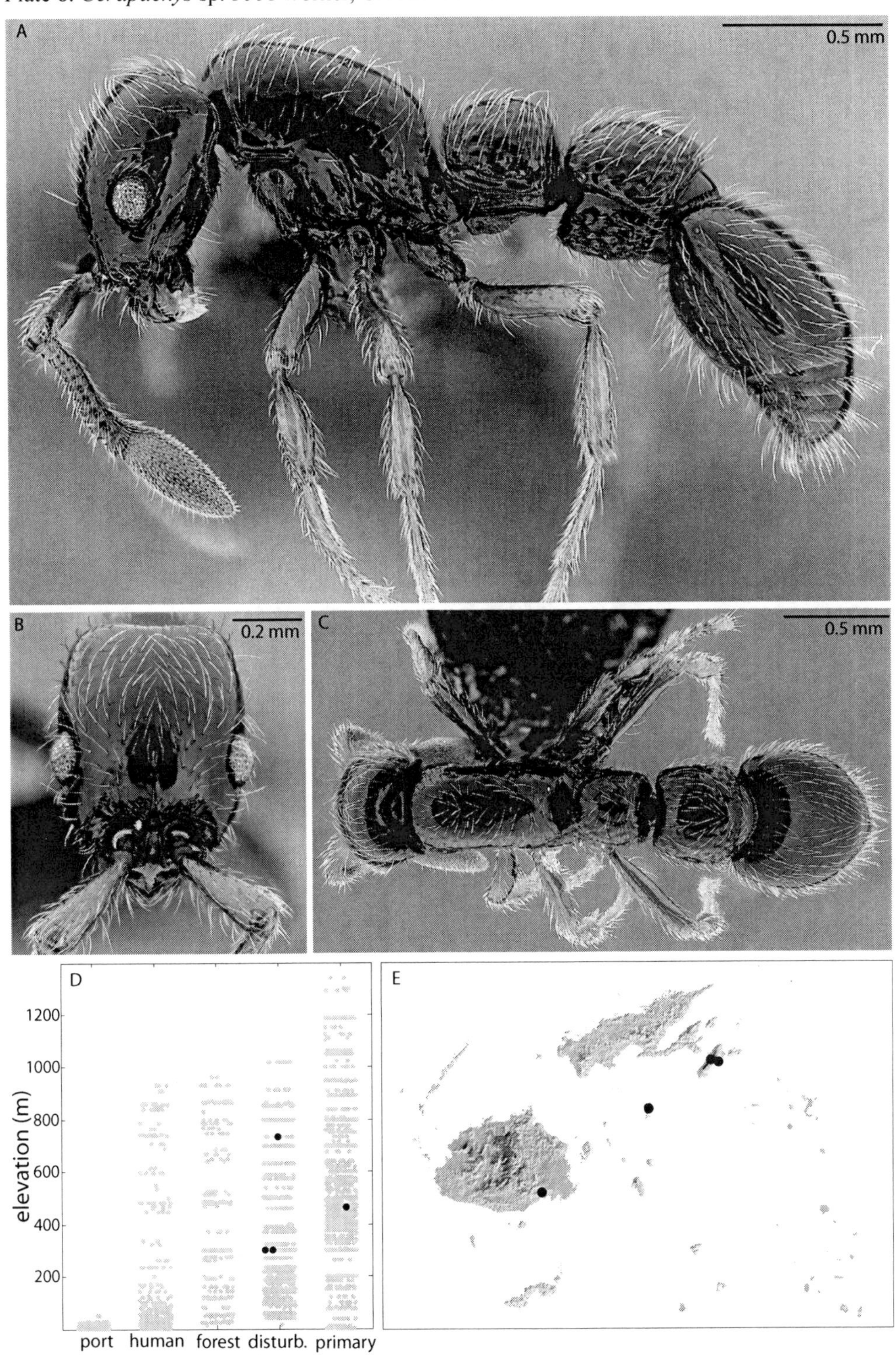

Plate 9. *Cerapachys majusculus* syntype worker, USNMENT00529628.

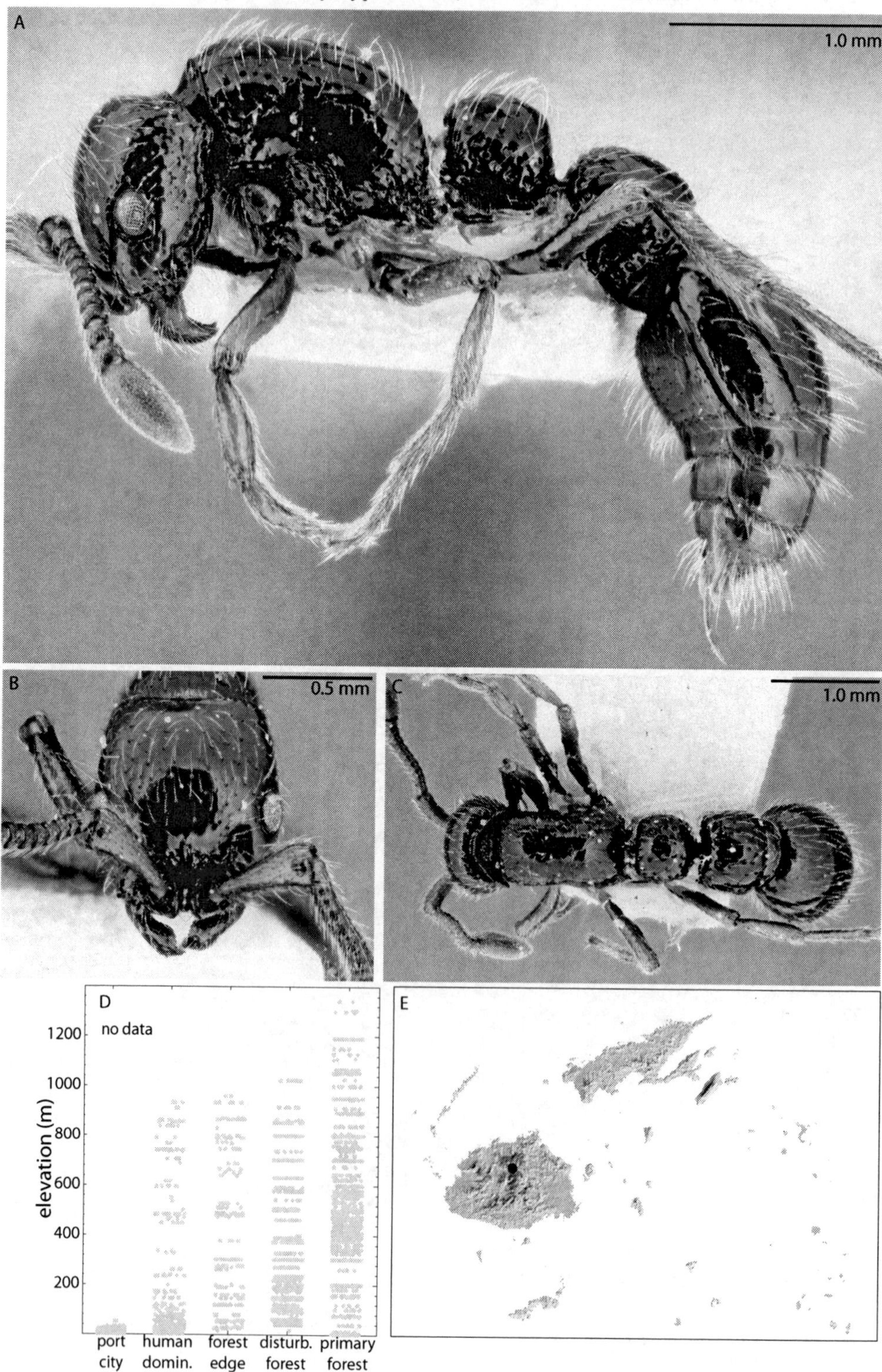

Plate 10. *Cerapachys sculpturatus* syntype worker, USNMENT00693041.

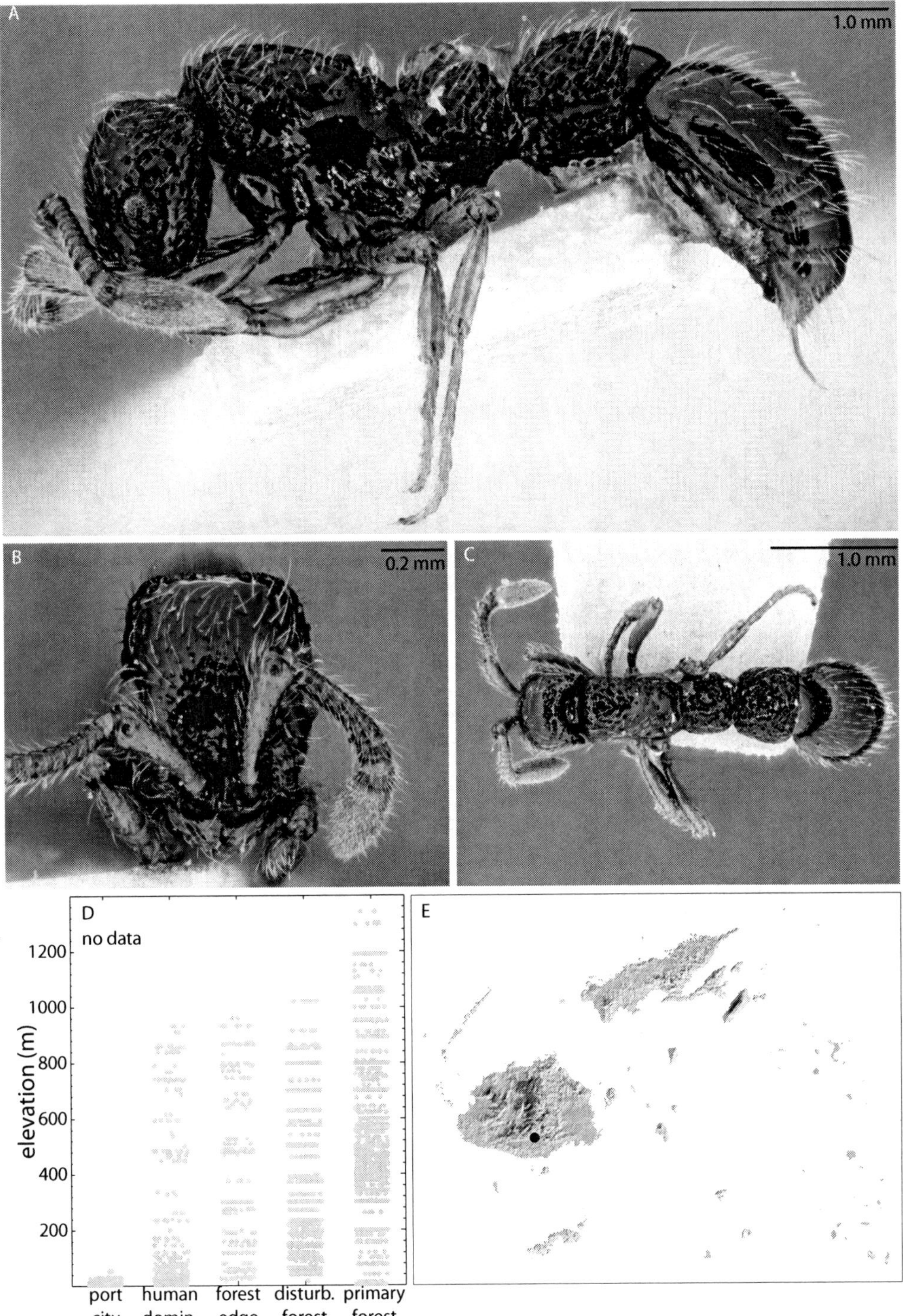

Plate 11. *Cerapachys vitiensis* worker, CASENT0175795.

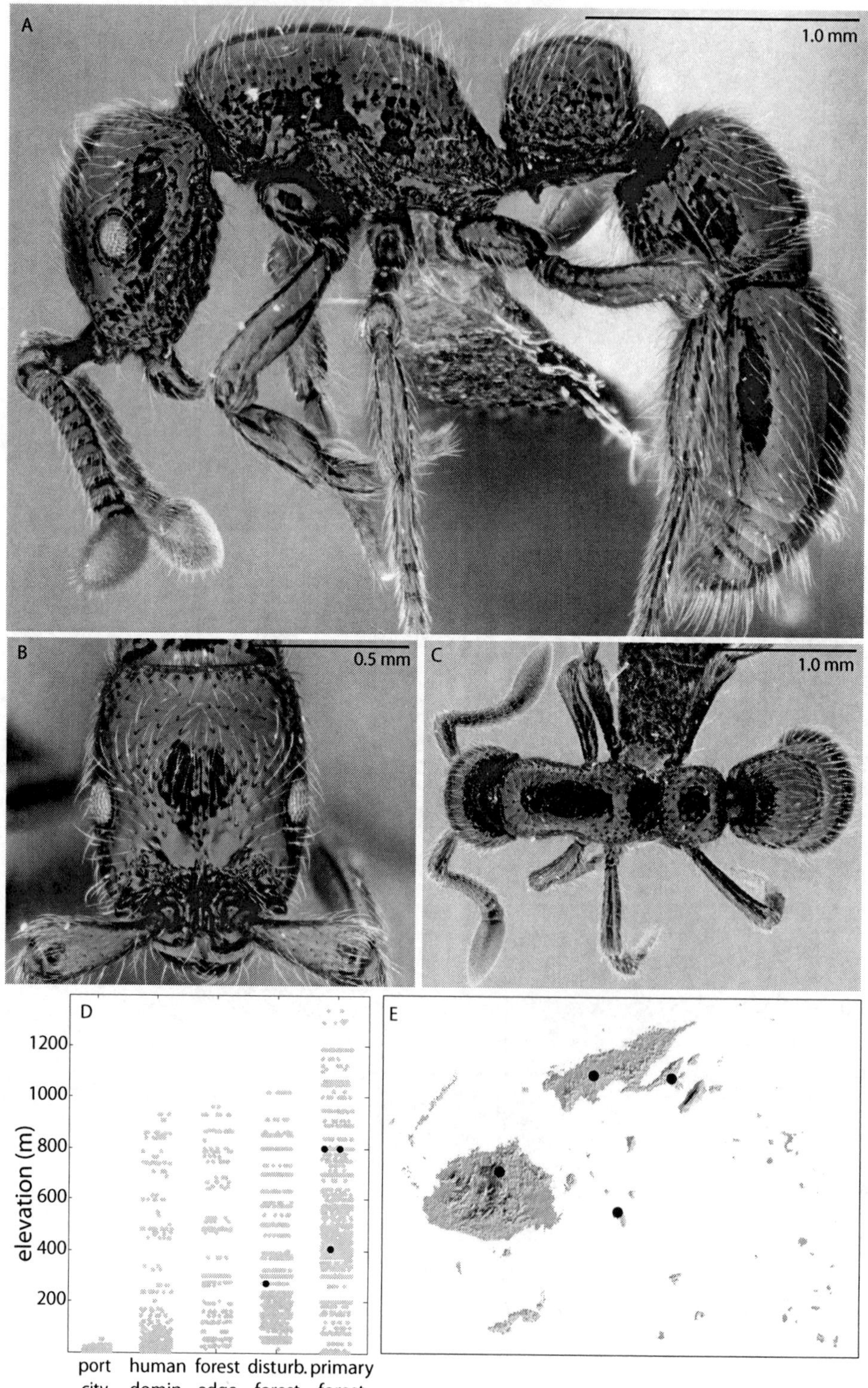

Plate 12. *Cerapachys* sp. FJ07 worker, CASENT0175790.

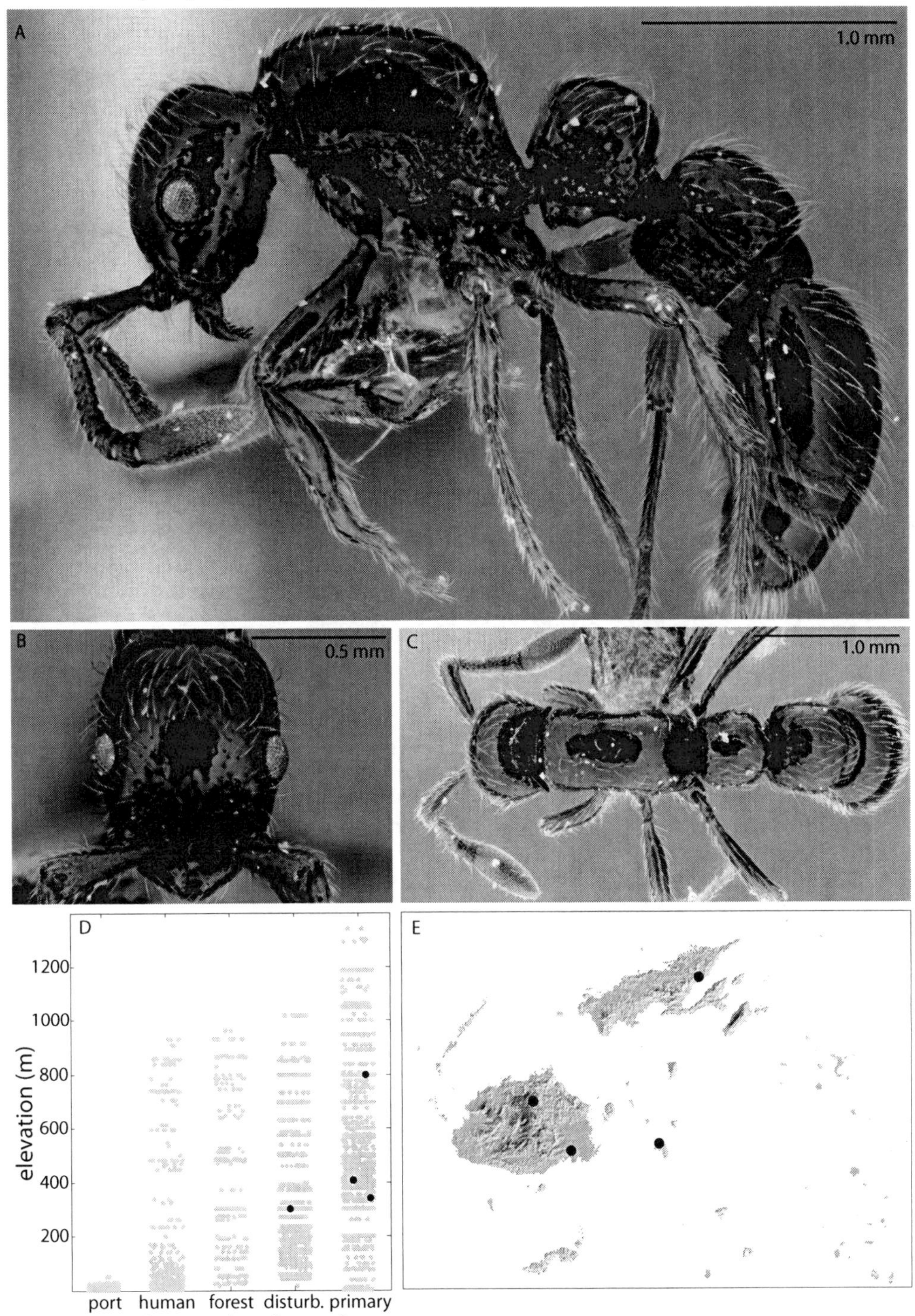

Plate 13. *Cerapachys* sp. FJ05 worker, CASENT0175808.

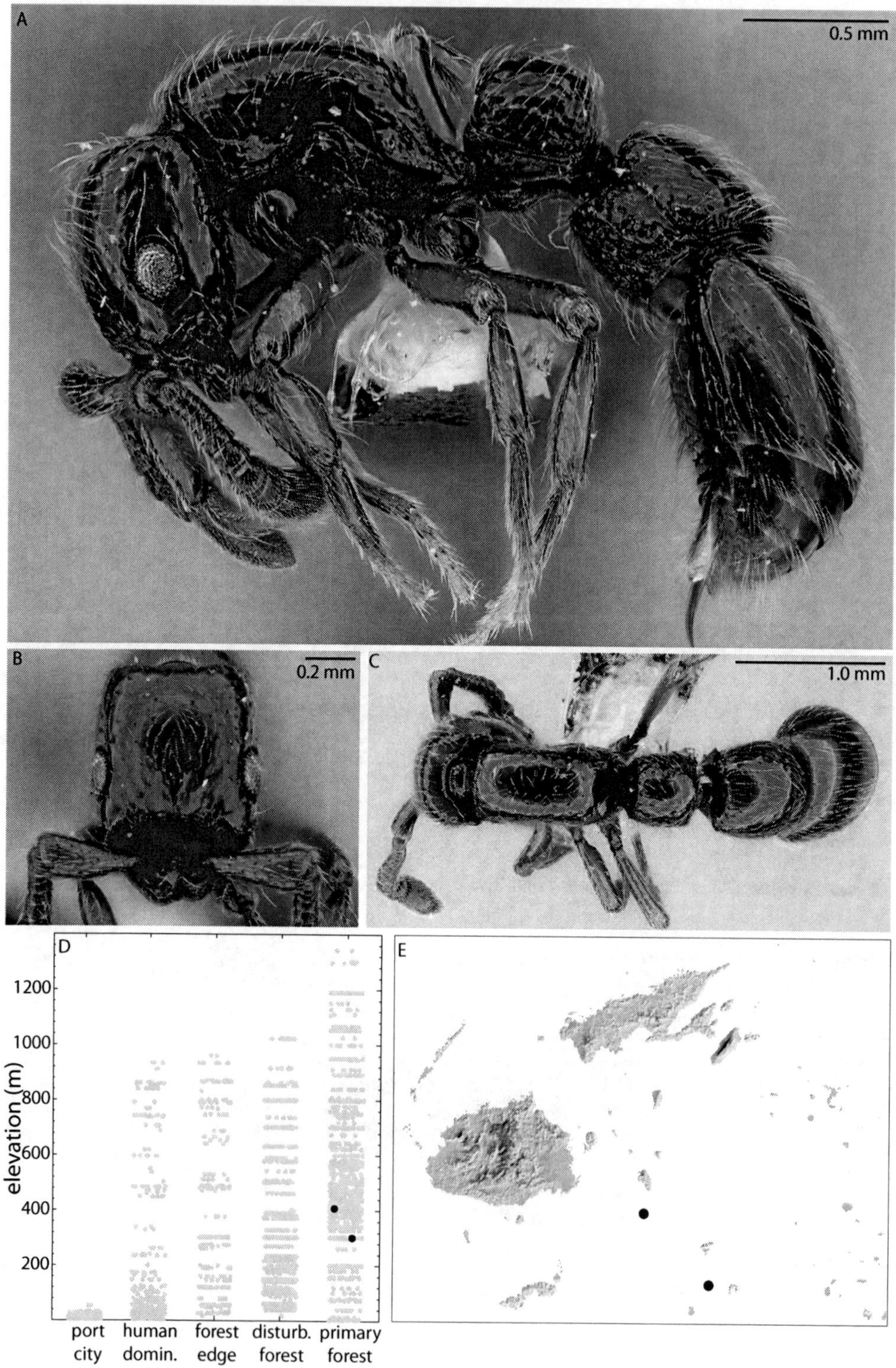

Plate 14. *Cerapachys* sp. FJ04 worker, CASENT0171150.

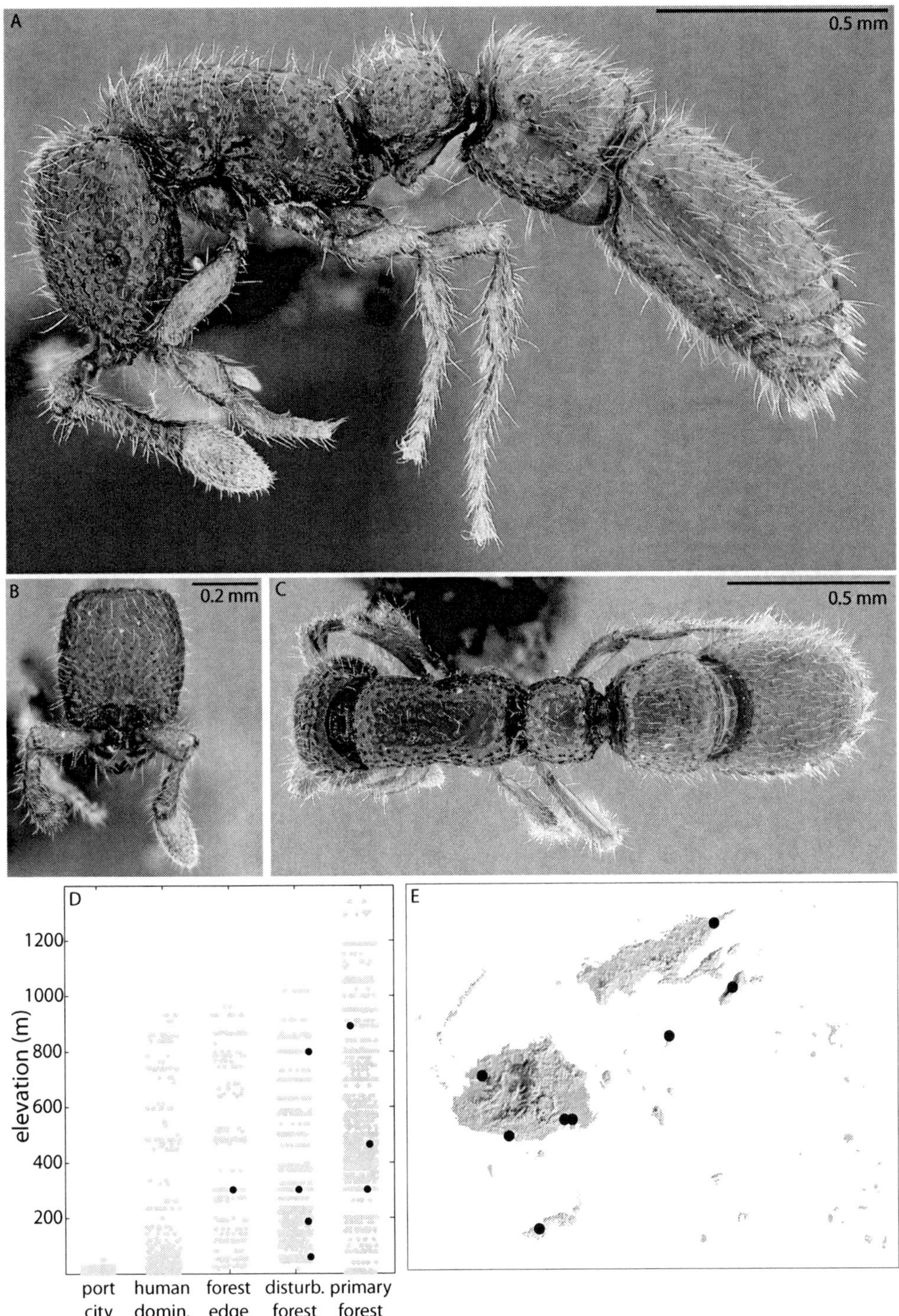

Plate 15. *Cerapachys* sp. FJ08 worker, CASENT0175805.

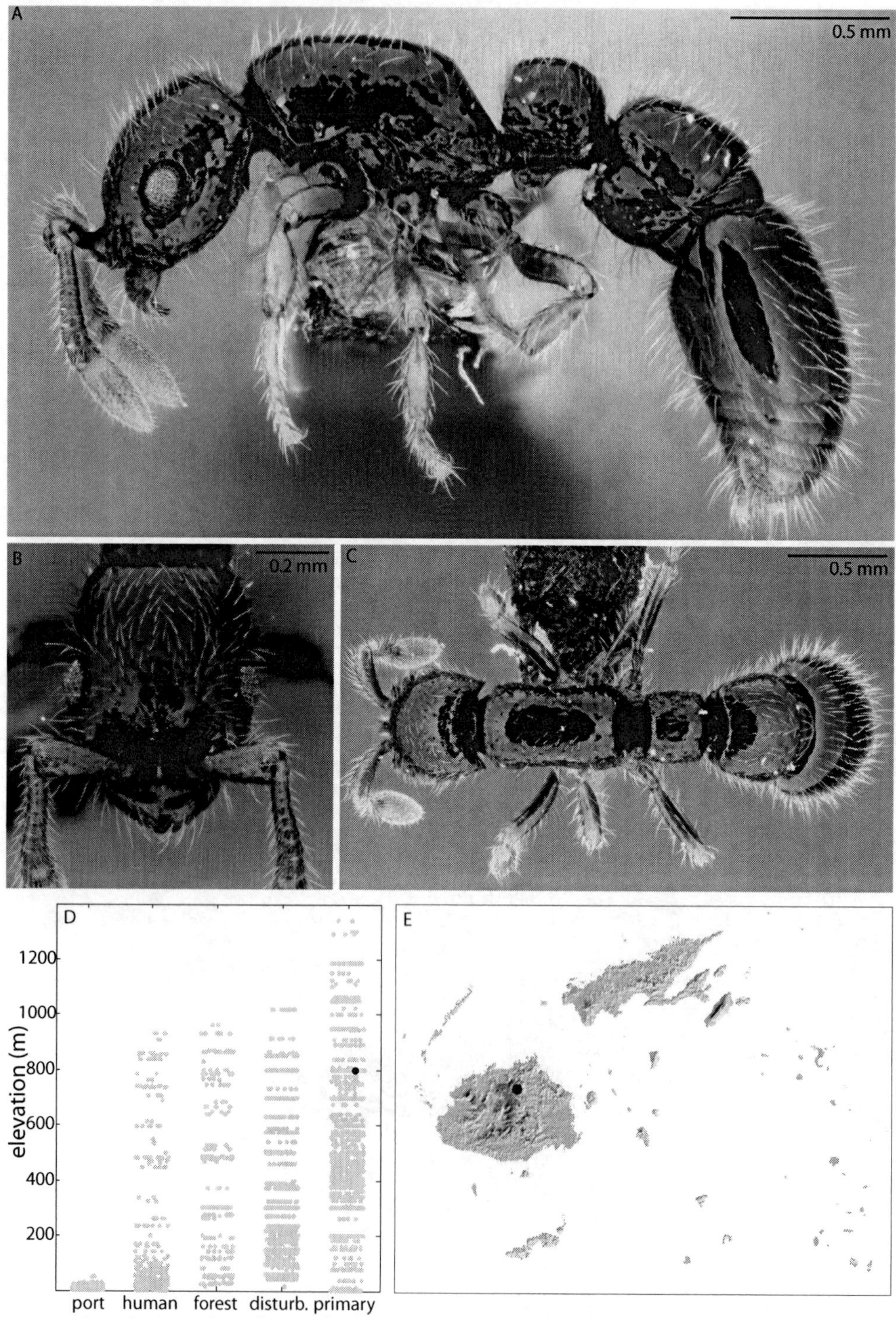

Plate 16. *Cerapachys* sp. FJ10 worker, CASENT0177223.

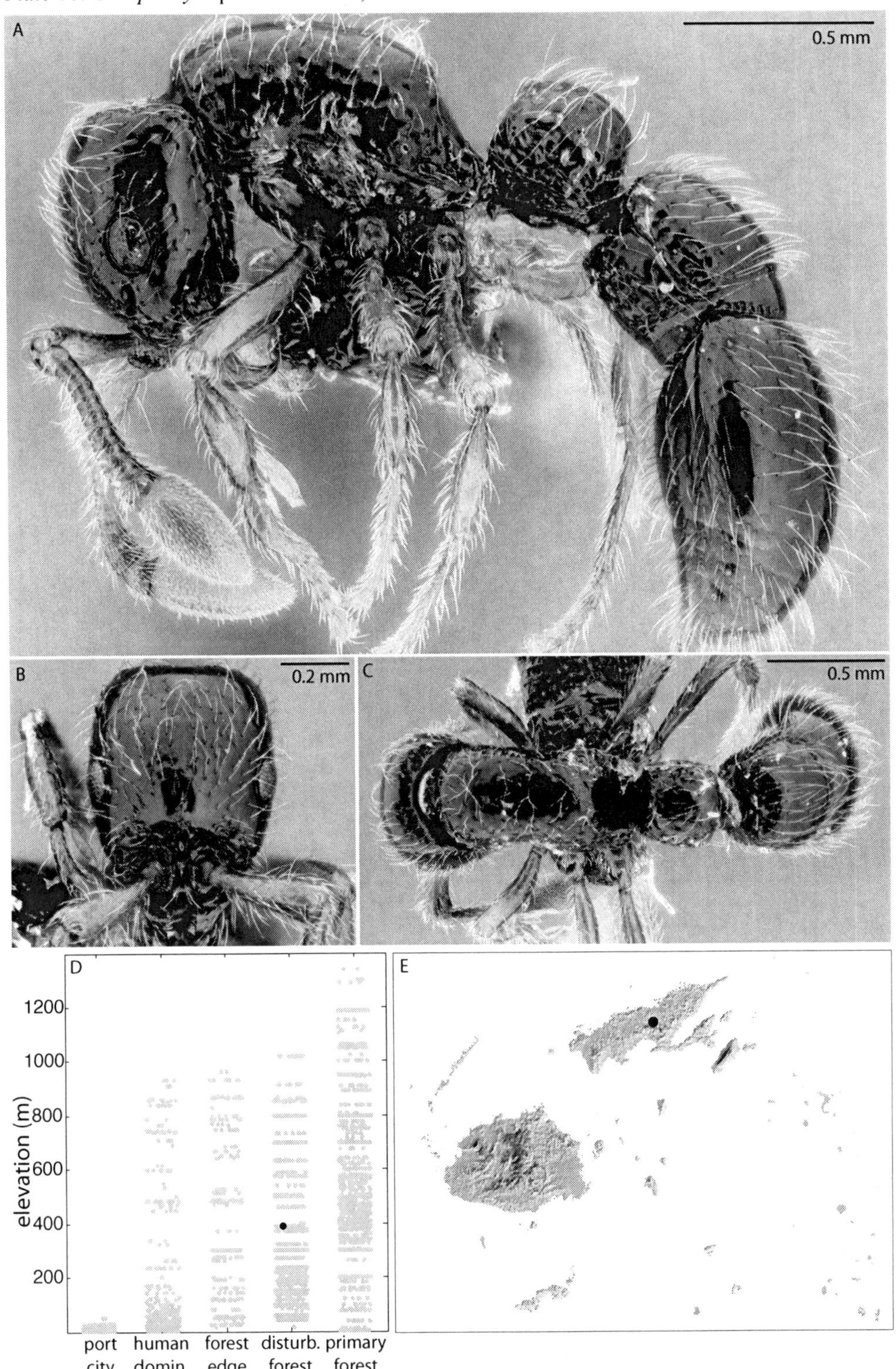

Plate 17. *Iridomyrmex anceps* worker, CASENT0171061.

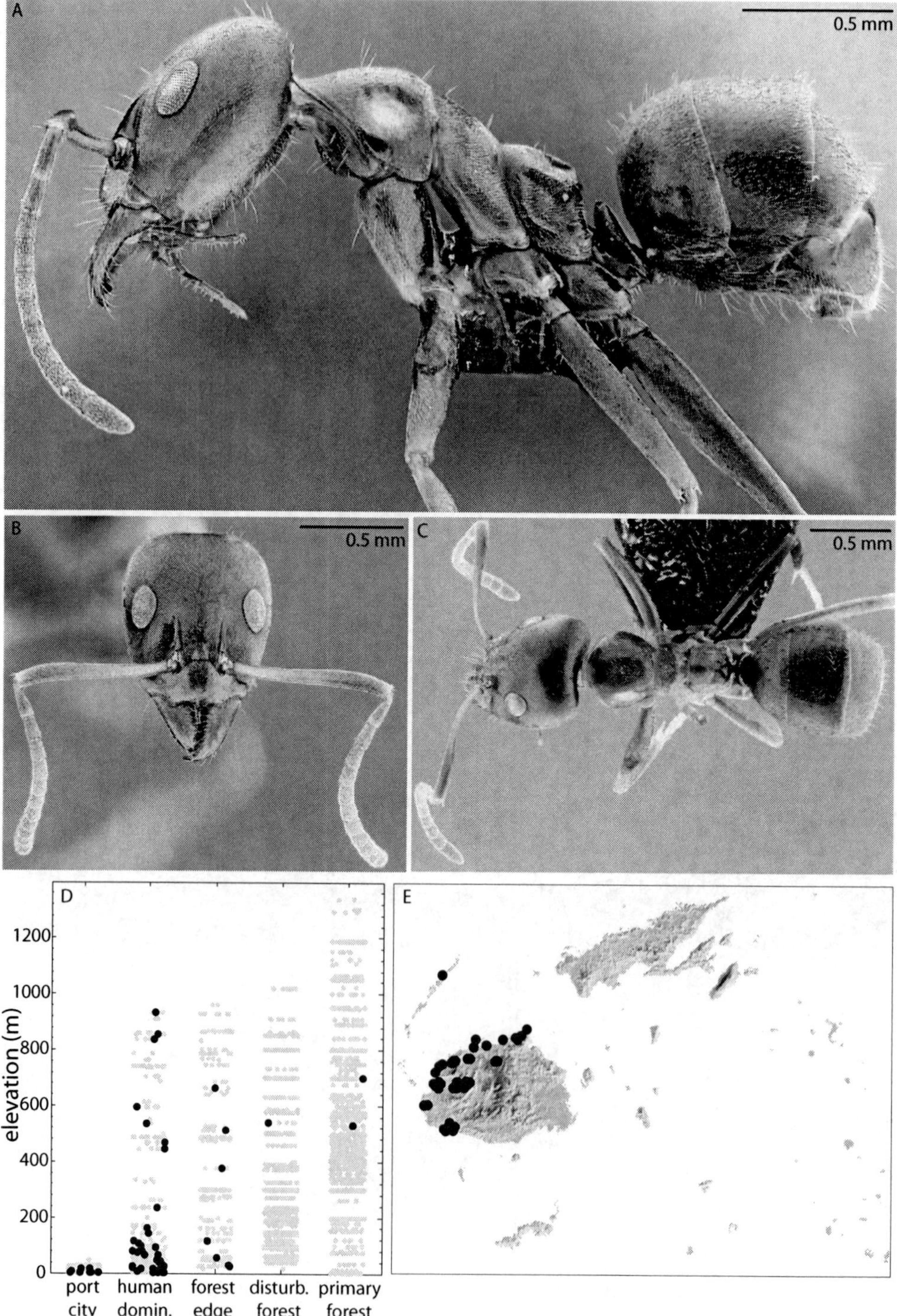

Plate 18. *Ochetellus sororis* worker, CASENT0171060.

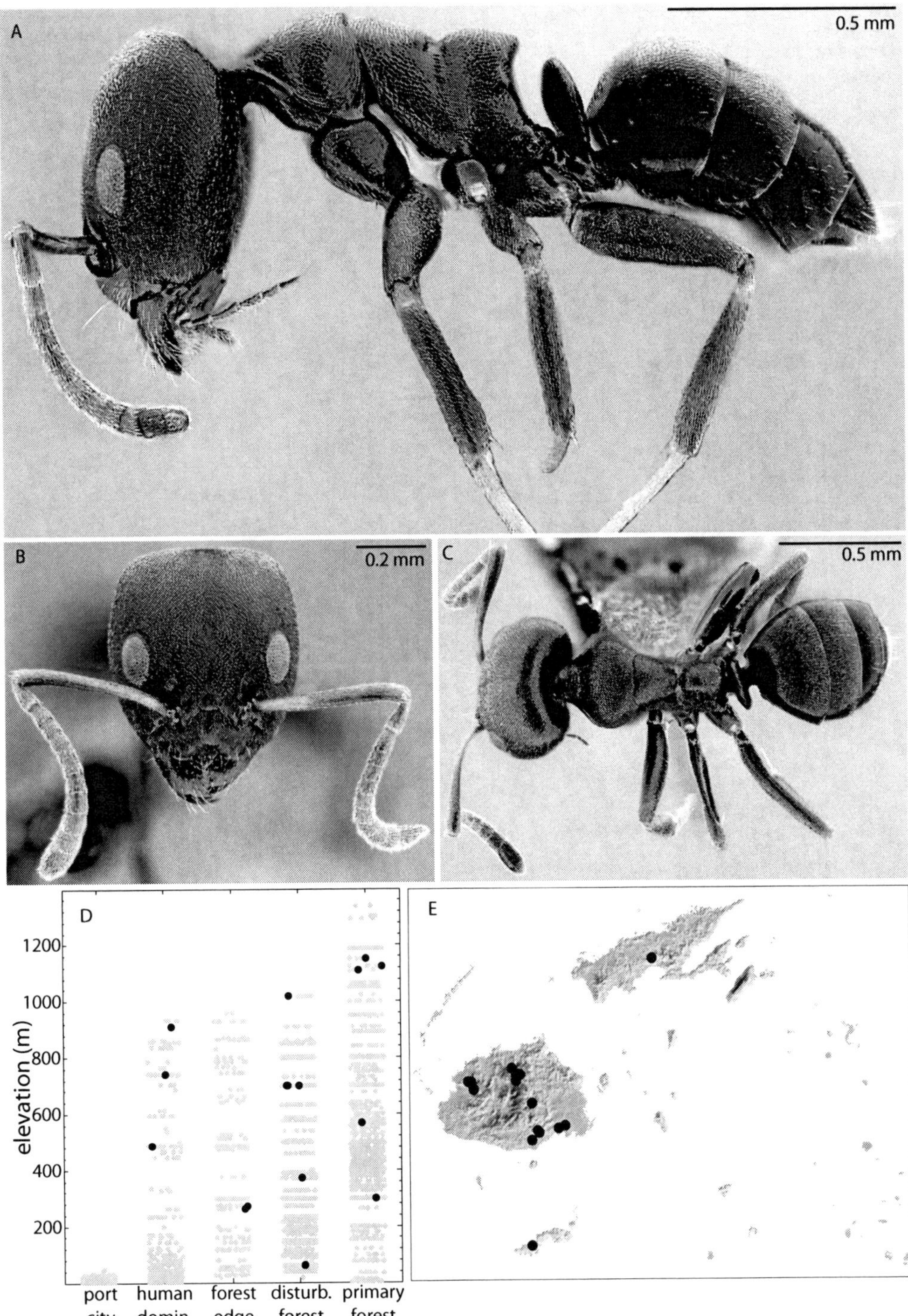

Plate 19. *Philidris nagasau* worker, CASENT0171058.

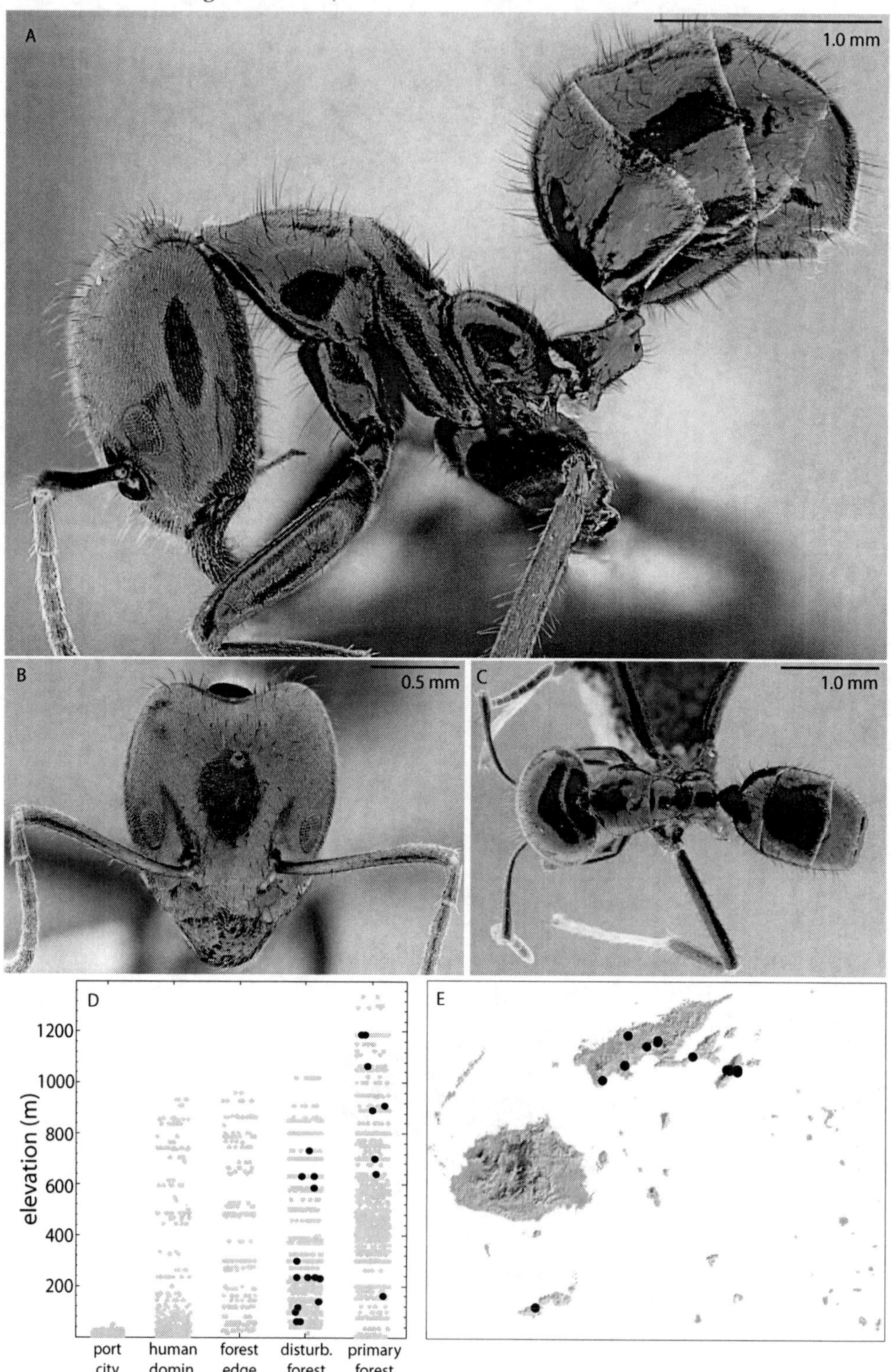

Plate 20. *Tapinoma melanocephalum* worker, CASENT0171078.

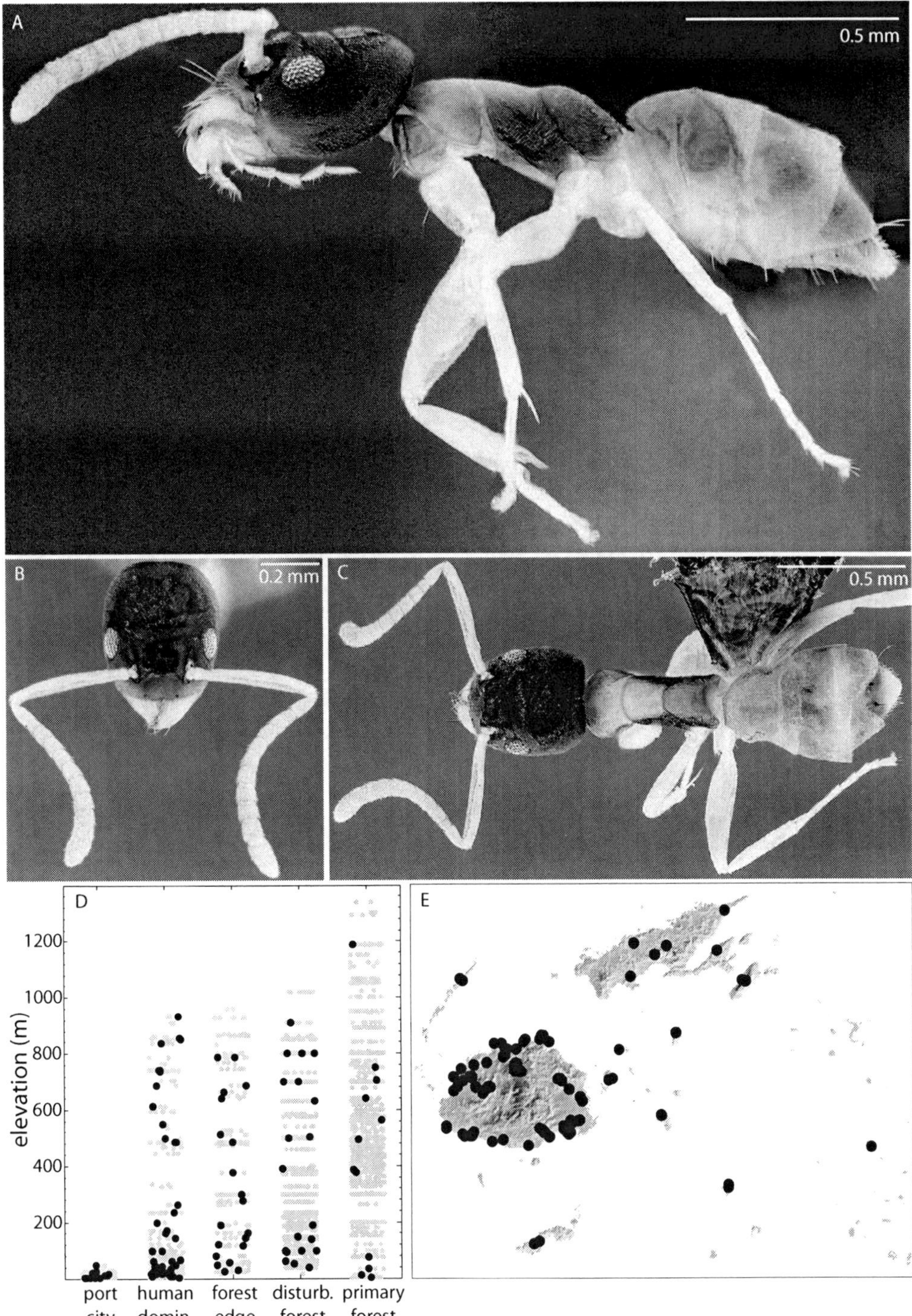

Plate 21. *Tapinoma minutum* worker, CASENT0177146.

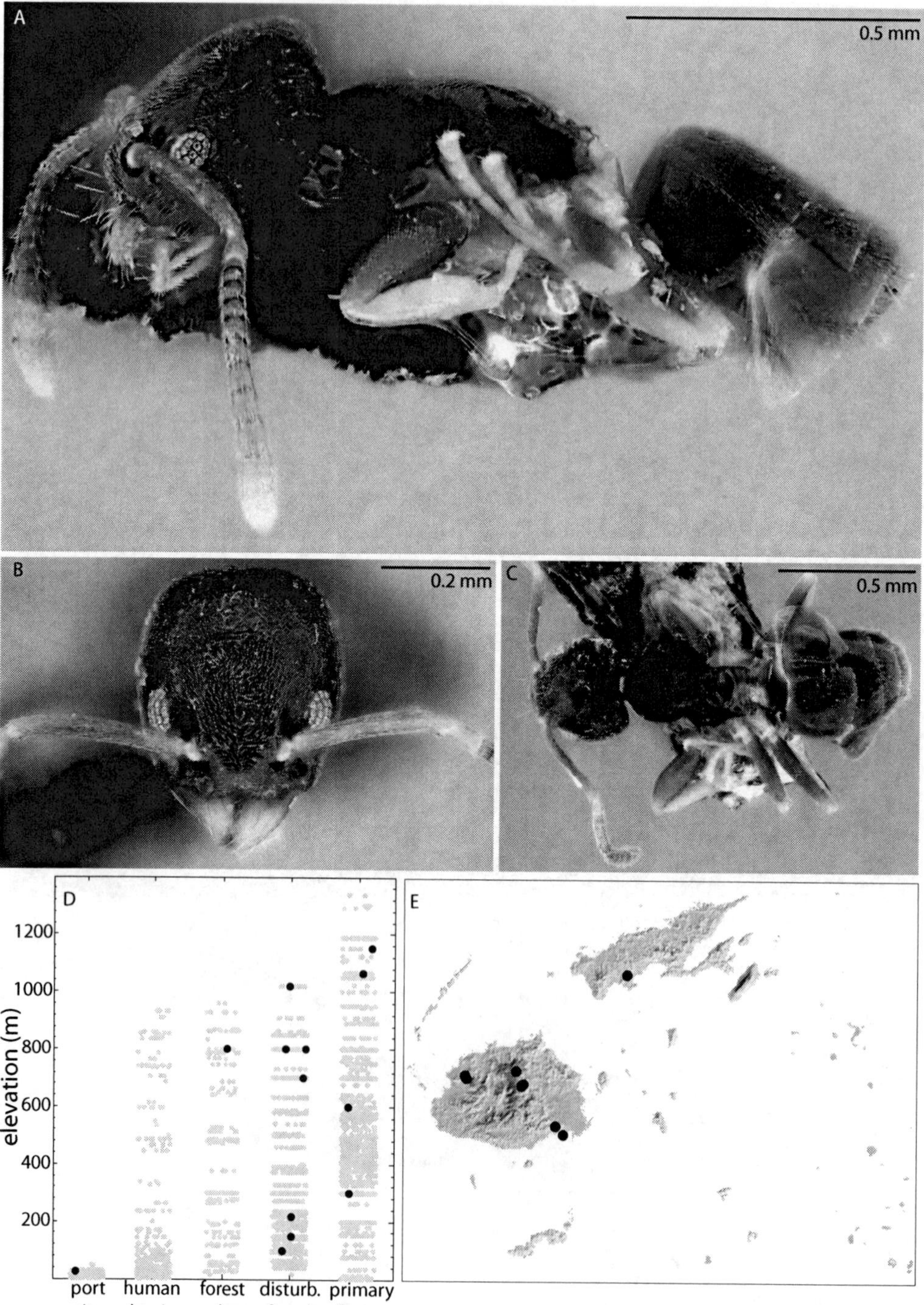

Plate 22. *Tapinoma* sp. FJ01 worker, CASENT0177206.

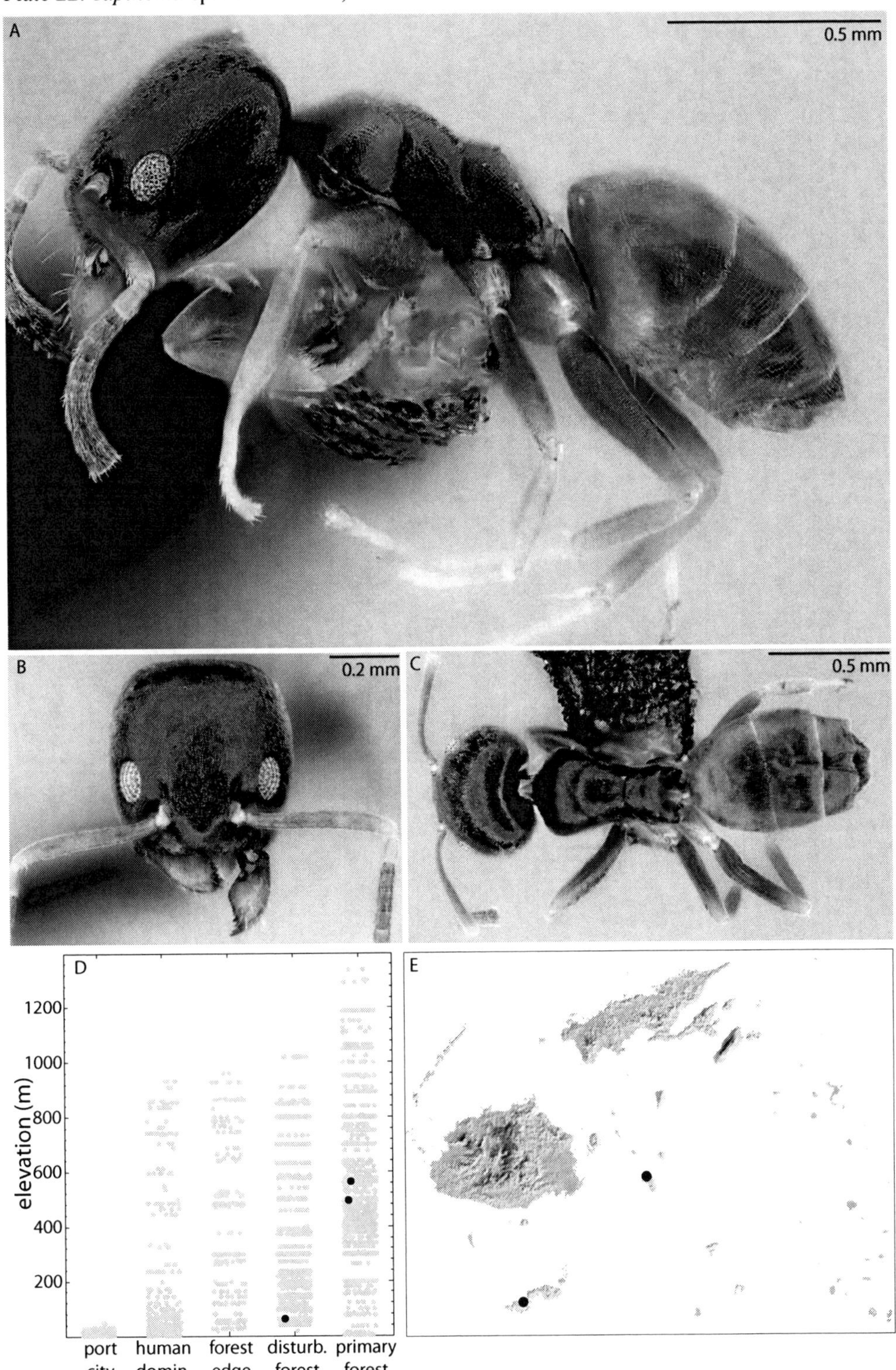

Plate 23. *Tapinoma* sp. FJ02 worker, CASENT0177104.

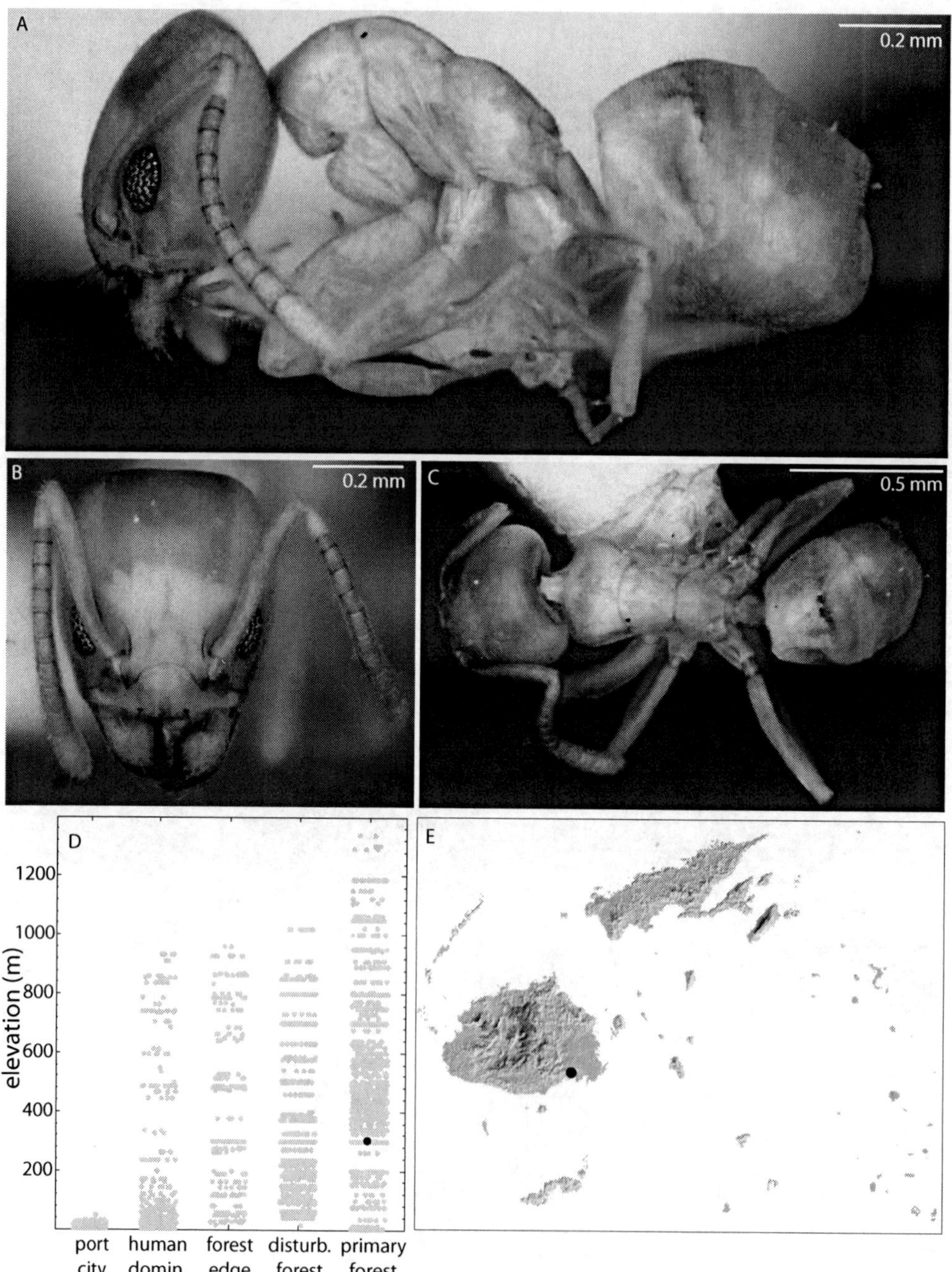

Plate 24. *Technomyrmex vitiensis* worker, CASENT0171057.

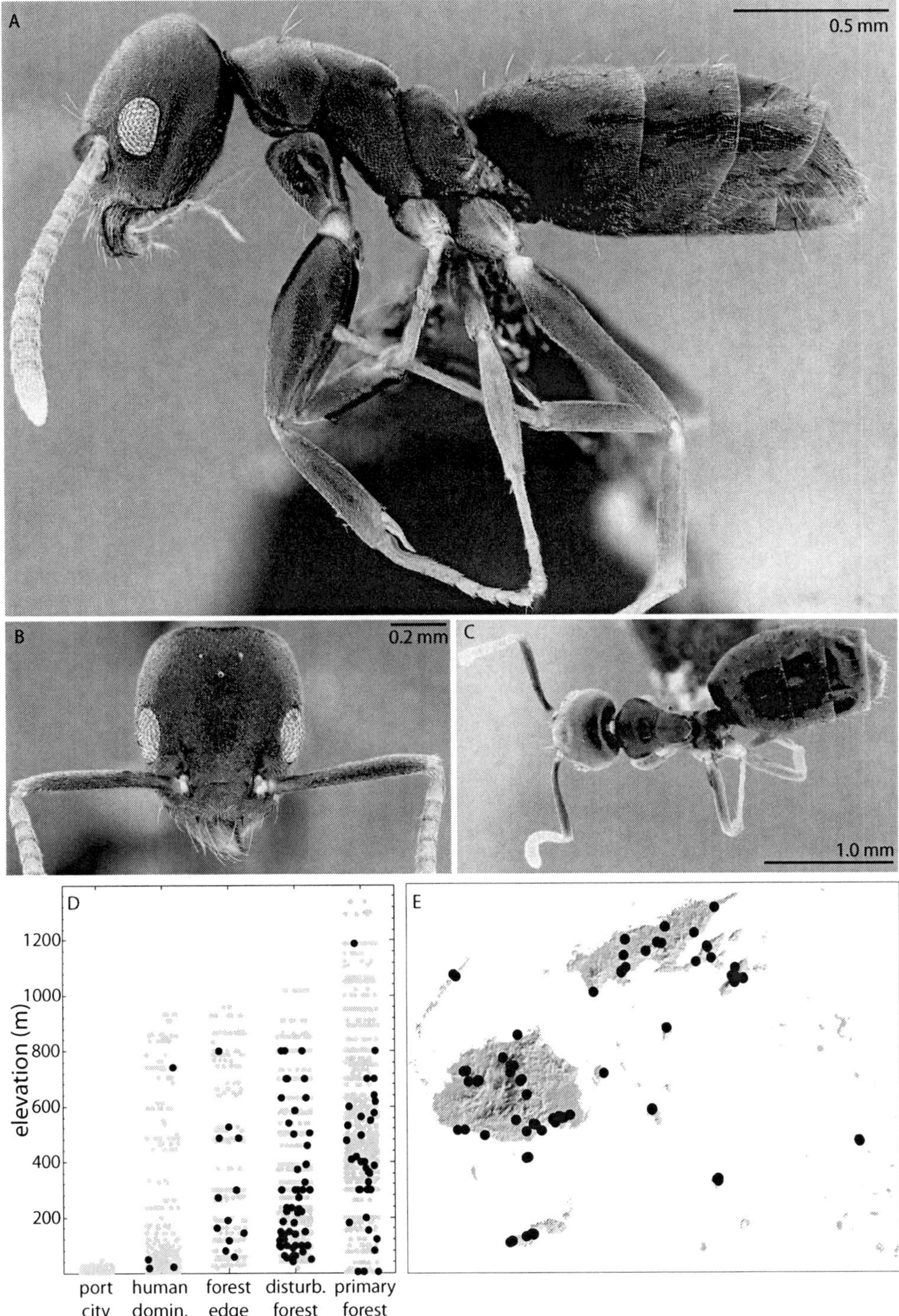

Plate 25. *Gnamptogenys aterrima* worker, CASENT0171062.

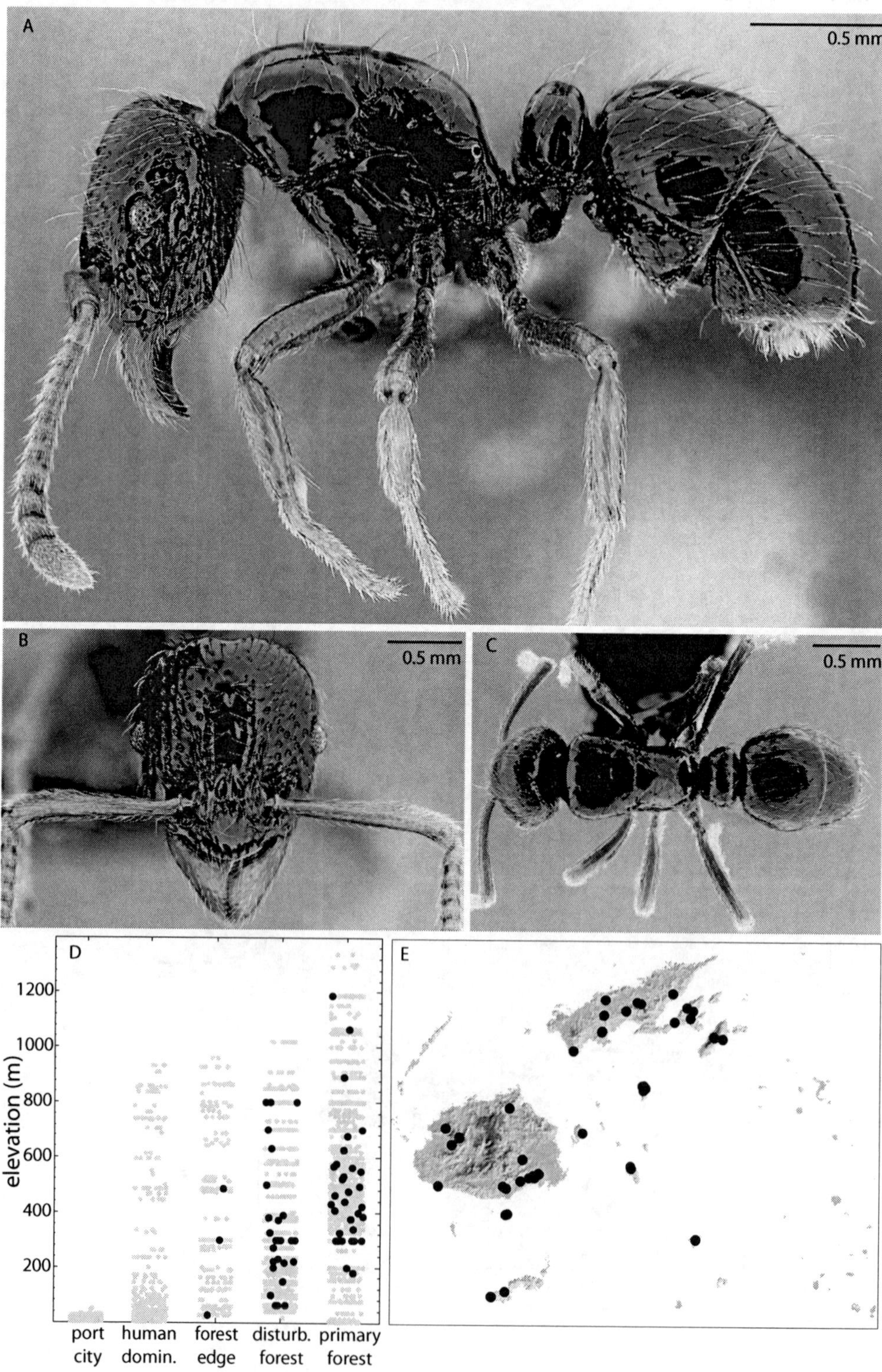

Plate 26. *Acropyga lauta* worker, CASENT0171066.

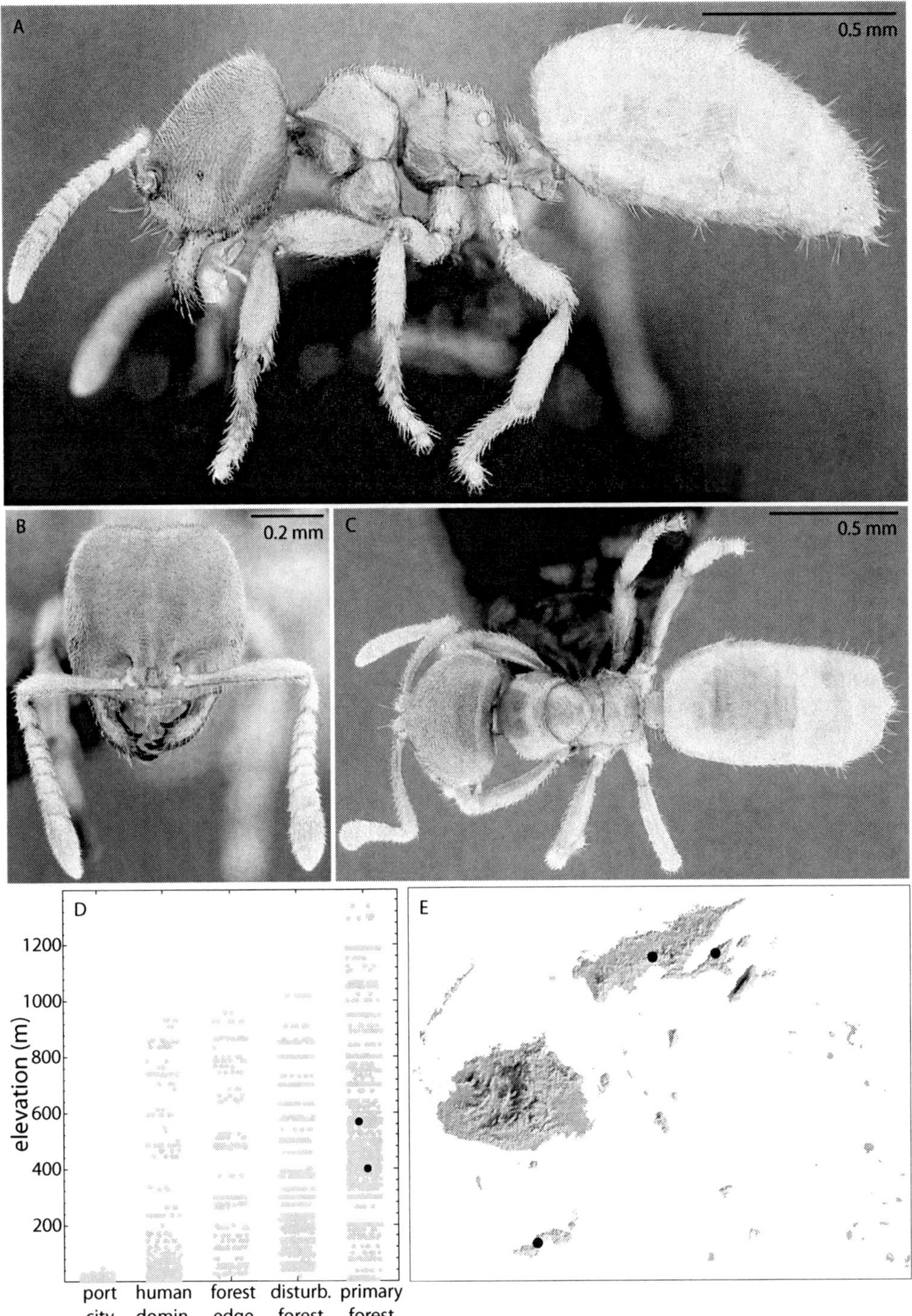

Plate 27. *Acropyga* sp. FJ02 worker, CASENT0127488.

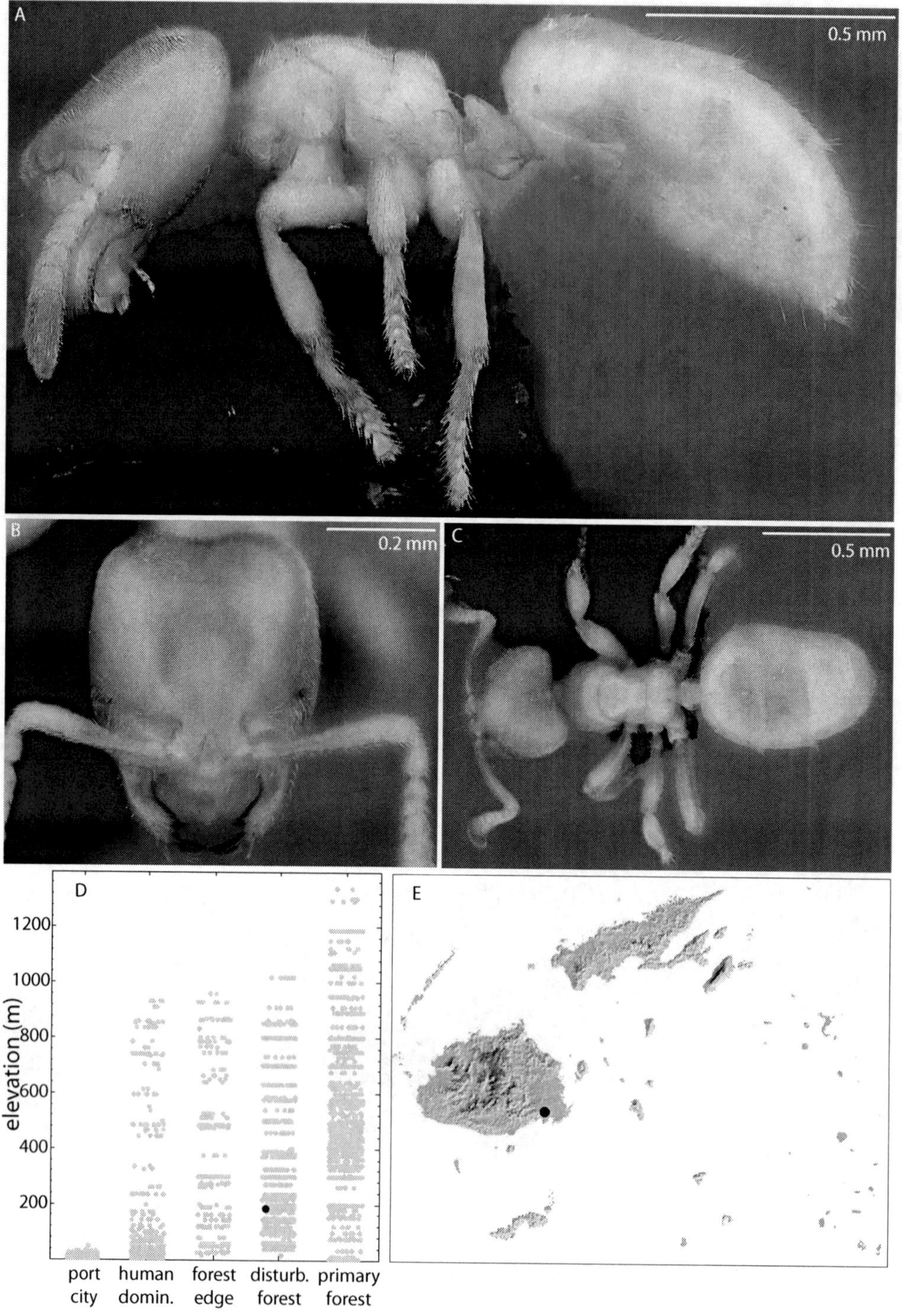

Plate 28. *Anoplolepis gracilipes* worker, CASENT0171031.

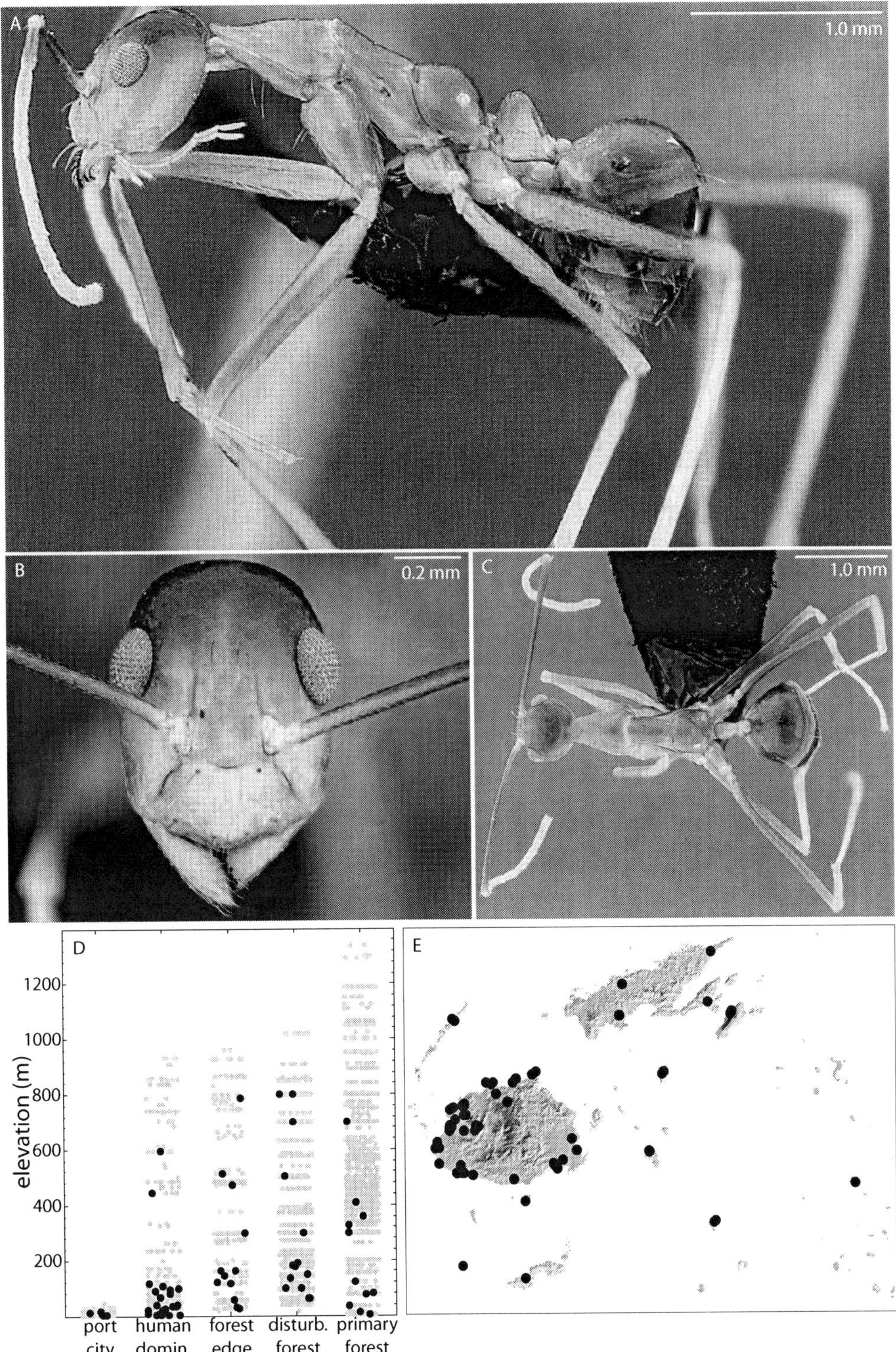

Plate 29. *Camponotus chloroticus* (minor, CASENT0171139; major, CASENT0171140).

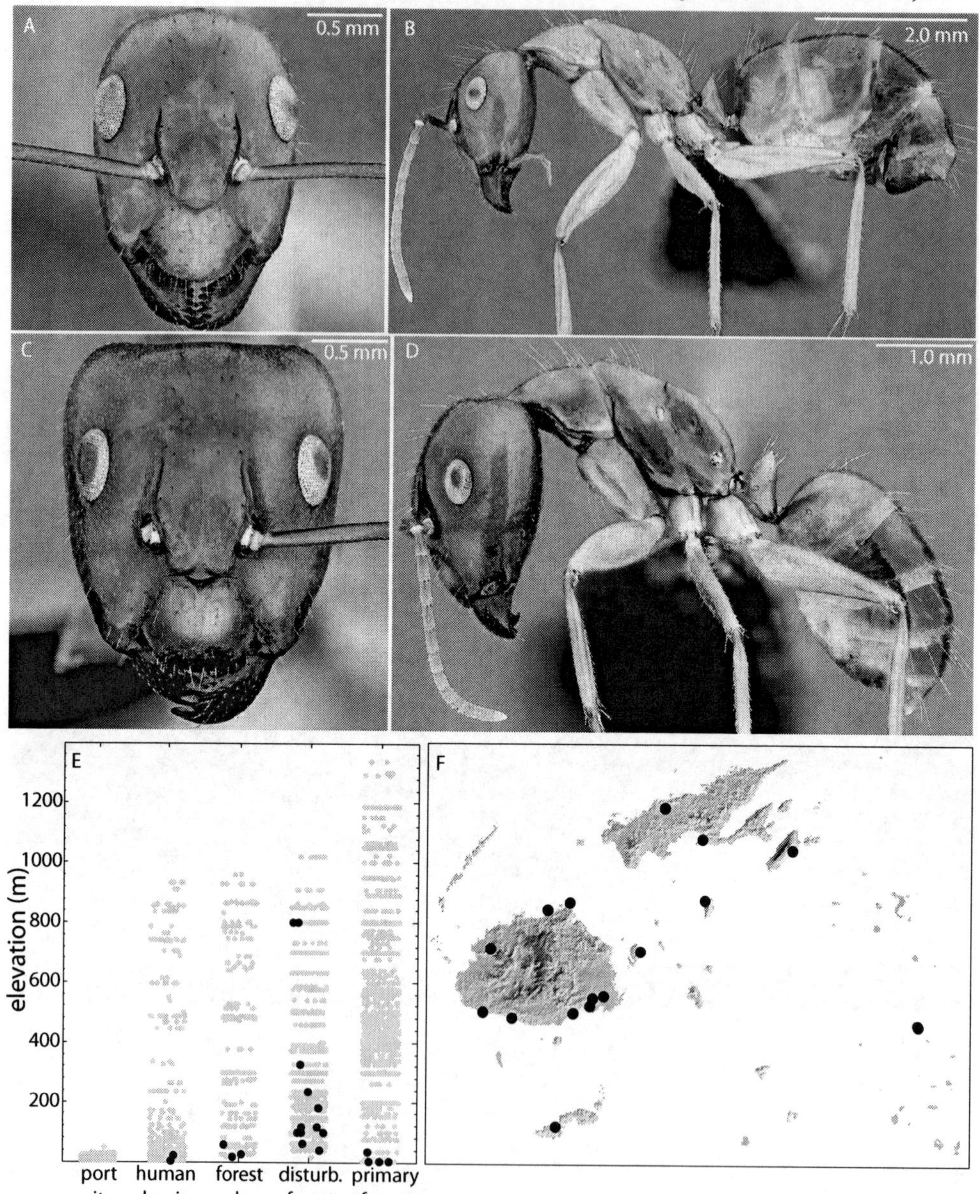

Plate 30. *Camponotus polynesicus* (minor, CASENT0187250; major, CASENT0187263).

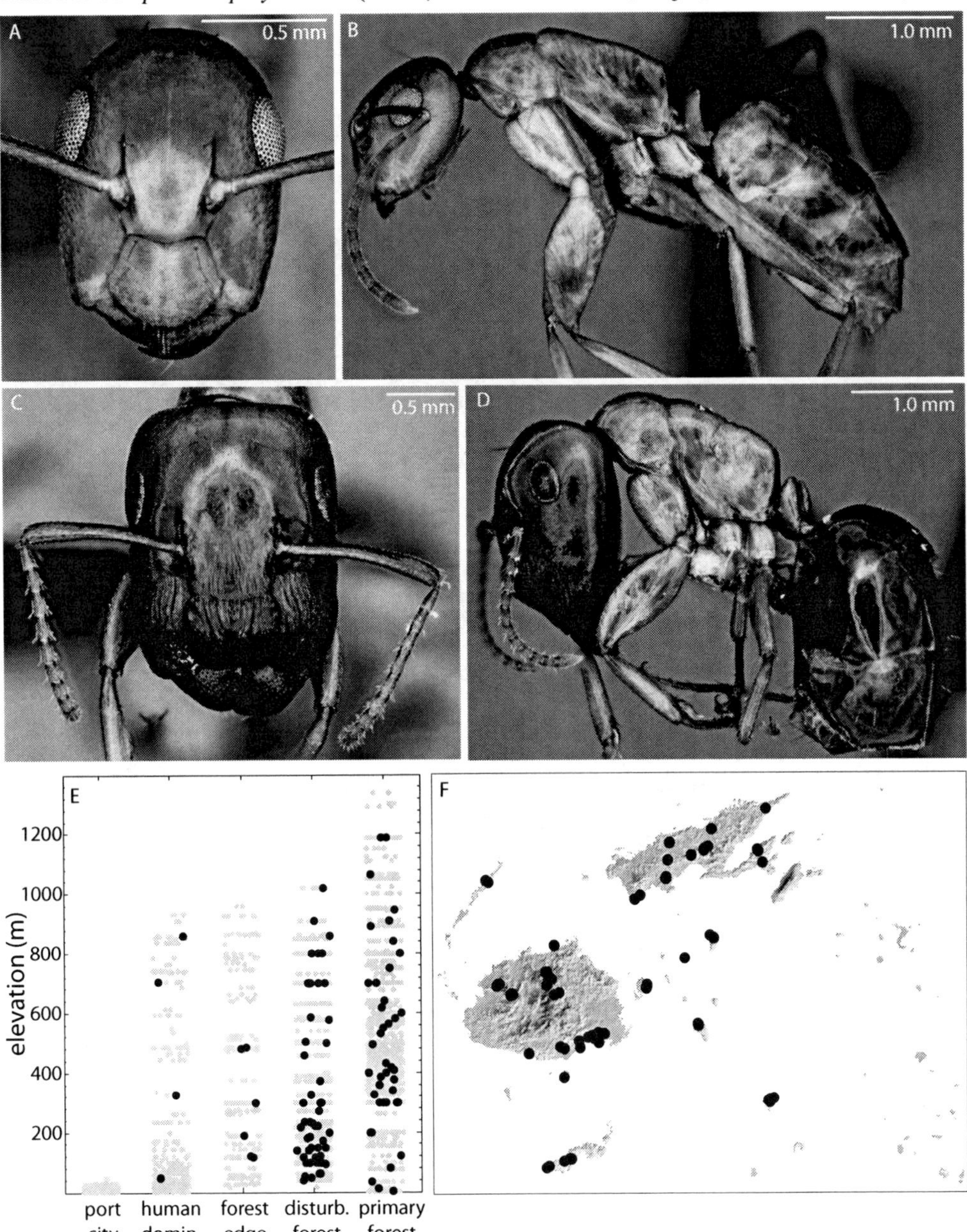

Plate 31. *Camponotus vitiensis* worker, CASENT0186859.

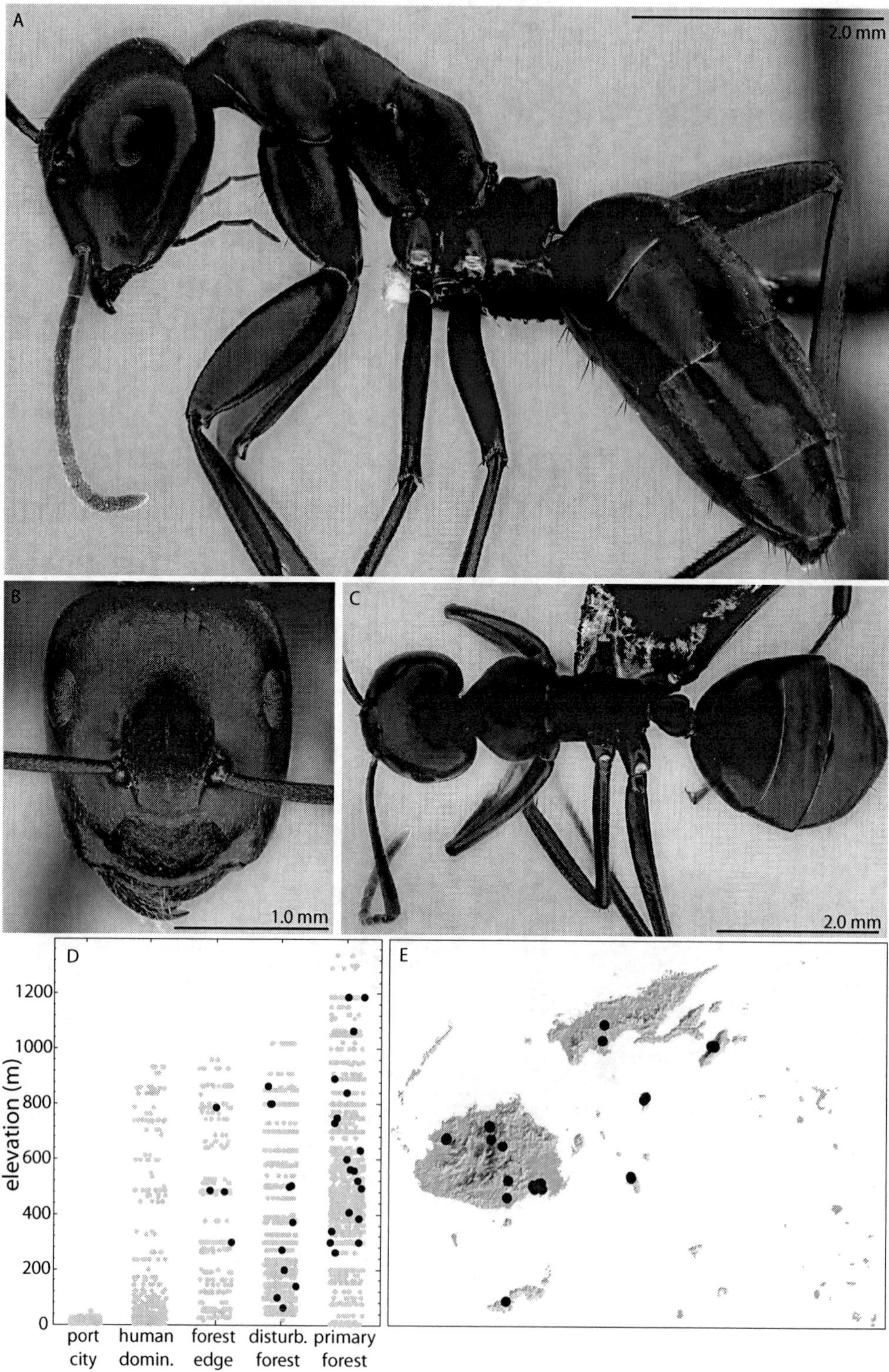

Plate 32. *Camponotus* sp. FJ04 worker, CASENT0177727.

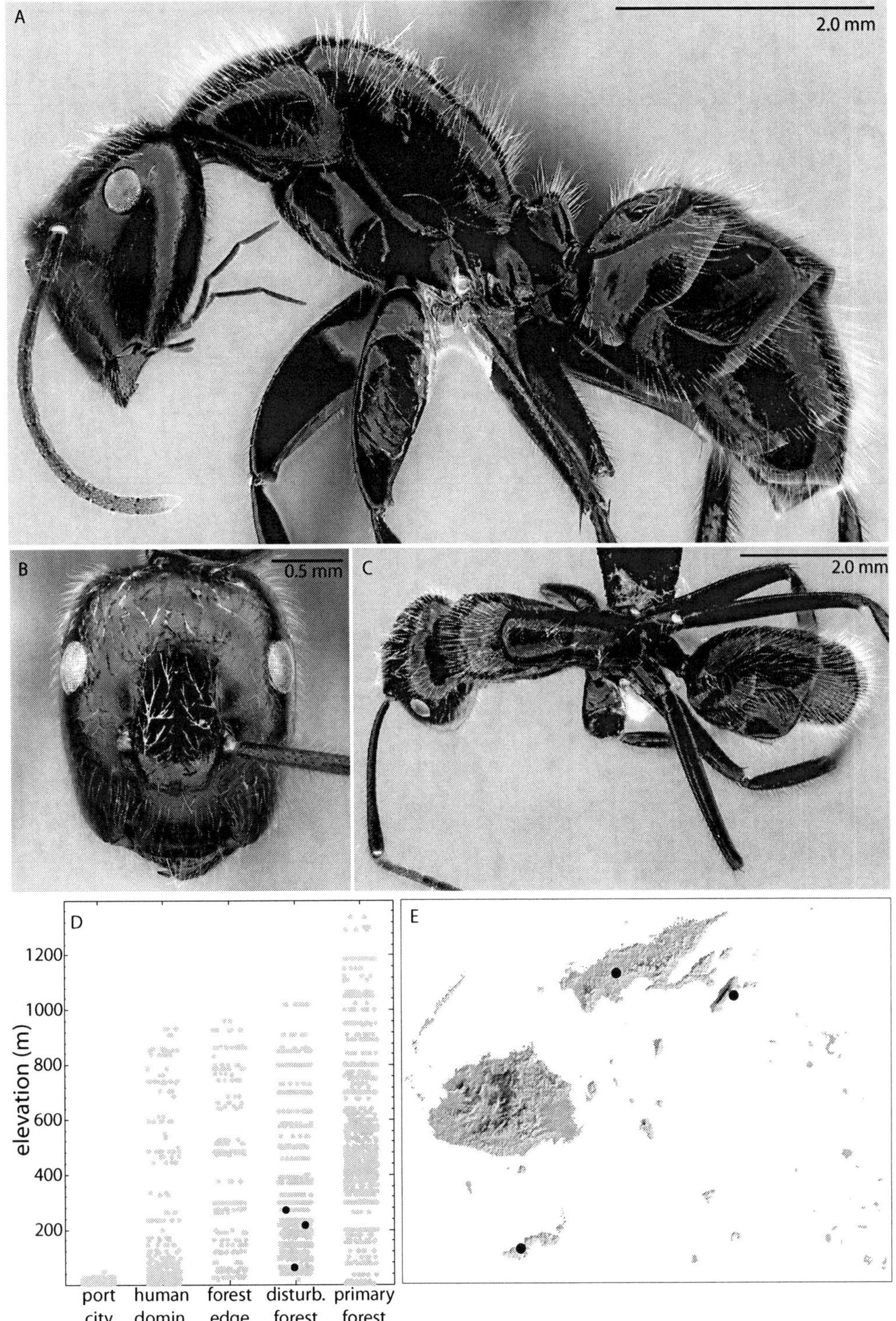

Plate 33. *Camponotus fijianus* worker, CASENT0177590.

Plate 34. *Camponotus dentatus* worker, CASENT0177557.

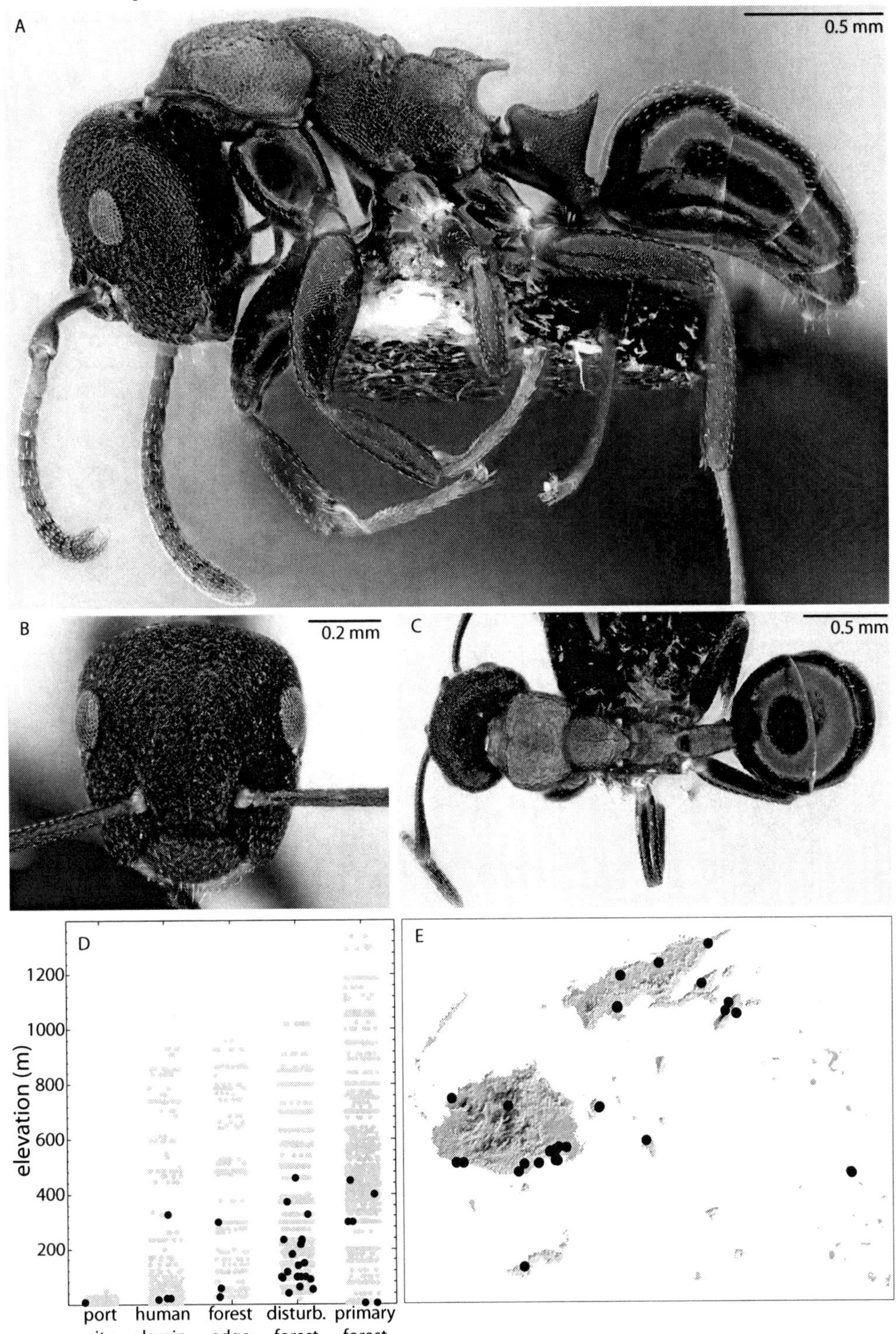

Plate 35. *Camponotus bryani* worker, CASENT0177563.

Plate 36. *Camponotus manni* worker, CASENT0177564.

Plate 37. *Camponotus umbratilis* worker, CASENT0177580.

Plate 38. *Camponotus* sp. FJ02 worker, CASENT0177581.

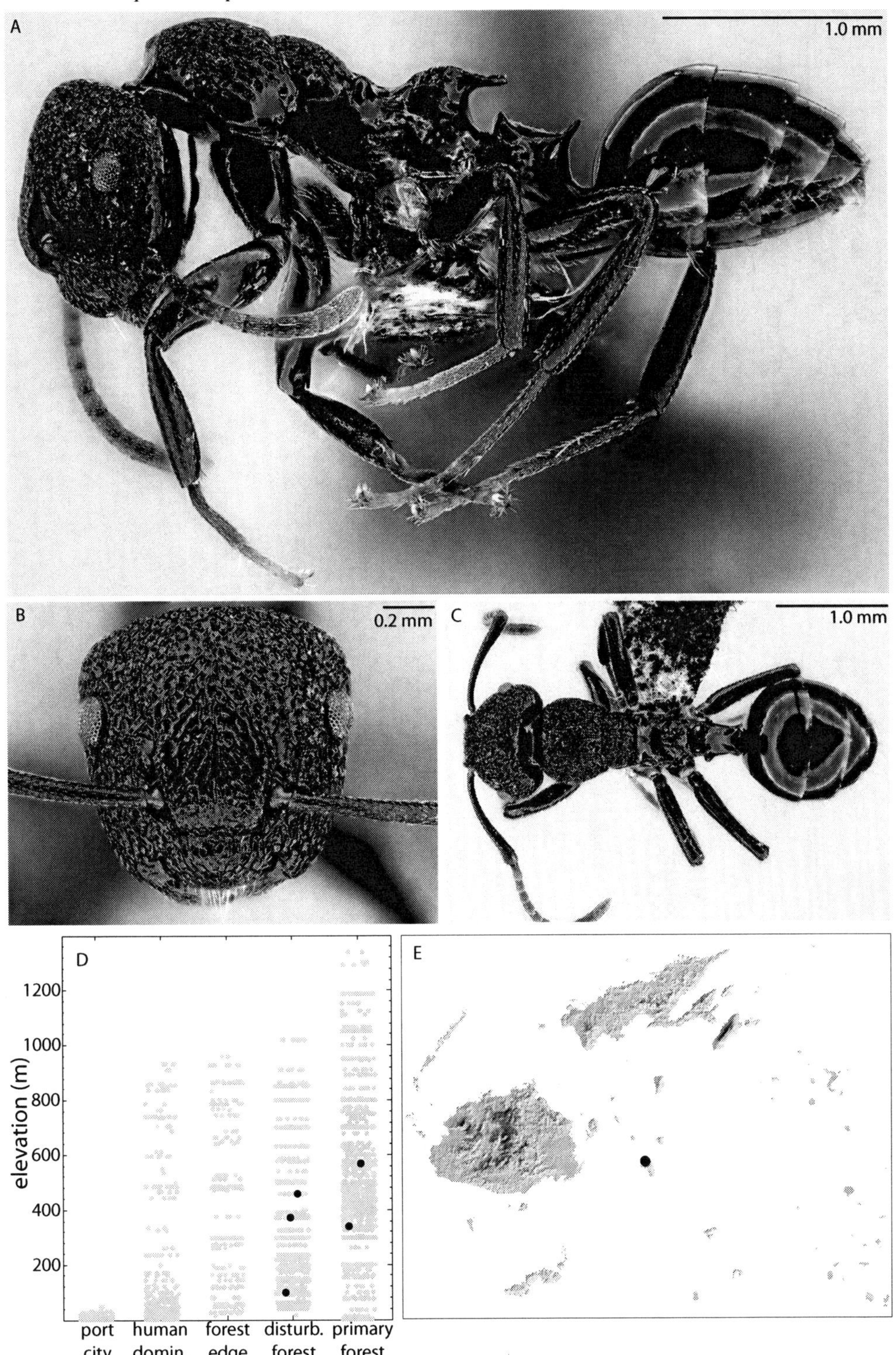

Plate 39. *Camponotus* sp. FJ03 worker, CASENT0177667.

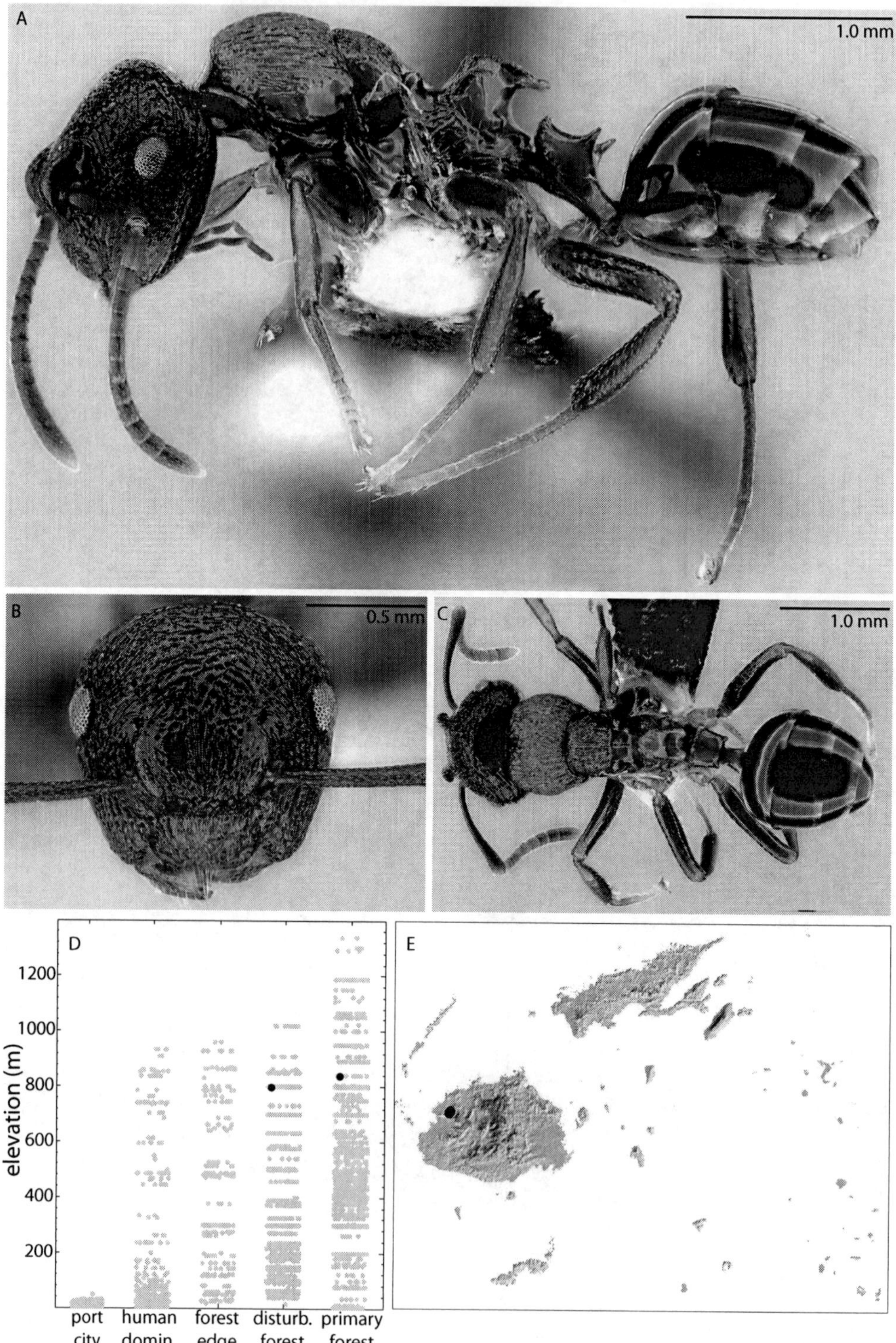

Plate 40. *Camponotus cristatus* worker, CASENT0180230.

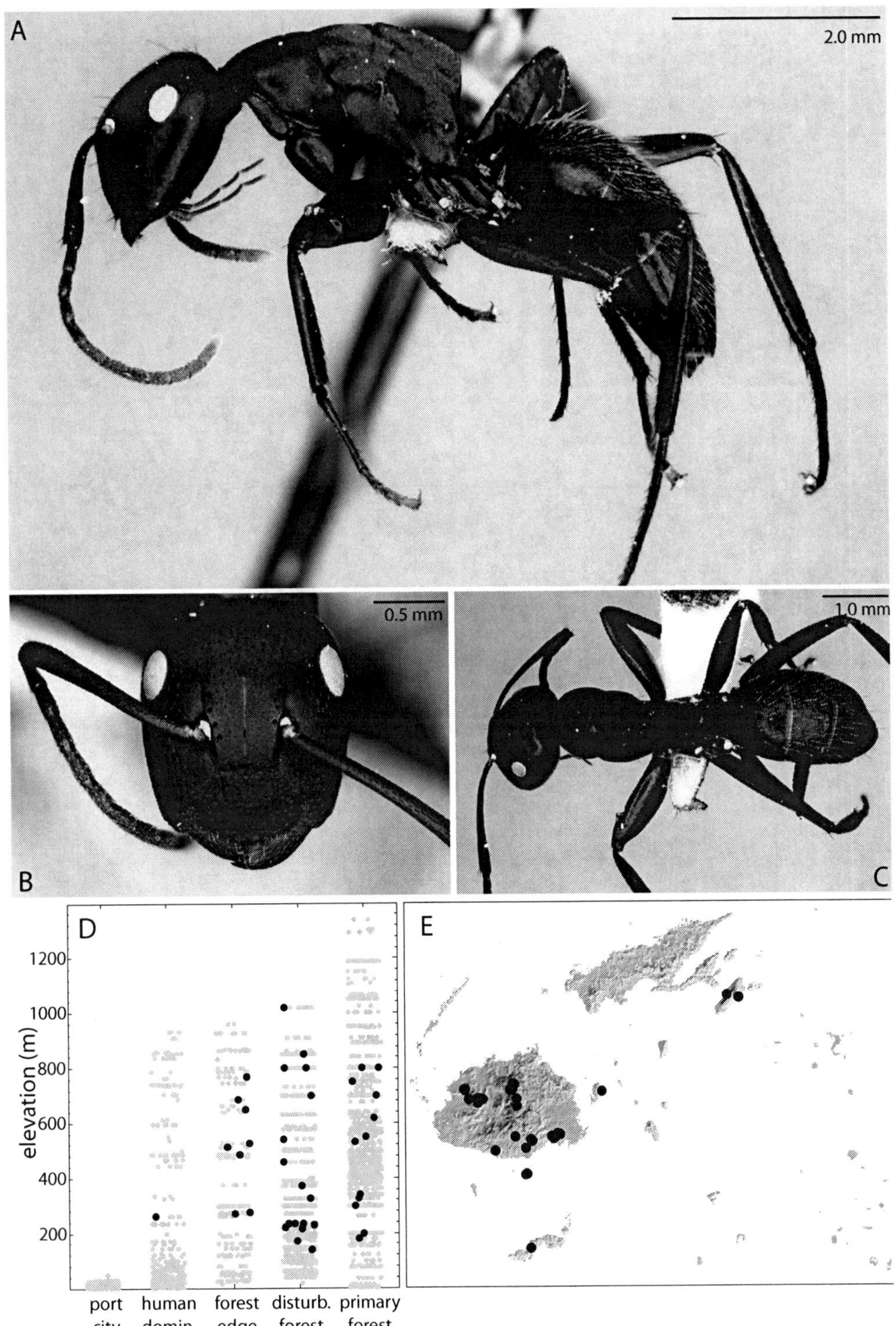

Plate 41. *Camponotus laminatus* red morph (minor, CASENT0180461; major, CASENT0180148).

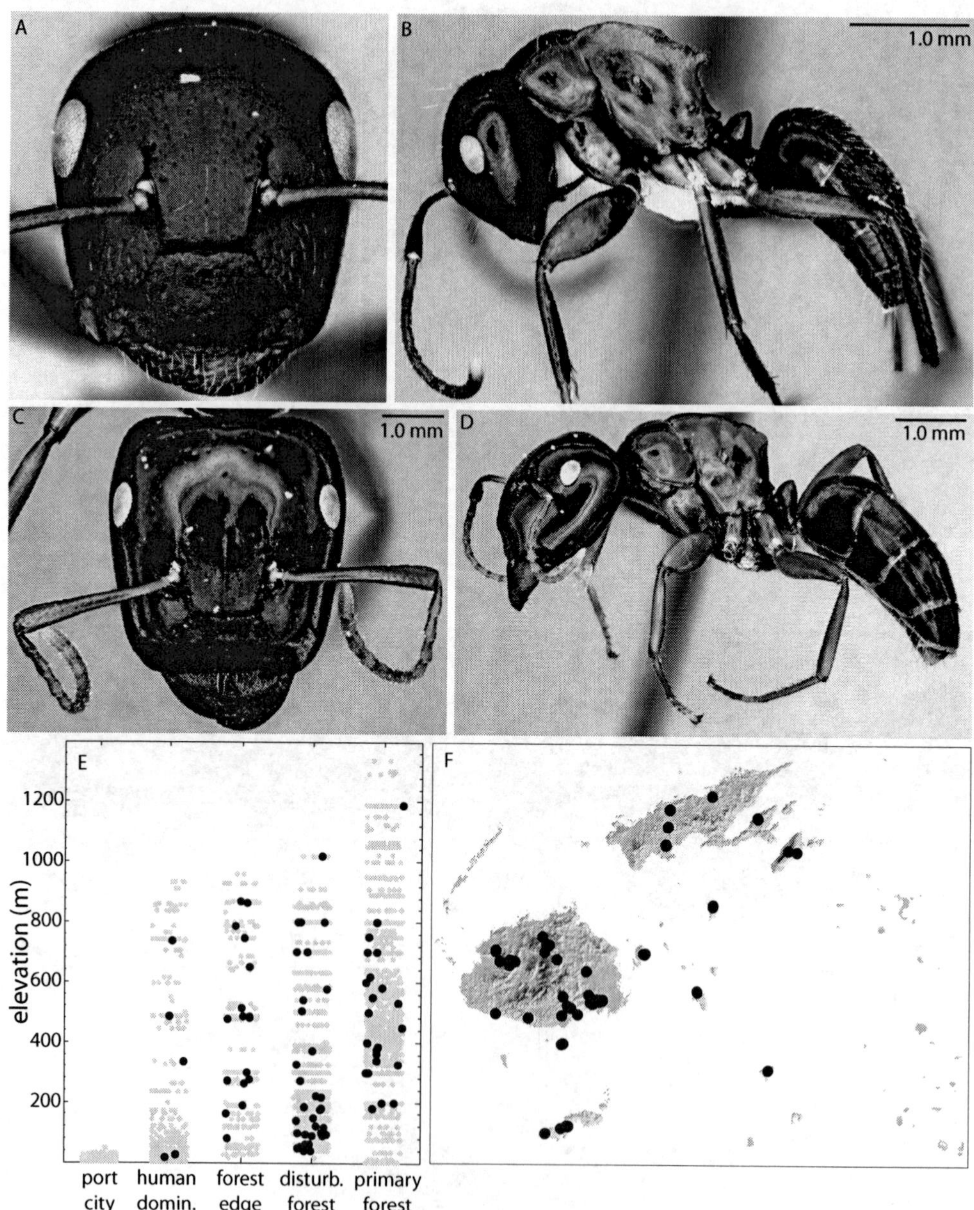

Plate 42. *Camponotus levuanus* black morph worker, CASENT0180271.

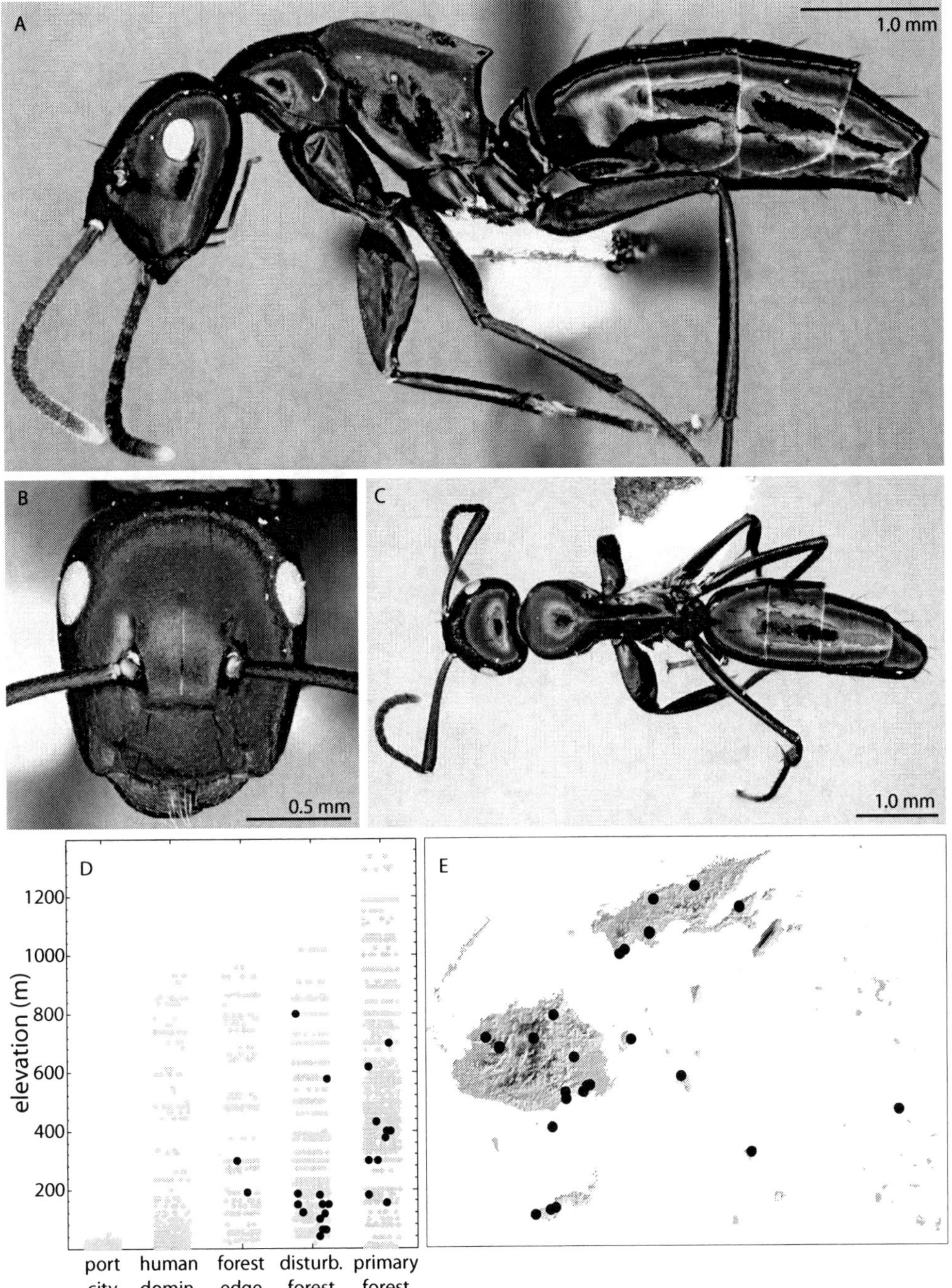

Plate 43. *Camponotus maafui* worker, CASENT0180410.

Plate 44. *Camponotus sadinus* (minor, CASENT0180304; major, CASENT0180333).

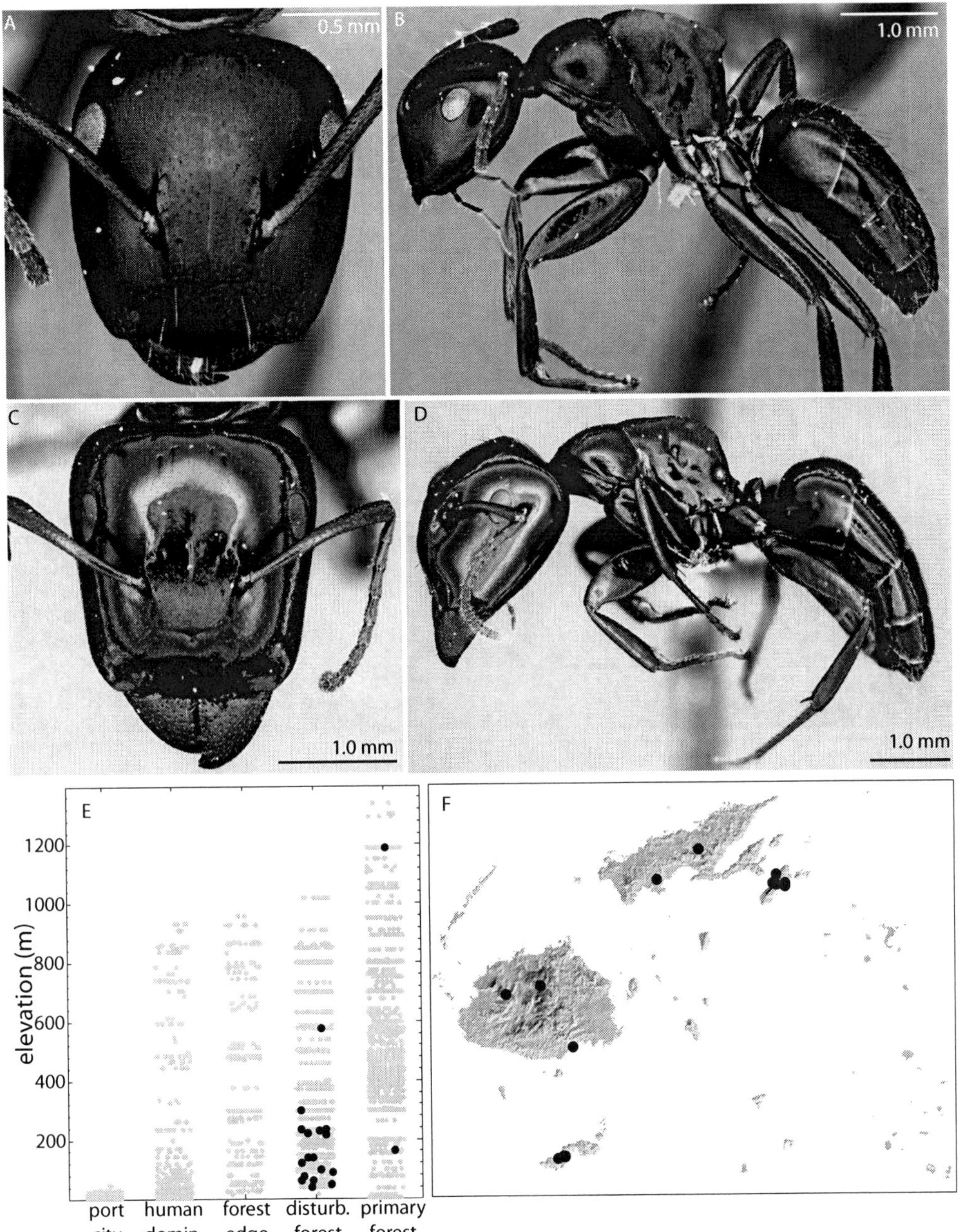

Plate 45. *Camponotus kadi* (minor, CASENT0180467; major, CASENT0180434).

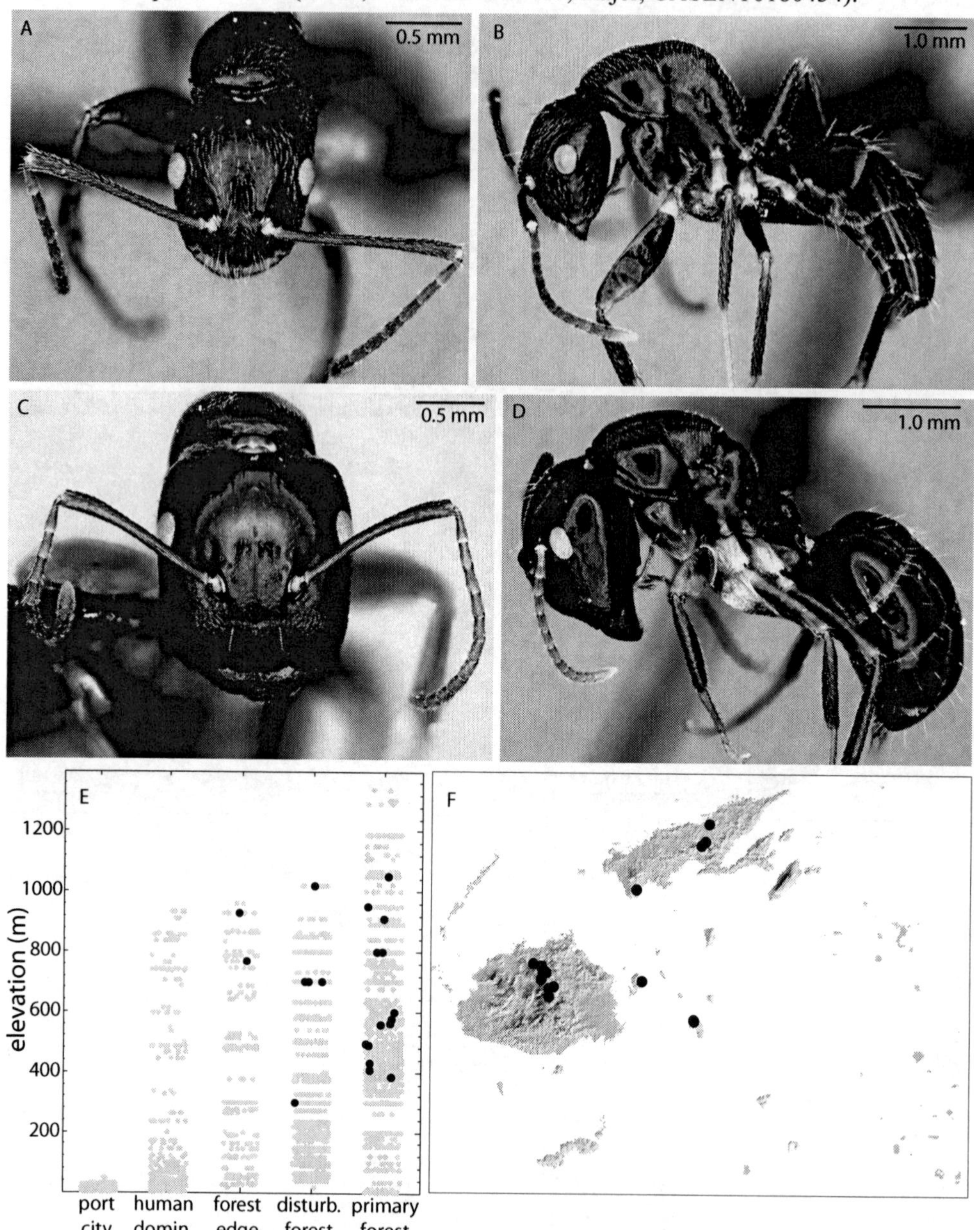

Plate 46. *Camponotus lauensis* syntype worker, CASENT0235853.

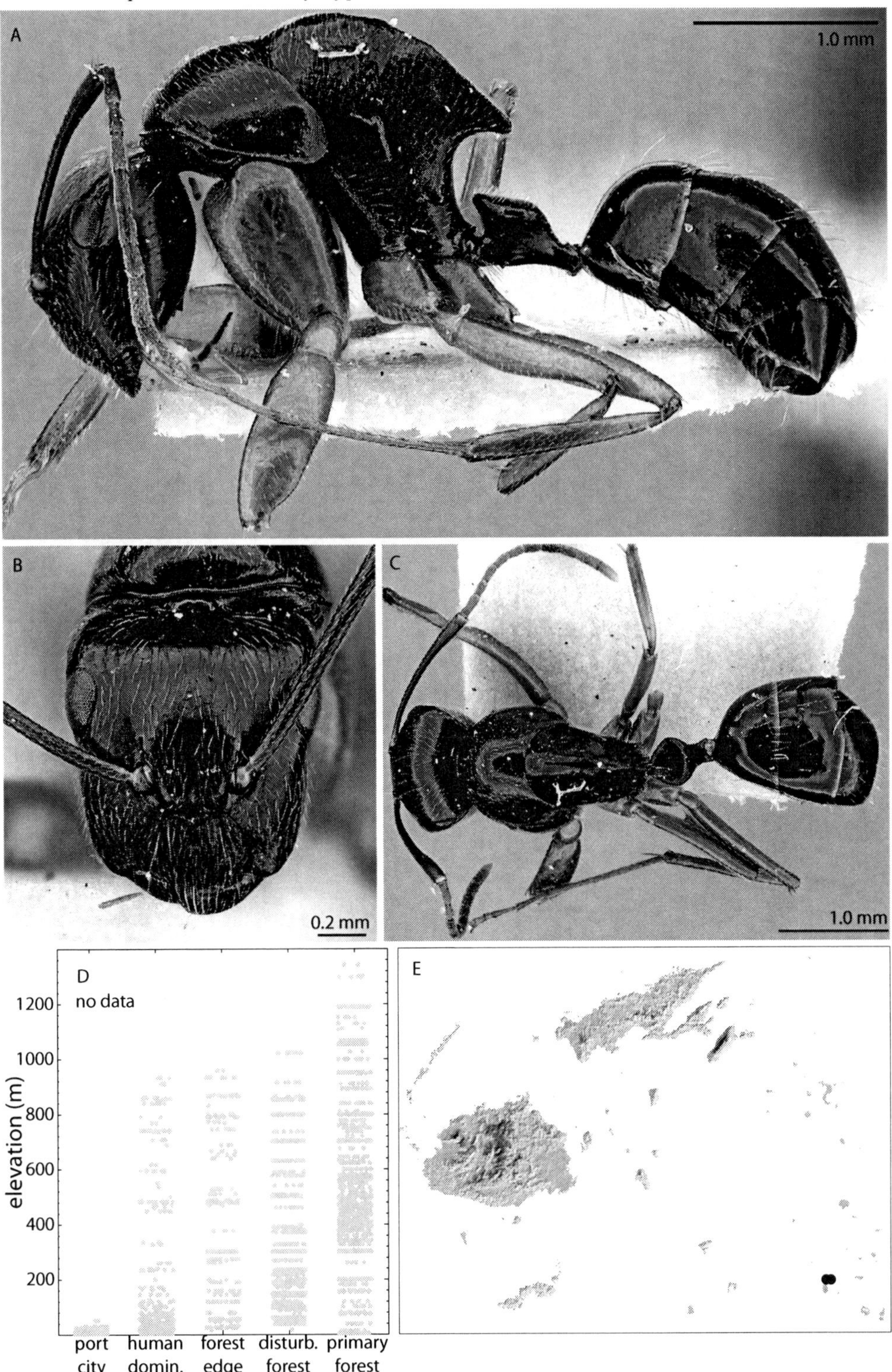

Plate 47. *Camponotus schmeltzi* (minor, CASENT0187073; major, CASENT0187063).

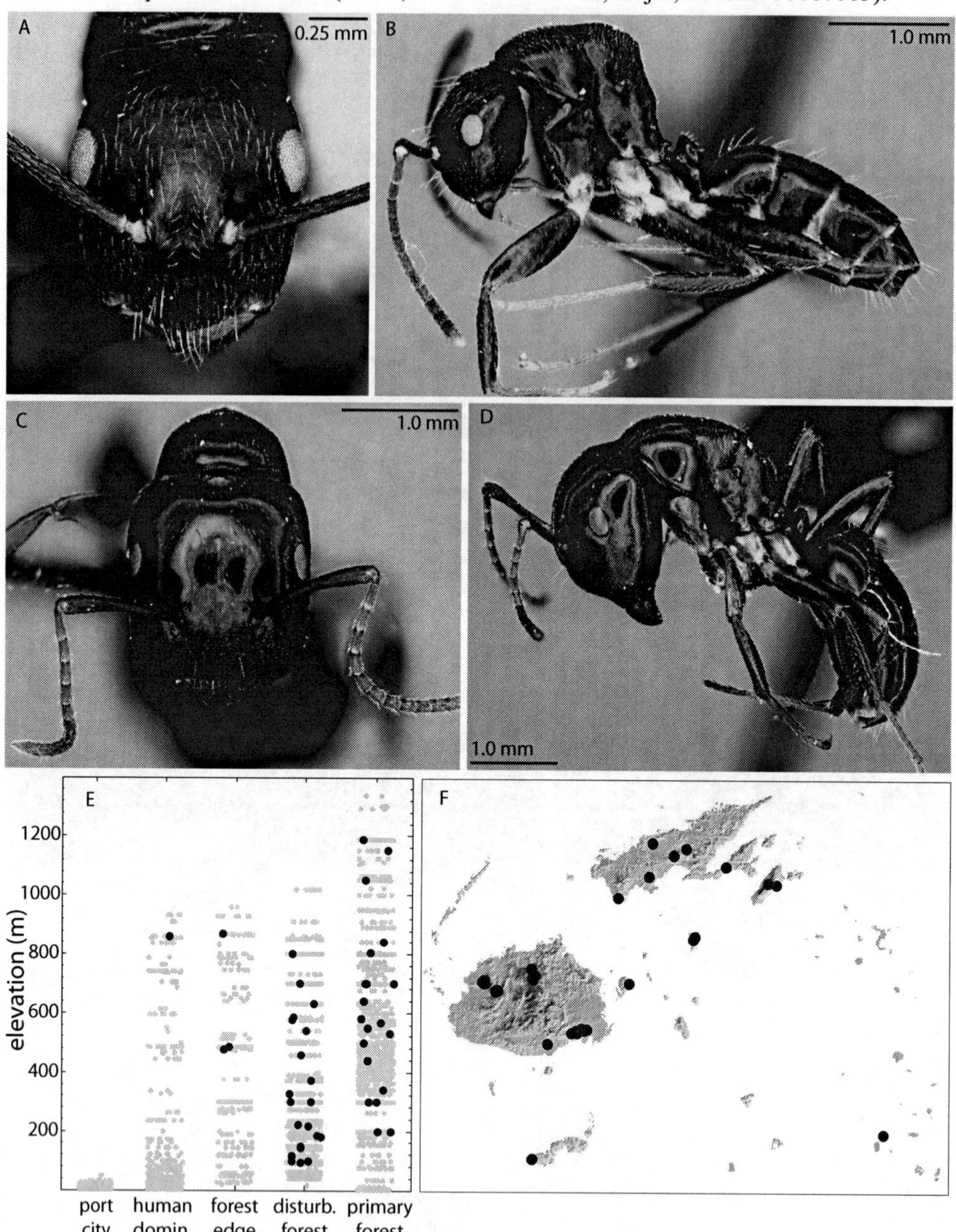

Plate 48. *Nylanderia glabrior* worker, CASENT0181474.

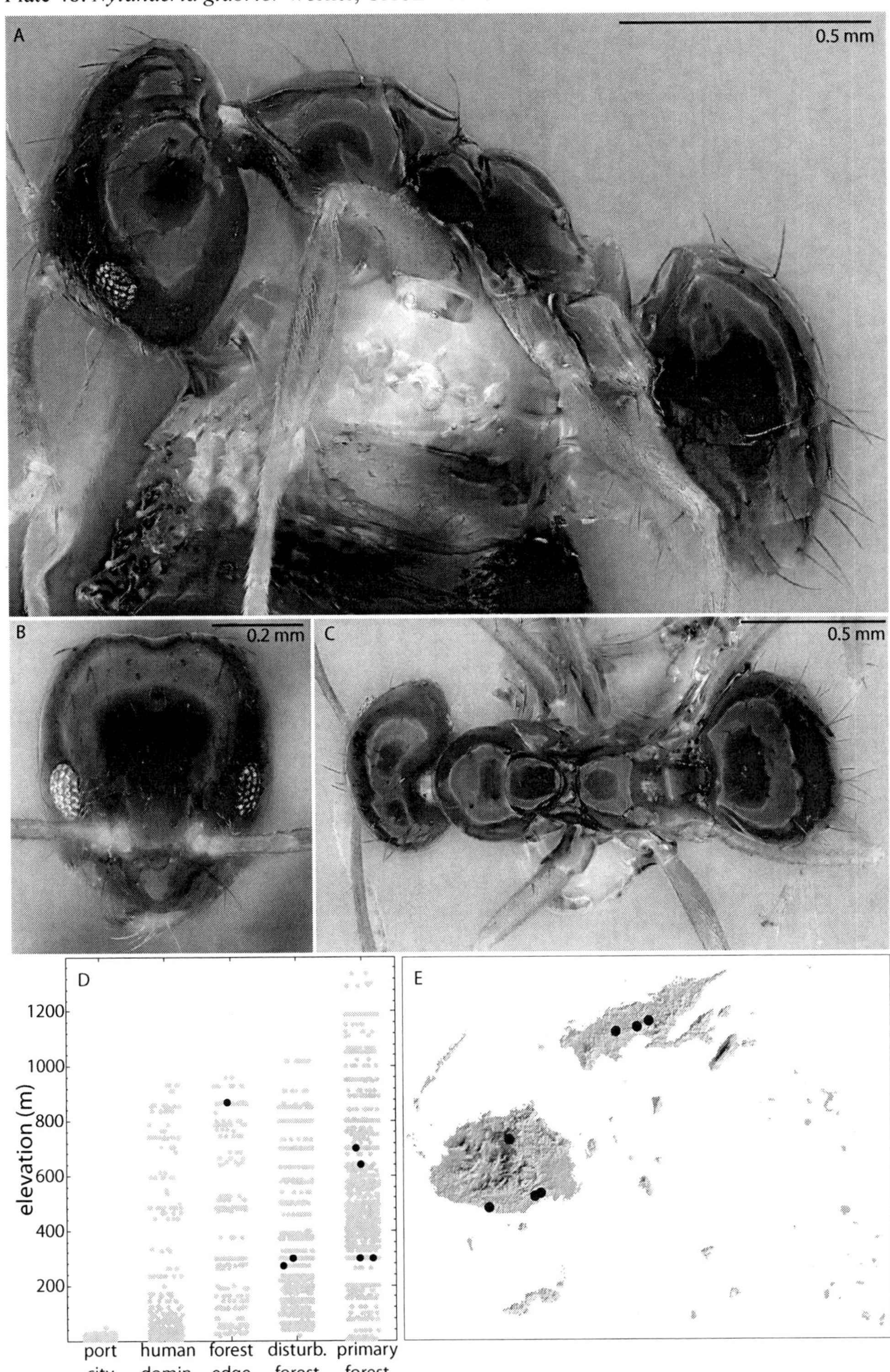

Plate 49. *Nylanderia vaga* worker, CASENT0171069.

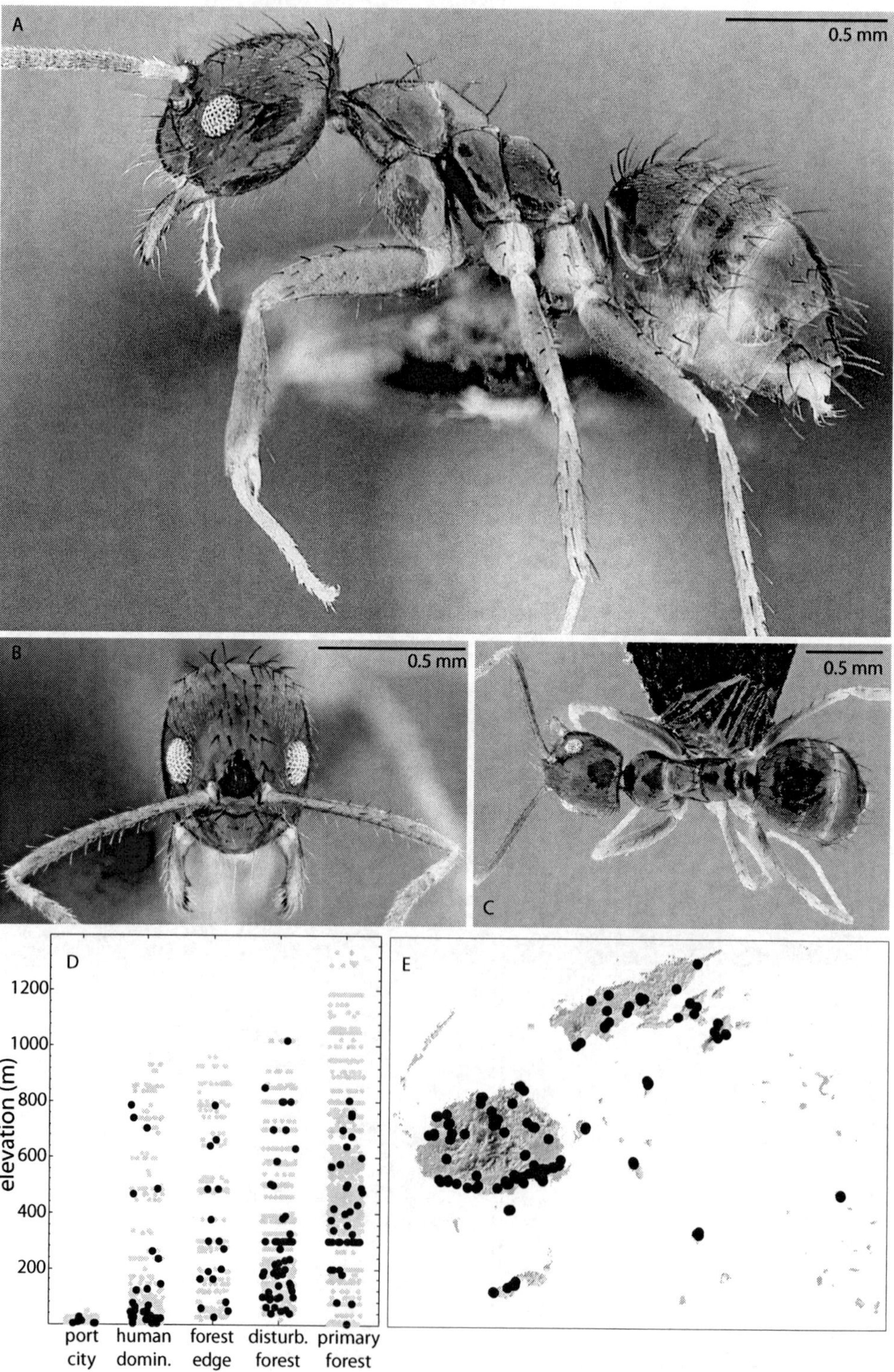

Plate 50. *Nylanderia vitiensis* worker, CASENT0181499.

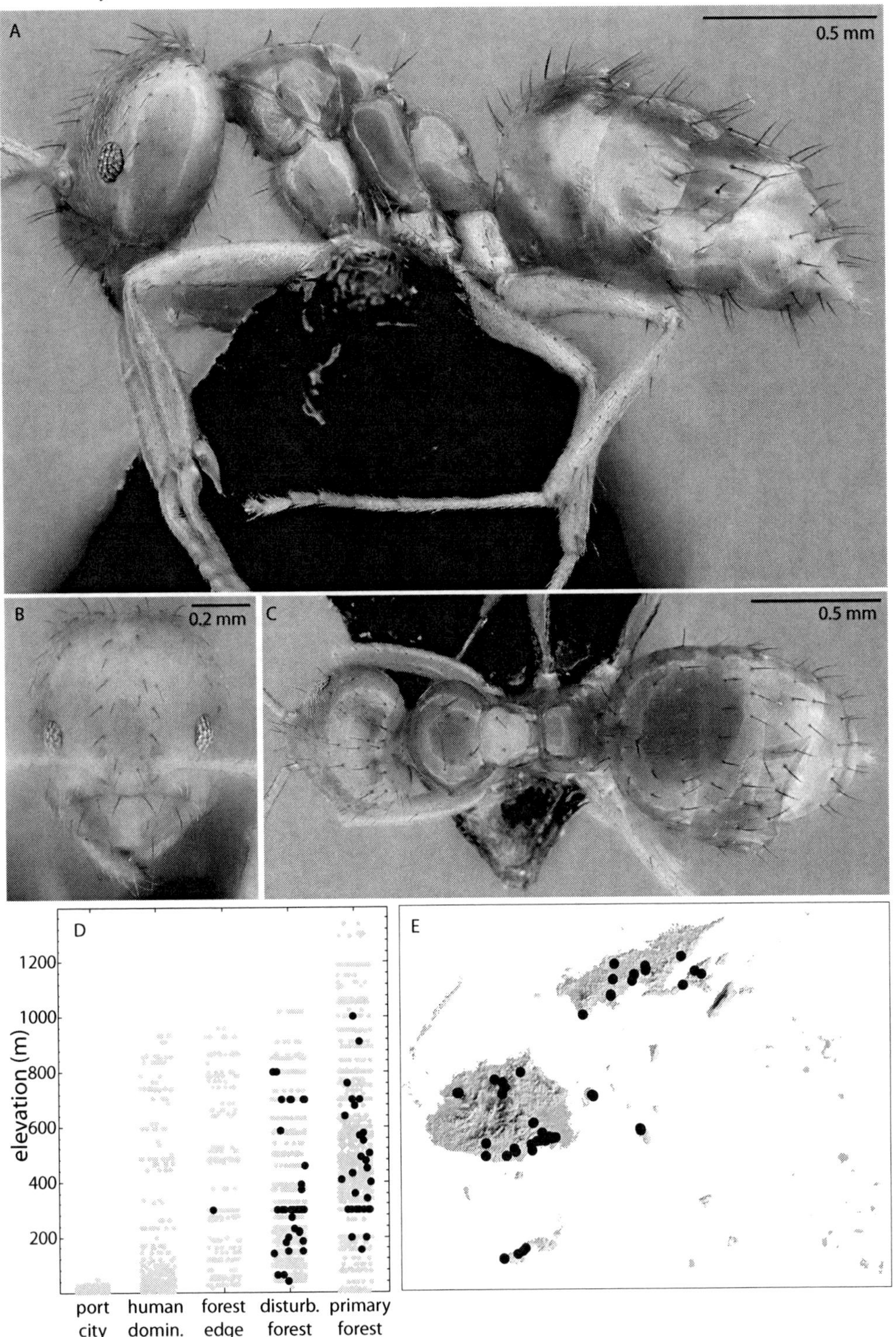

Plate 51. *Nylanderia* sp. FJ03 worker, CASENT0181303.

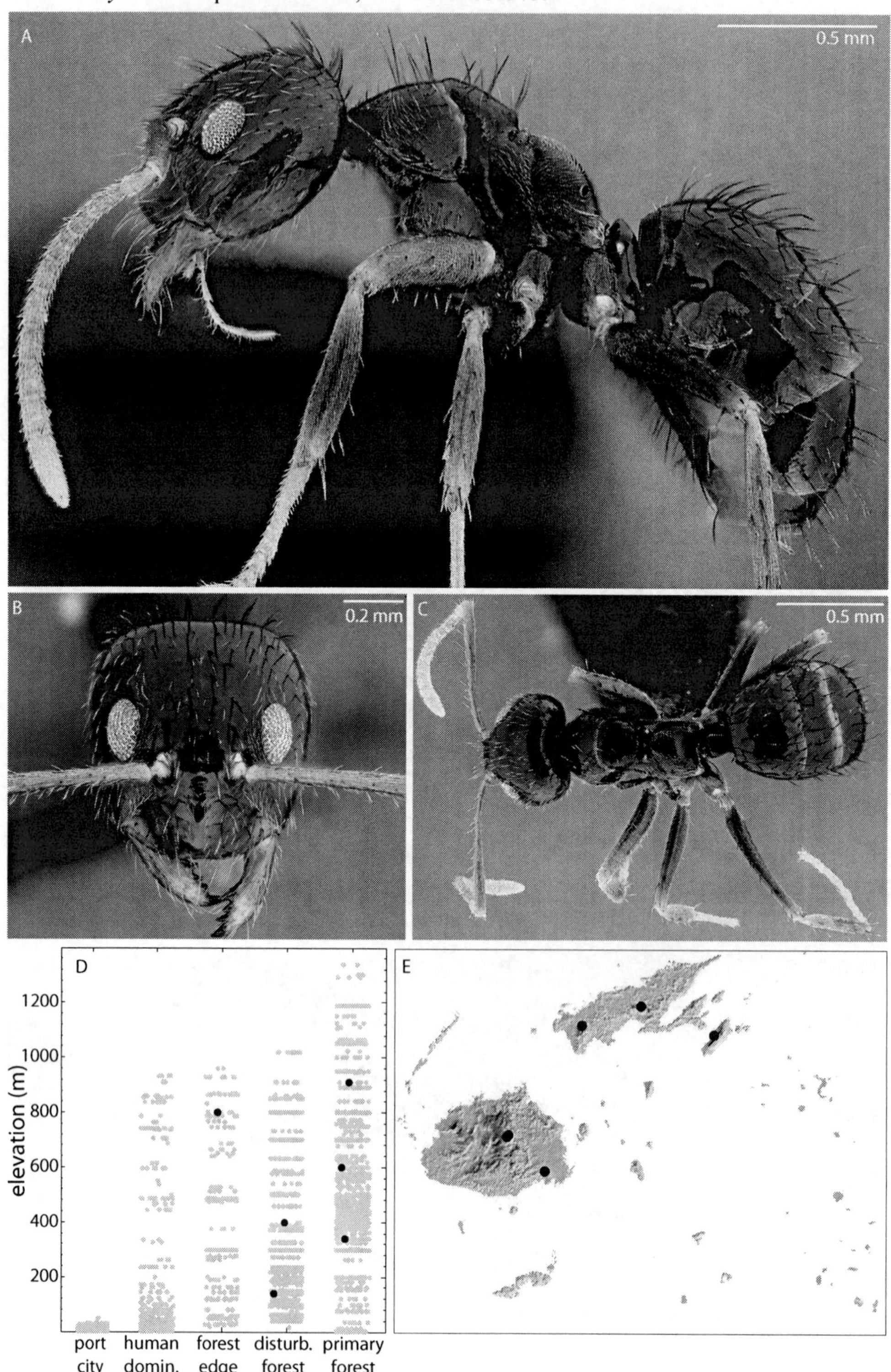

Plate 52. *Paraparatrechina oceanica* worker, CASENT0181303.

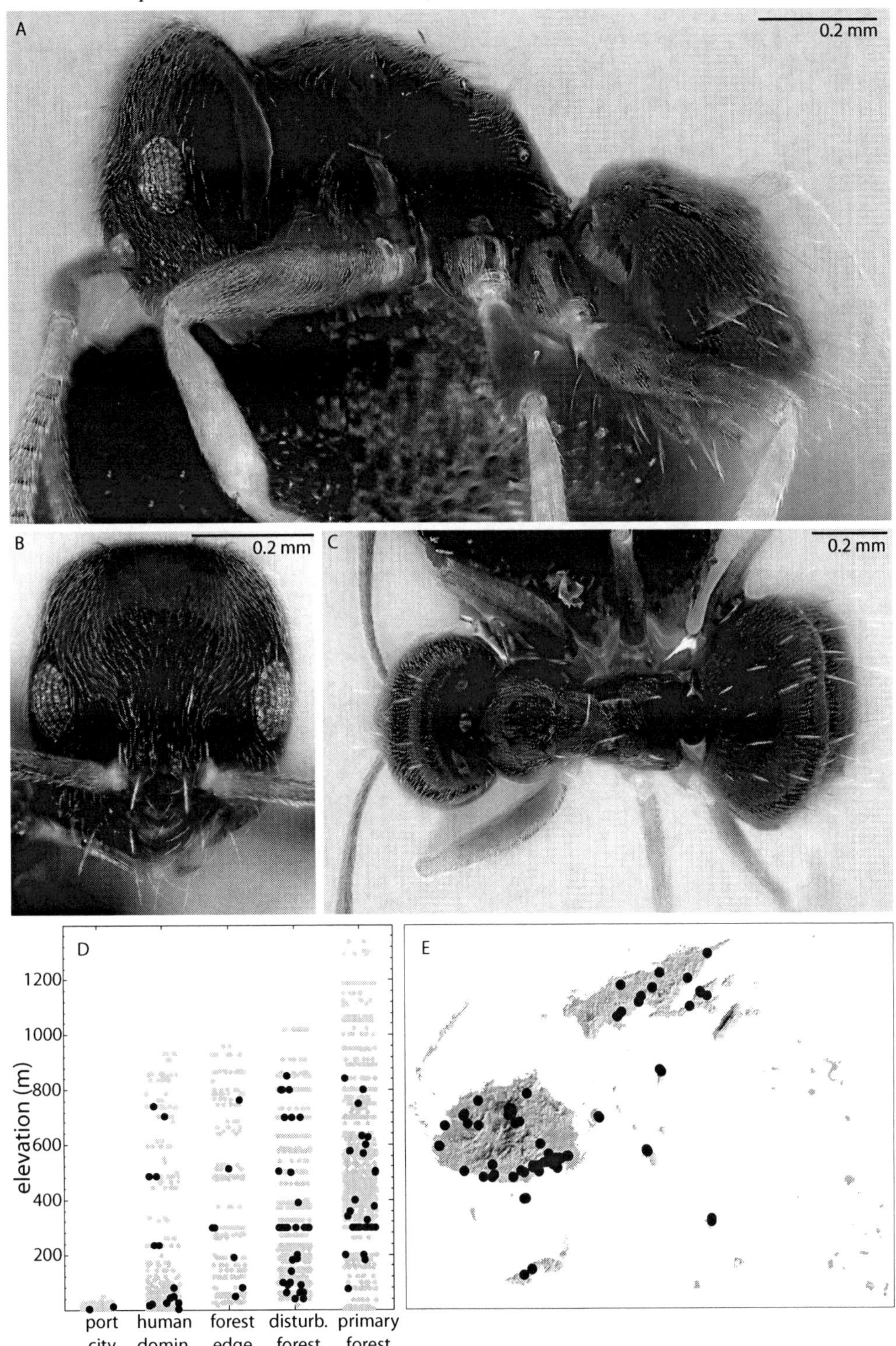

Plate 53. *Paratrechina longicornis* worker, CASENT0171073 (head and profile), CASENT0171035 (dorsal).

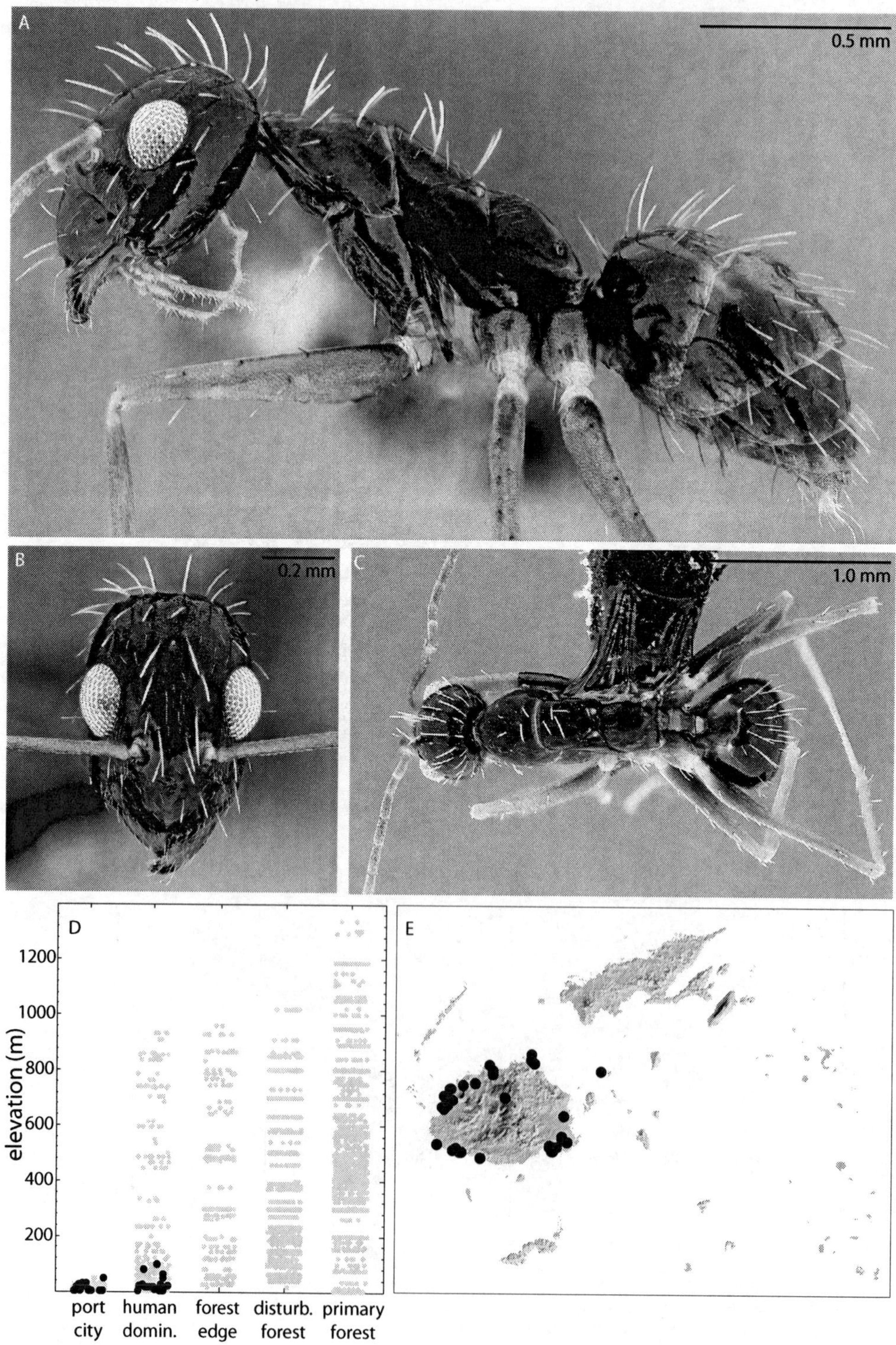

Plate 54. *Plagiolepis alluaudi* worker, CASENT0171065.

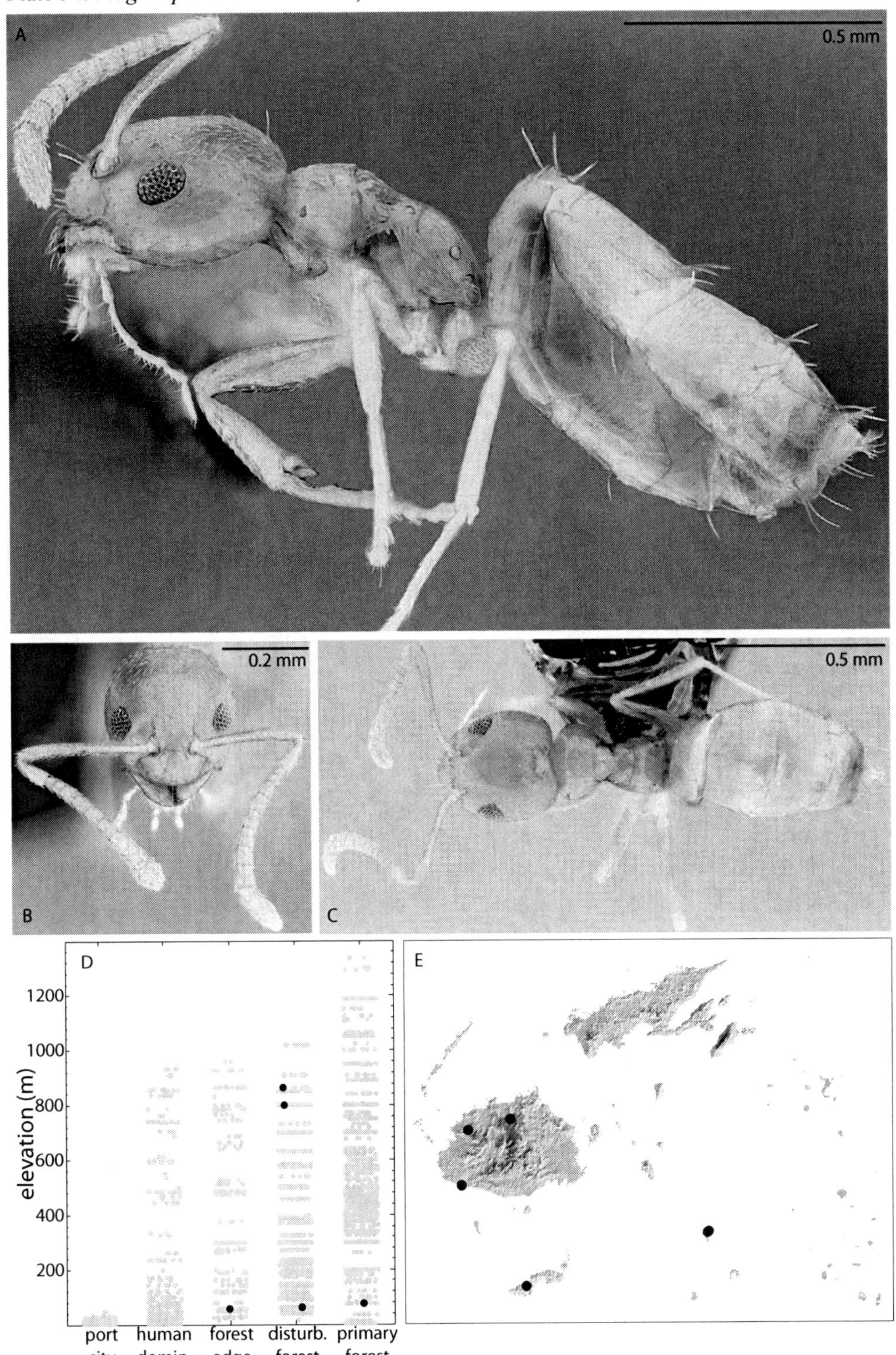

Plate 55. *Adelomyrmex hirsutus* worker, CASENT0181559.

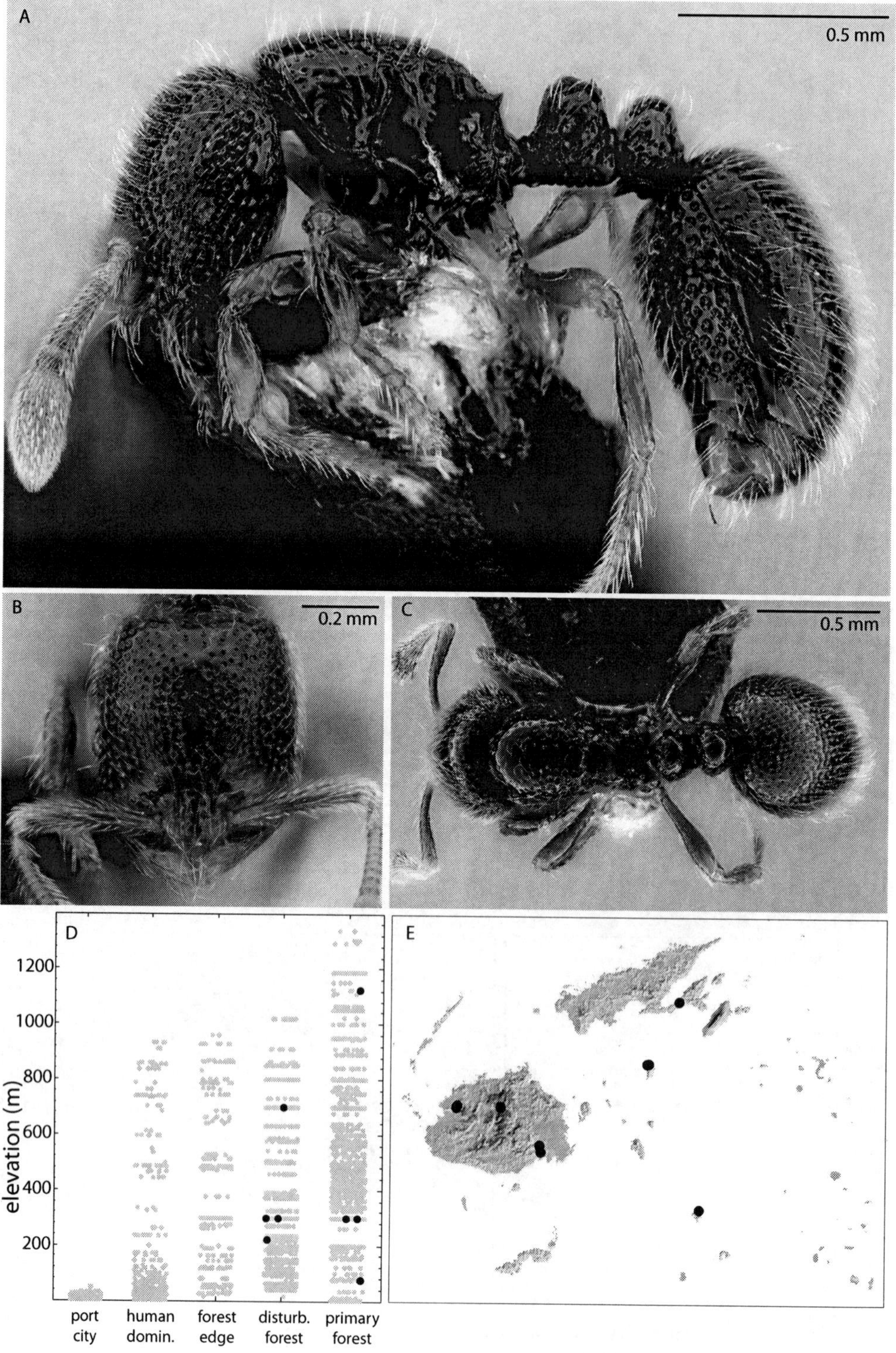

Plate 56. *Adelomyrmex samoanus* worker, CASENT0171037.

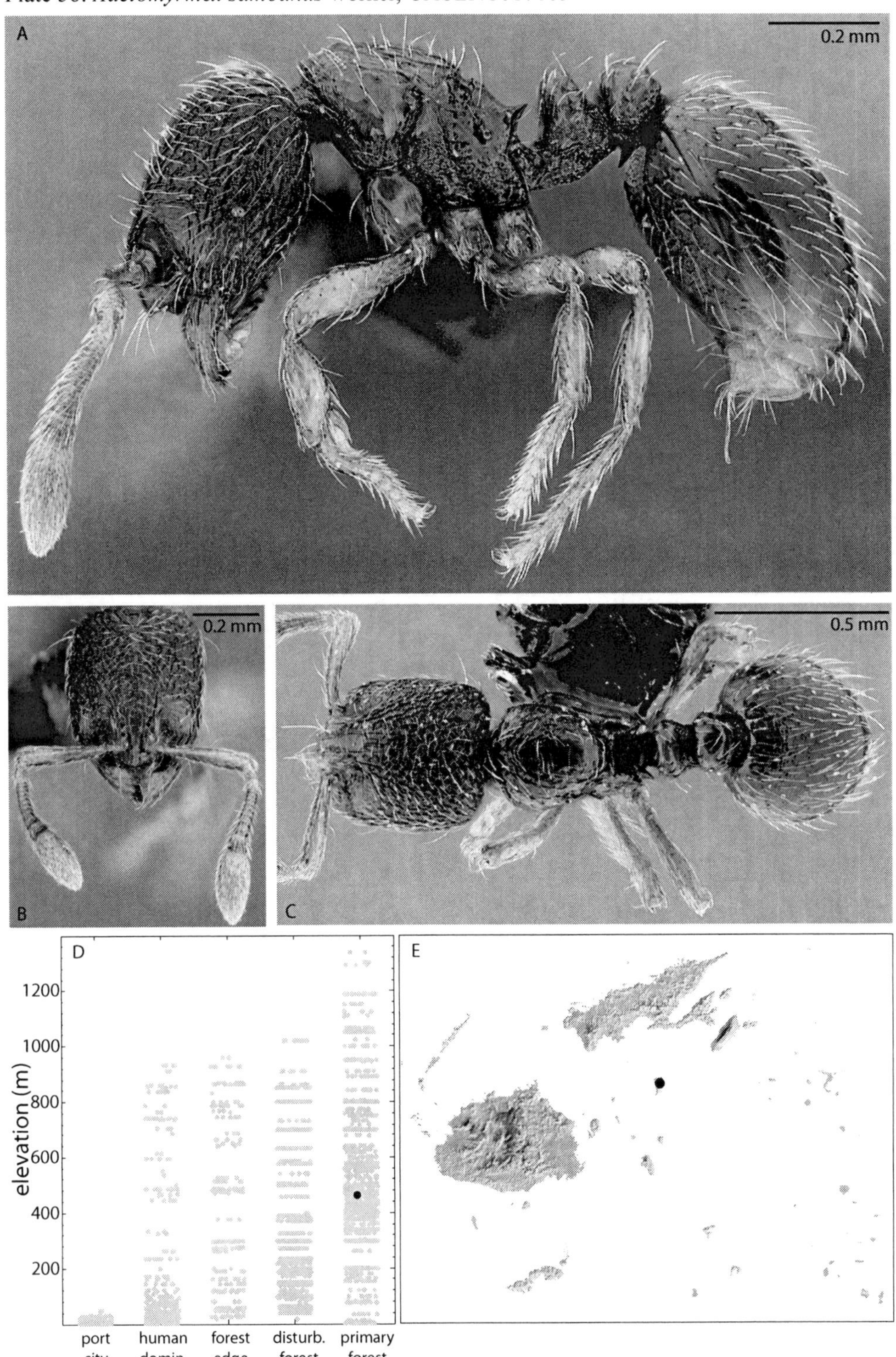

Plate 57. *Cardiocondyla emeryi* worker, CASENT0171087.

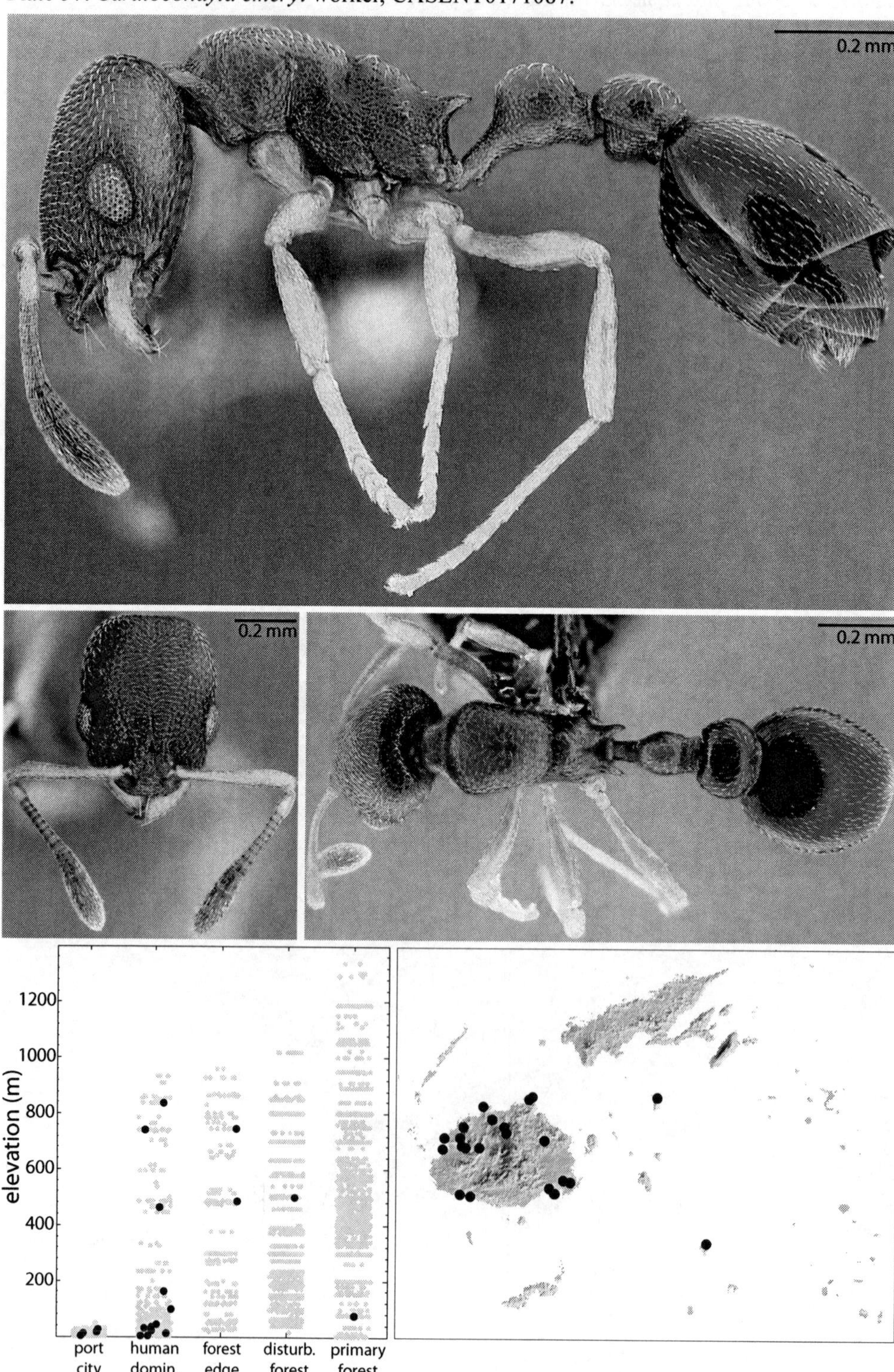

Plate 58. *Cardiocondyla kagutsuchi* worker, CASENT0171071.

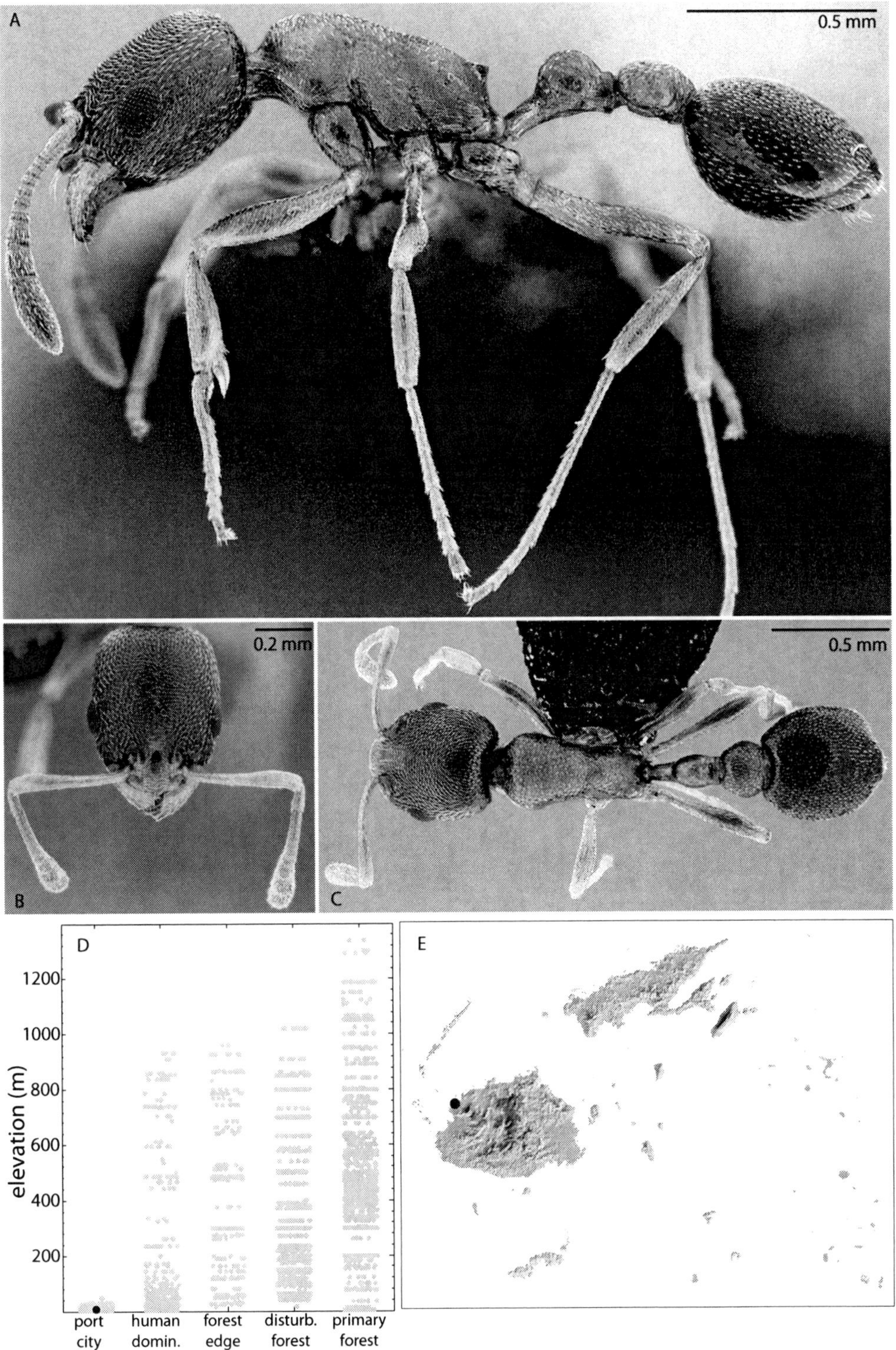

Plate 59. *Cardiocondyla minutior* v, CASENT0171077.

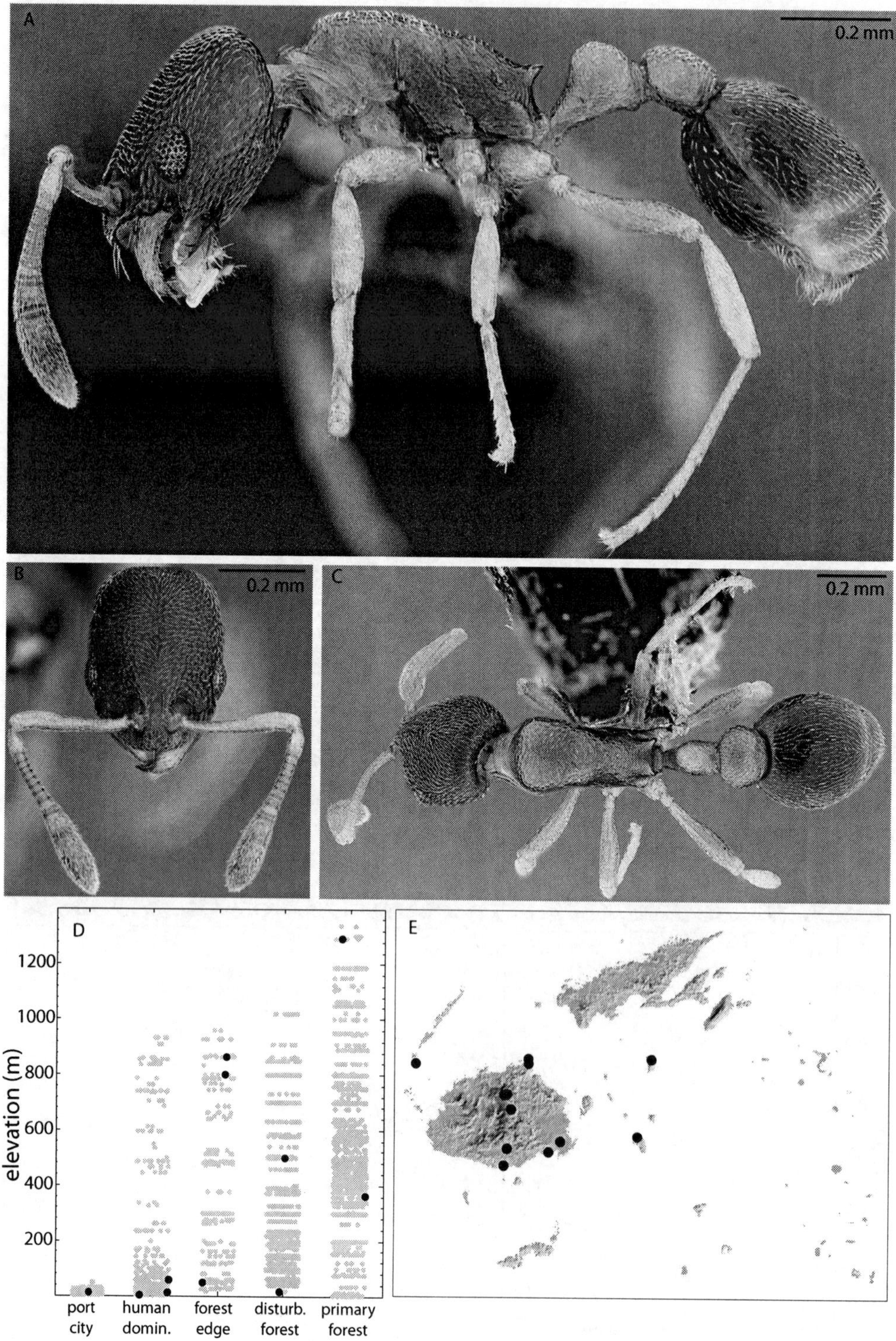

Plate 60. *Cardiocondyla nuda* worker, CASENT0181806.

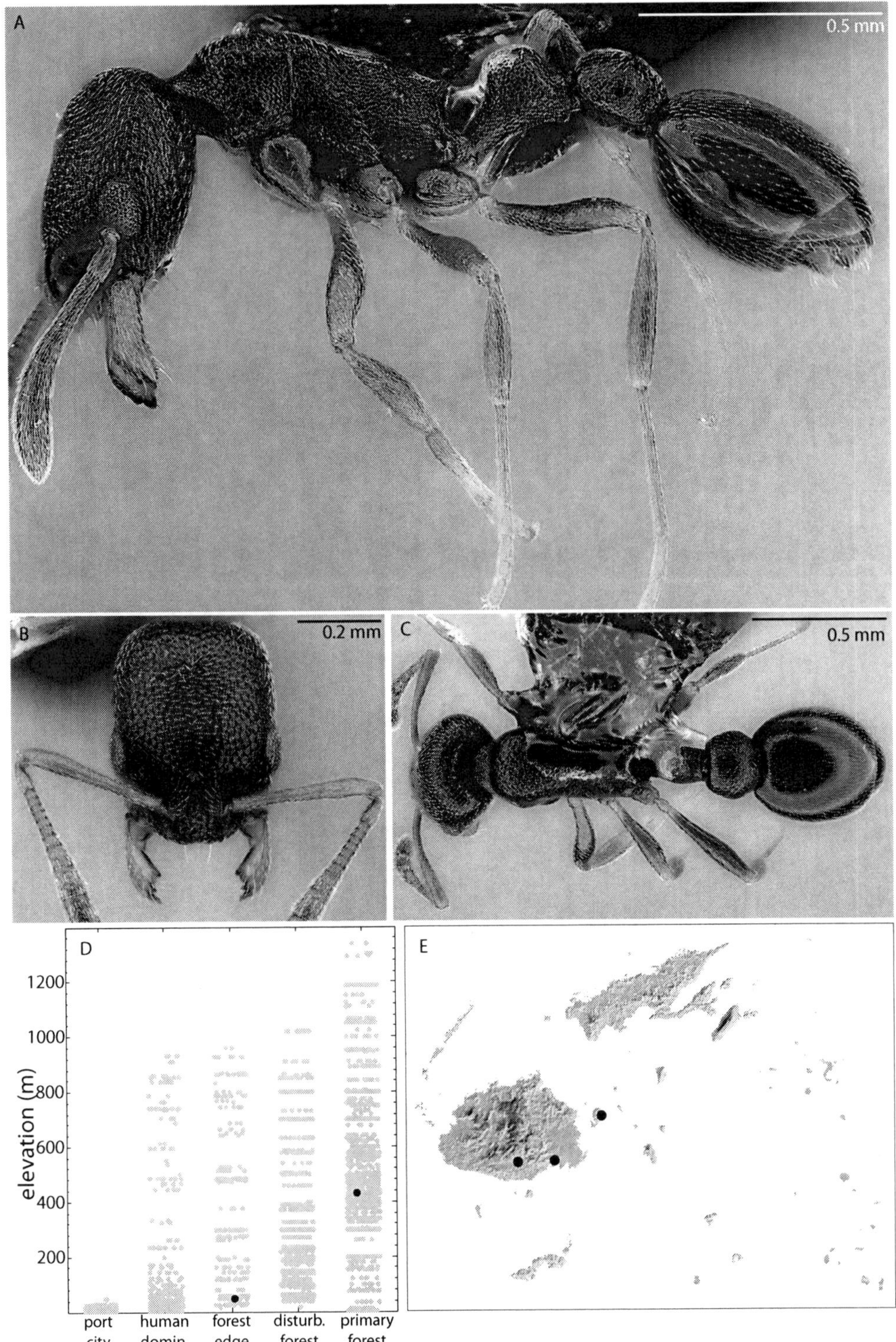

Plate 61. *Cardiocondyla obscurior* worker, CASENT0171038.

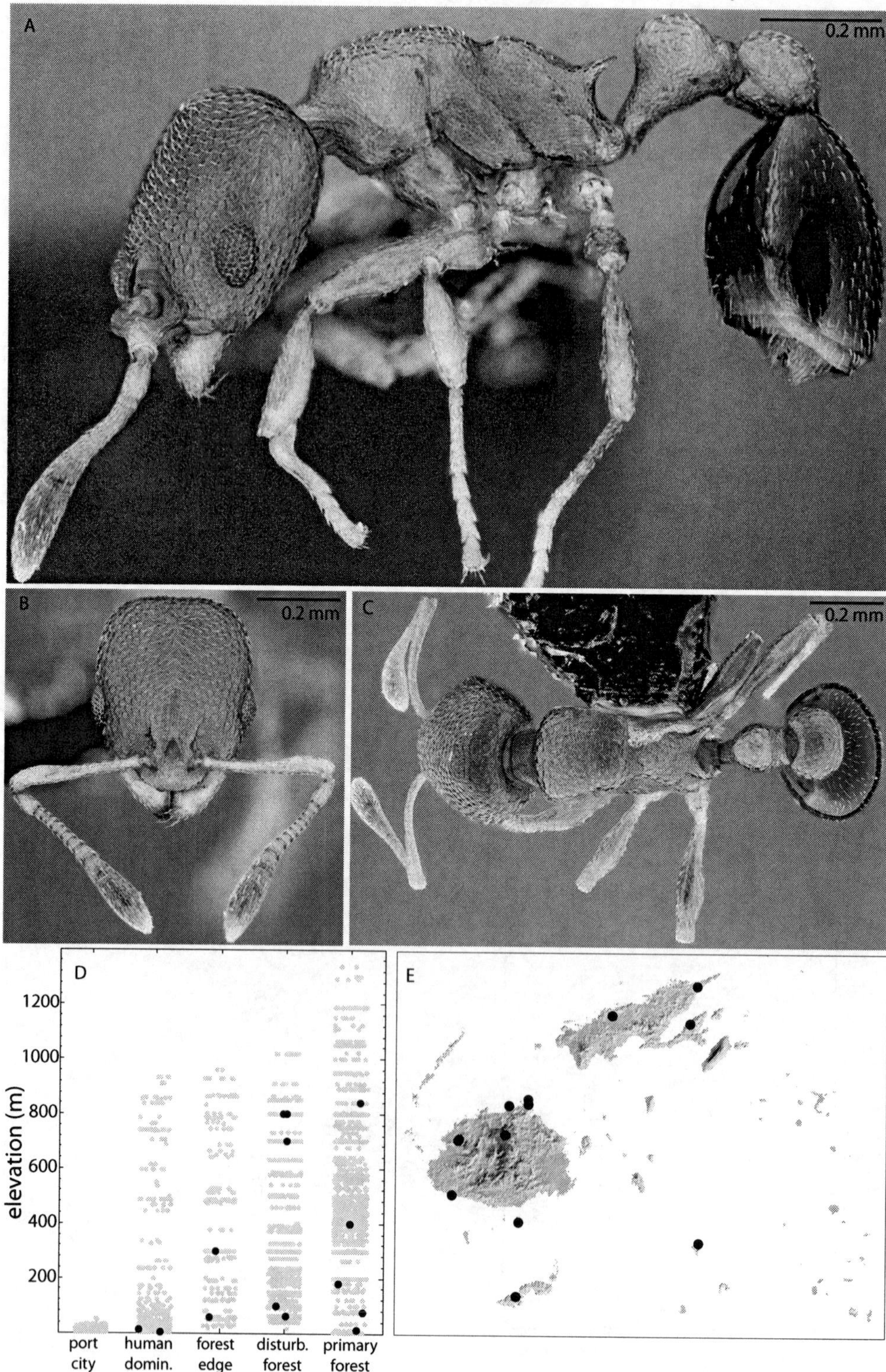

Plate 62. *Carebara atoma* (minor, CASENT0101462; major, CASENT0171039).

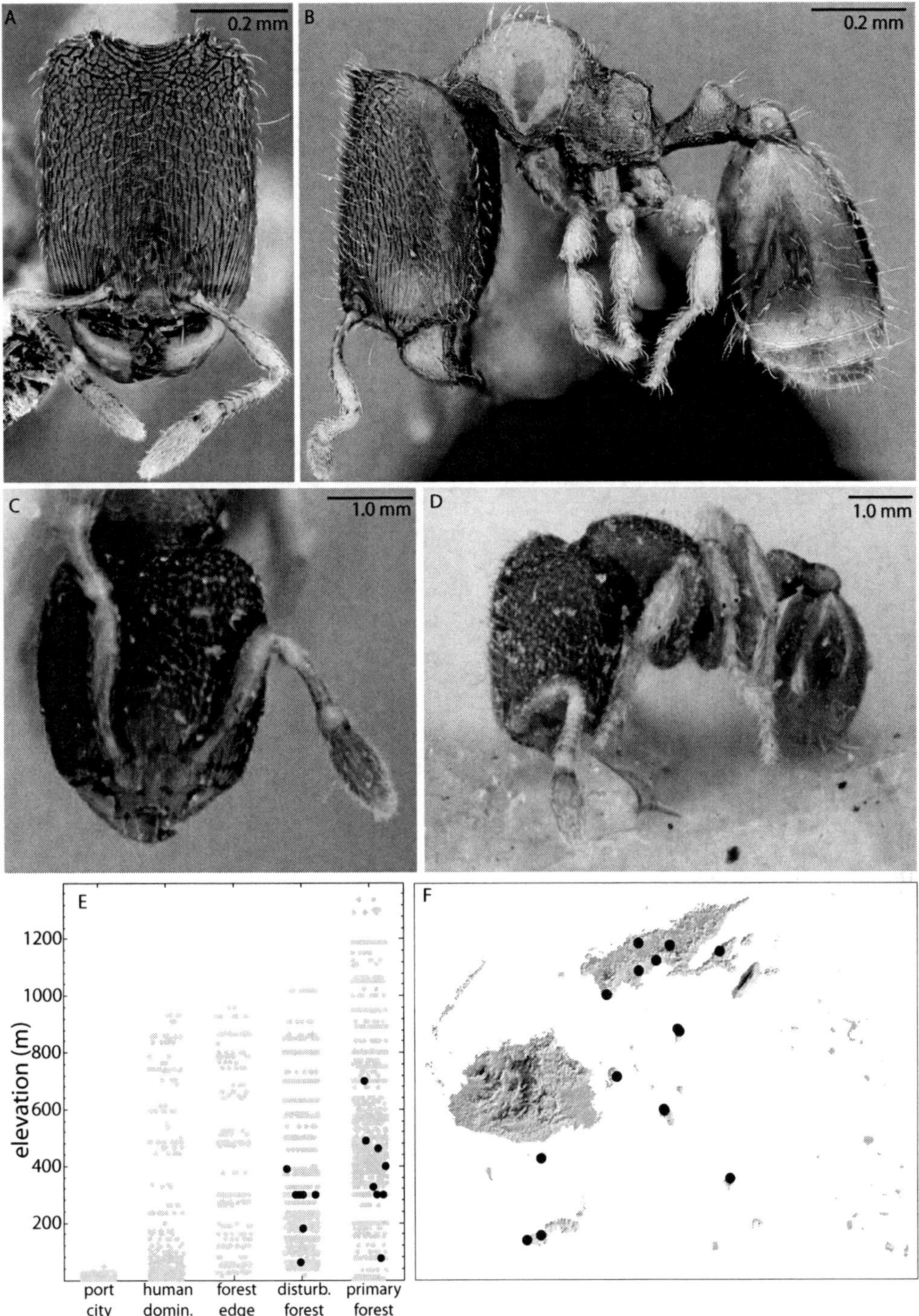

Plate 63. *Eurhopalothrix emeryi* worker, CASENT0181833.

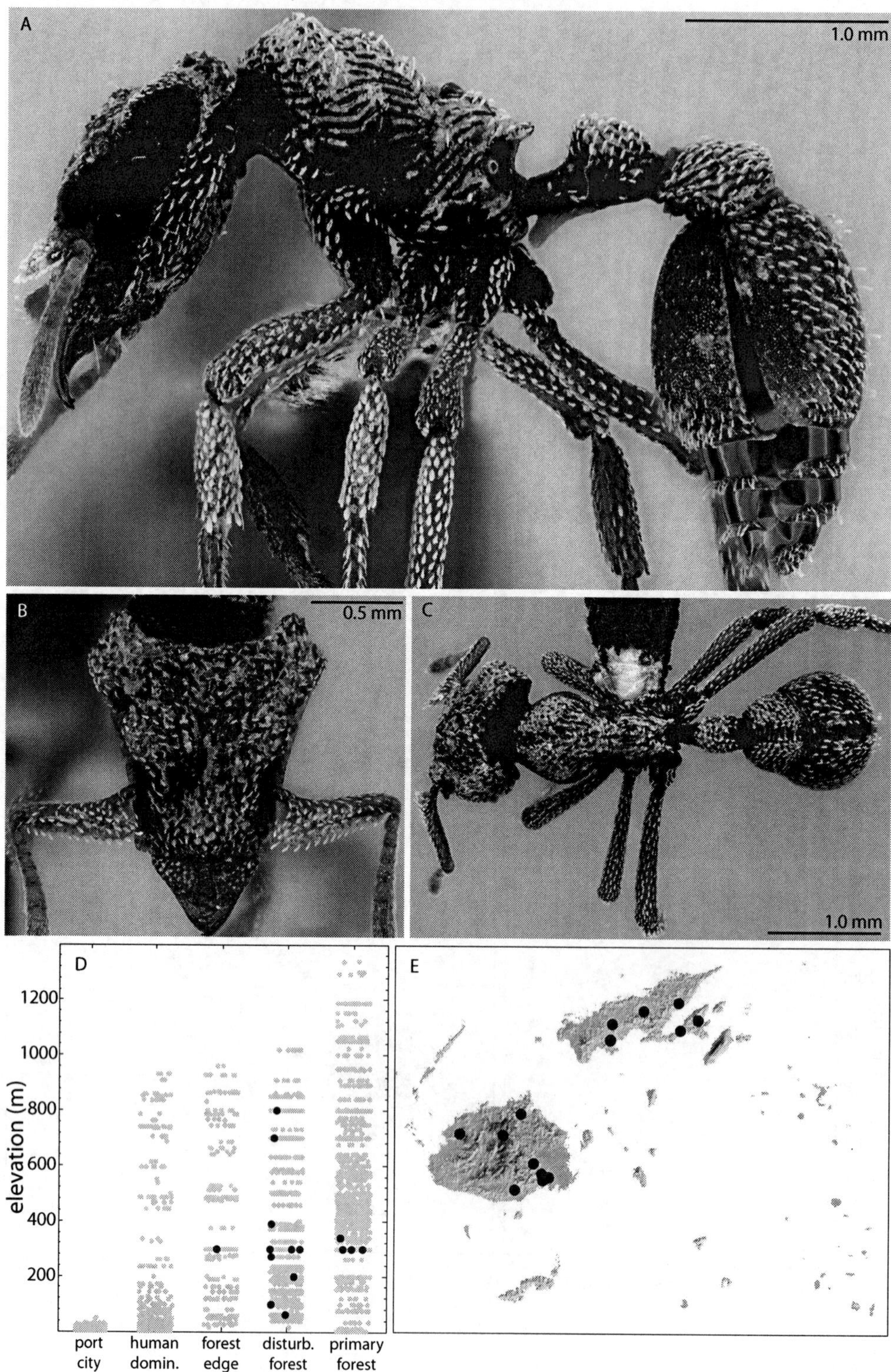

Plate 64. *Eurhopalothrix insidiatrix* worker, CASENT0171052.

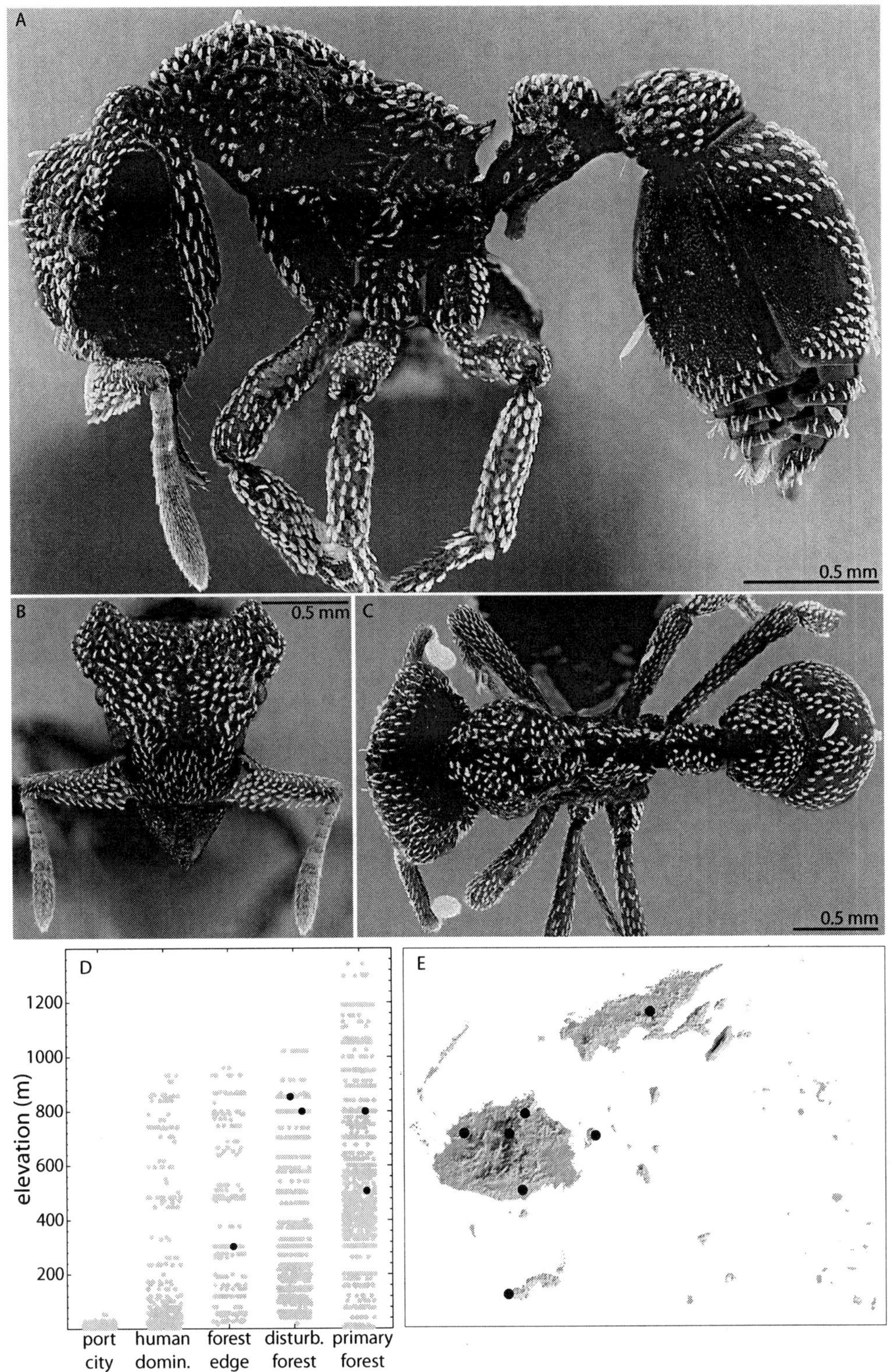

Plate 65. *Eurhopalothrix* sp. FJ52 male, CASENT0194603.

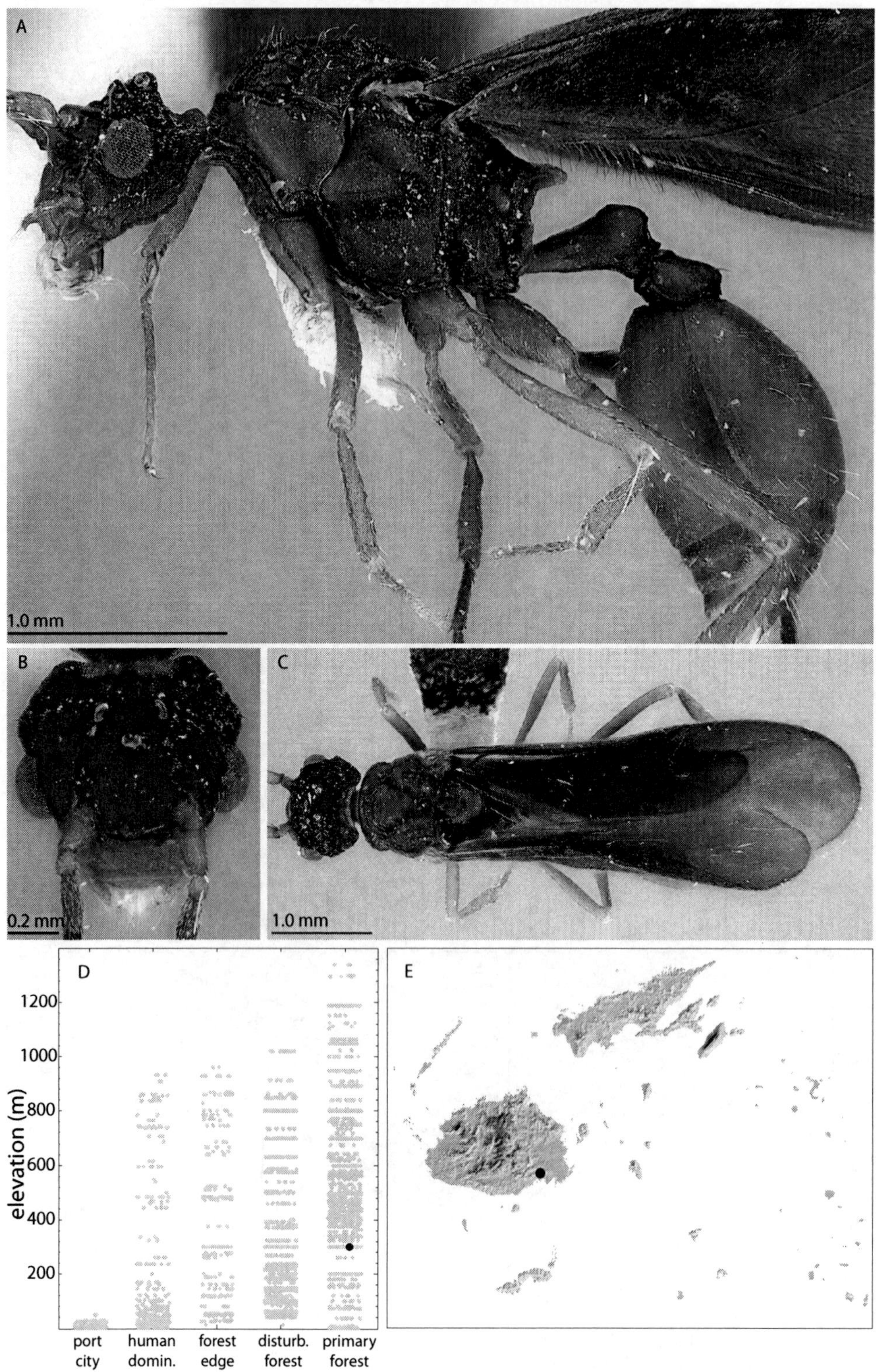

Plate 66. *Lordomyrma curvata* holotype worker, CASENT0171008.

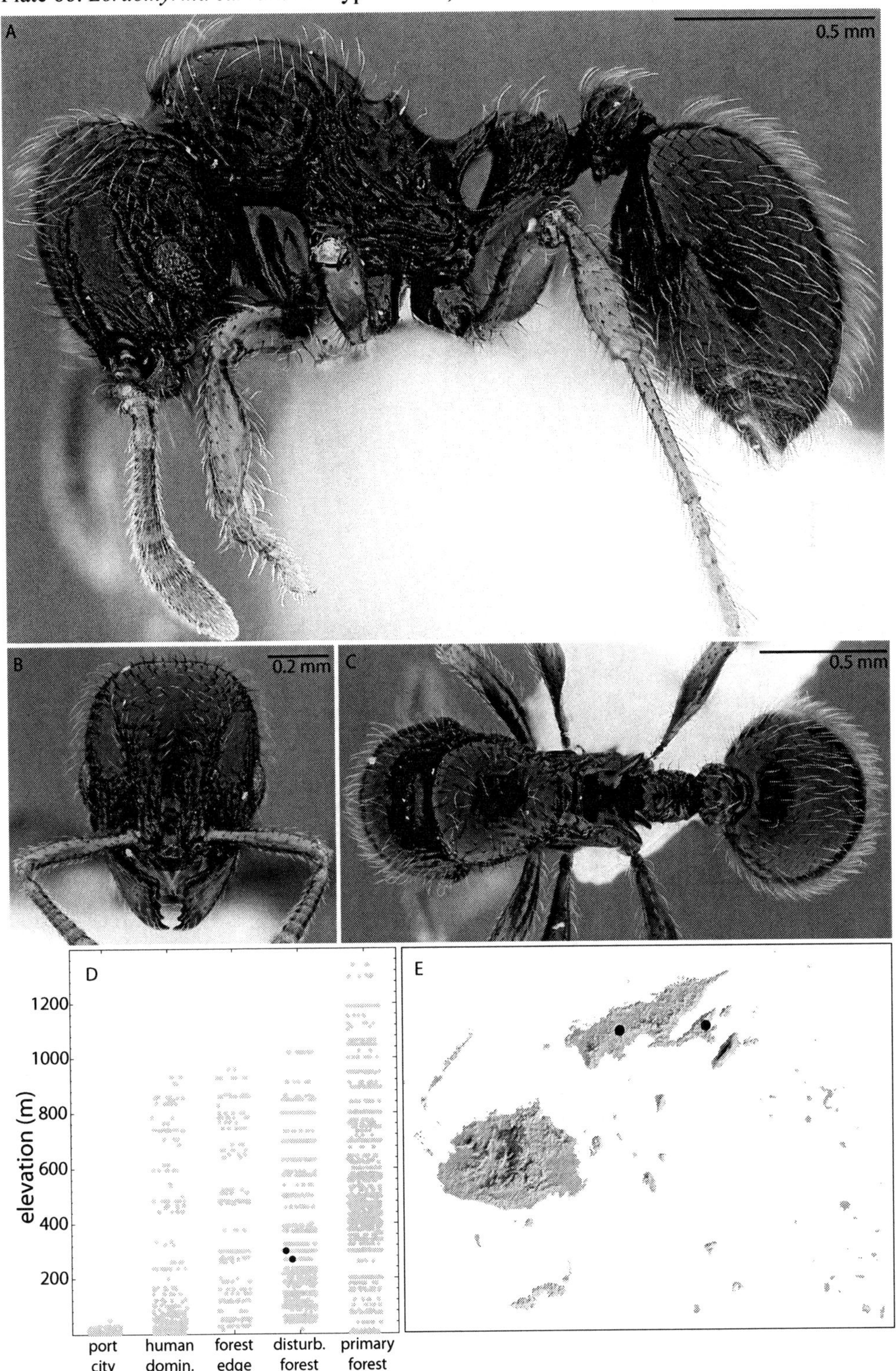

Plate 67. *Lordomyrma desupra* worker, CASENT0171002.

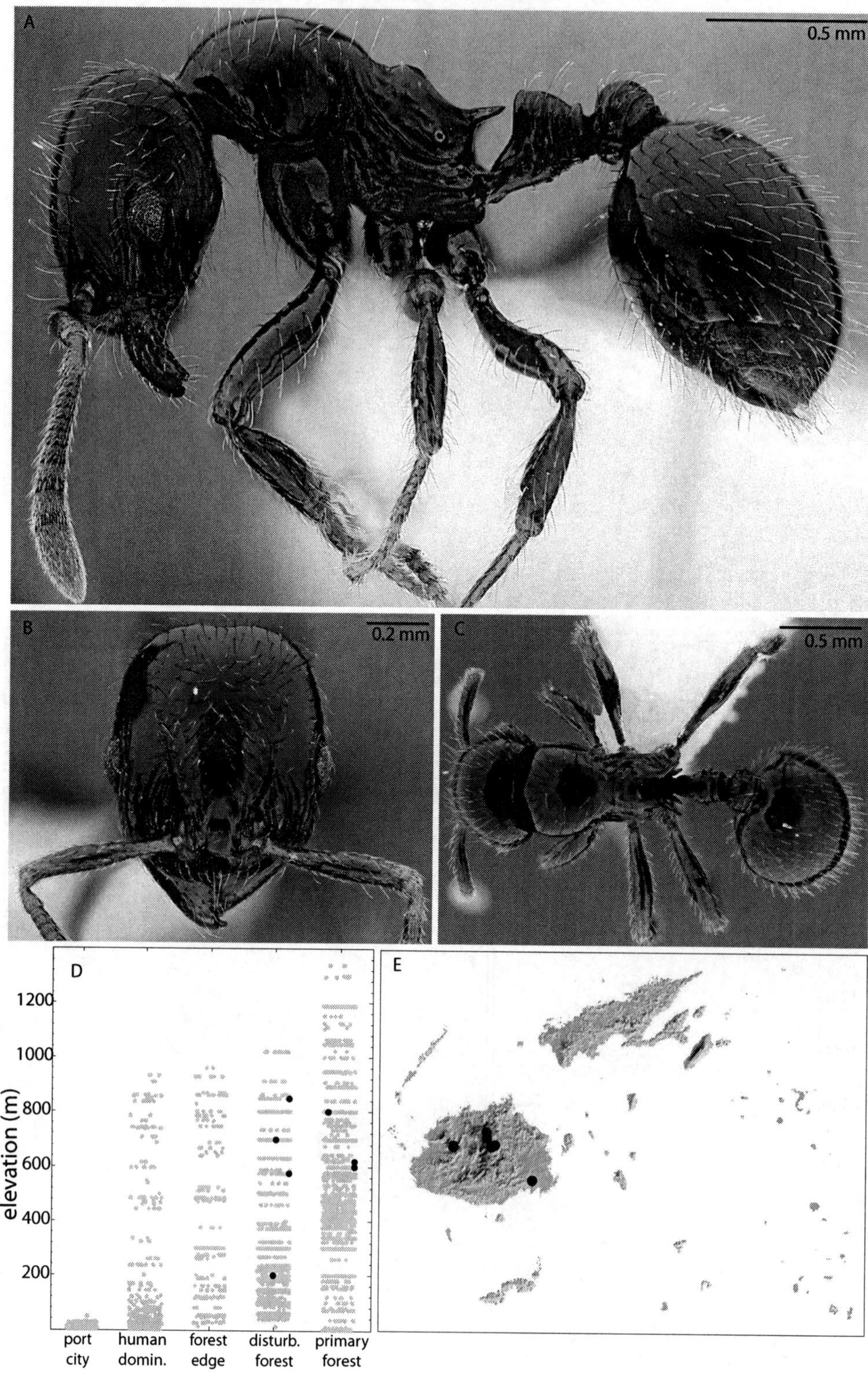

Plate 68. *Lordomyrma levifrons* syntype worker, CASENT0171004.

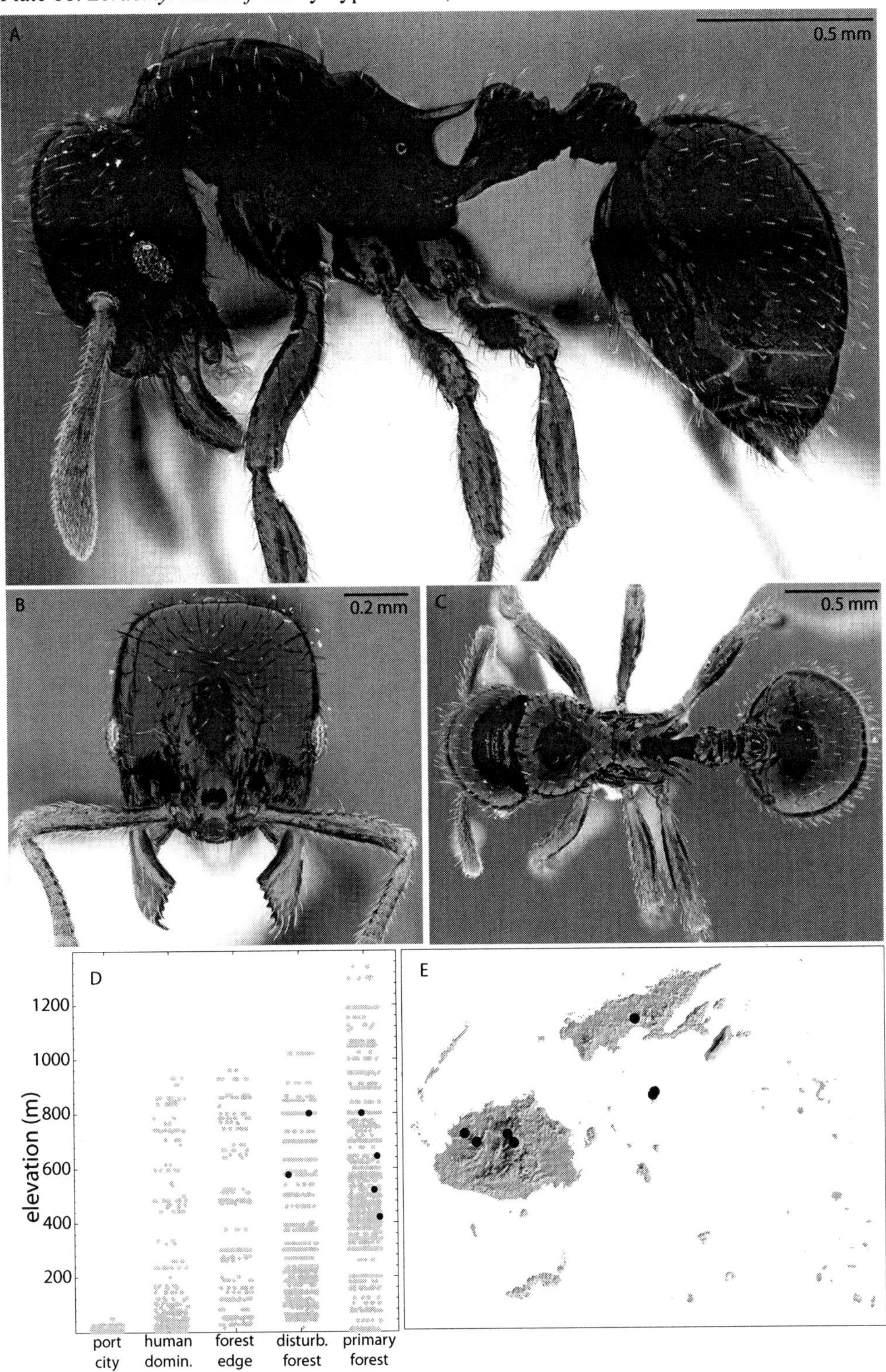

Plate 69. *Lordomyrma polita* worker, CASENT0171007.

Plate 70. *Lordomyrma rugosa* worker, CASENT0171009.

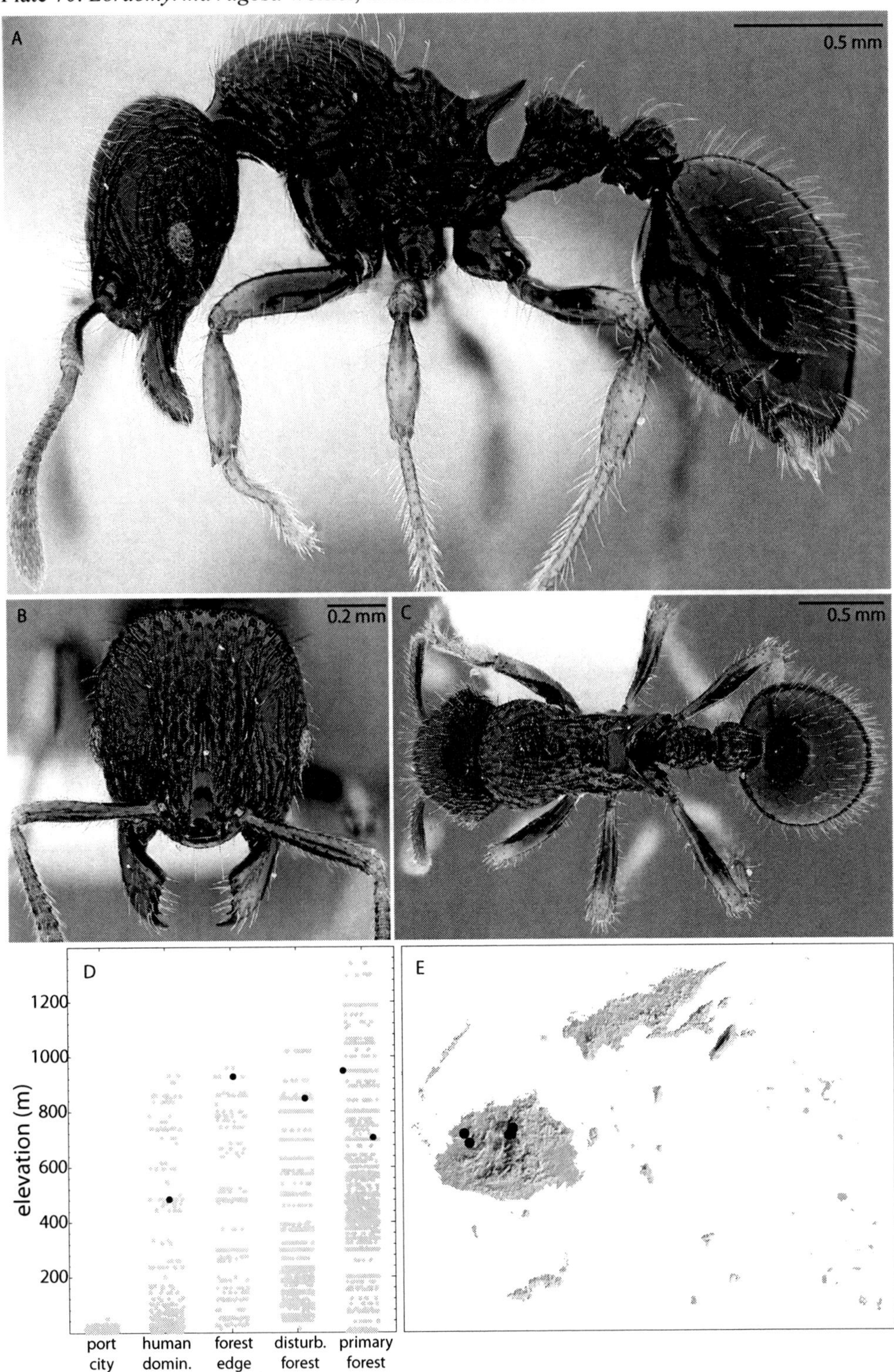

Plate 71. *Lordomyrma stoneri* worker, CASENT0171014.

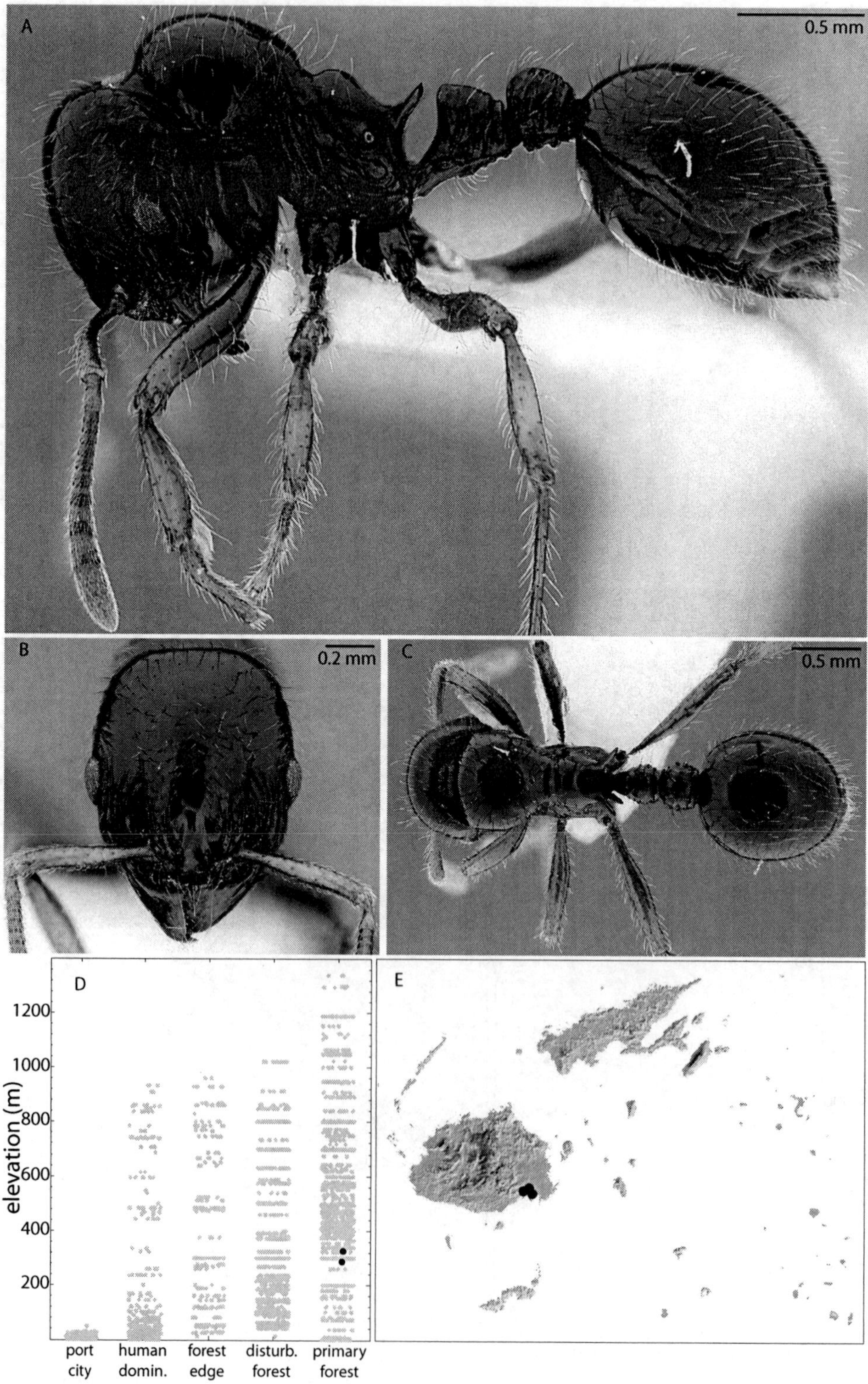

Plate 72. *Lordomyrma striatella* worker, CASENT0171010.

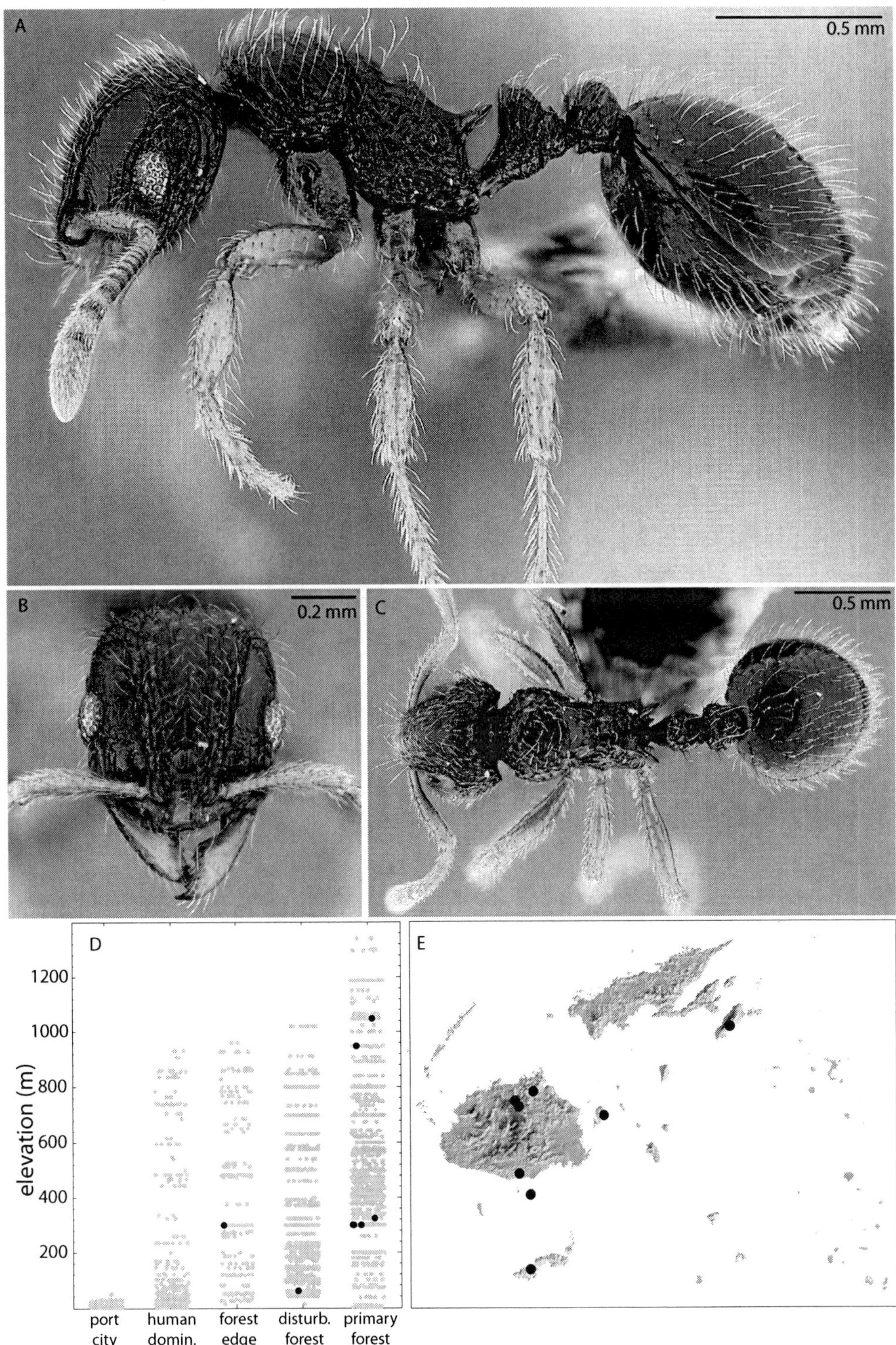

Plate 73. *Lordomyrma sukuna* worker, CASENT0171011.

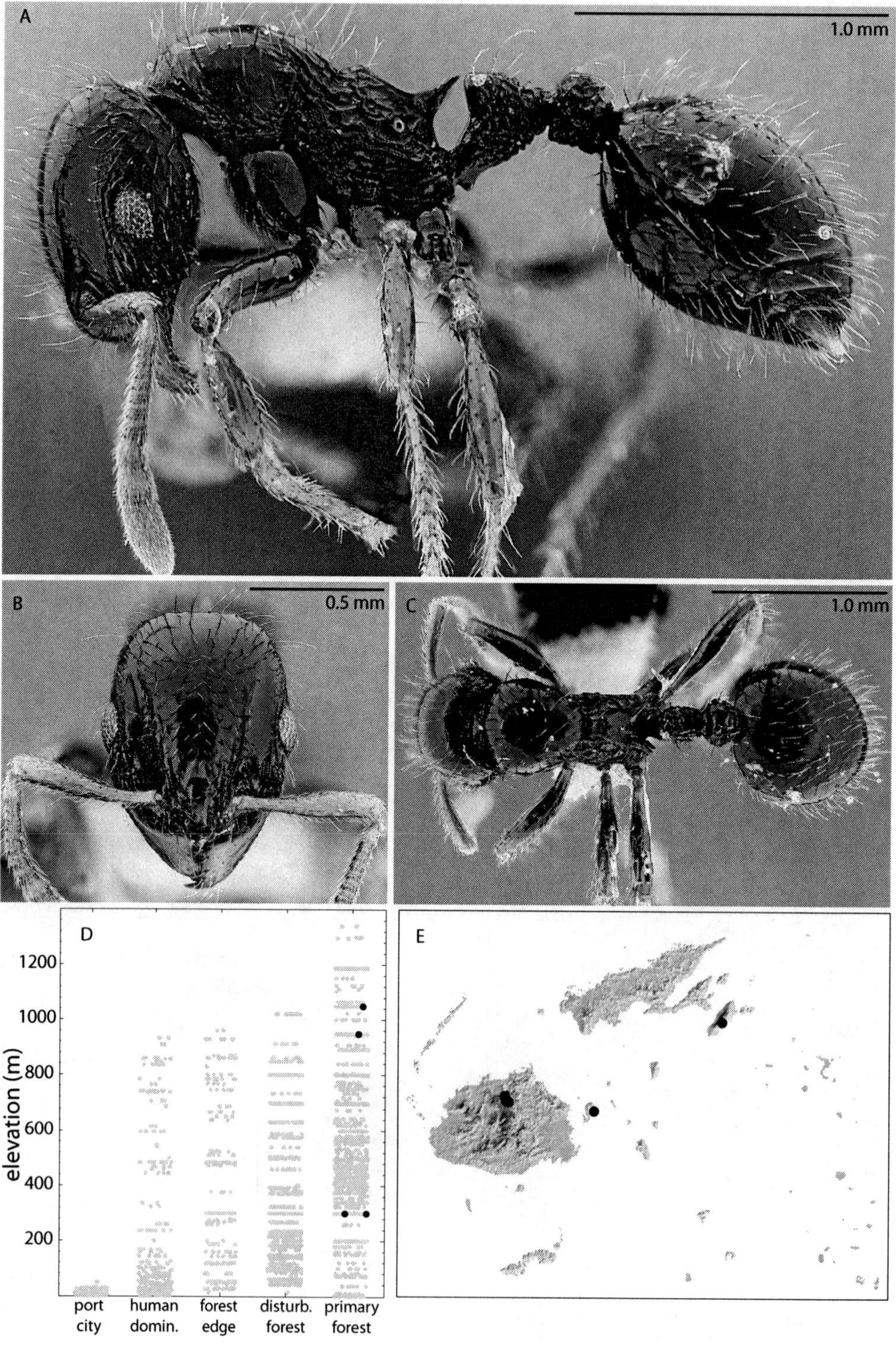

Plate 74. *Lordomyrma tortuosa* worker, CASENT0171000.

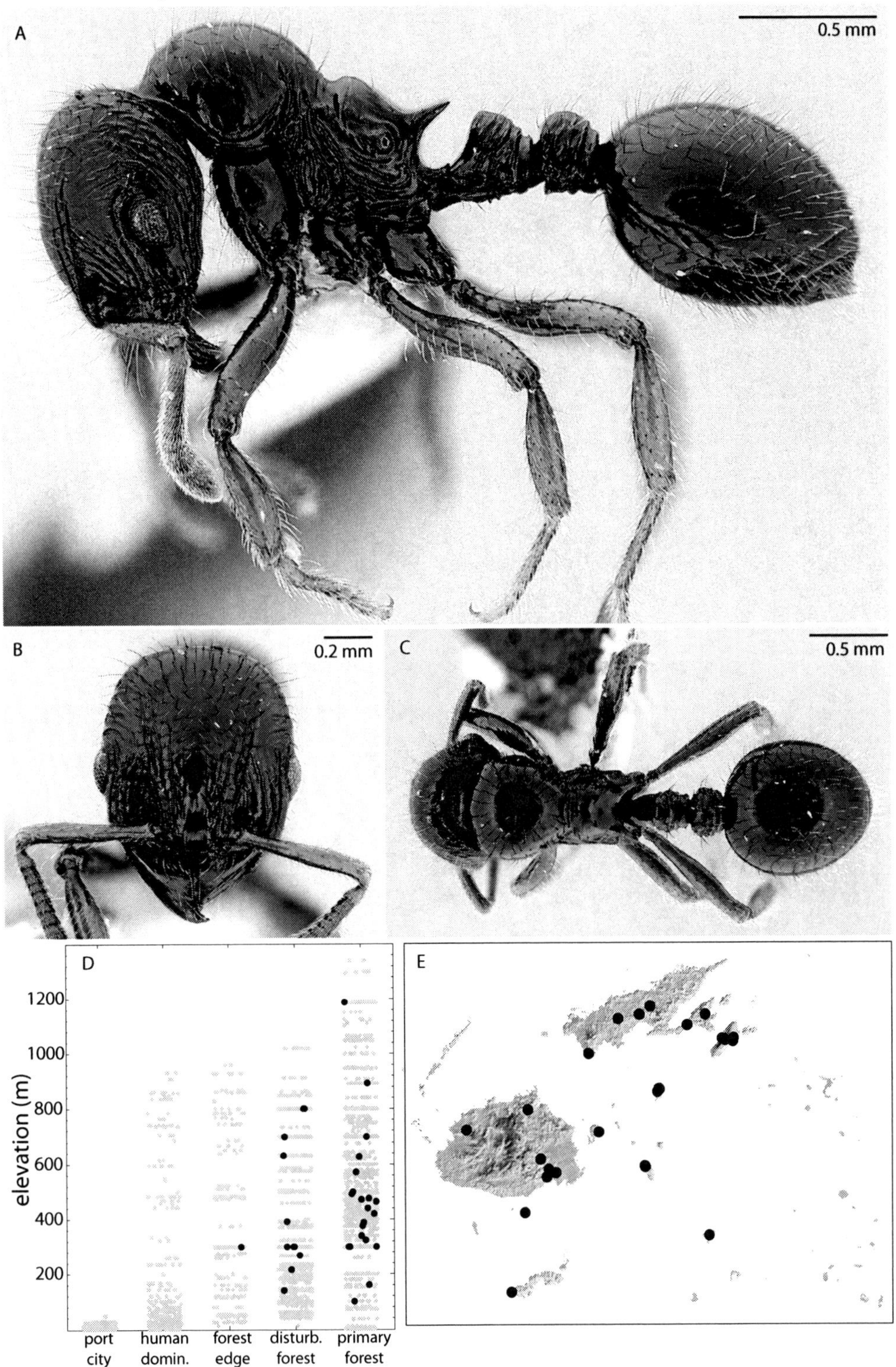

Plate 75. *Lordomyrma vanua* worker, CASENT0171051.

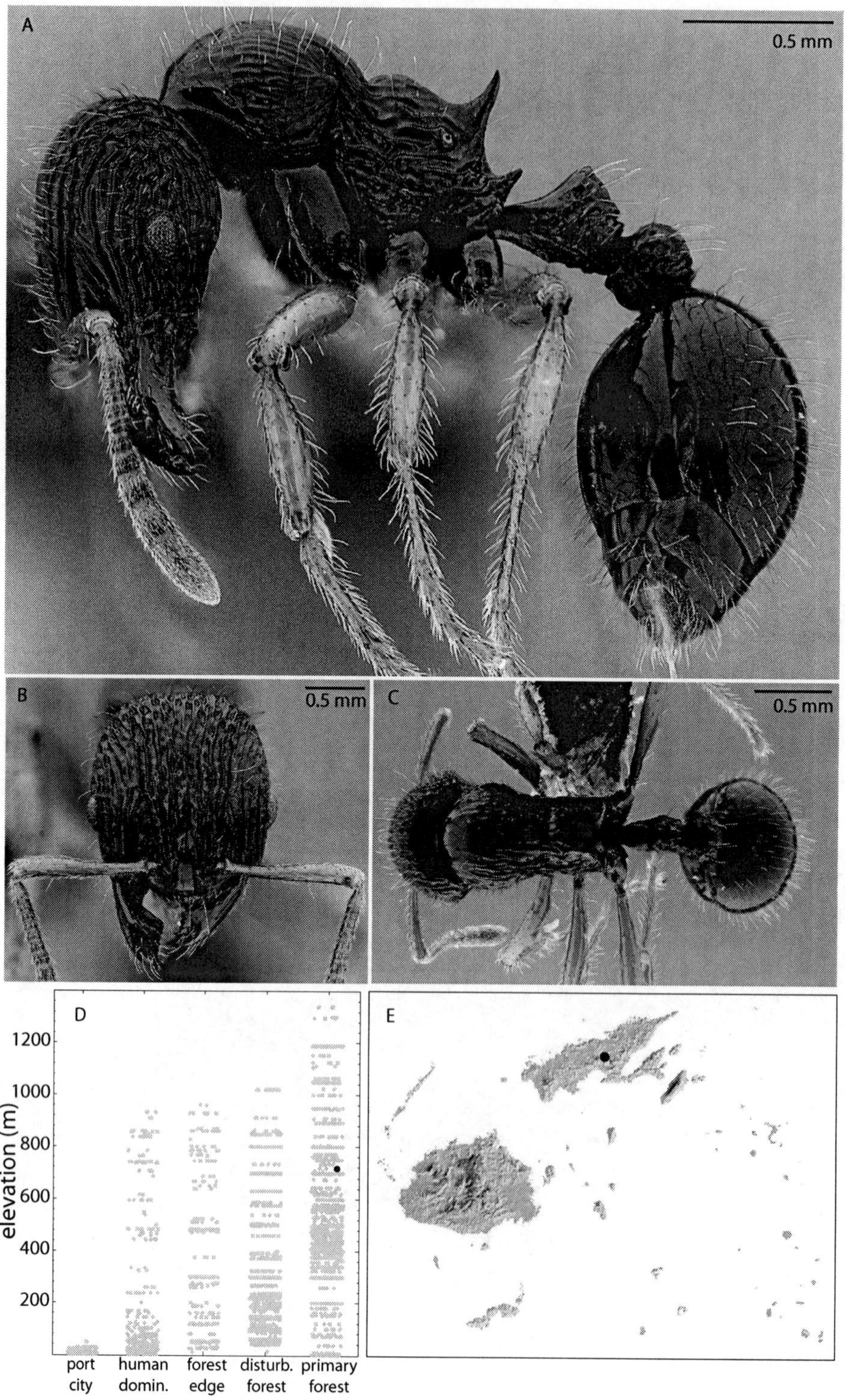

Plate 76. *Lordomyrma vuda* holotype worker, CASENT0171018.

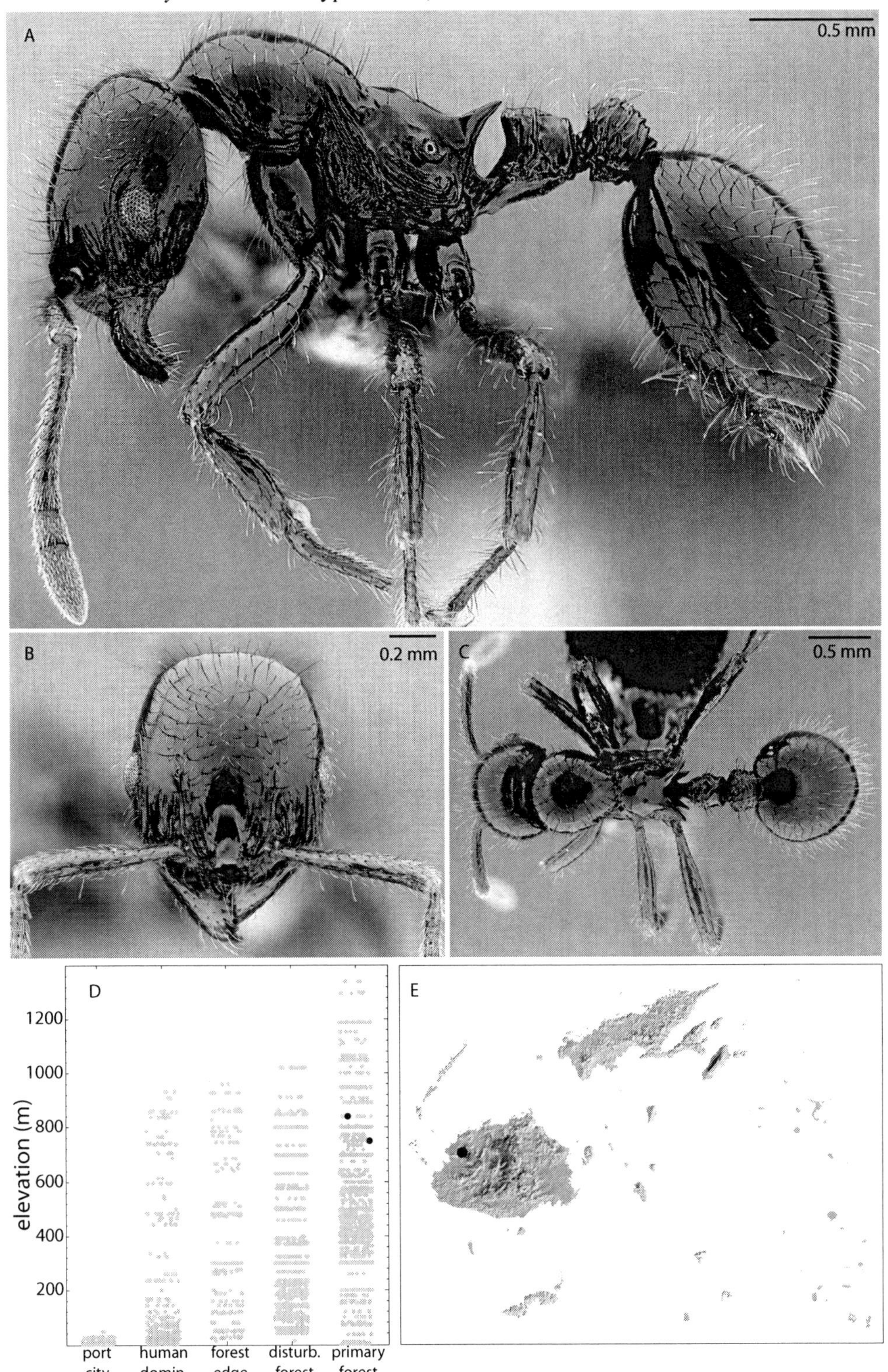

Plate 77. *Metapone* sp. FJ01 queen, CASENT0181802.

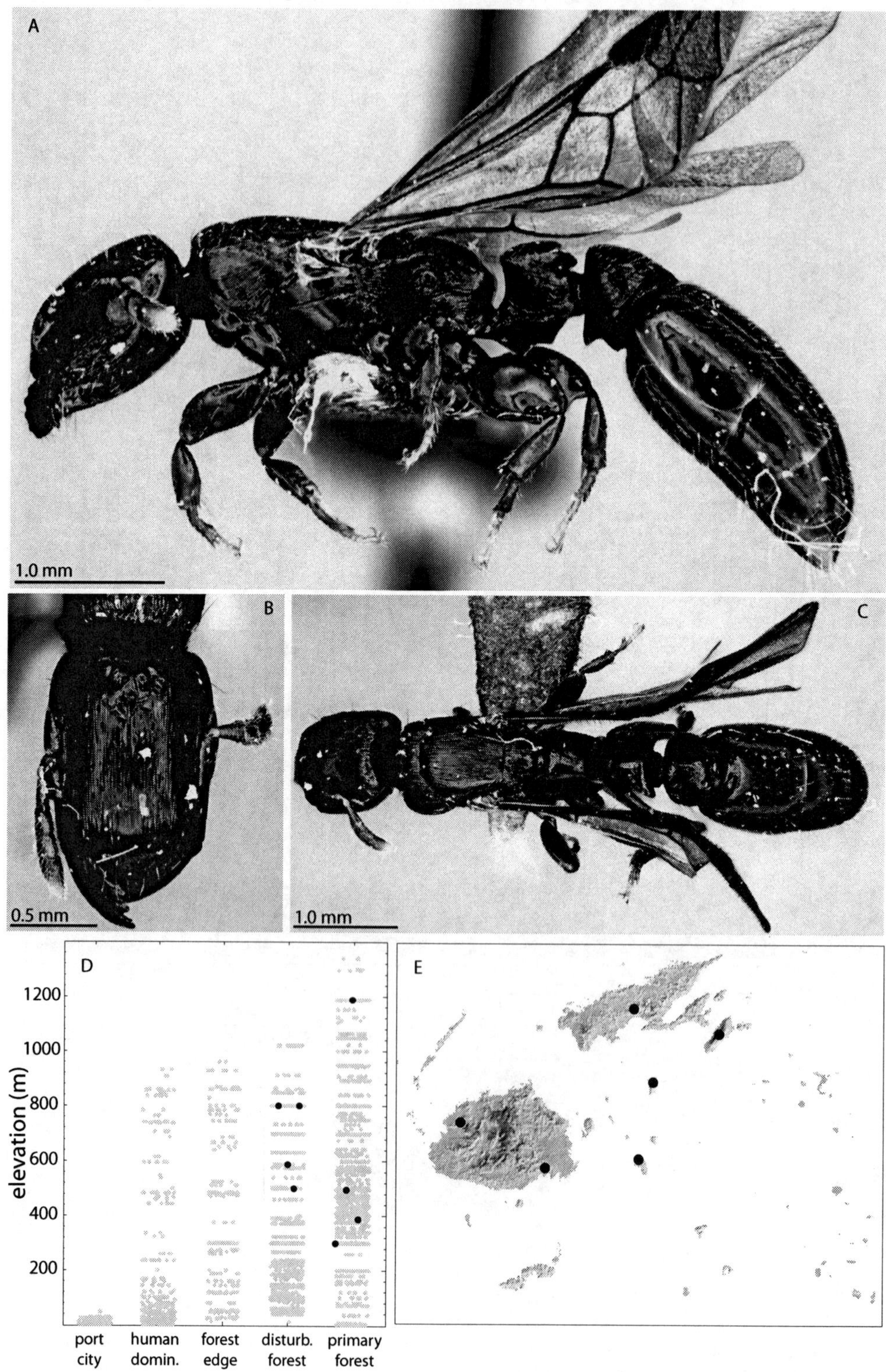

Plate 78. *Monomorium destructor* worker, CASENT0171088.

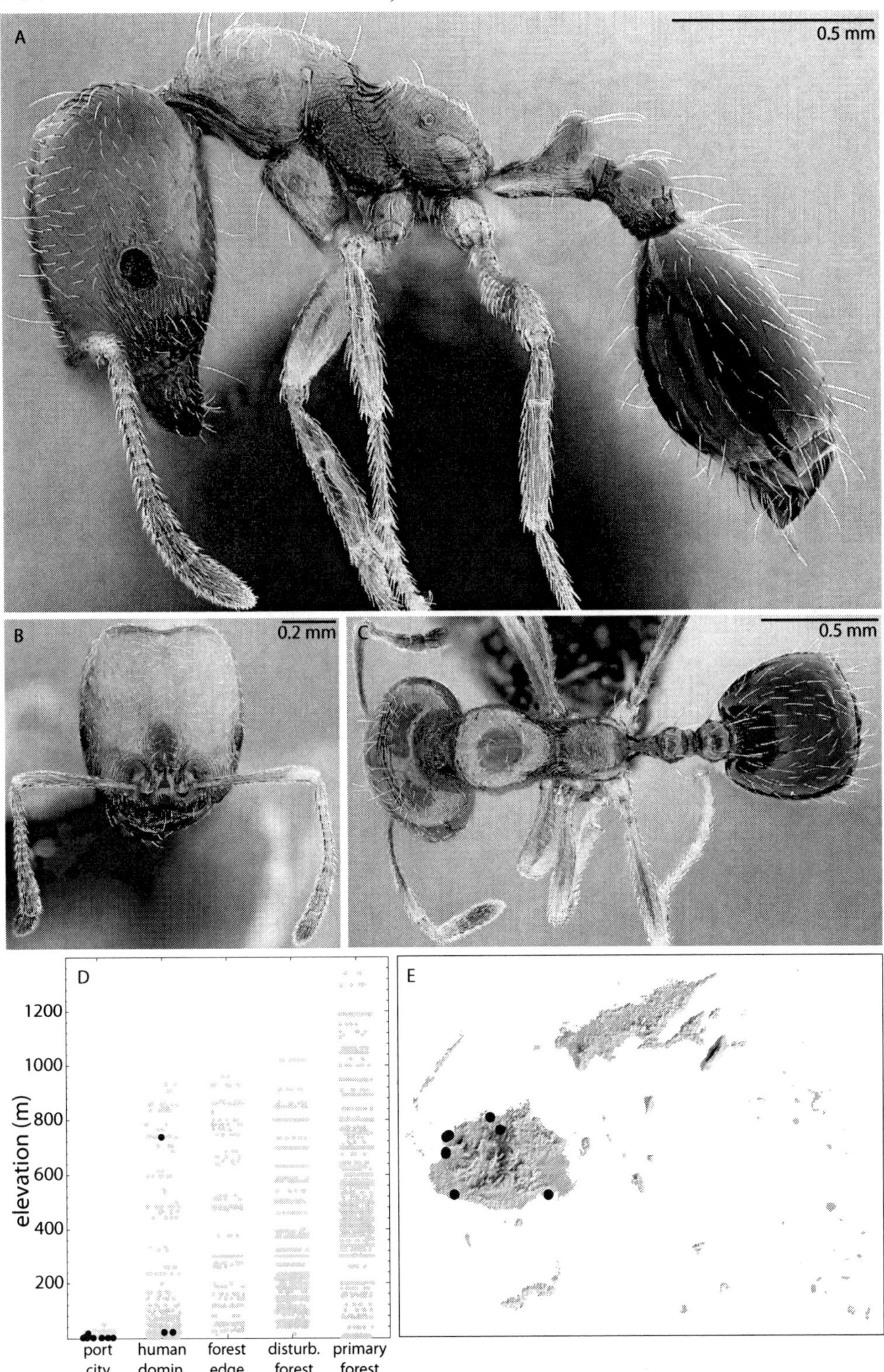

Plate 79. *Monomorium floricola* worker, CASENT0171072.

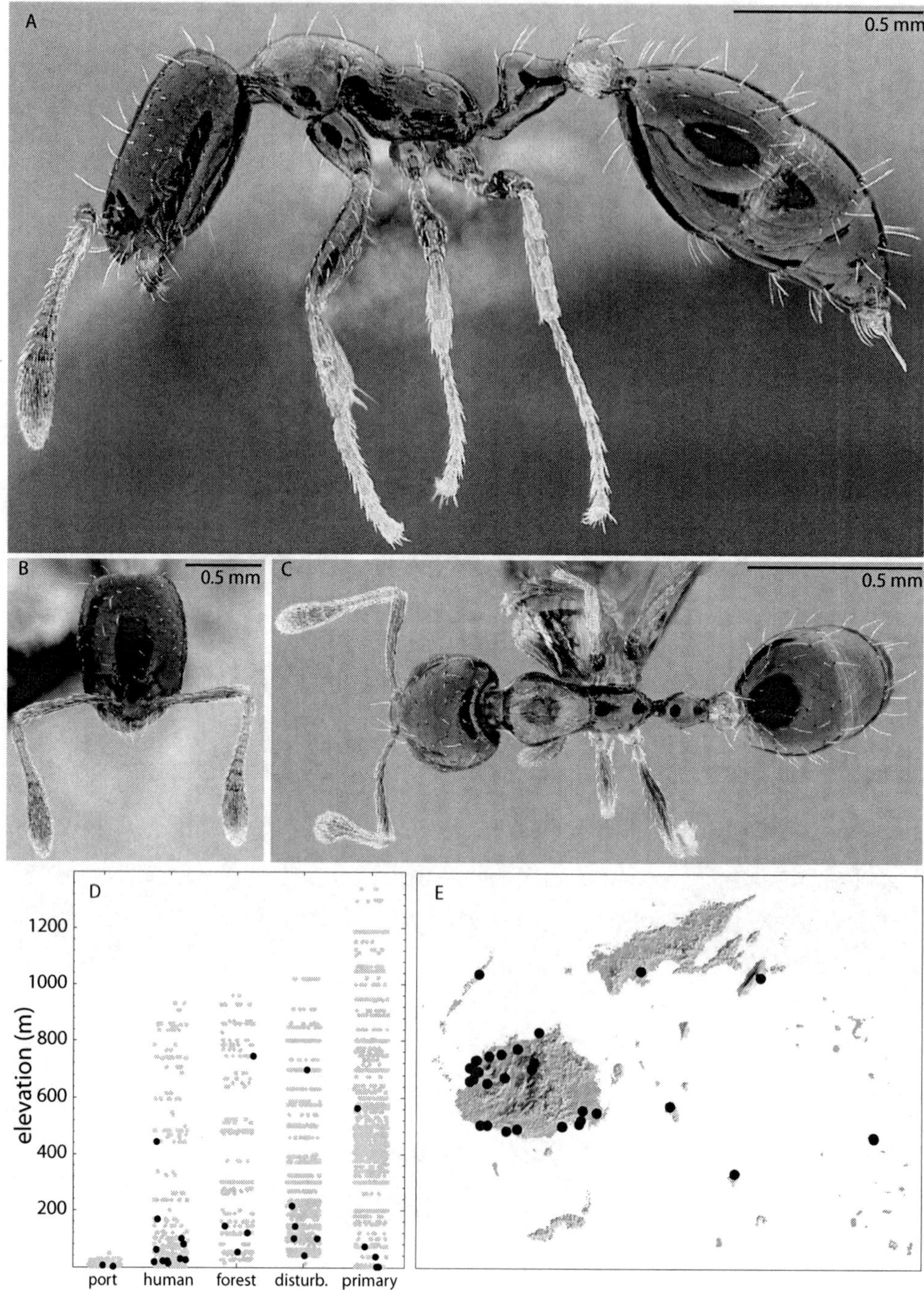

Plate 80. *Monomorium pharaonis* worker, CASENT0171086.

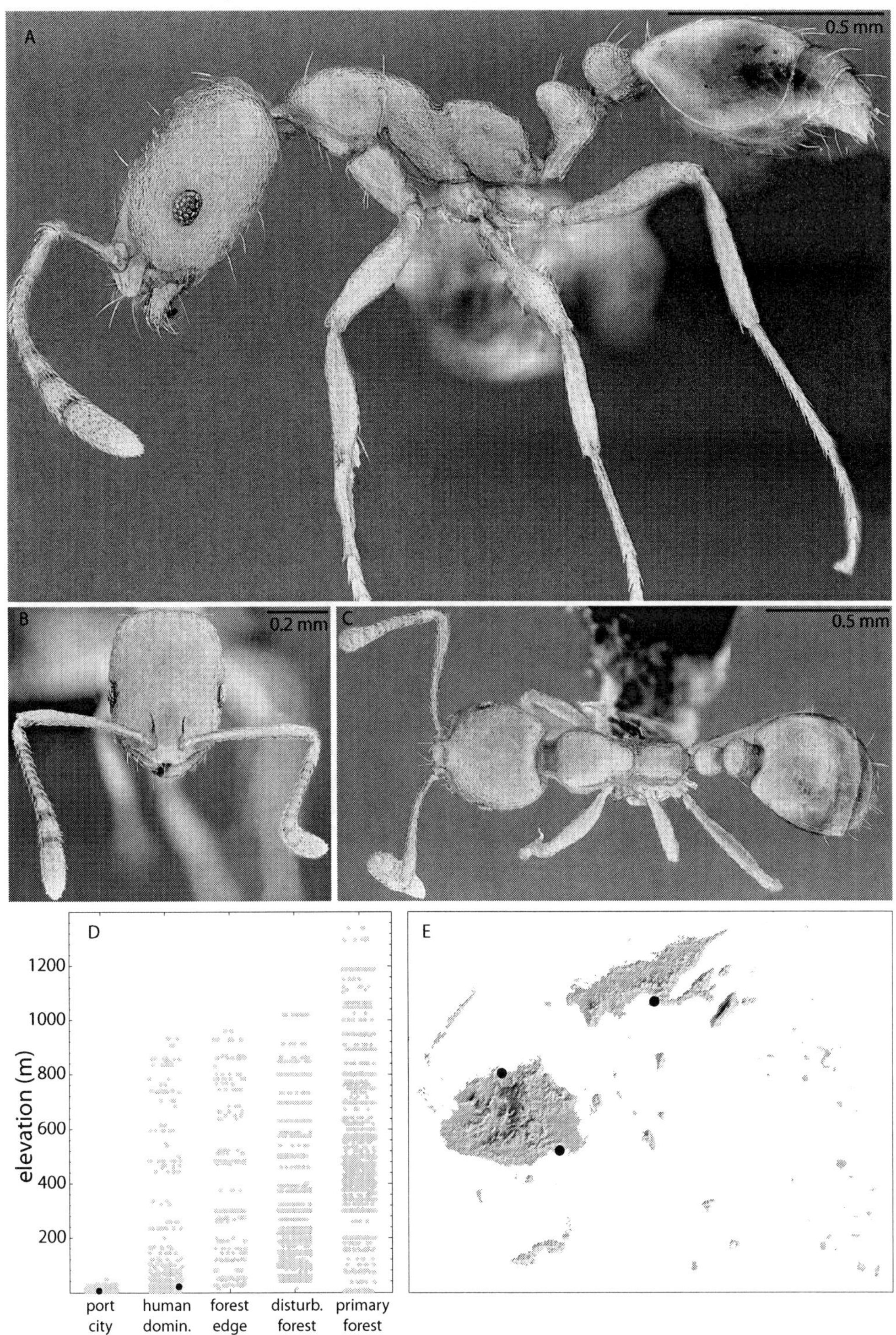

Plate 81. *Monomorium sechellense* worker, CASENT0181877.

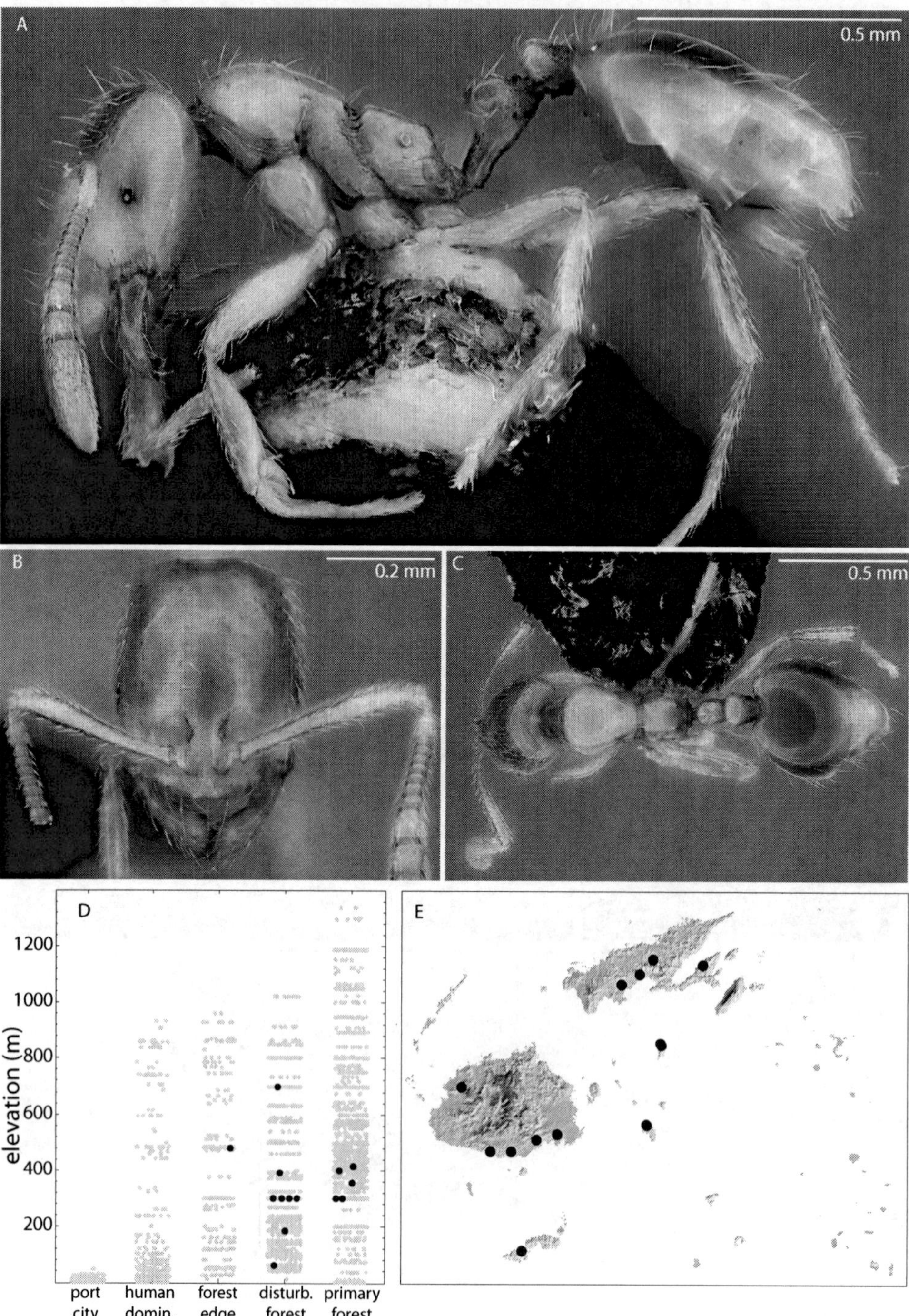

Plate 82. *Monomorium vitiense* worker, CASENT0181872.

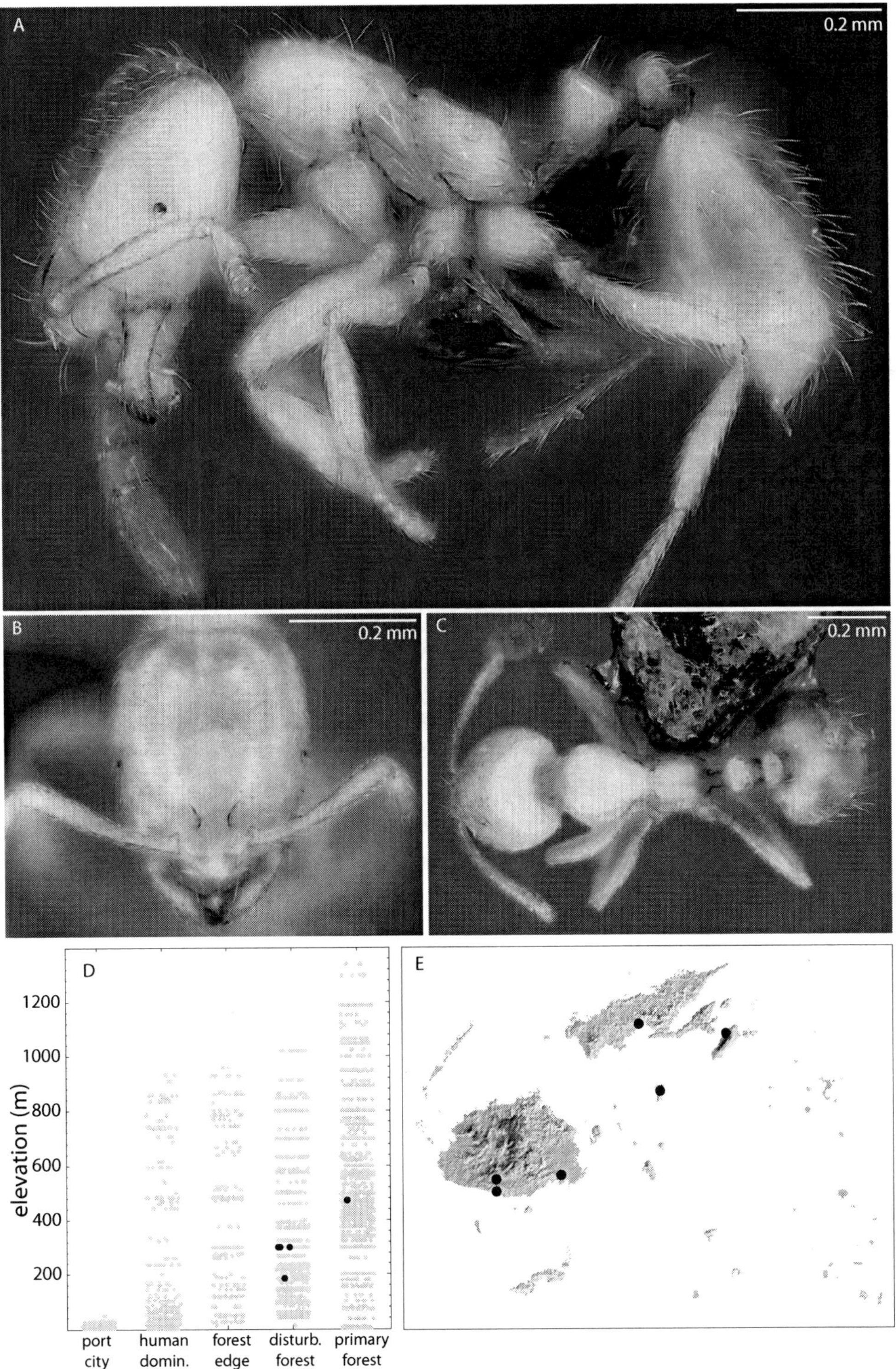

Plate 83. *Monomorium* sp. FJ02 worker, CASENT0181926.

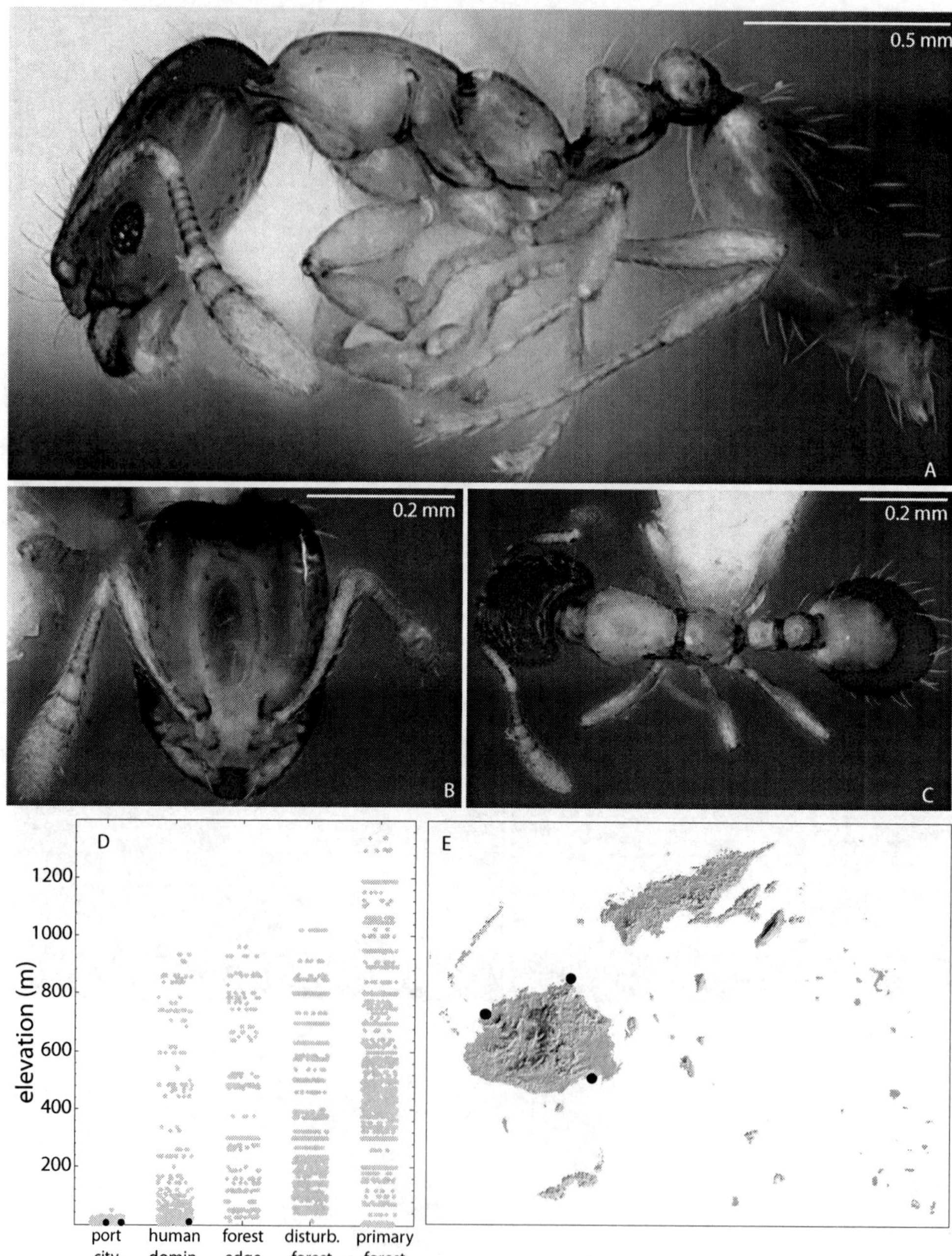

Plate 84. *Myrmecina cacabau* holotype worker, CASENT0260381.

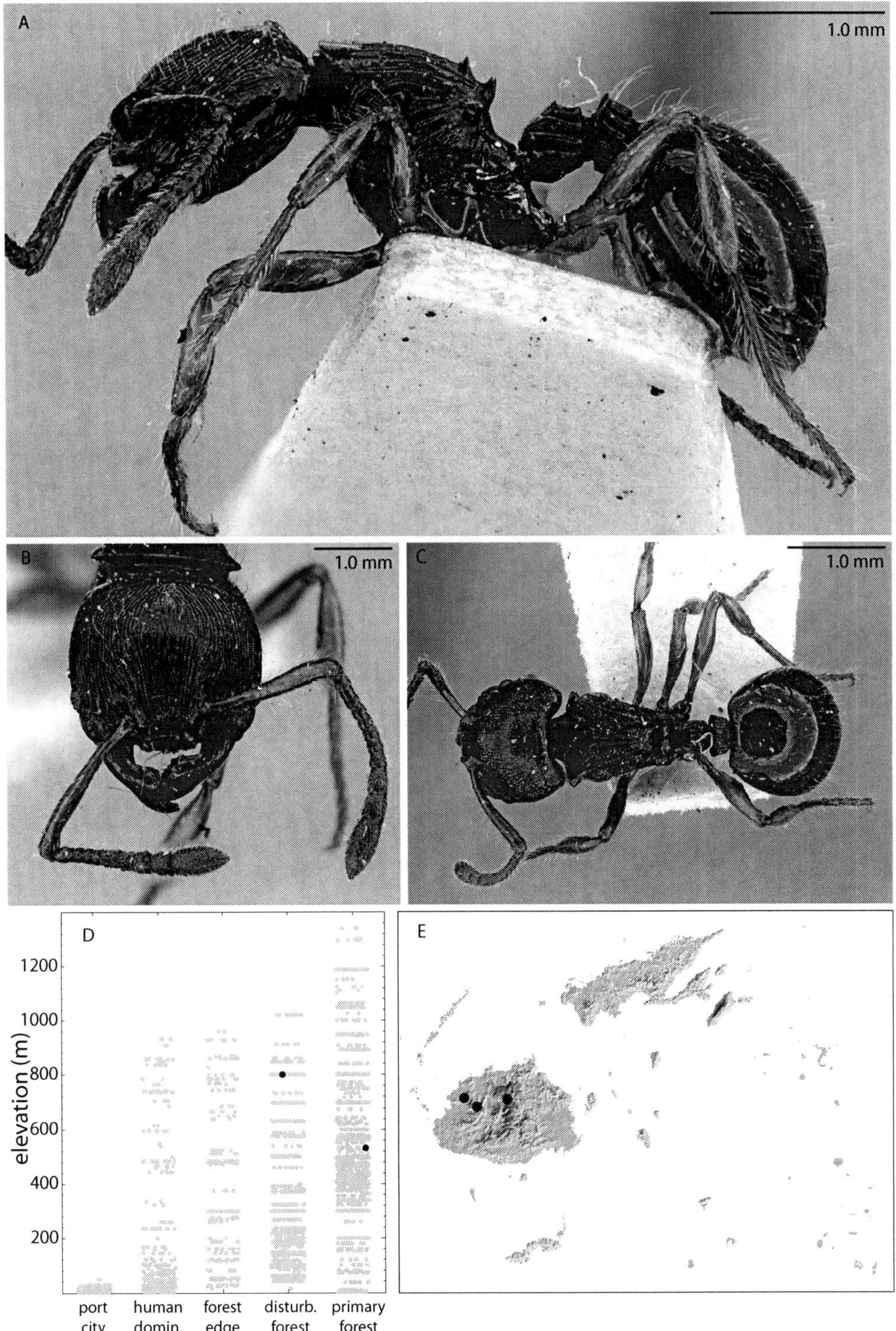

Plate 85. *Myrmecina* sp. FJ01 worker, CASENT0182571.

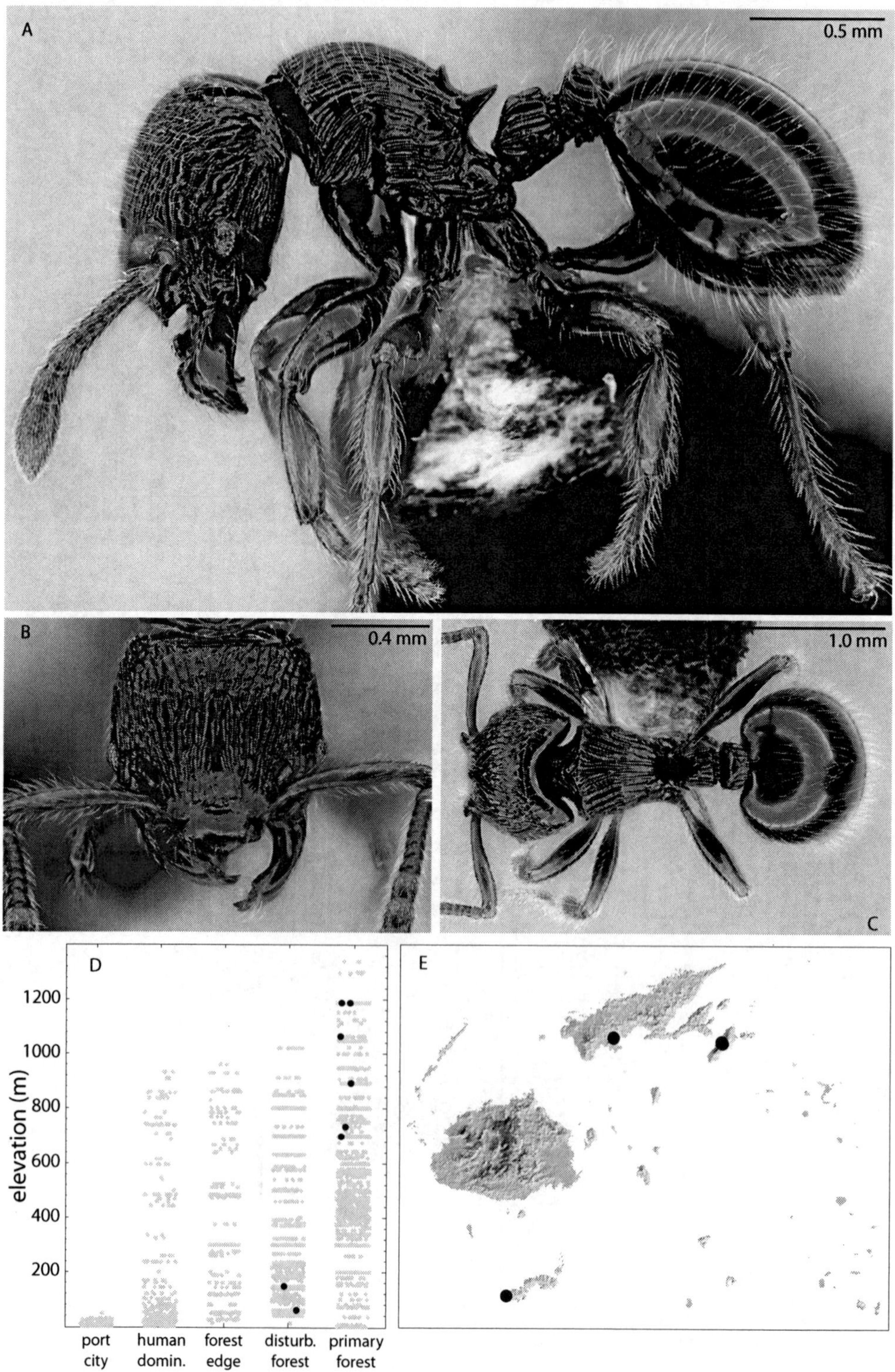

Plate 86. *Pheidole caldwelli* worker, CASENT0185147.

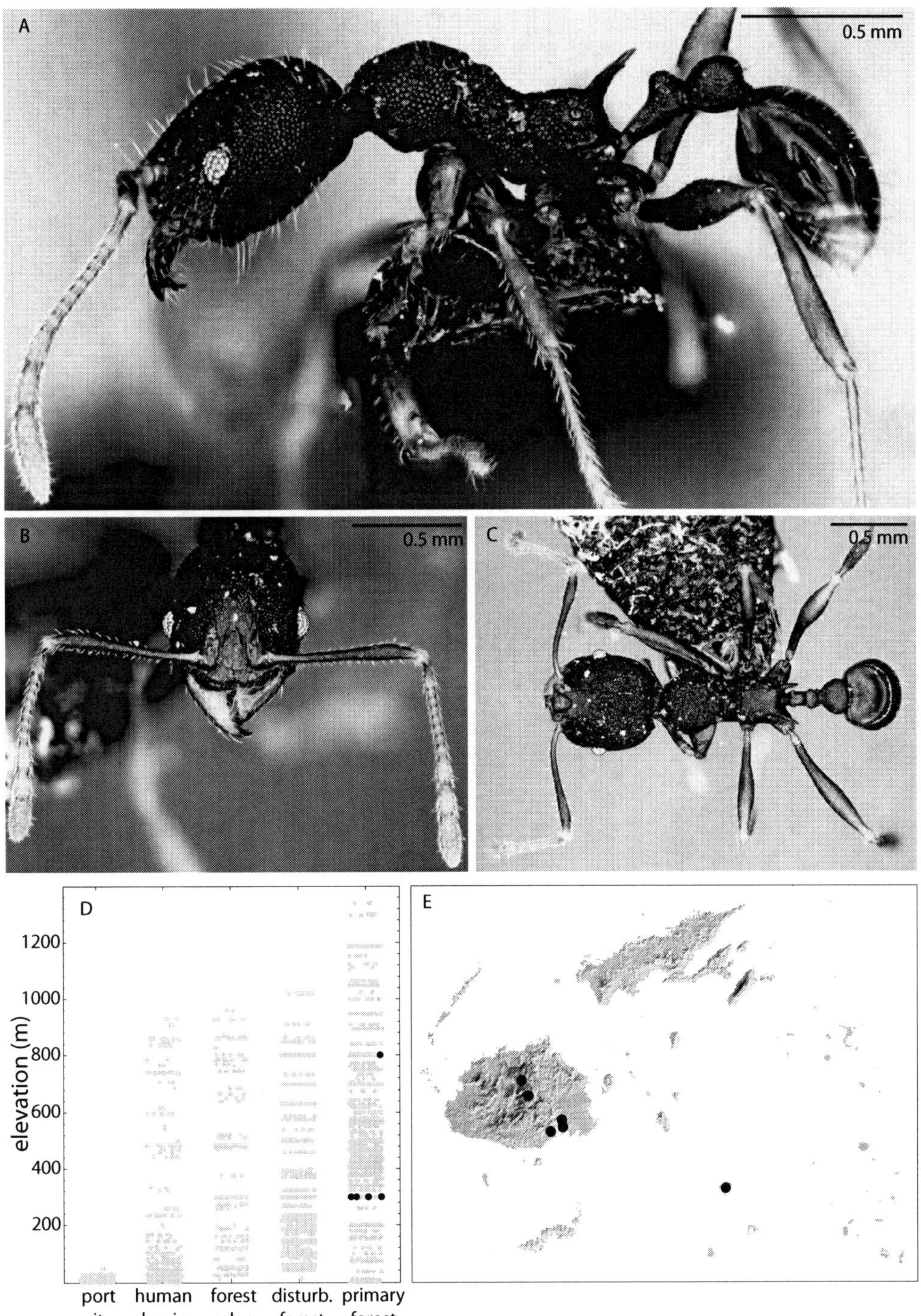

Plate 87. *Pheidole fervens* (minor, CASENT0171076; major CASENT0171099).

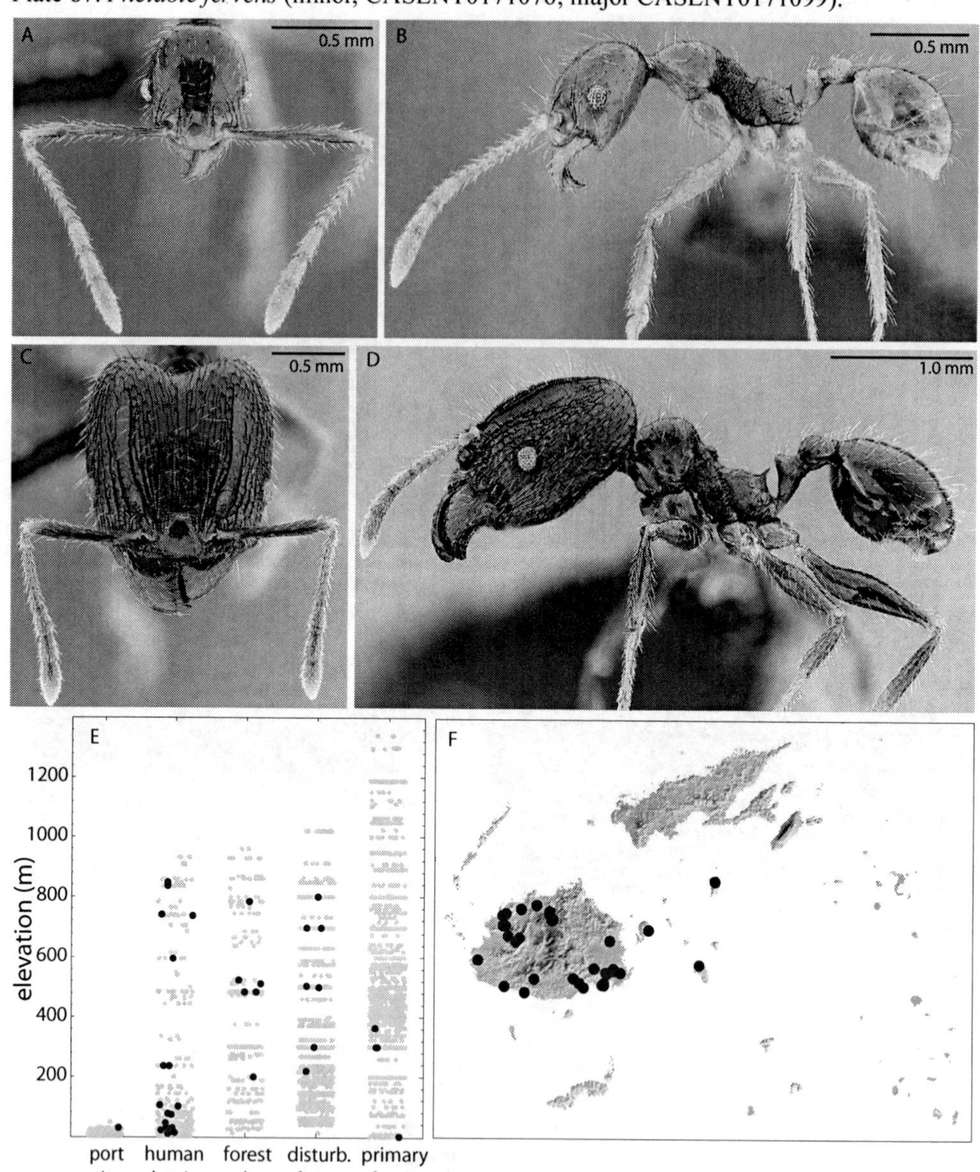

Plate 88. *Pheidole knowlesi* (minor, CASENT0171041; major CASENT0171097).

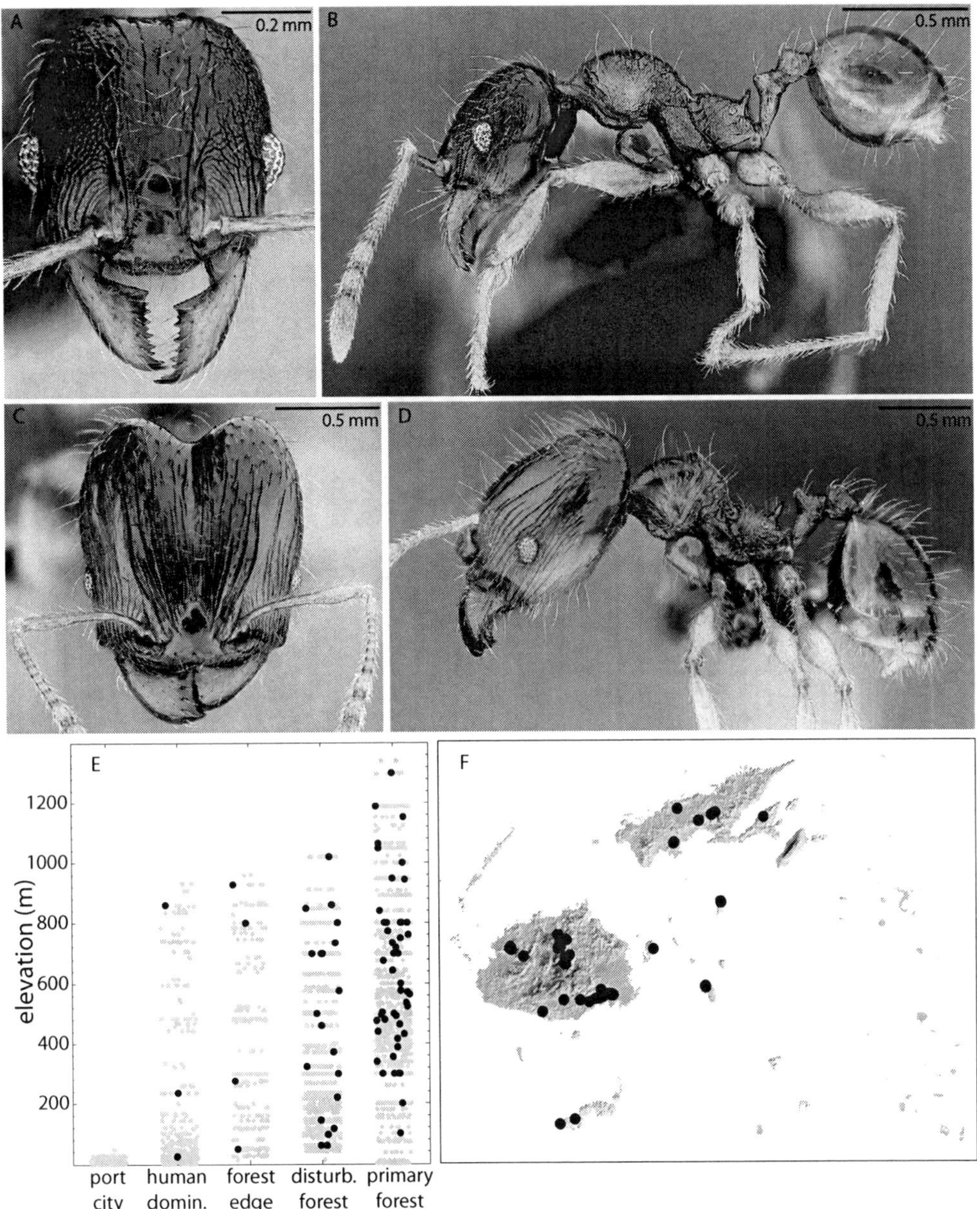

Plate 89. *Pheidole megacephala* (minor, CASENT0171092; major, CASENT0171036).

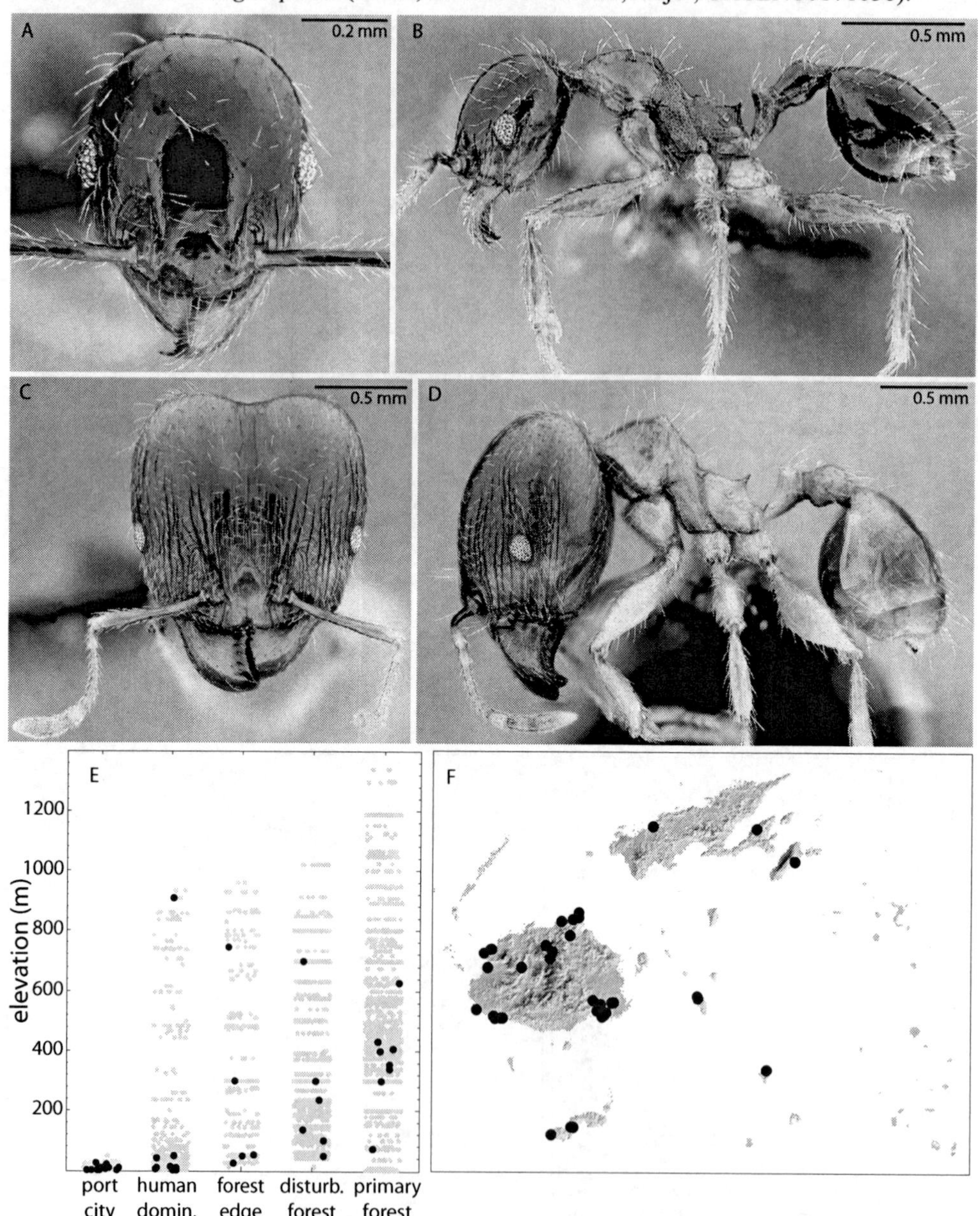

Plate 90. *Pheidole oceanica* (minor, CASENT0171127, major, CASENT0171126).

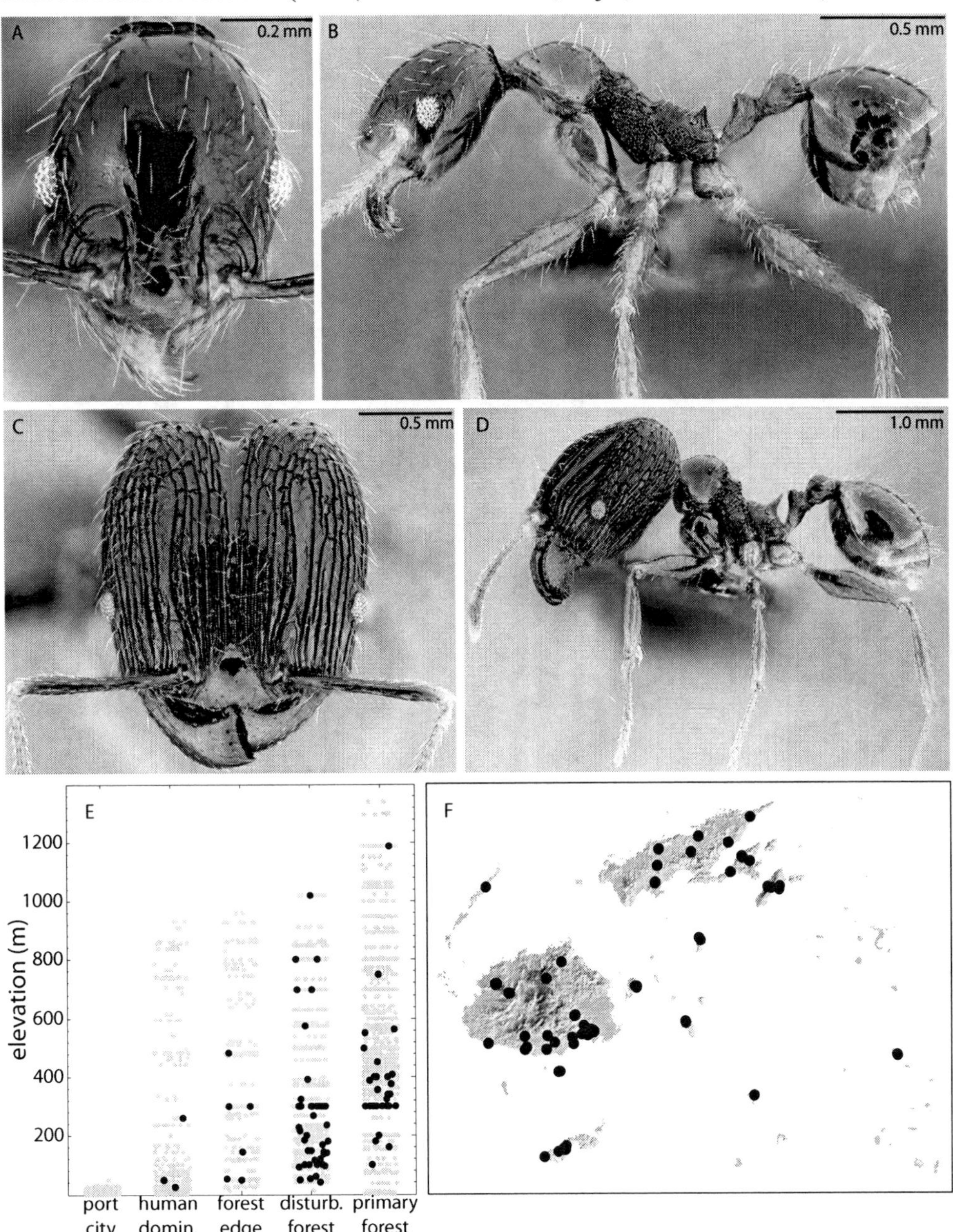

Plate 91. *Pheidole onifera* (minor, CASENT0185425; major, CASENT0185253).

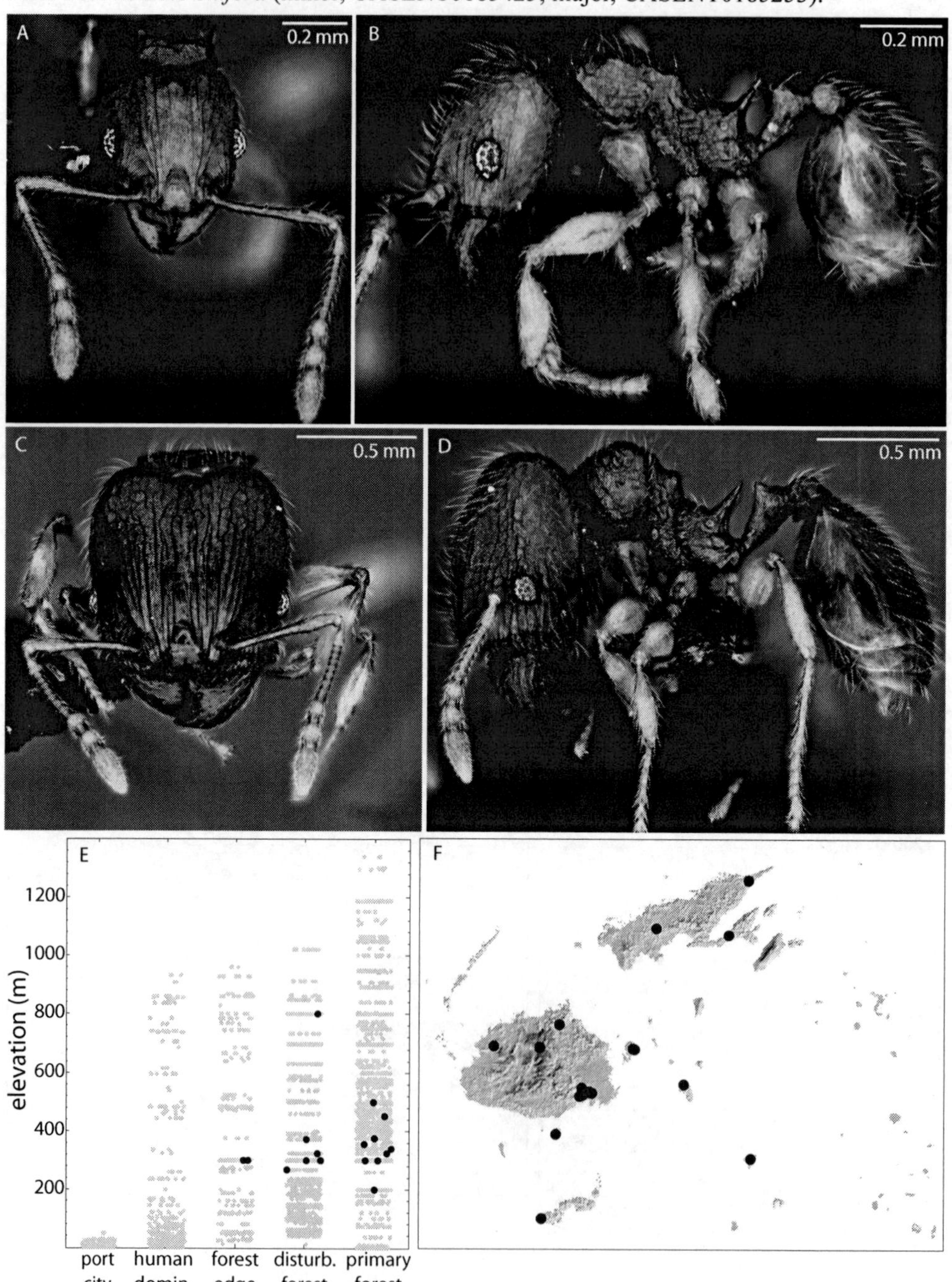

Plate 92. *Pheidole sexspinosa* worker, CASENT0194651.

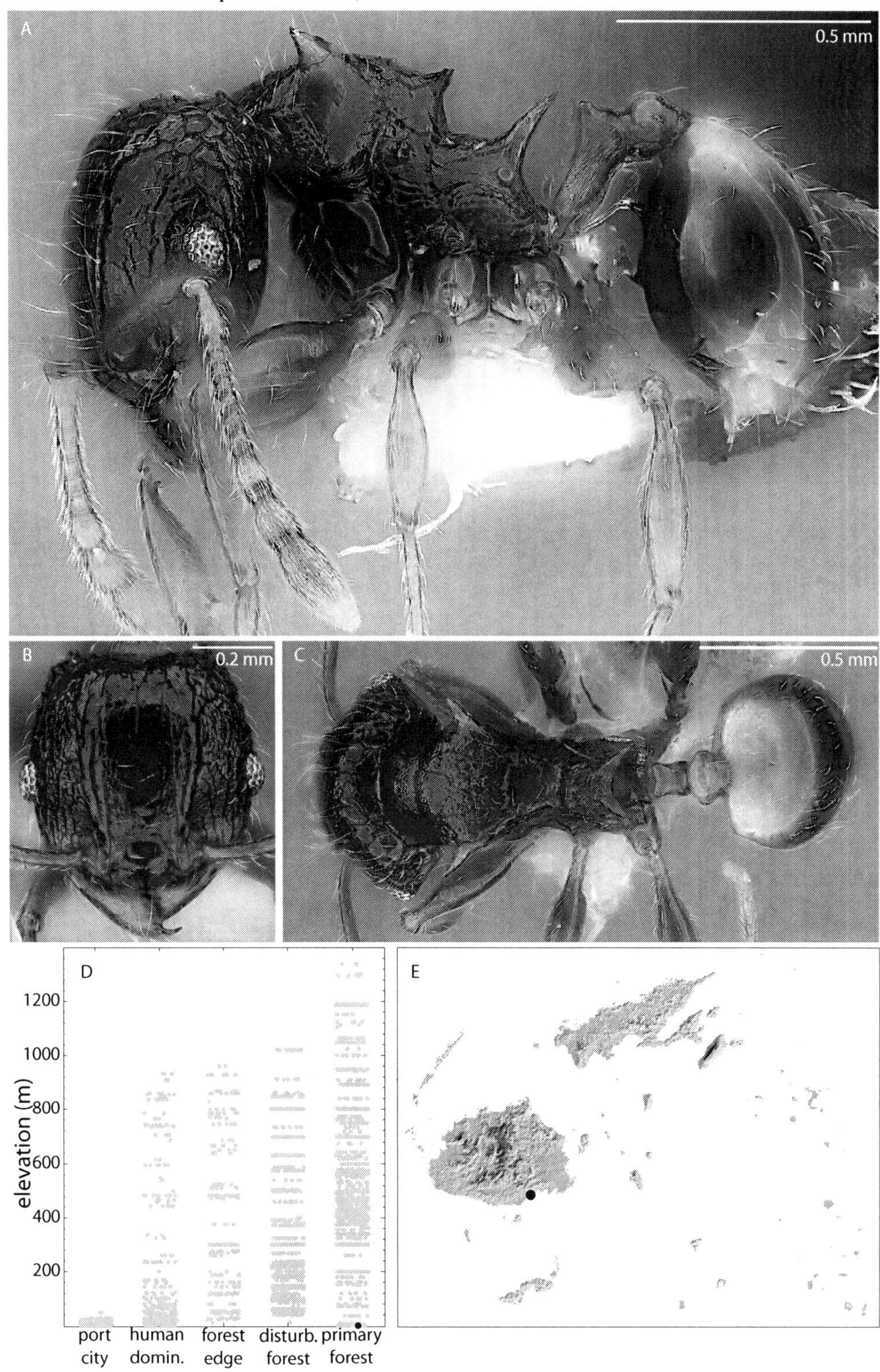

Plate 93. *Pheidole umbonata* (minor, CASENT0171136, minor CASENT0171135).

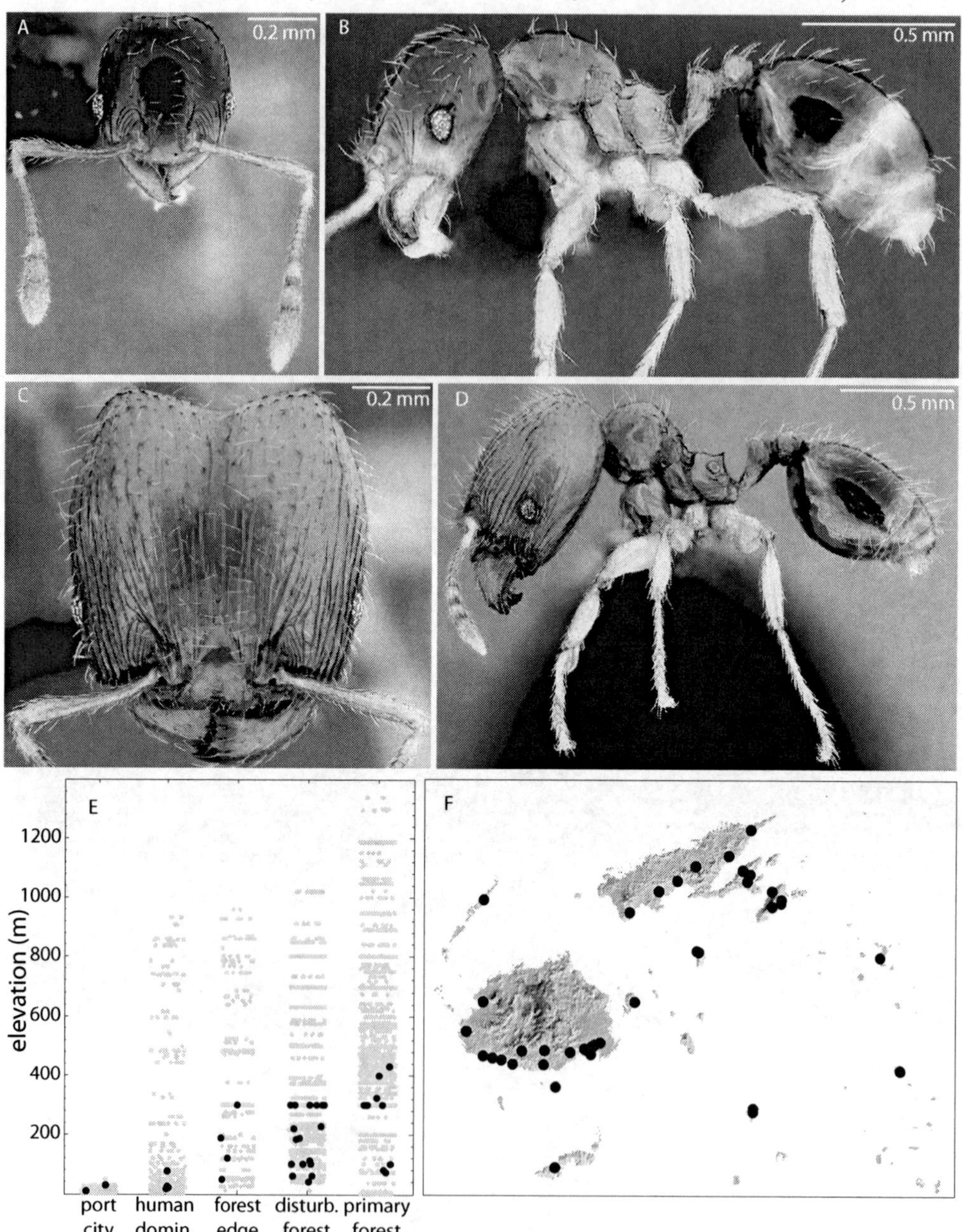

Plate 94. *Pheidole vatu* (minor, CASENT0185673; major CASENT0185525).

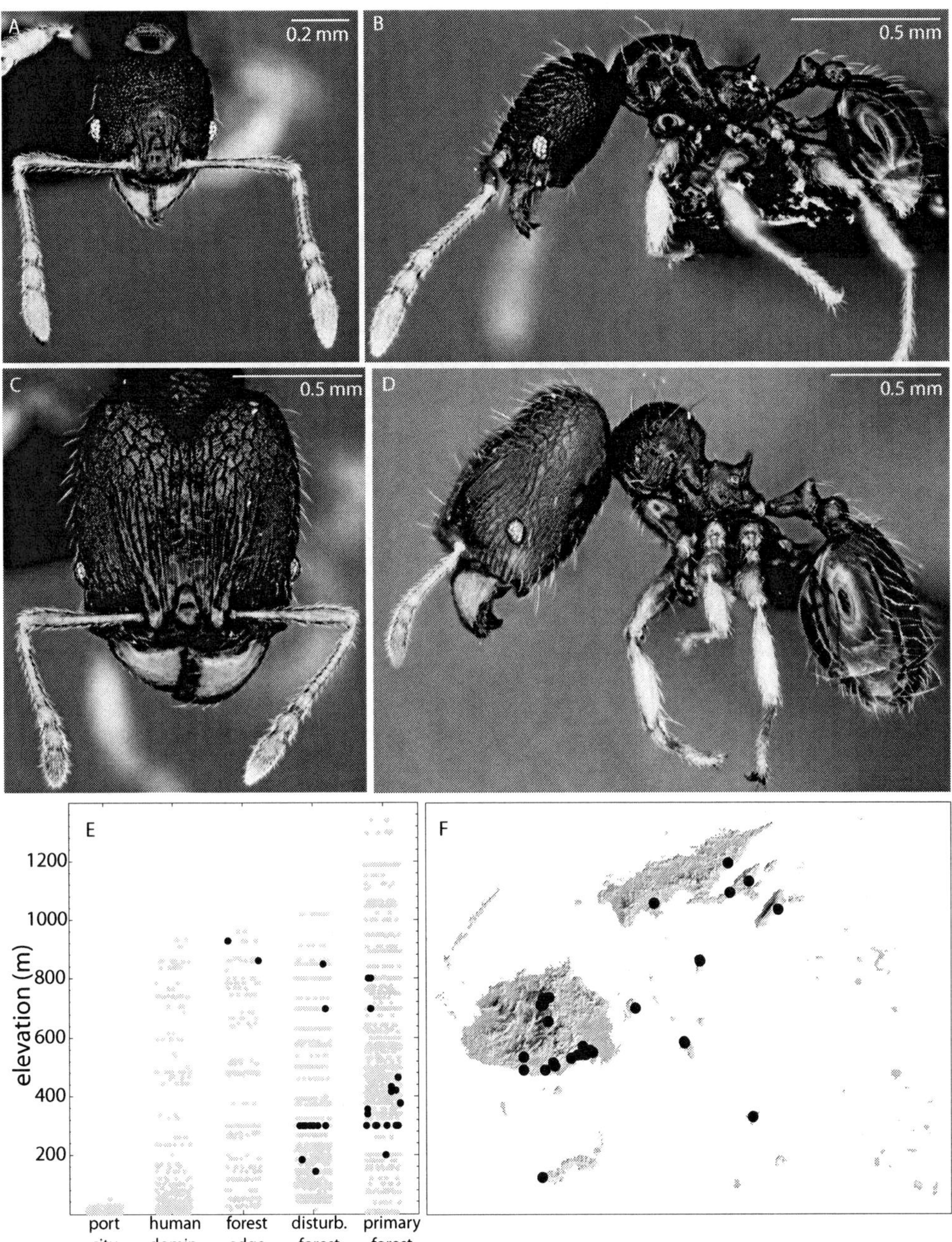

Plate 95. *Pheidole wilsoni* (minor, CASENT0183603; major, CASENT0183622).

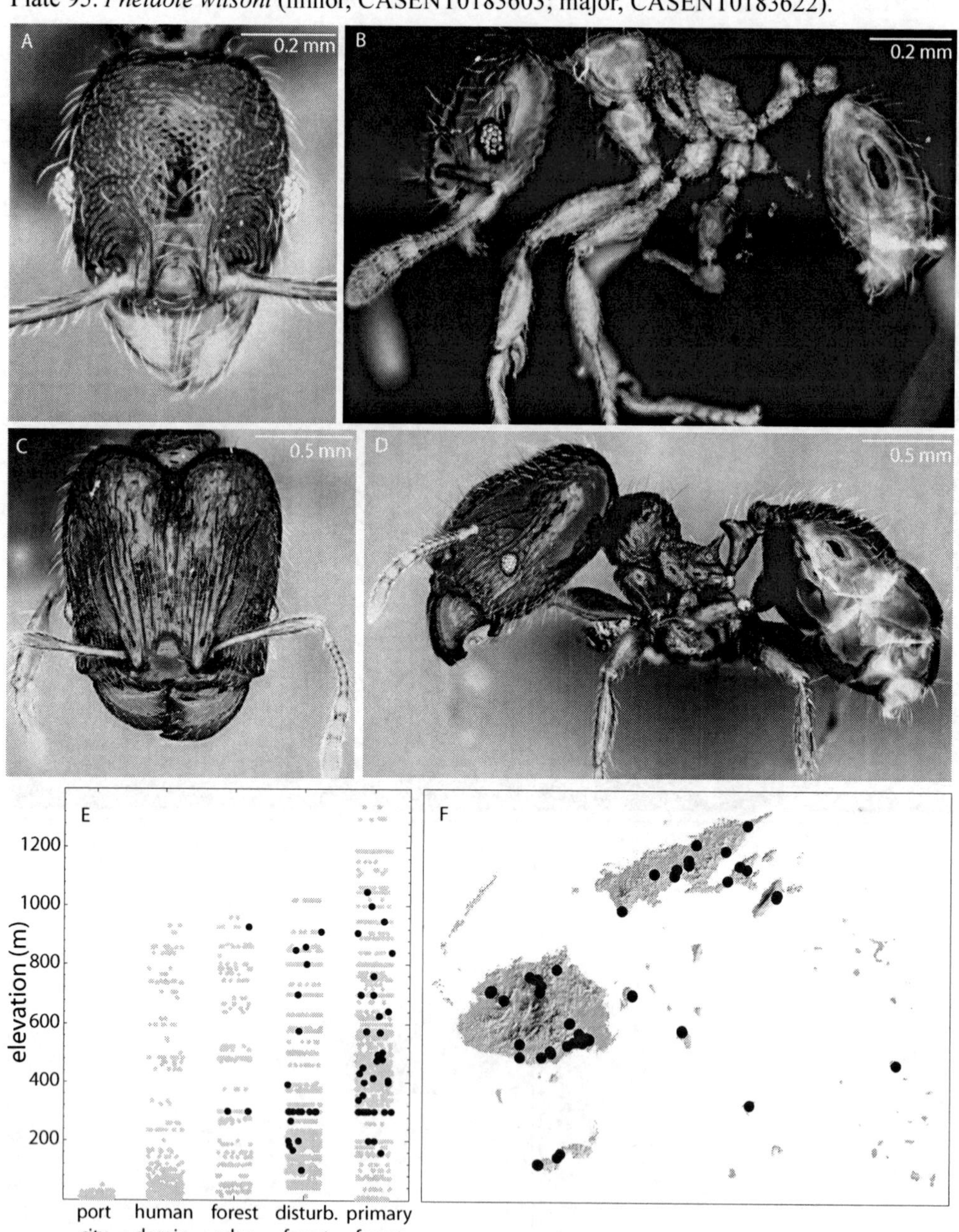

Plate 96. *Pheidole* sp. FJ05 (minor, CASENT0183979; major, CASENT0183626).

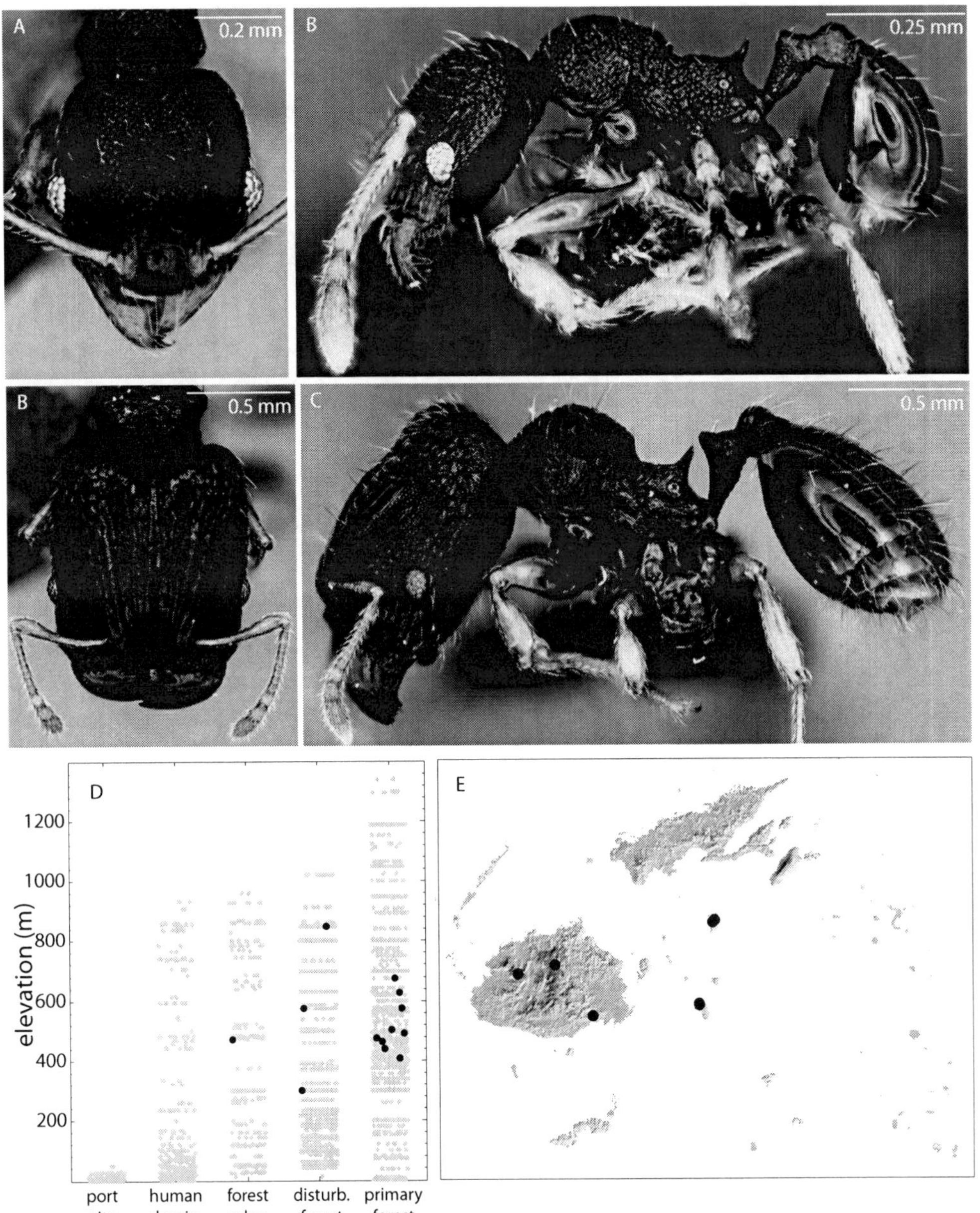

Plate 97. *Pheidole* sp. FJ09 (minor, CASENT0183387; major, CASENT0183927).

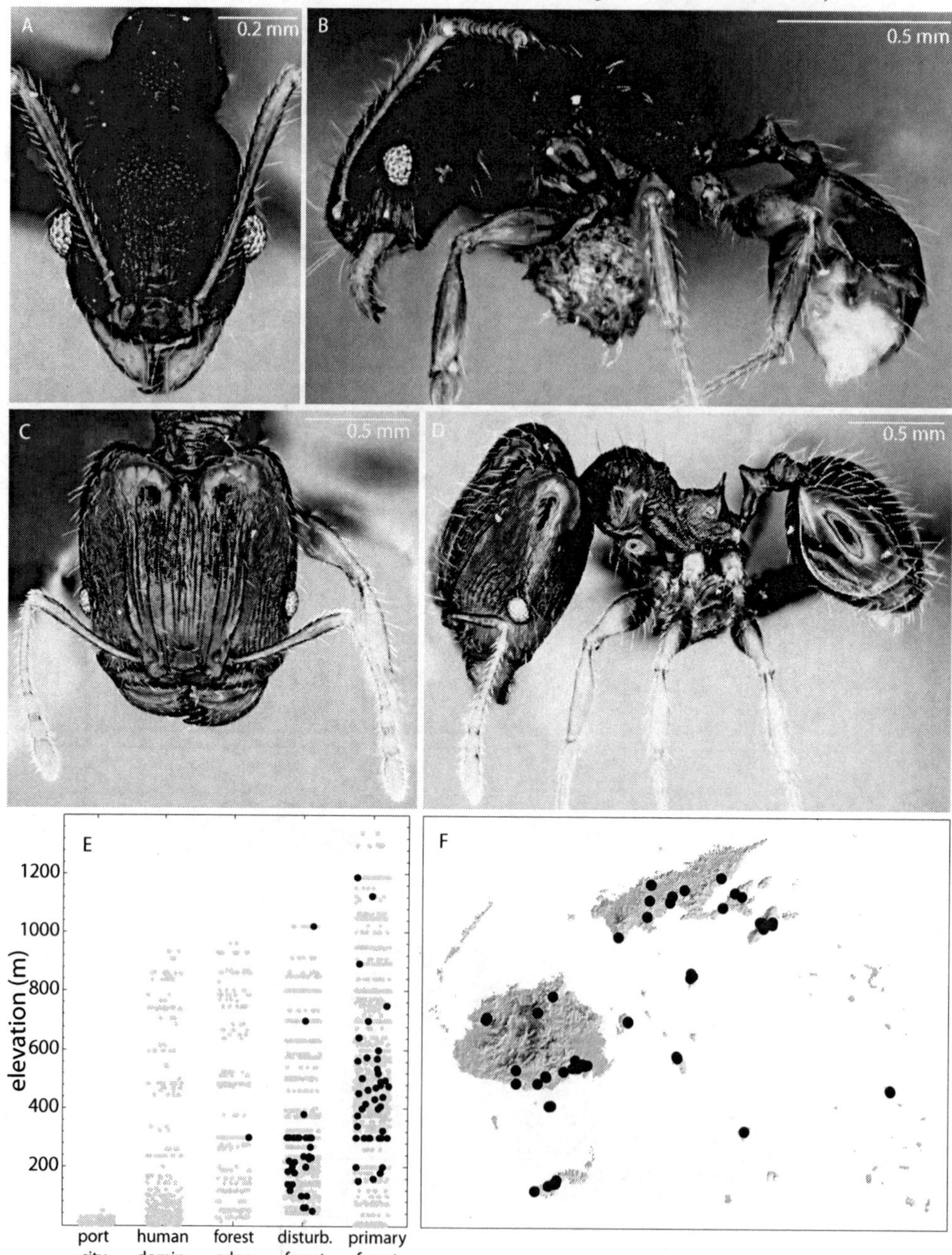

Plate 98. *Pheidole bula* (paratype minor, CASENT0171017; holotype major, CASENT0171113).

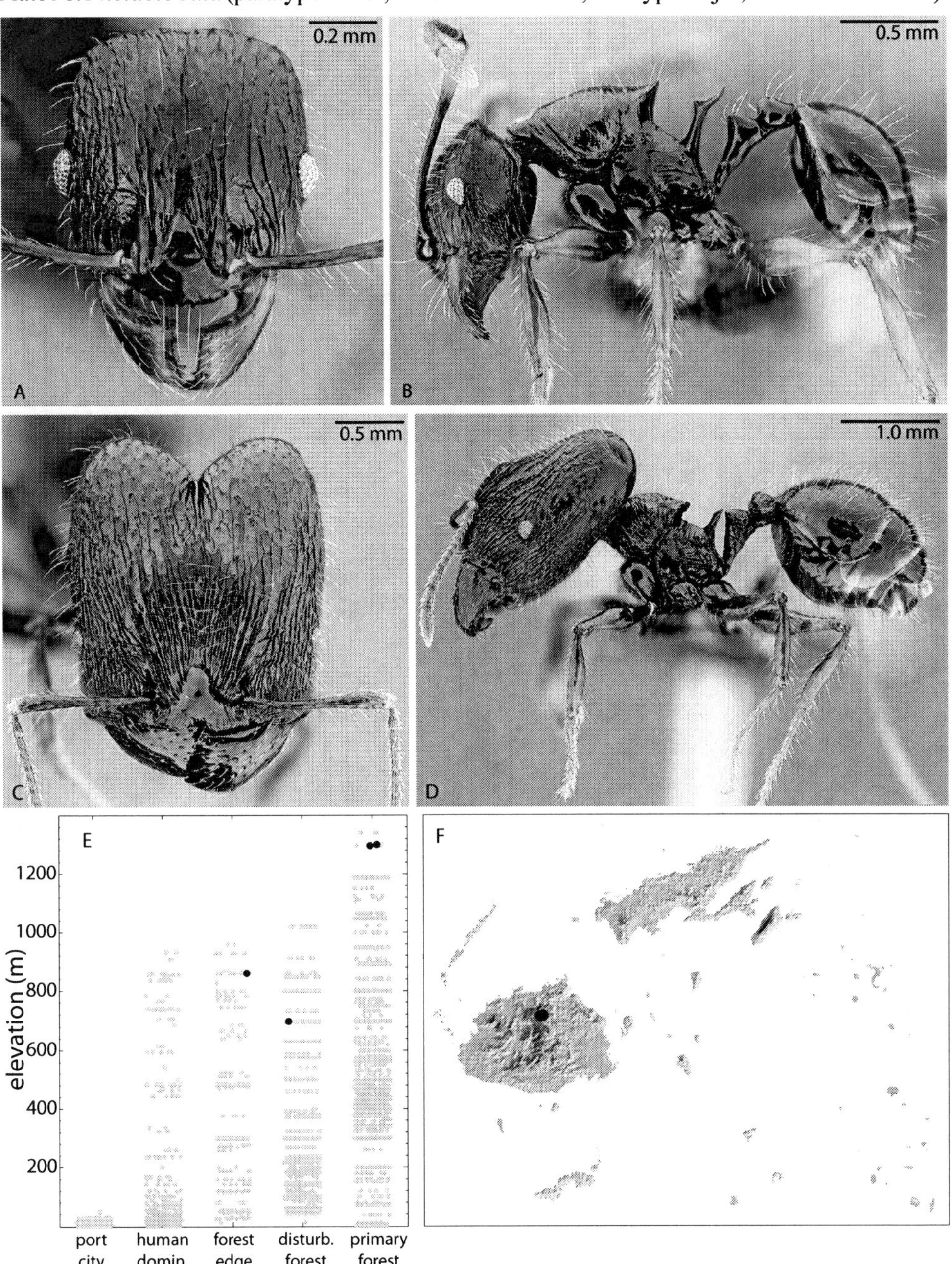

Plate 99. *Pheidole colaensis* (minor, CASENT0171020; major, CASENT0171103).

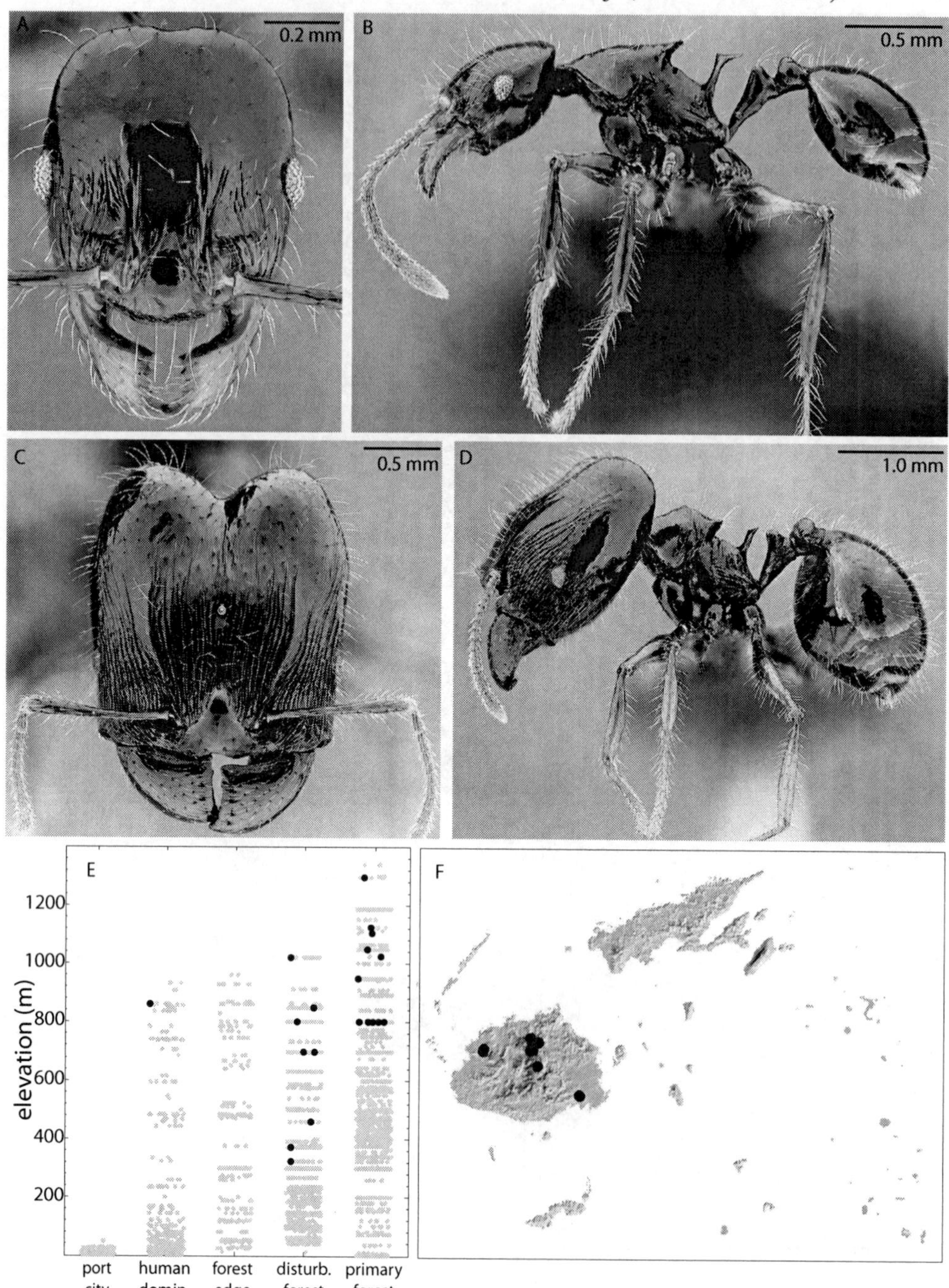

Plate 100. *Pheidole furcata* (paratype minor, CASENT0171025; holotype major, CASENT0171111).

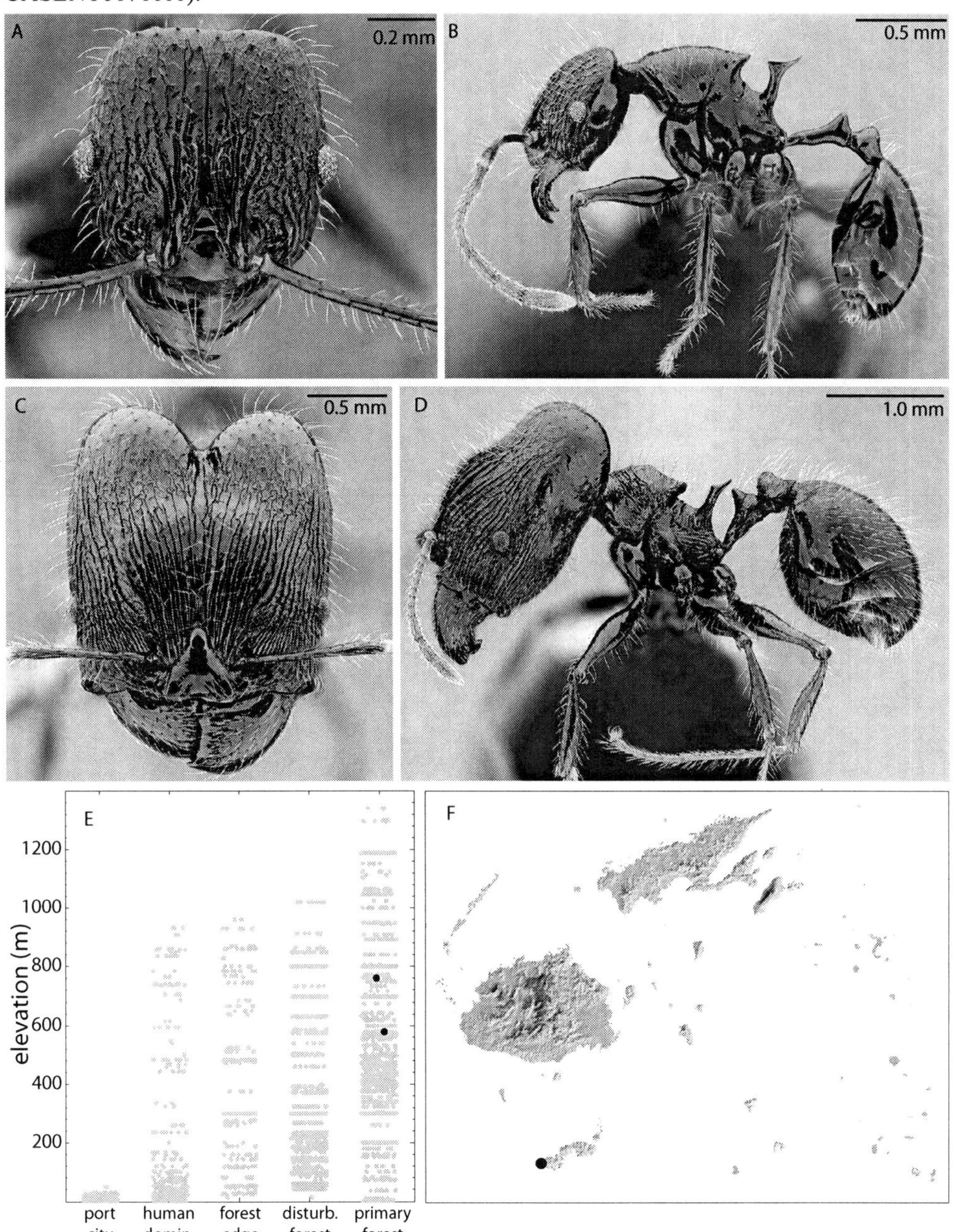

Plate 101. *Pheidole pegasus* (paratype minor, CASENT0171024; holotype major, CASENT0171108).

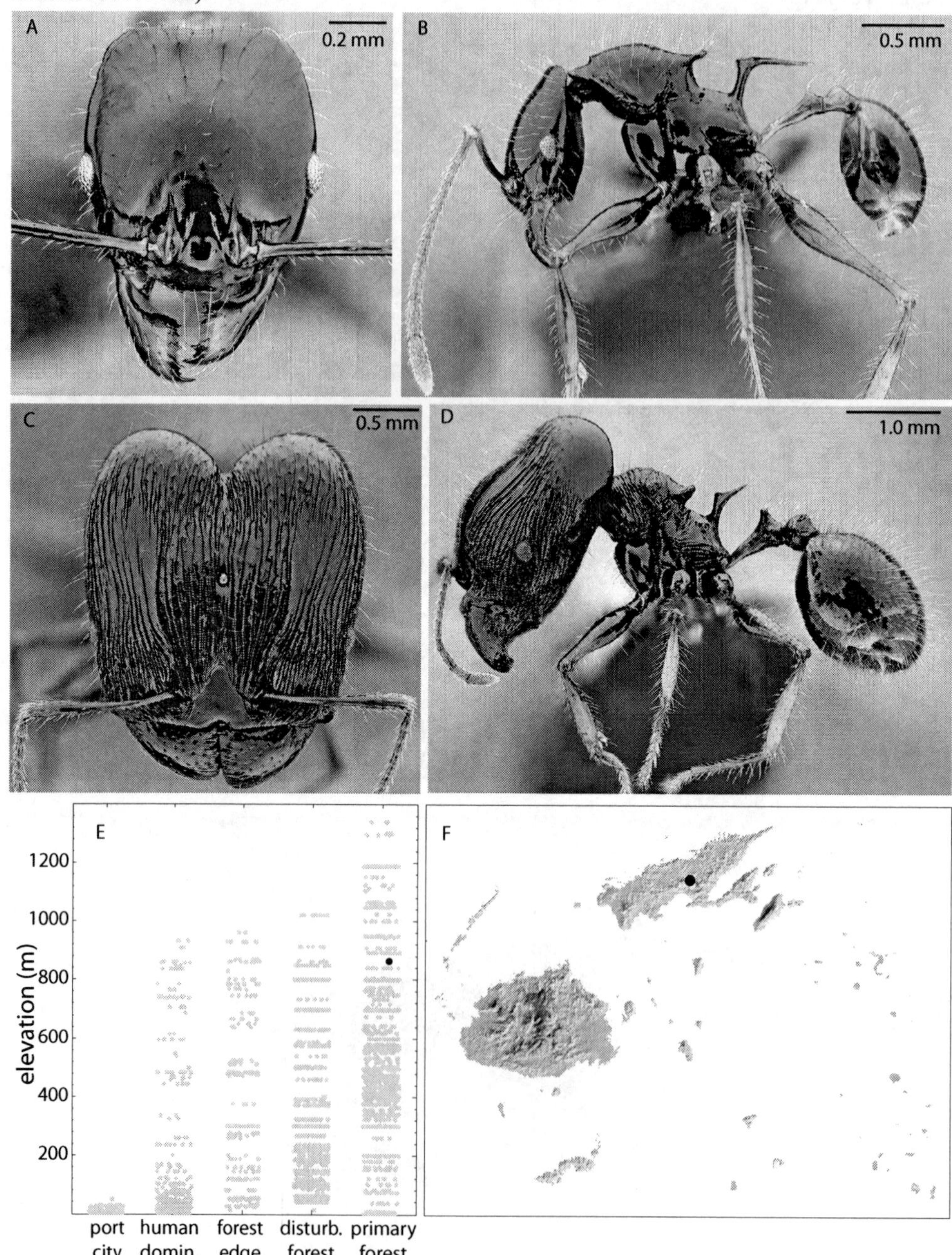

Plate 102. *Pheidole roosevelti* (minor, CASENT0171023; major, CASENT0171027).

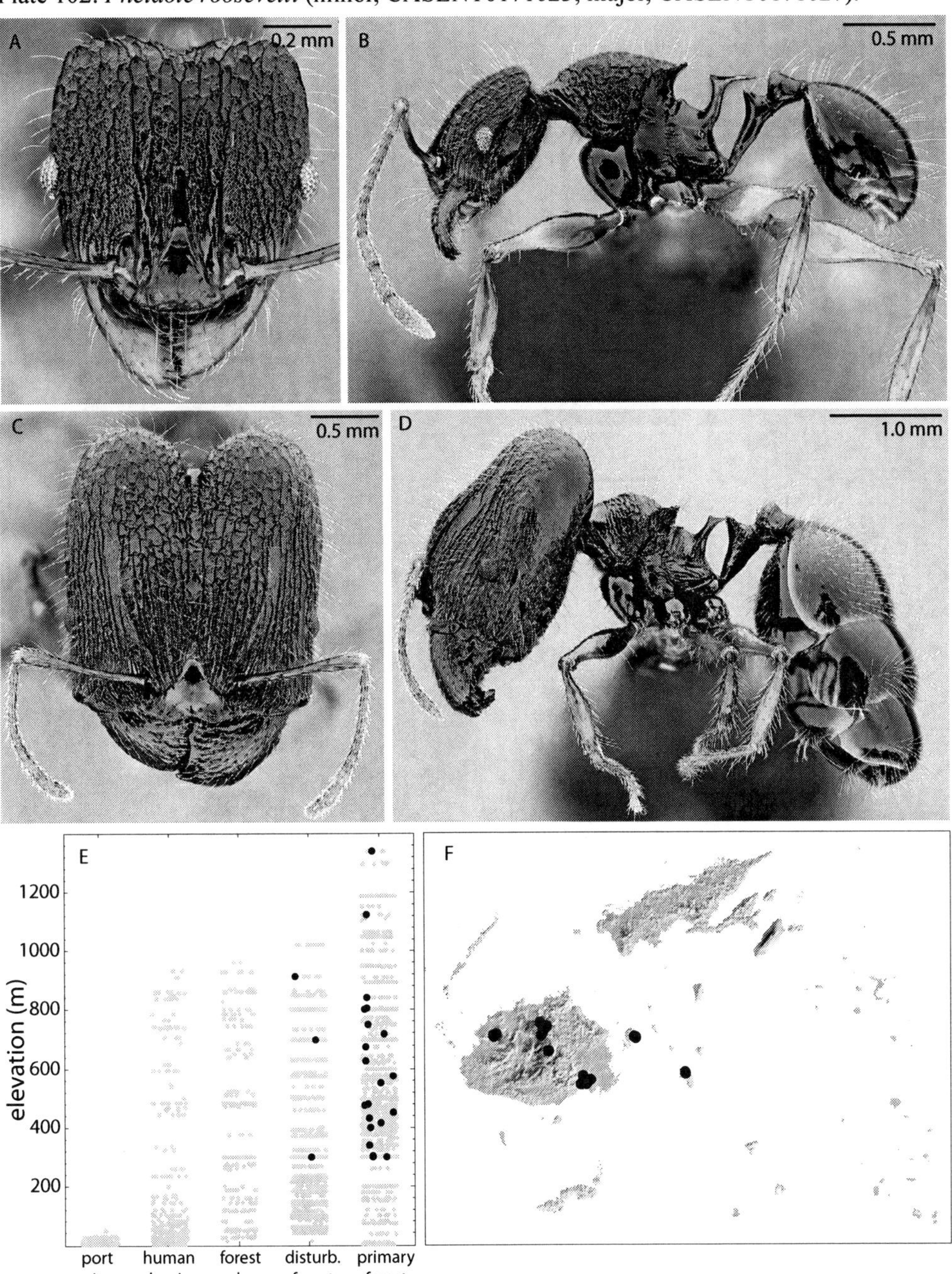

Plate 103. *Pheidole simplispinosa* (paratype minor, CASENT0171022; holotype major, CASENT0171106).

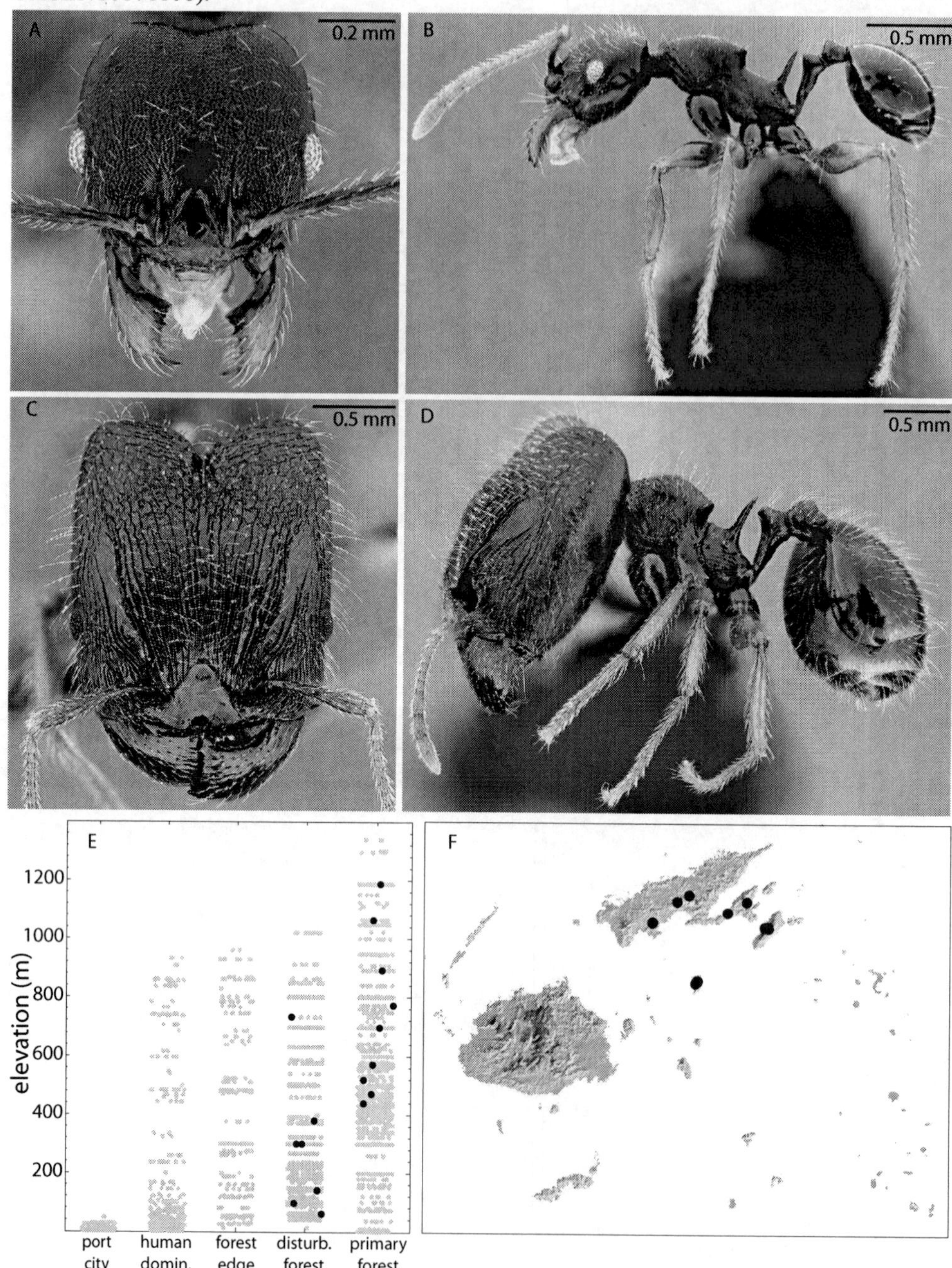

Plate 104. *Pheidole uncagena* (paratype minor, CASENT0171026; holotype major CASENT0171110).

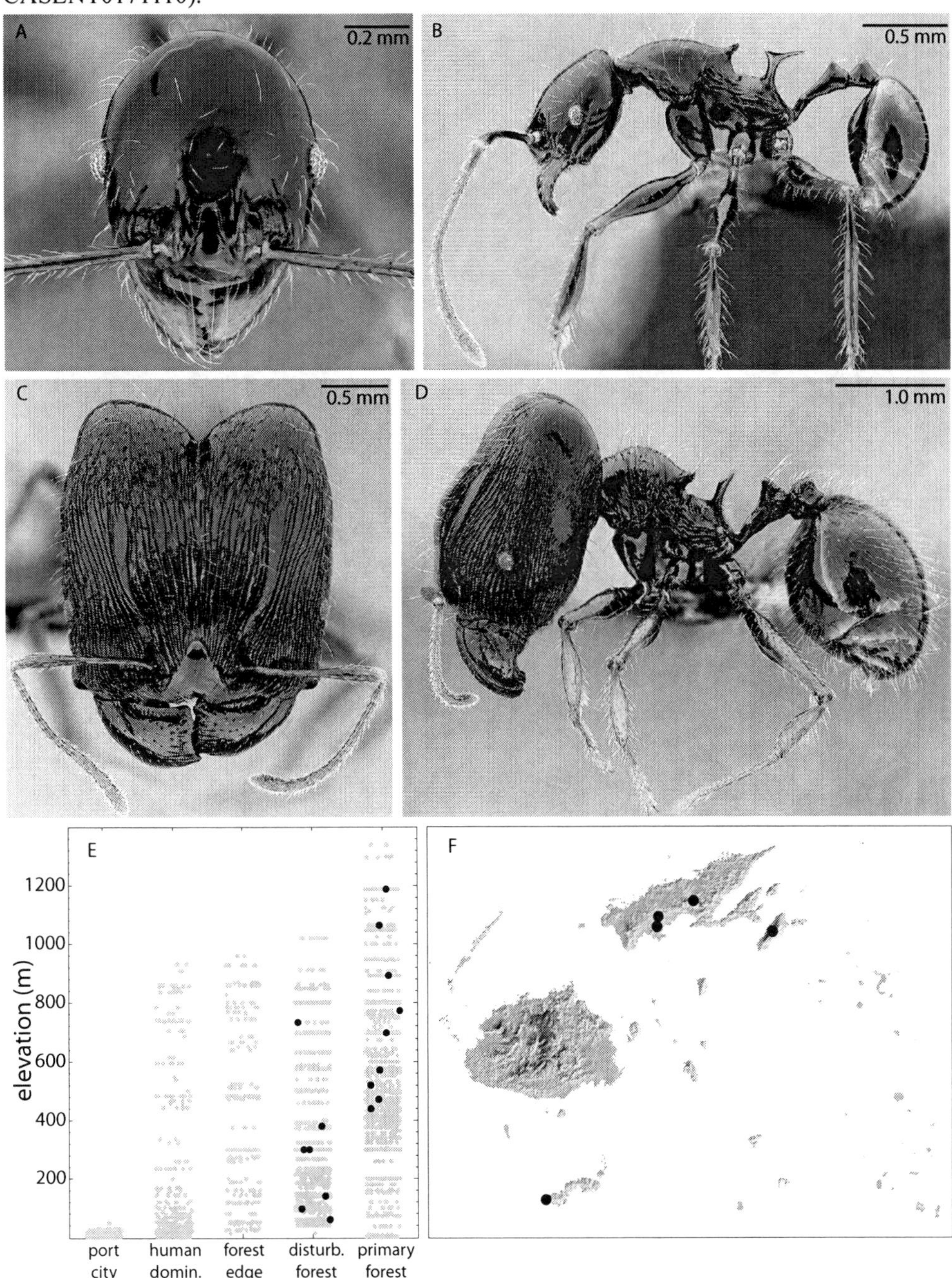

Plate 105. *Poecilomyrma myrmecodiae* worker, CASENT0217355.

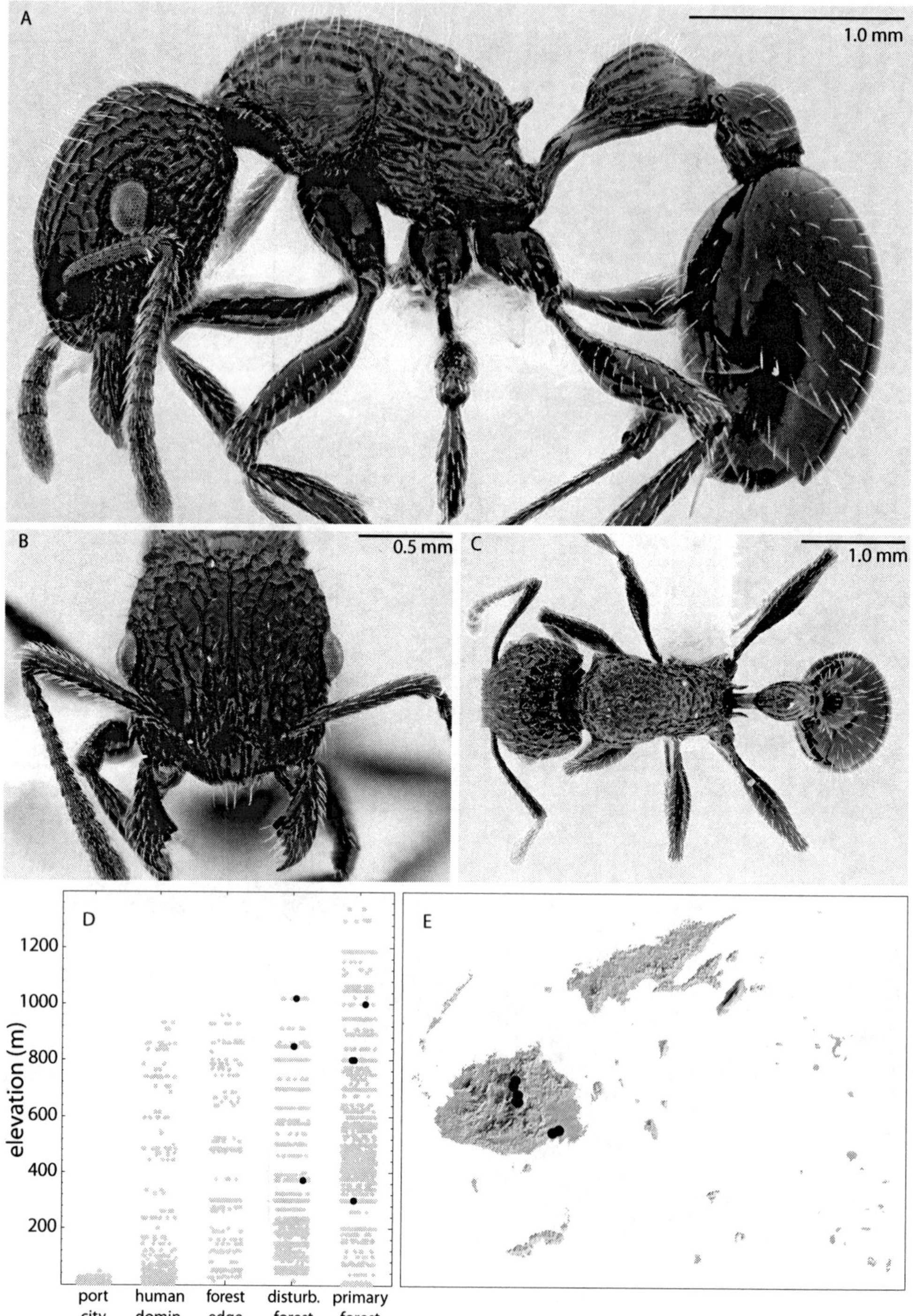

Plate 106. *Poecilomyrma senirewae* worker, CASENT0181574.

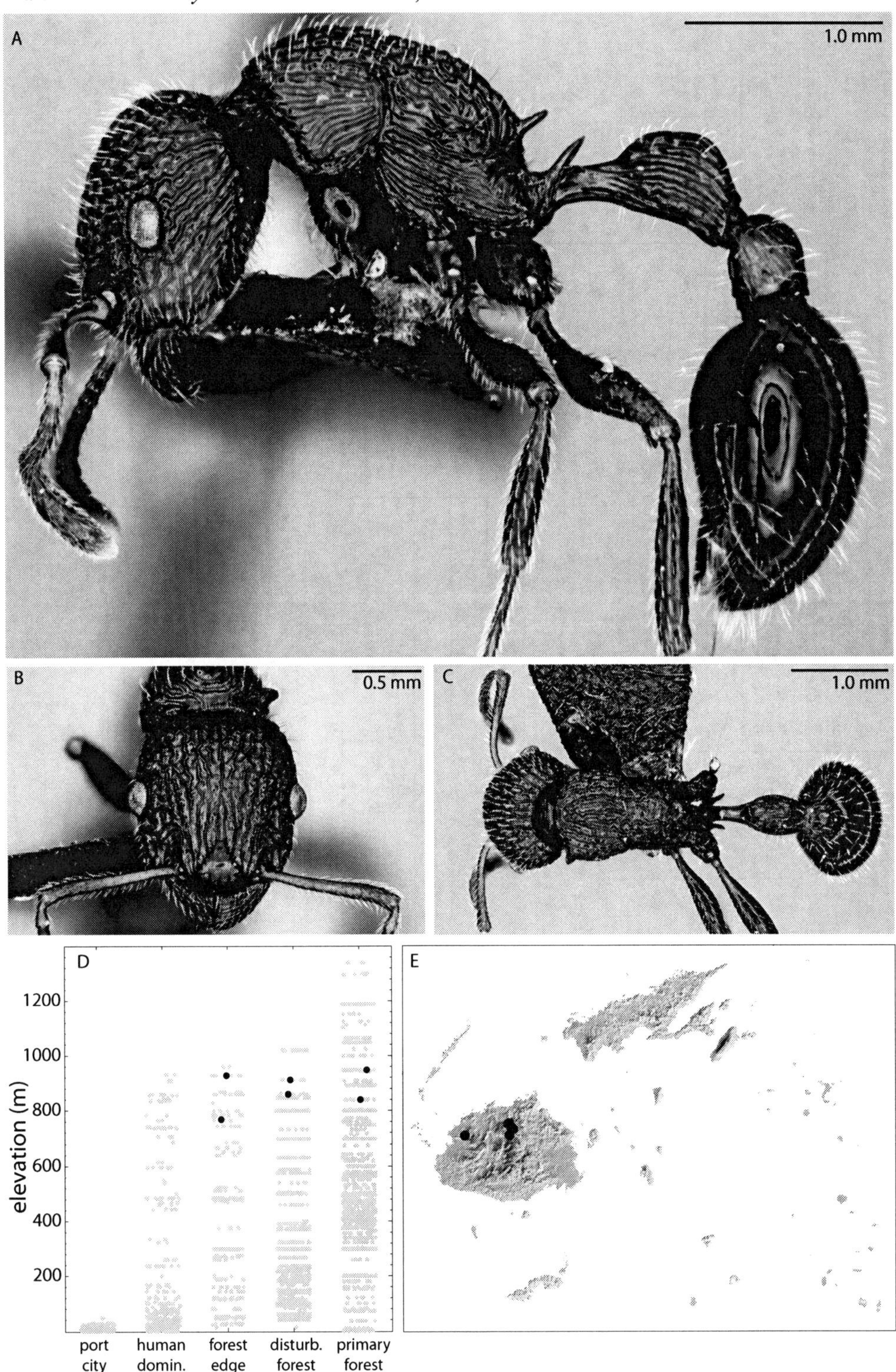

Plate 107. *Poecilomyrma* sp. FJ03 worker, CASENT0181454.

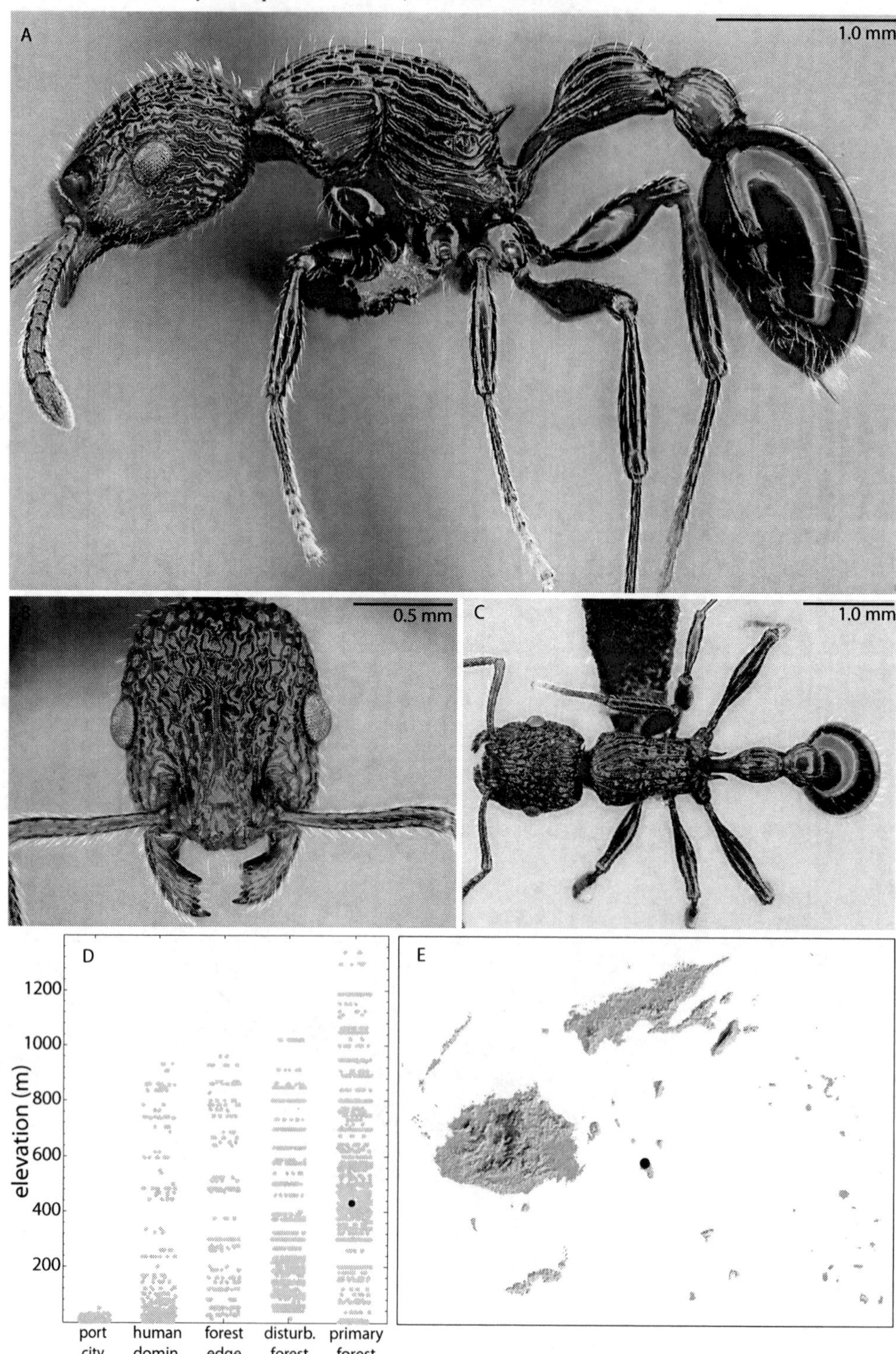

Plate 108. *Poecilomyrma* sp. FJ05 worker, CASENT0217355.

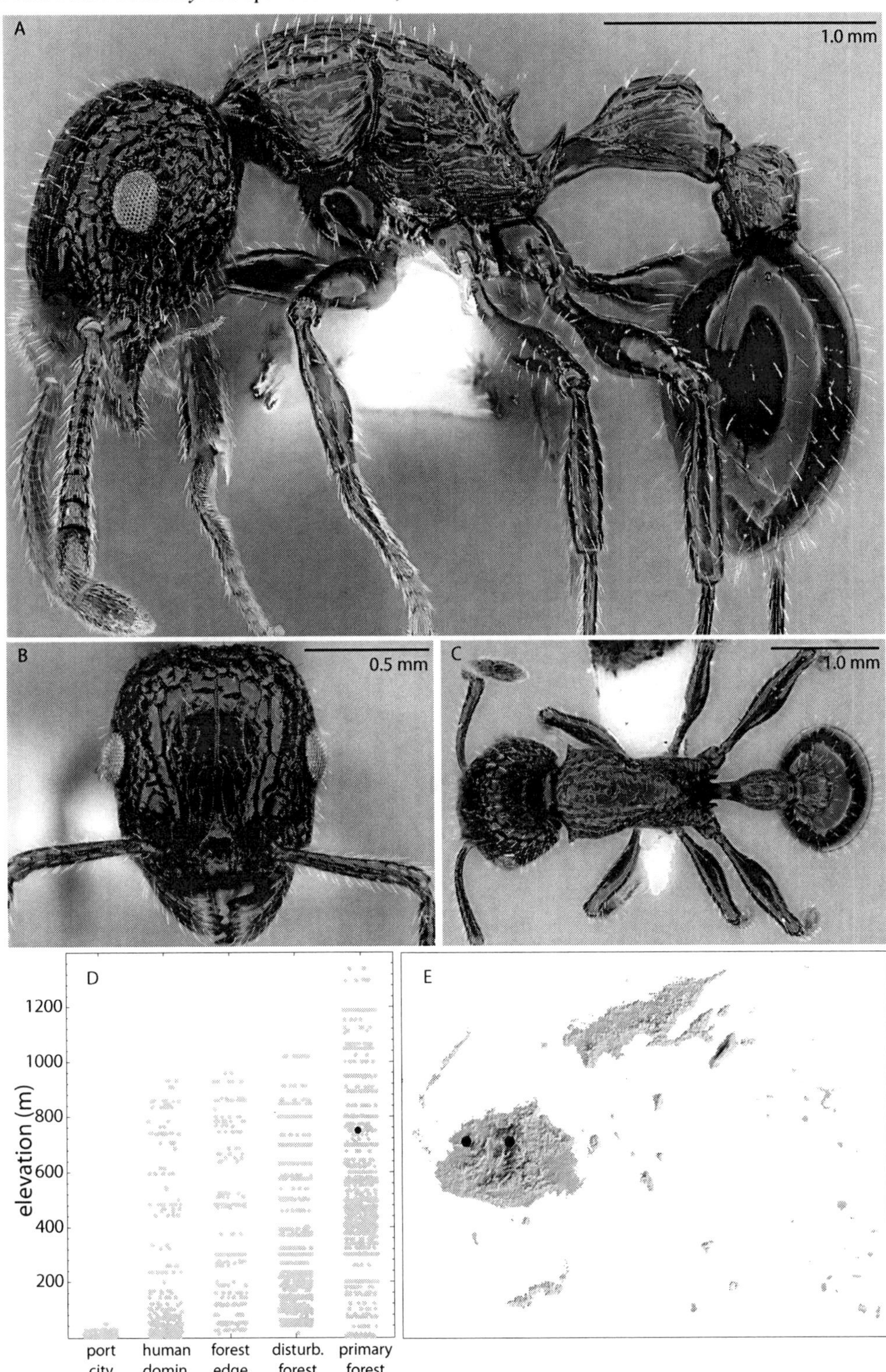

Plate 109. *Poecilomyrma* sp. FJ06 worker, CASENT0181607.

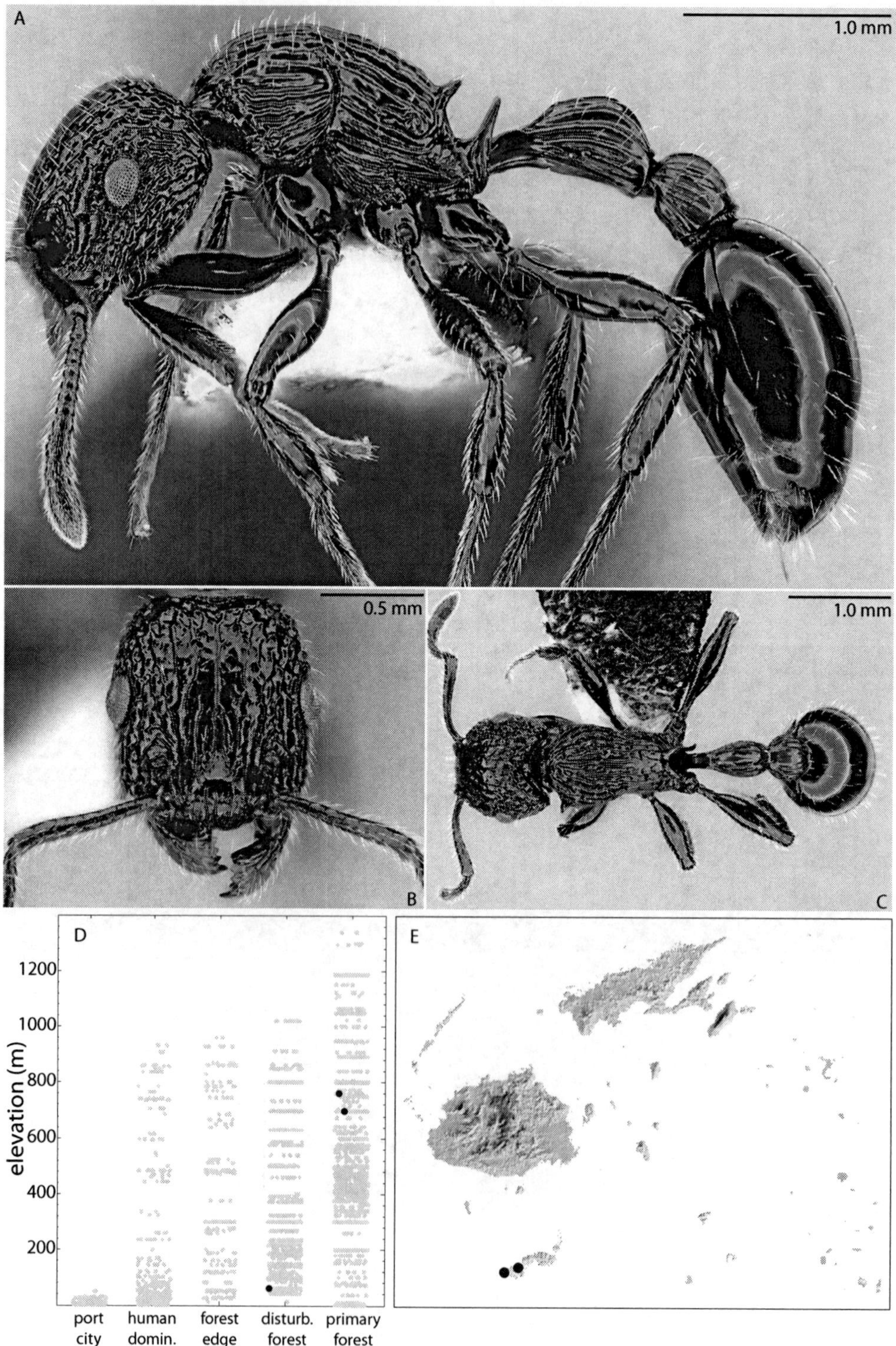

Plate 110. *Poecilomyrma* sp. FJ07 worker, CASENT0181539.

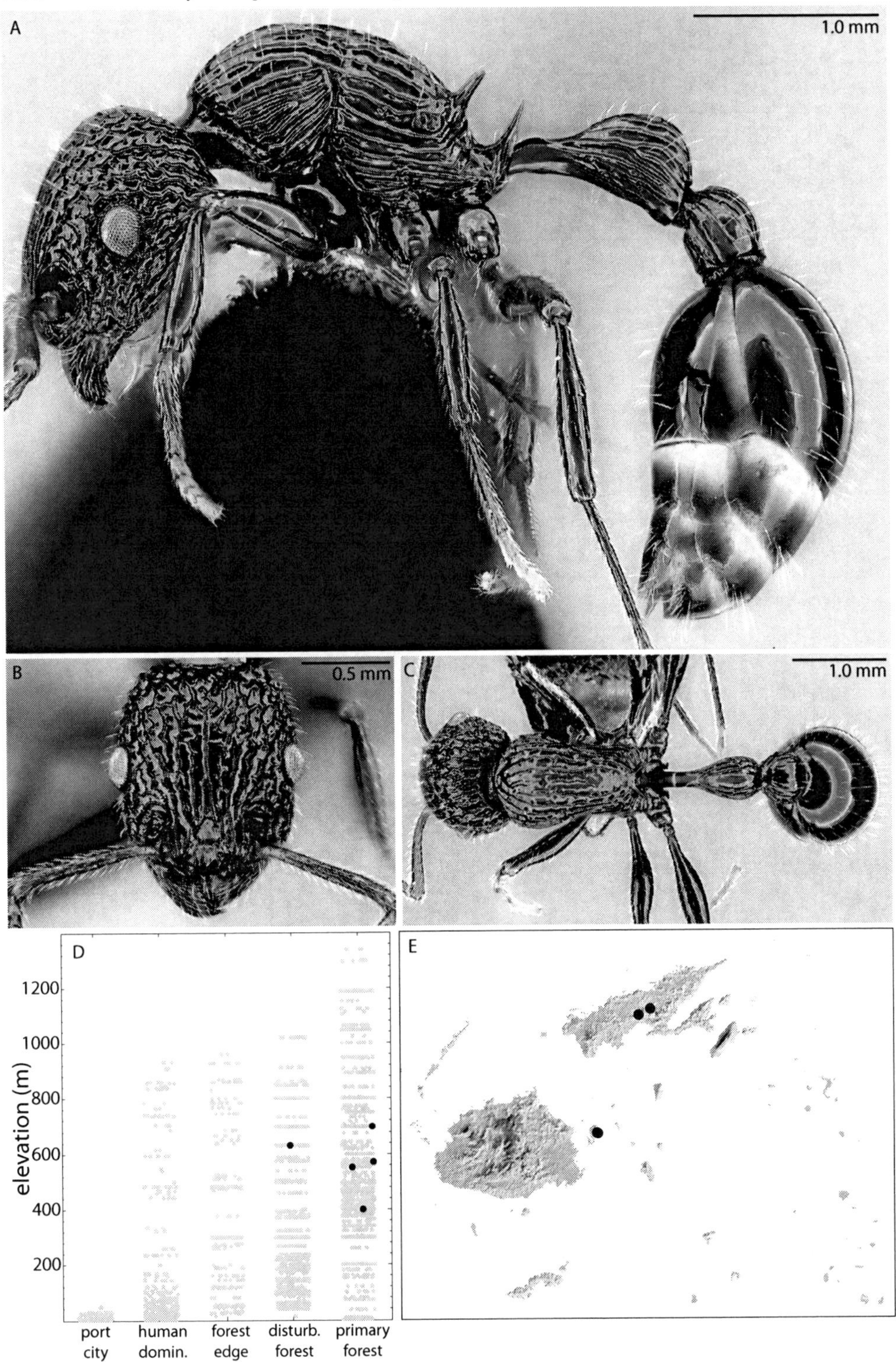

Plate 111. *Poecilomyrma* sp. FJ08 worker, CASENT0181455.

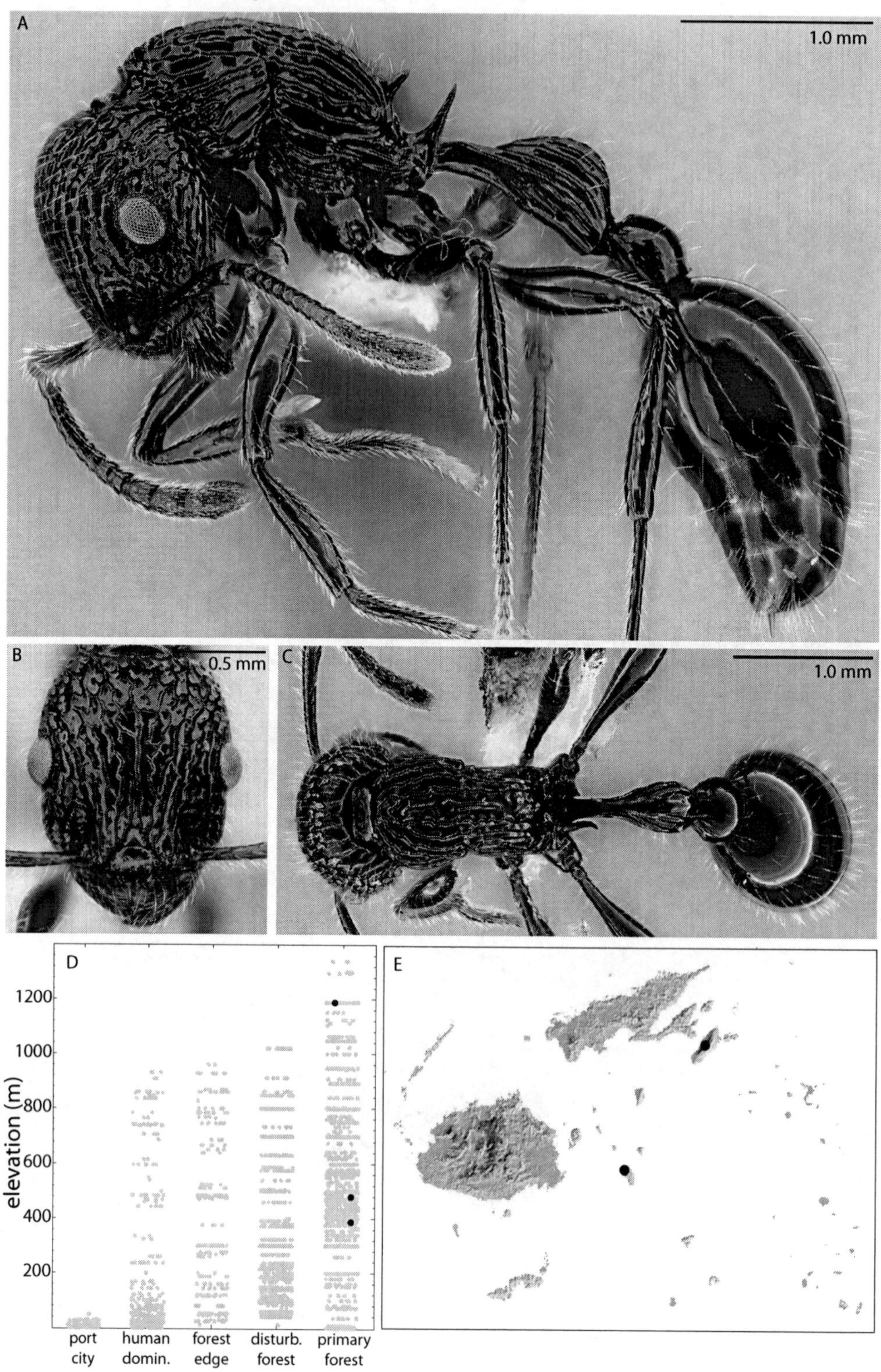

Plate 112. *Pristomyrmex mandibularis* worker, CASENT0171044.

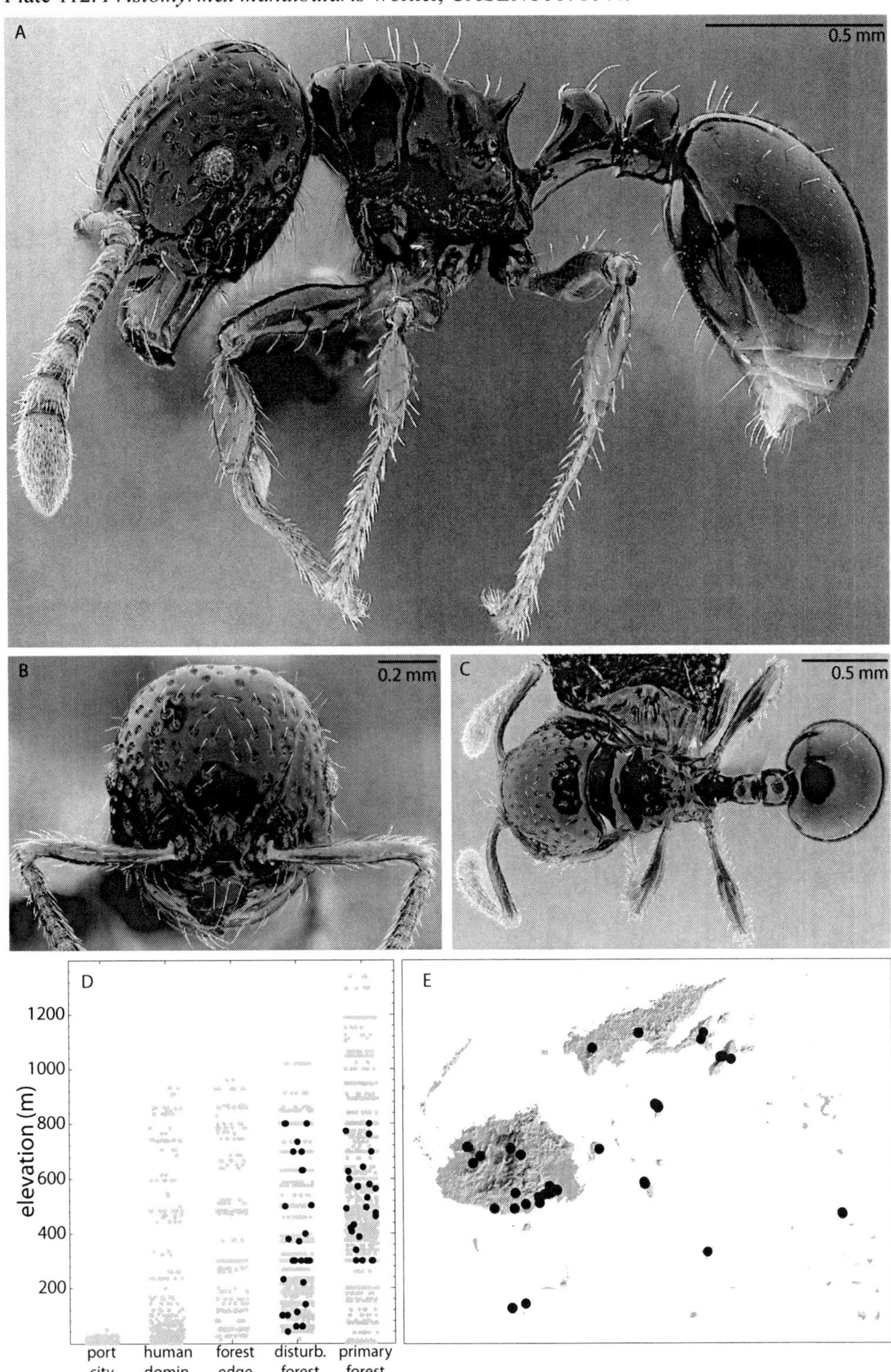

Plate 113. *Pristomyrmex* sp. FJ02 worker, CASENT0171144.

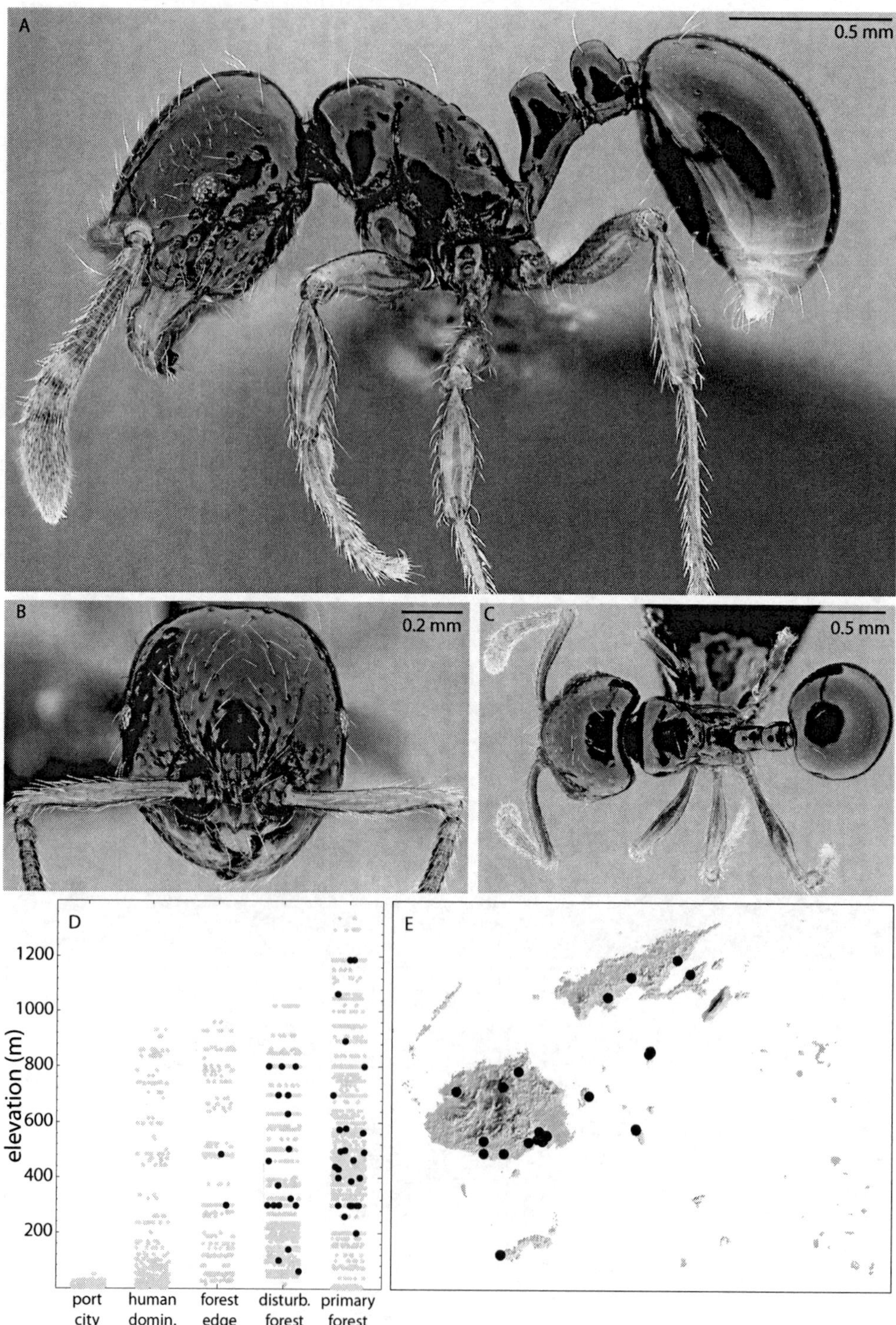

Plate 114. *Pyramica membranifera* worker, CASENT0171134.

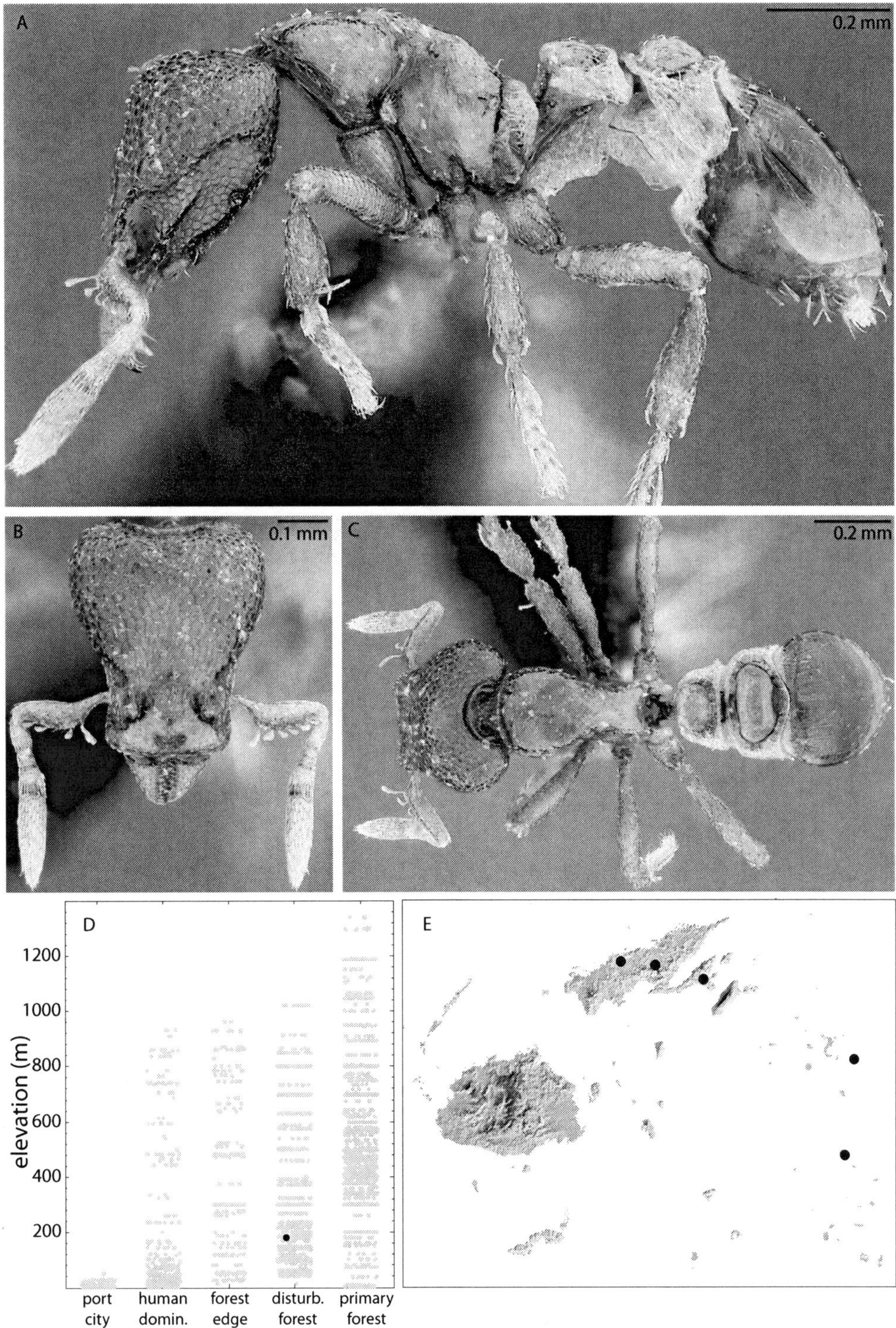

Plate 115. *Pyramica trauma* worker, CASENT0184729.

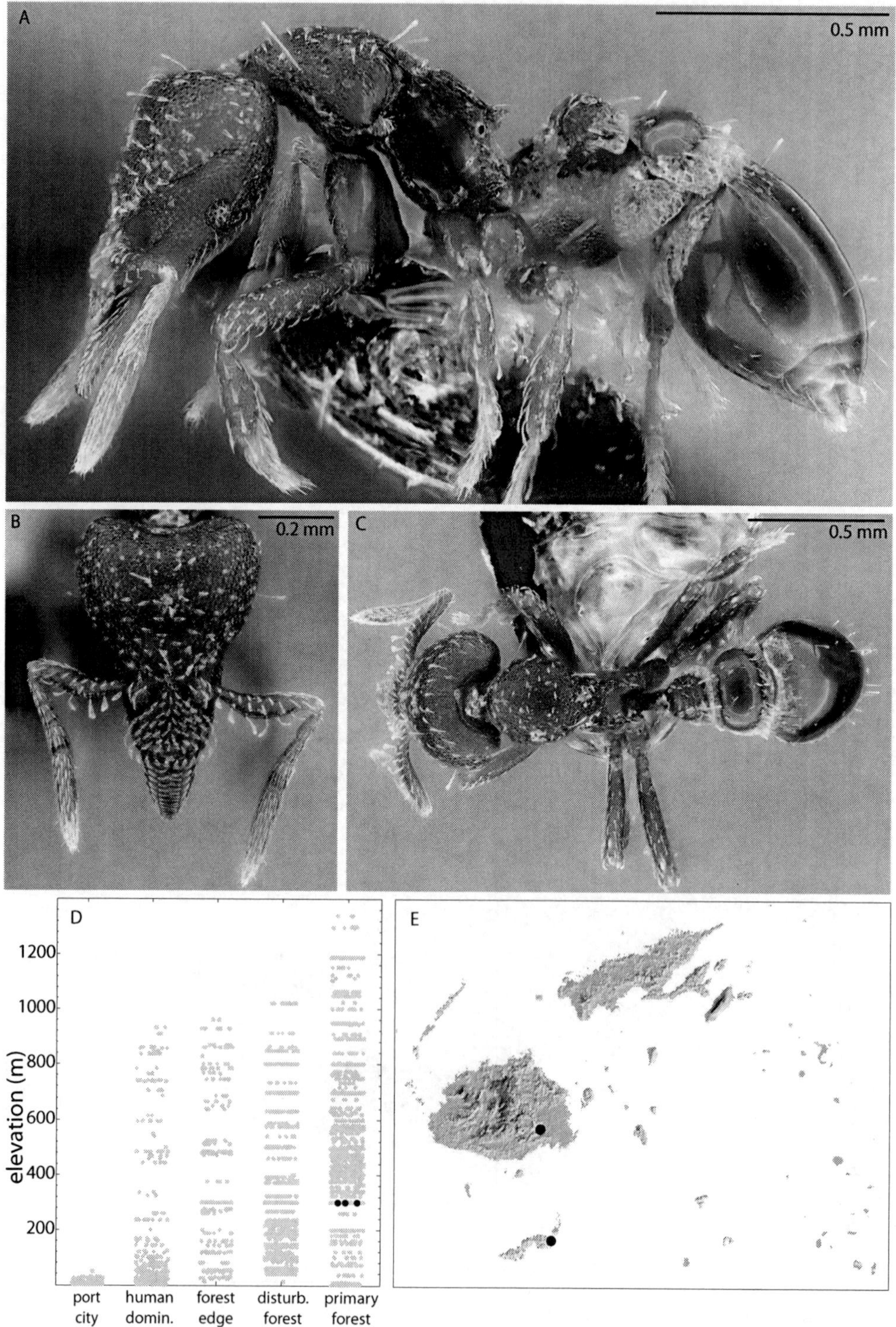

Plate 116. *Pyramica* sp. FJ02 worker, CASENT0184976.

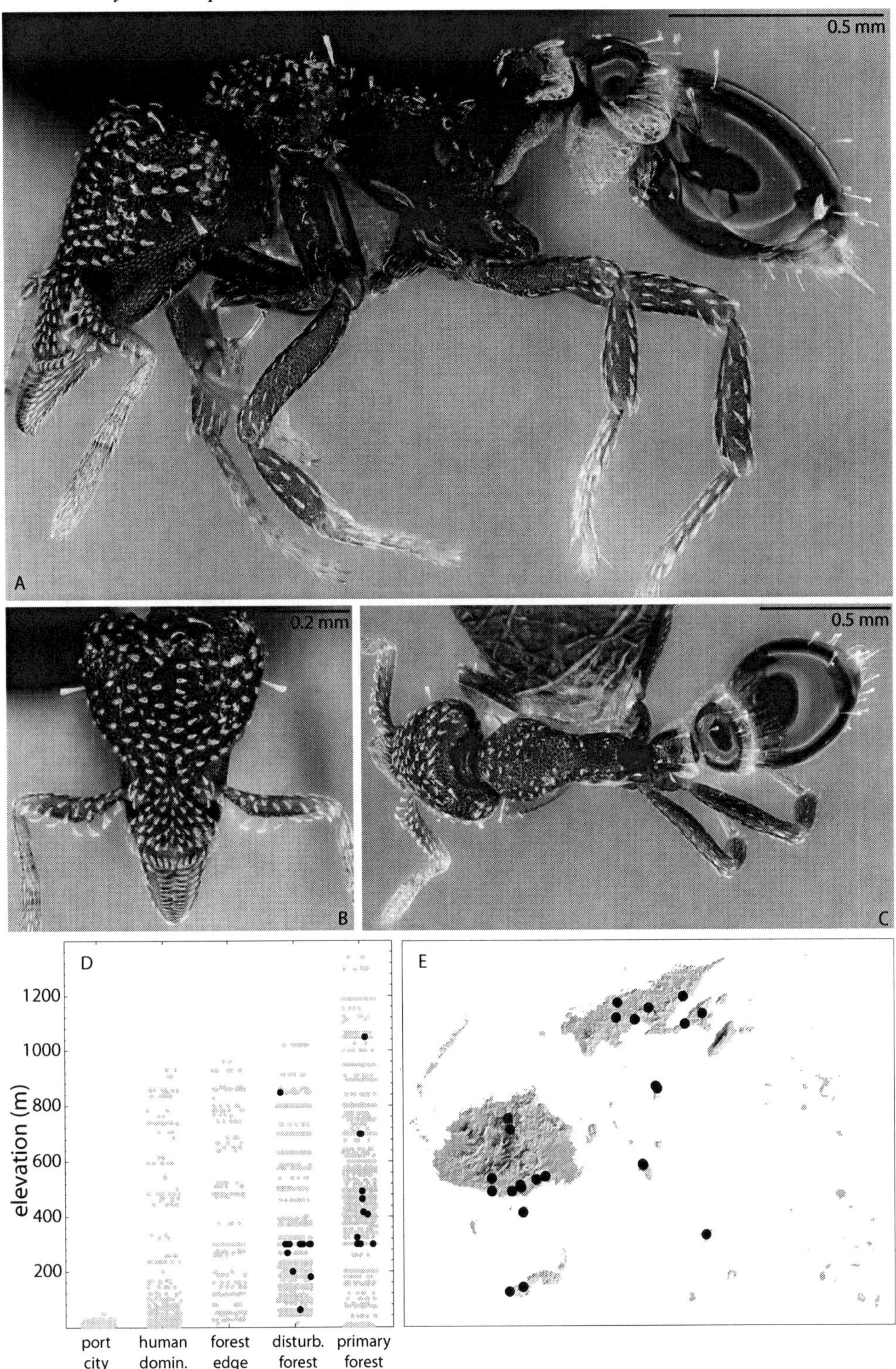

Plate 117. *Rogeria stigmatica* worker, CASENT0171046.

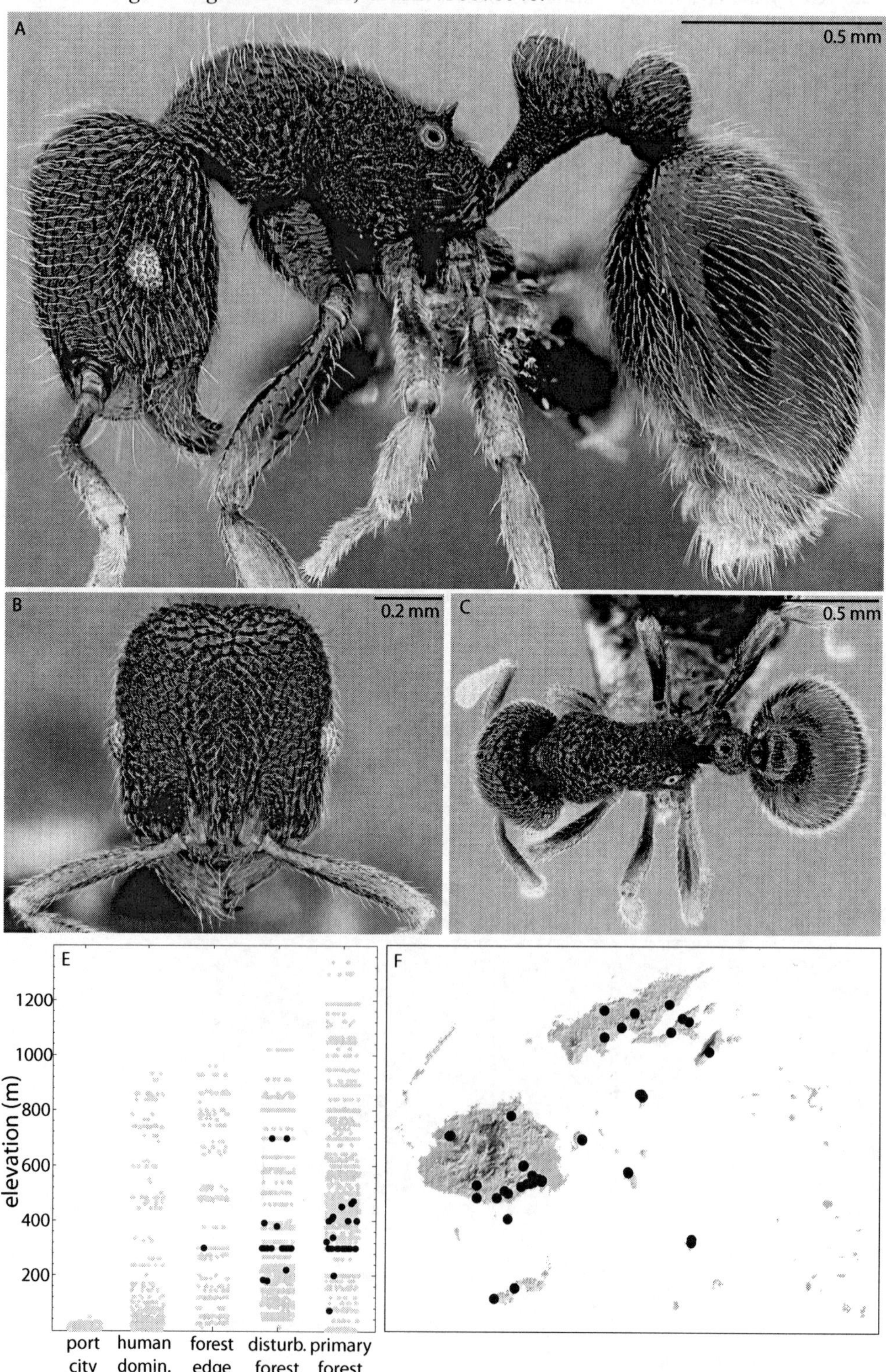

Plate 118. *Romblonella liogaster* holotype worker, CASENT0013155.

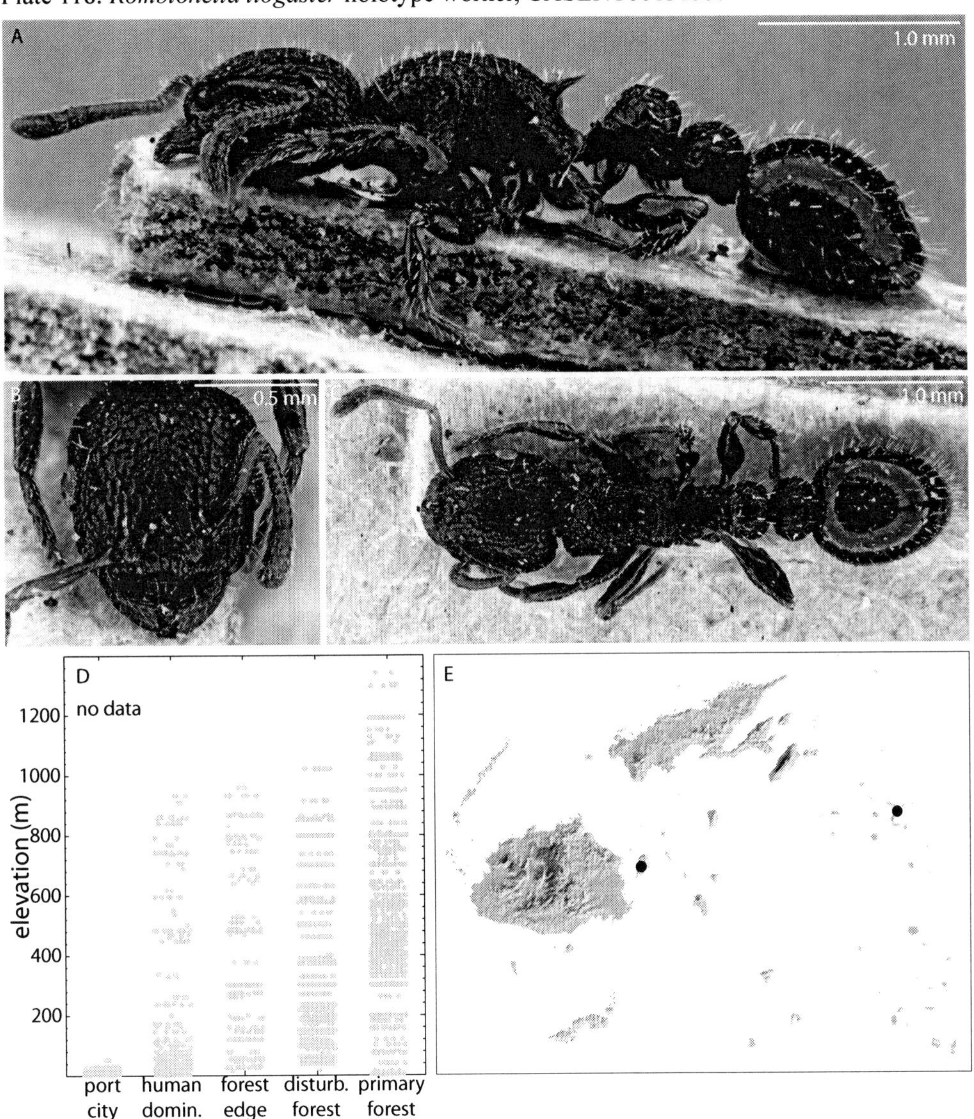

Plate 119. *Solenopsis geminata* worker, CASENT0171029.

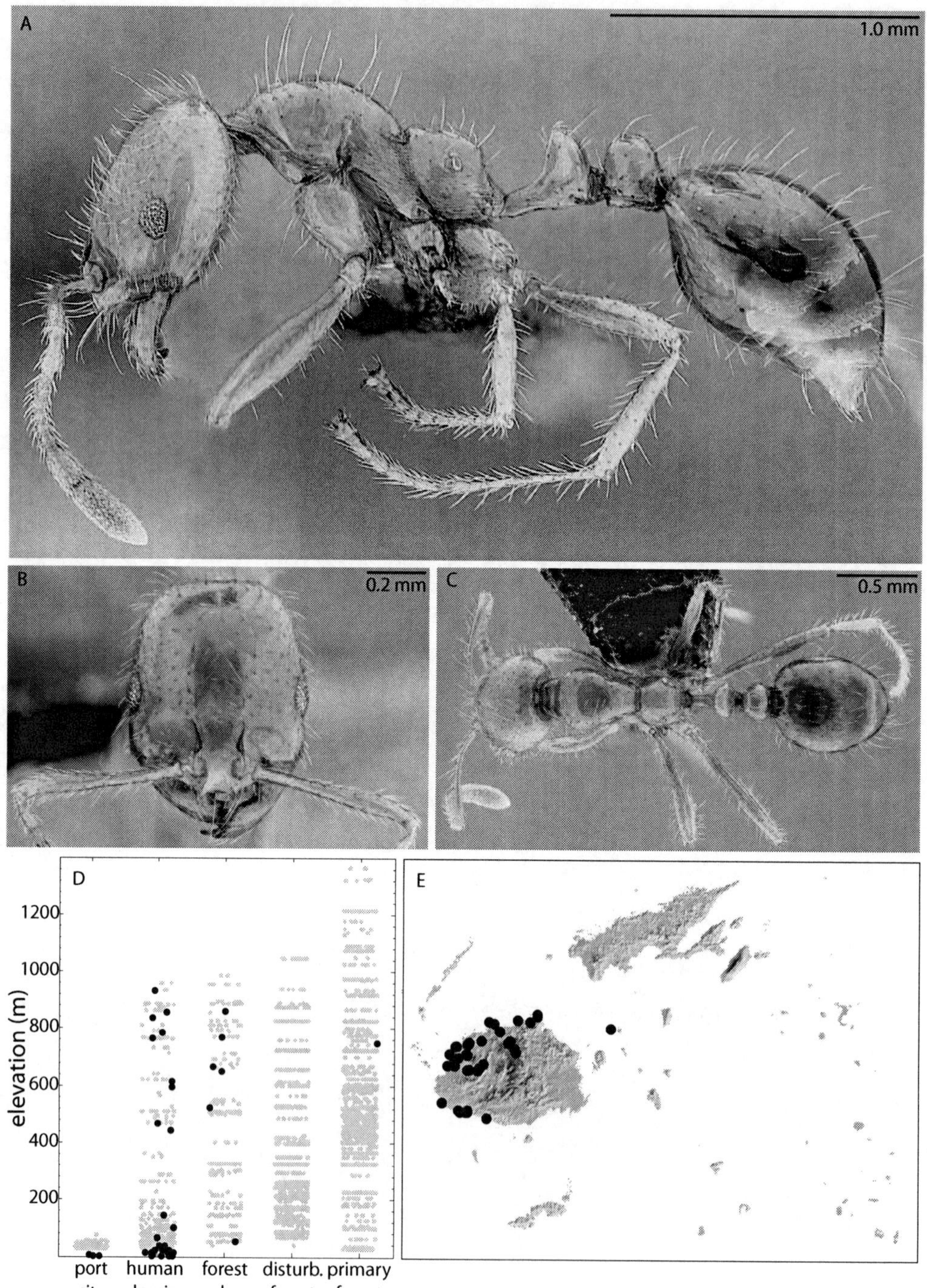

Plate 120. *Solenopsis papuana* worker, CASENT0171089.

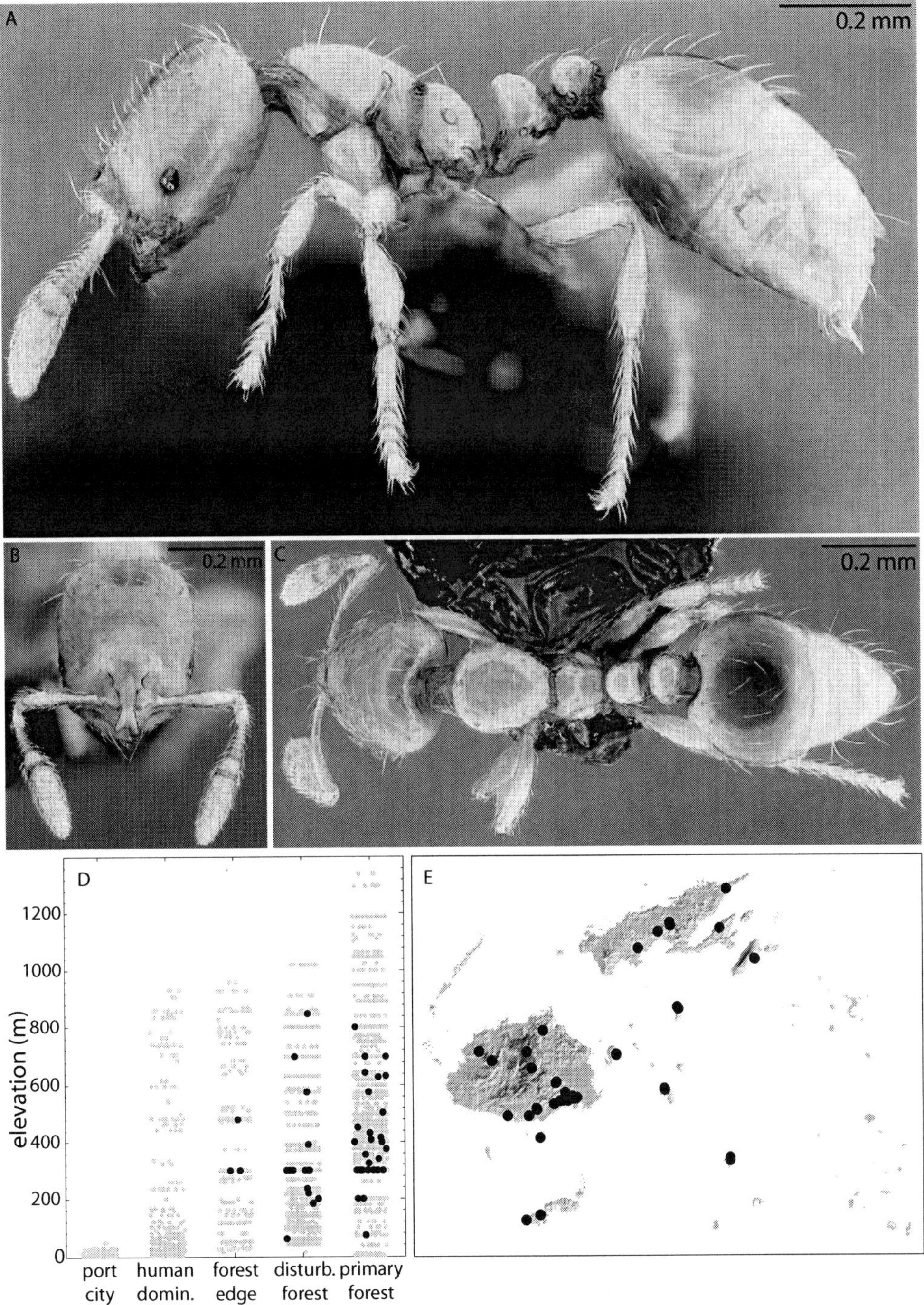

Plate 121. *Strumigenys basiliska* worker, CASENT0185864.

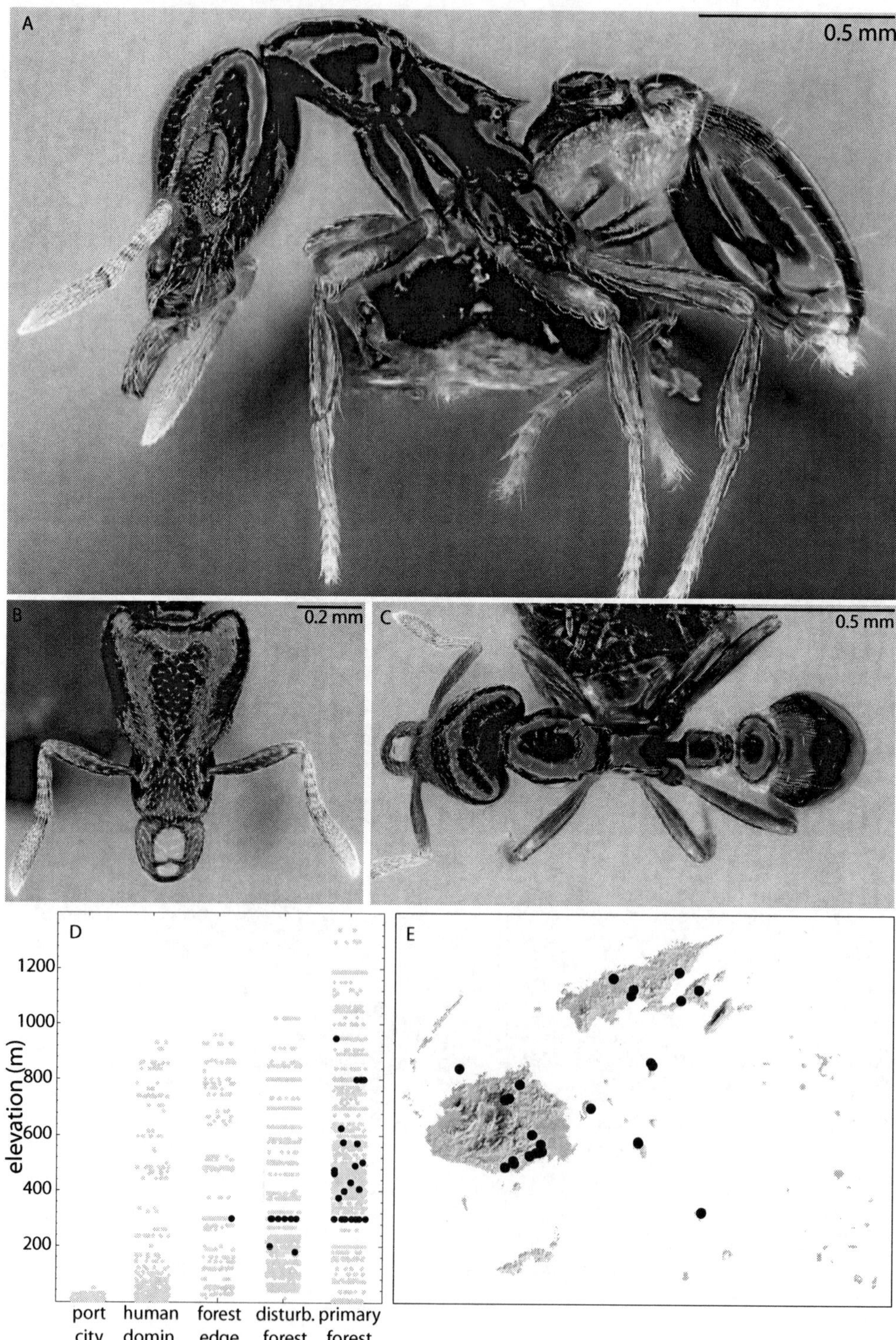

Plate 122. *Strumigenys chernovi* worker, CASENT0184914.

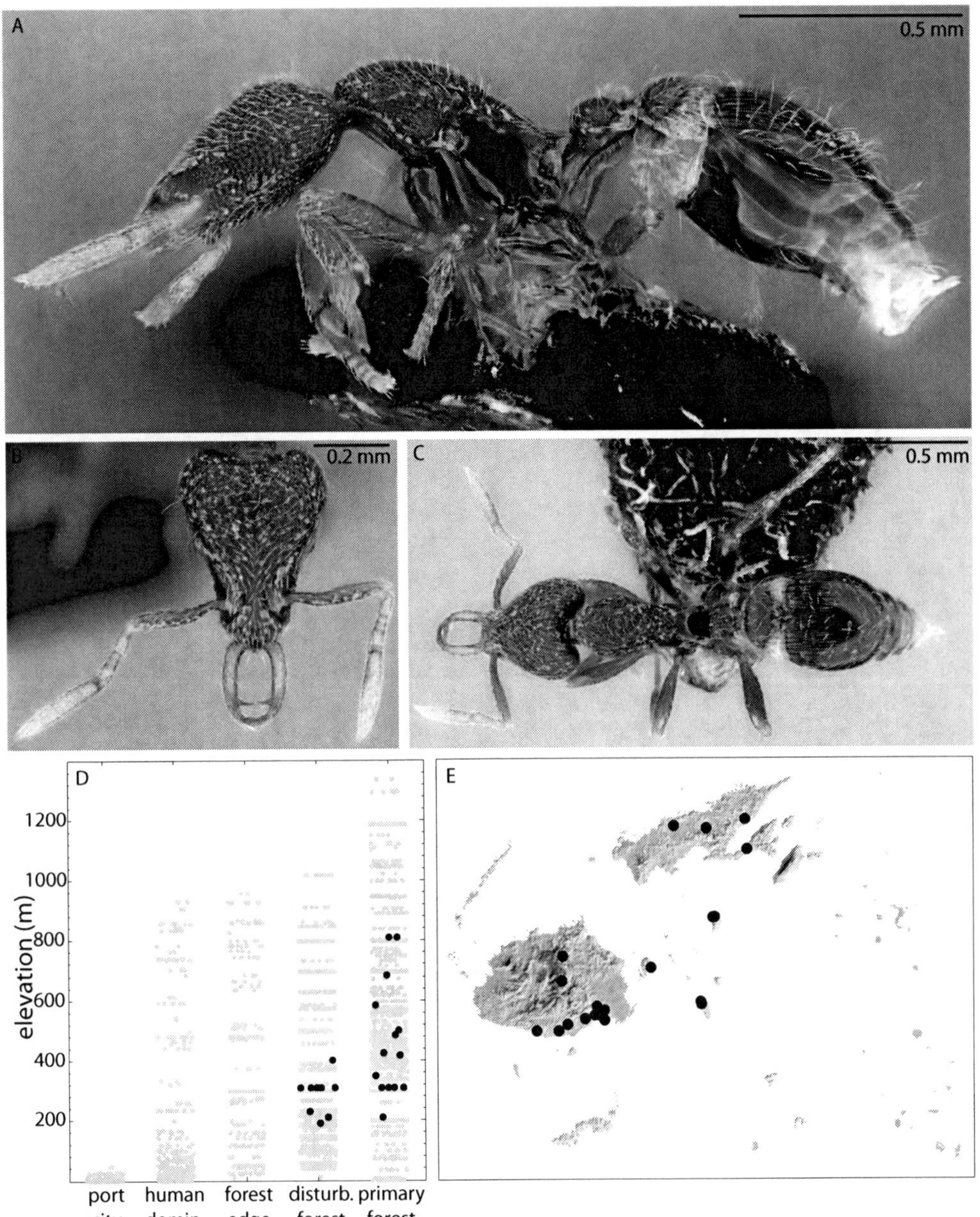

Plate 123. *Strumigenys daithma* worker, ANIC32-017716.

Plate 124. *Strumigenys ekasura* worker, CASENT0184678.

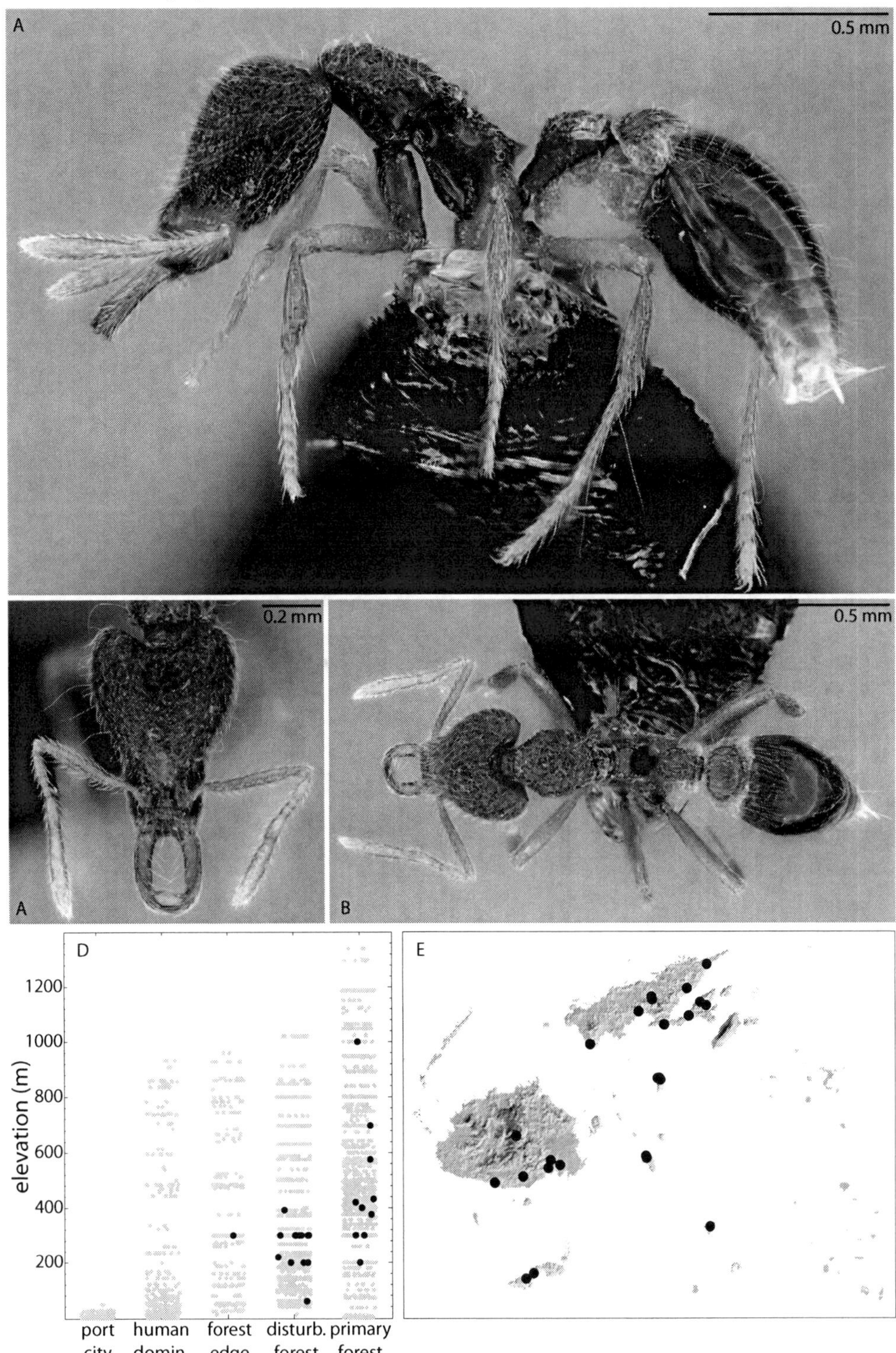

Plate 125. *Strumigenys frivola* worker, CASENT0186989.

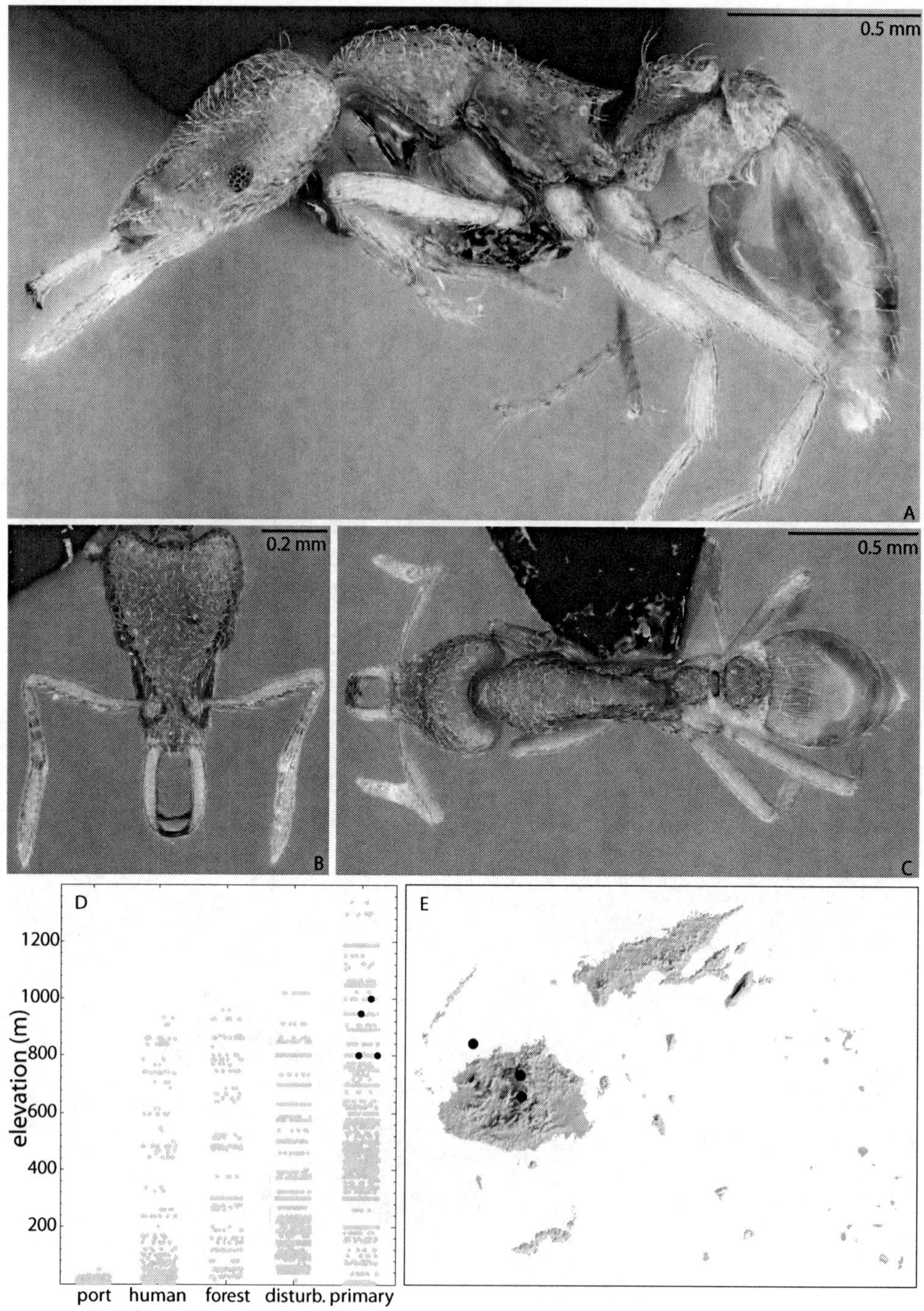

Plate 126. *Strumigenys godeffroyi* worker, CASENT0171155.

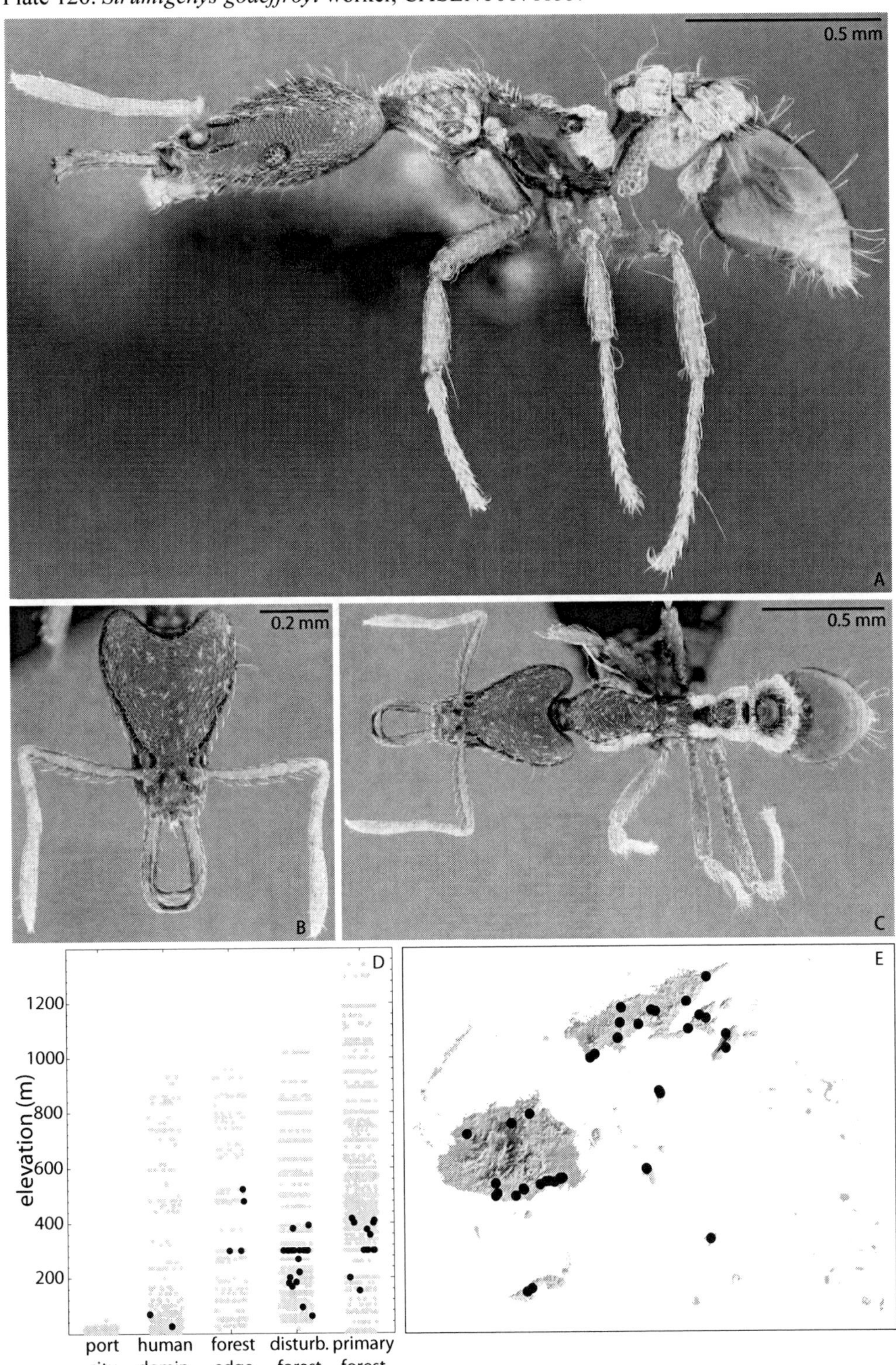

Plate 127. *Strumigenys jepsoni* worker, CASENT0235852.

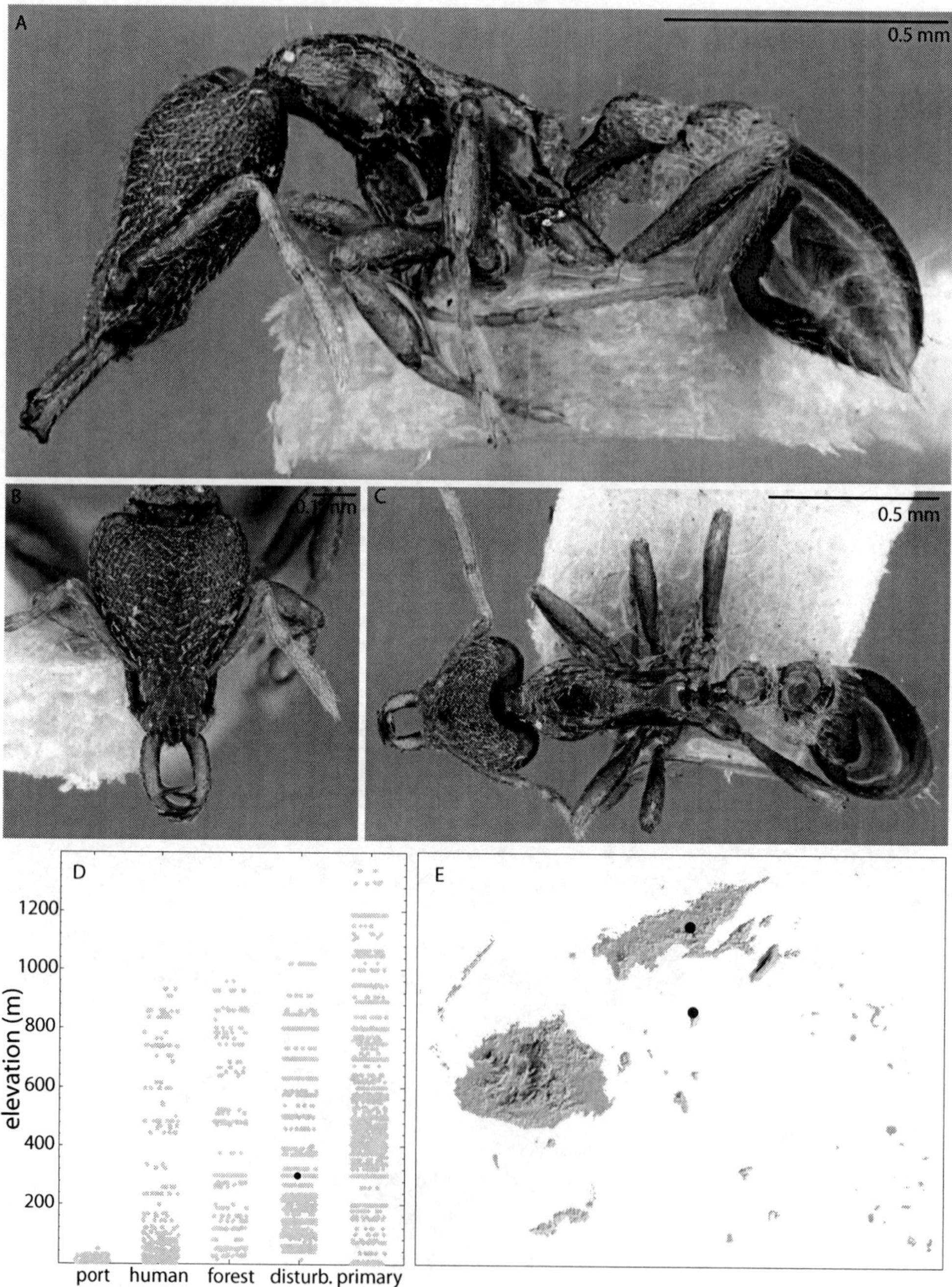

Plate 128. *Strumigenys mailei* worker, CASENT0186741.

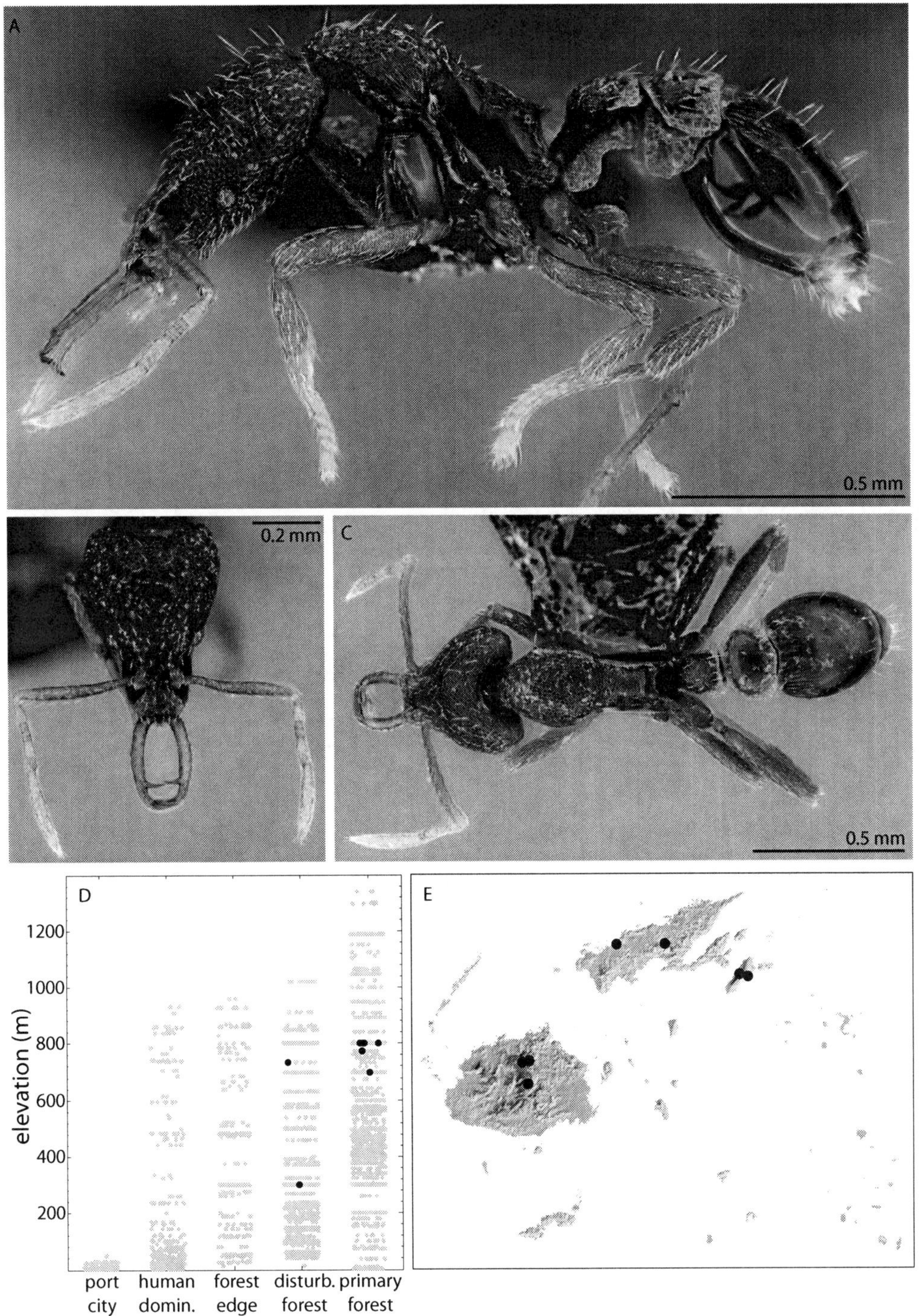

Plate 129. *Strumigenys nidifex* worker, CASENT0171047.

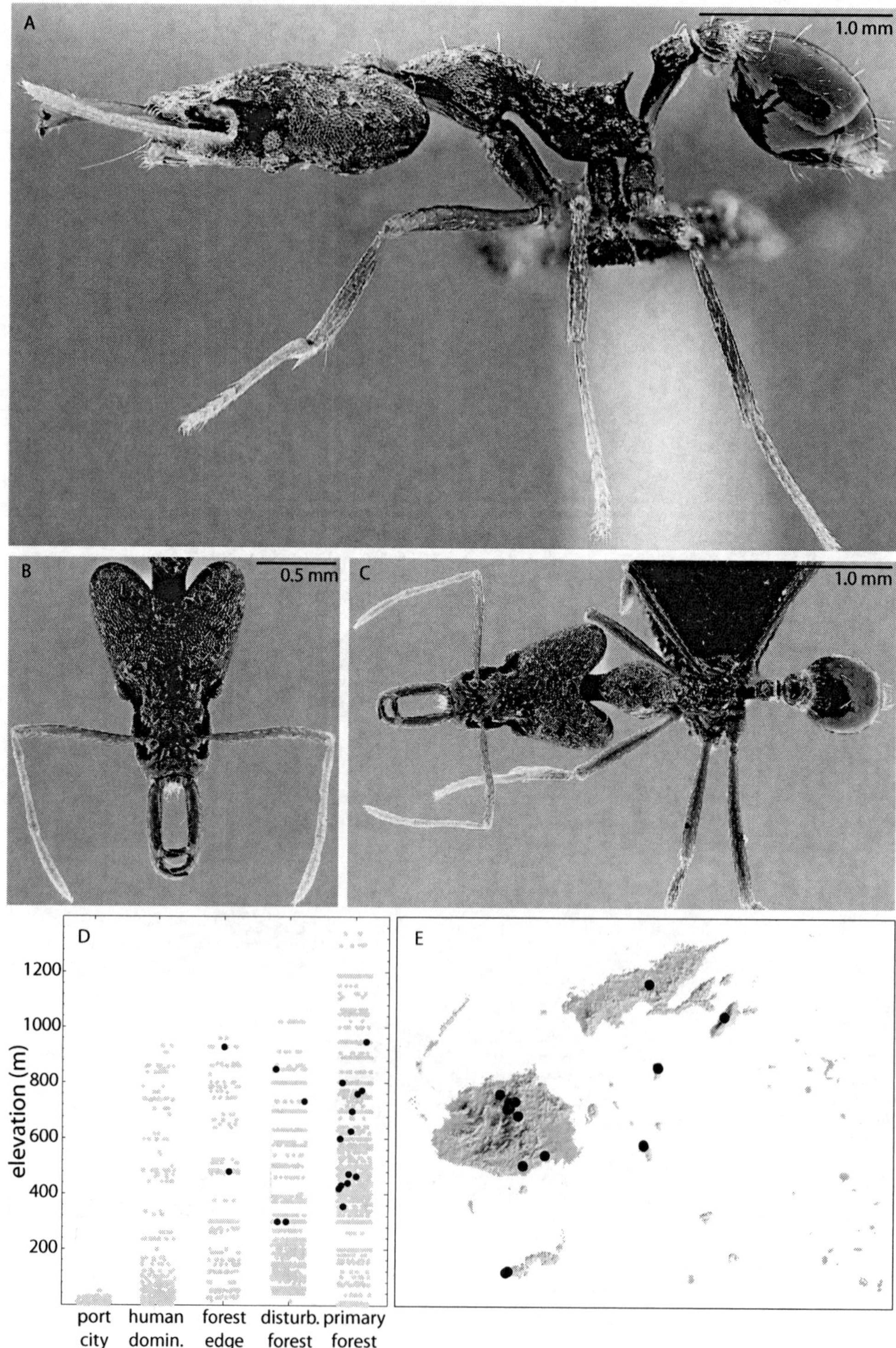

Plate 130. *Strumigenys panaulax* worker, CASENT0186960.

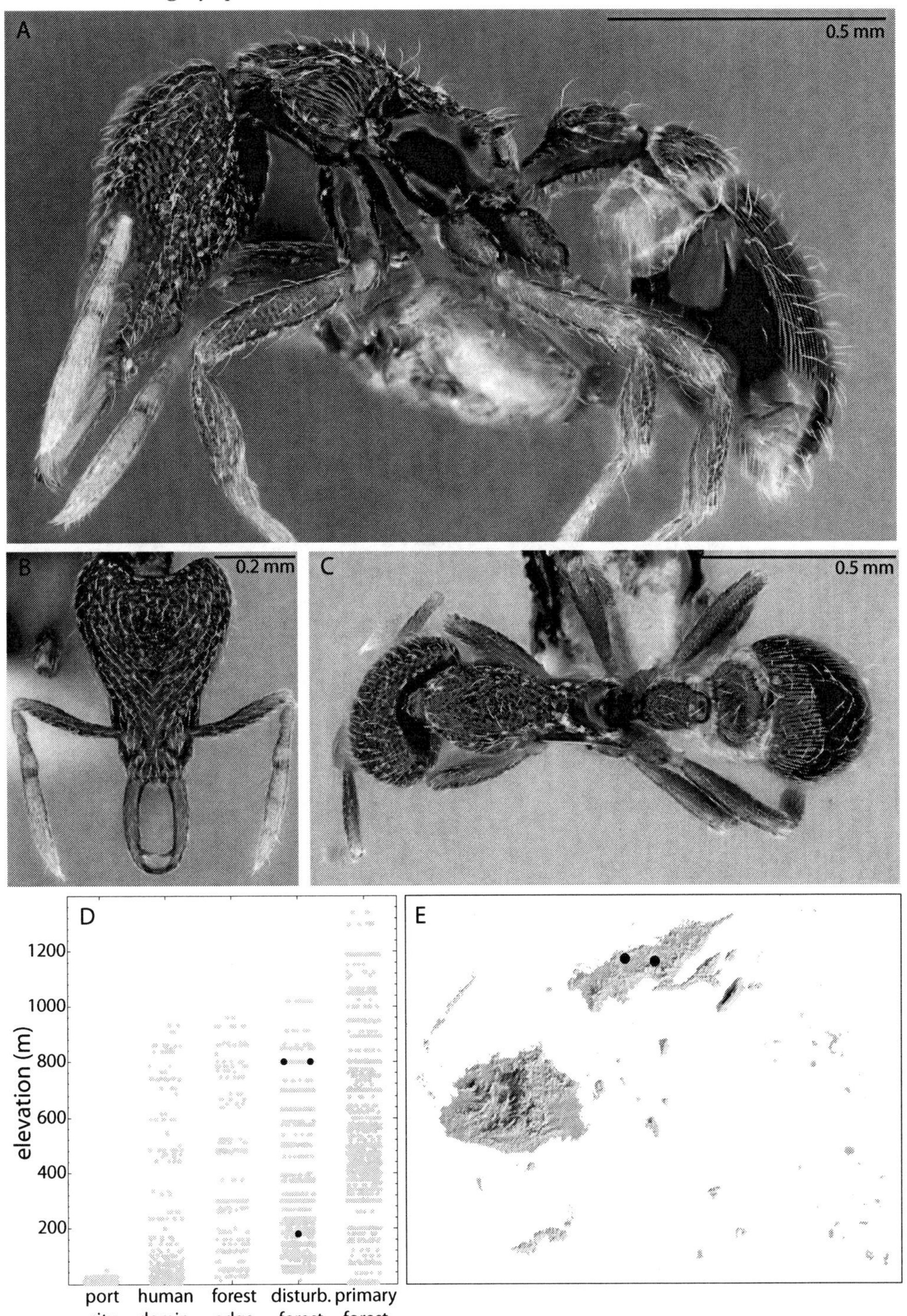

Plate 131. *Strumigenys praefecta* worker, CASENT0186986.

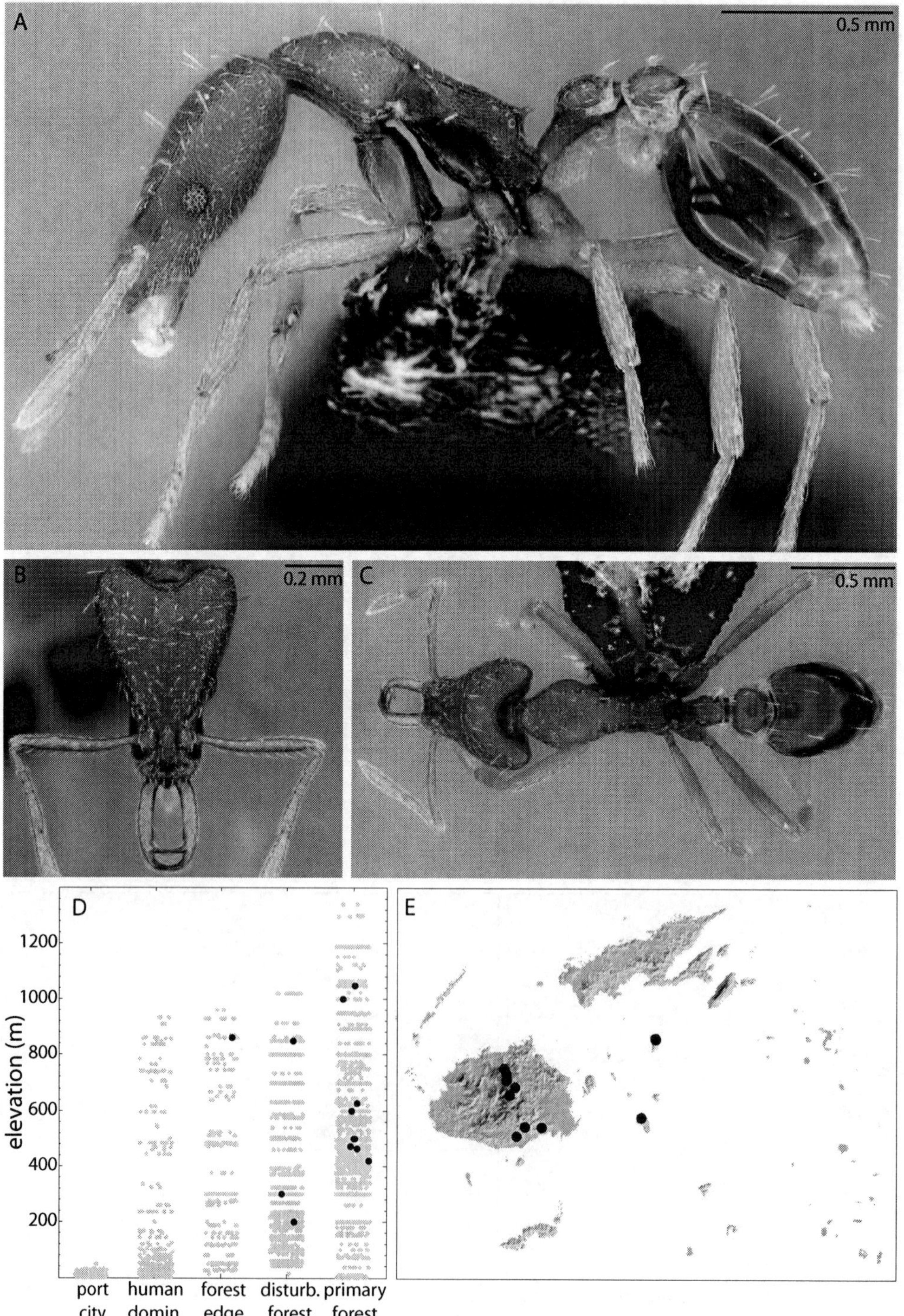

Plate 132. *Strumigenys rogeri* worker, CASENT0171133.

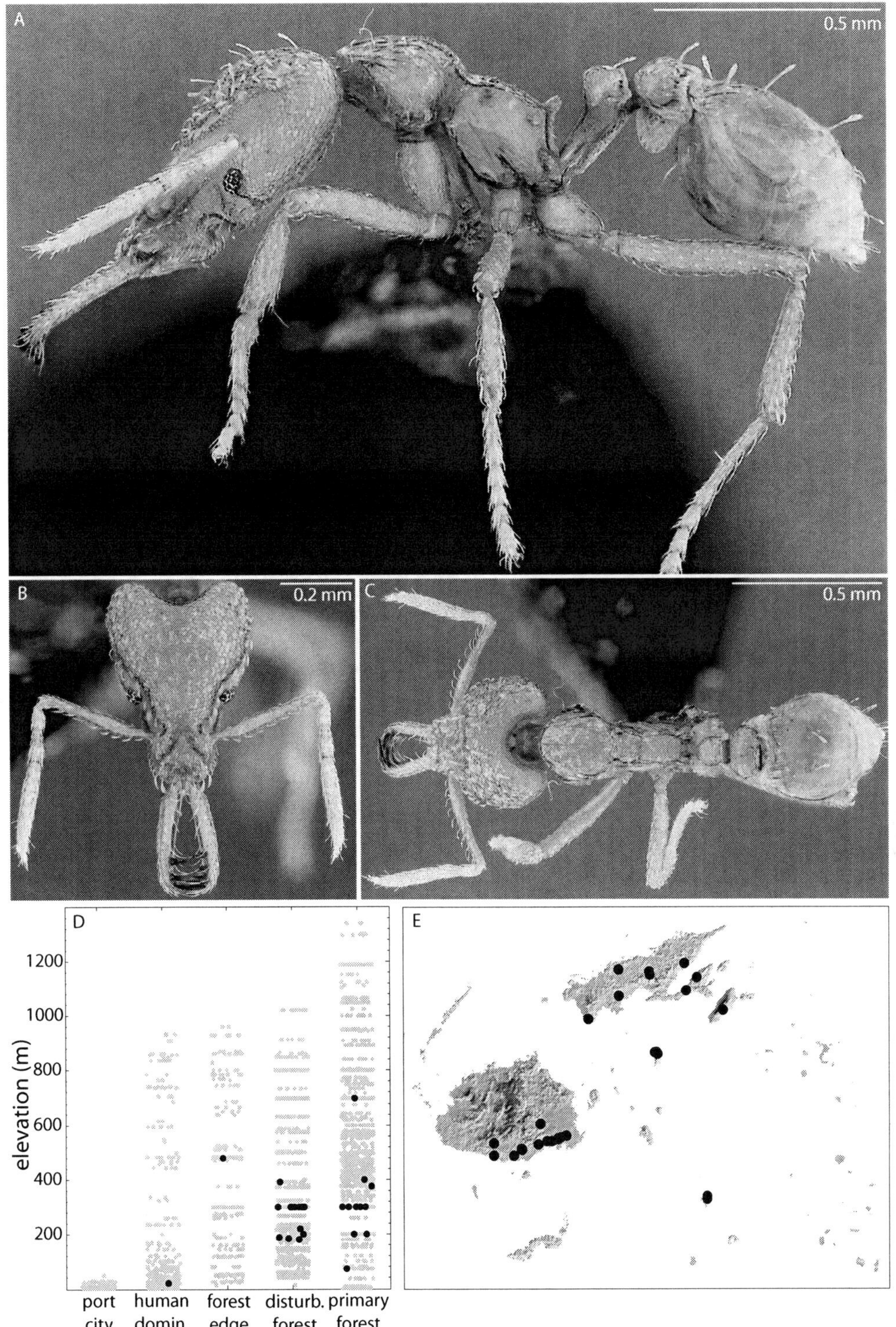

Plate 133. *Strumigenys scelesta* worker, CASENT0186833.

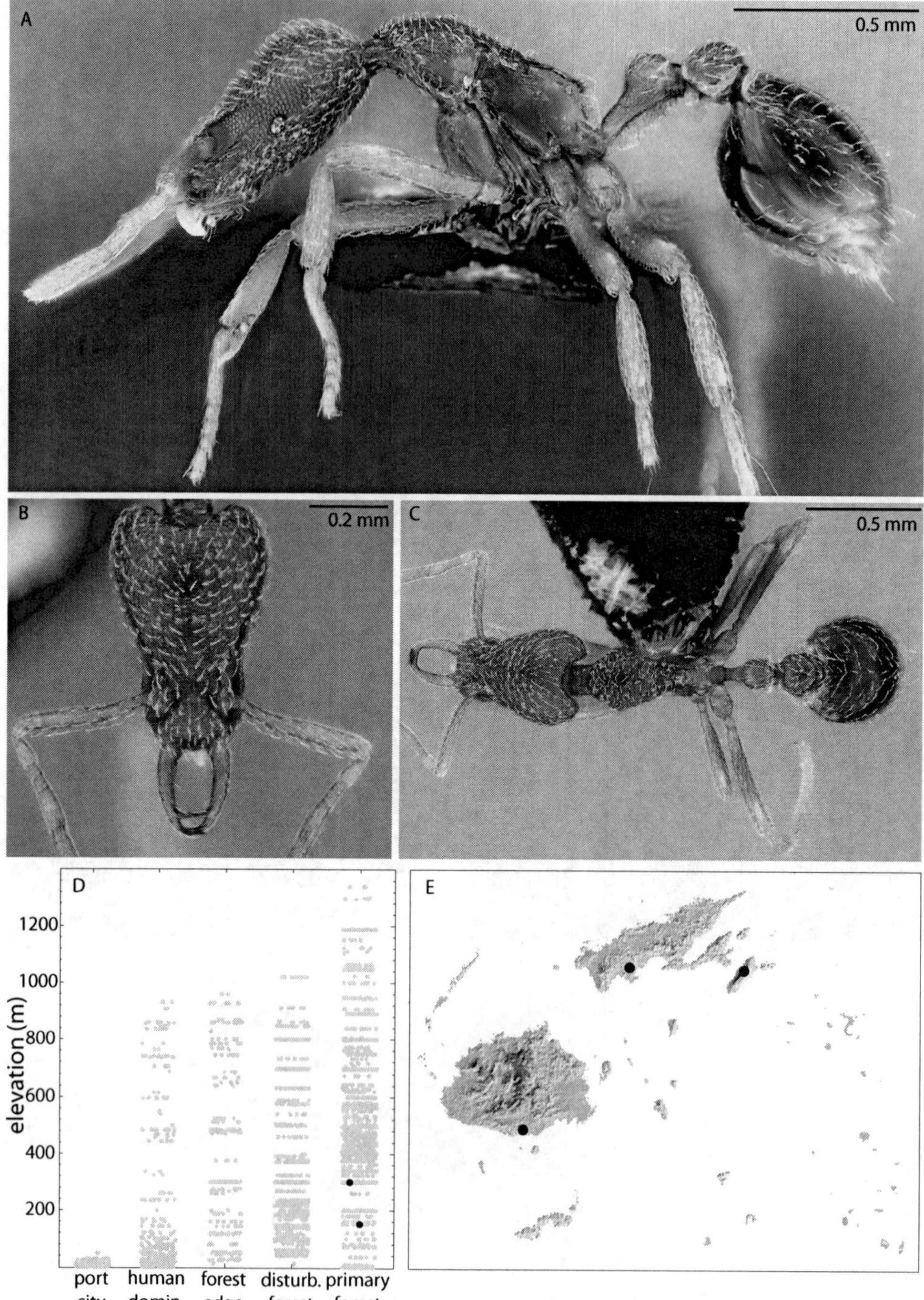

Plate 134. *Strumigenys sulcata* worker, CASENT0185962.

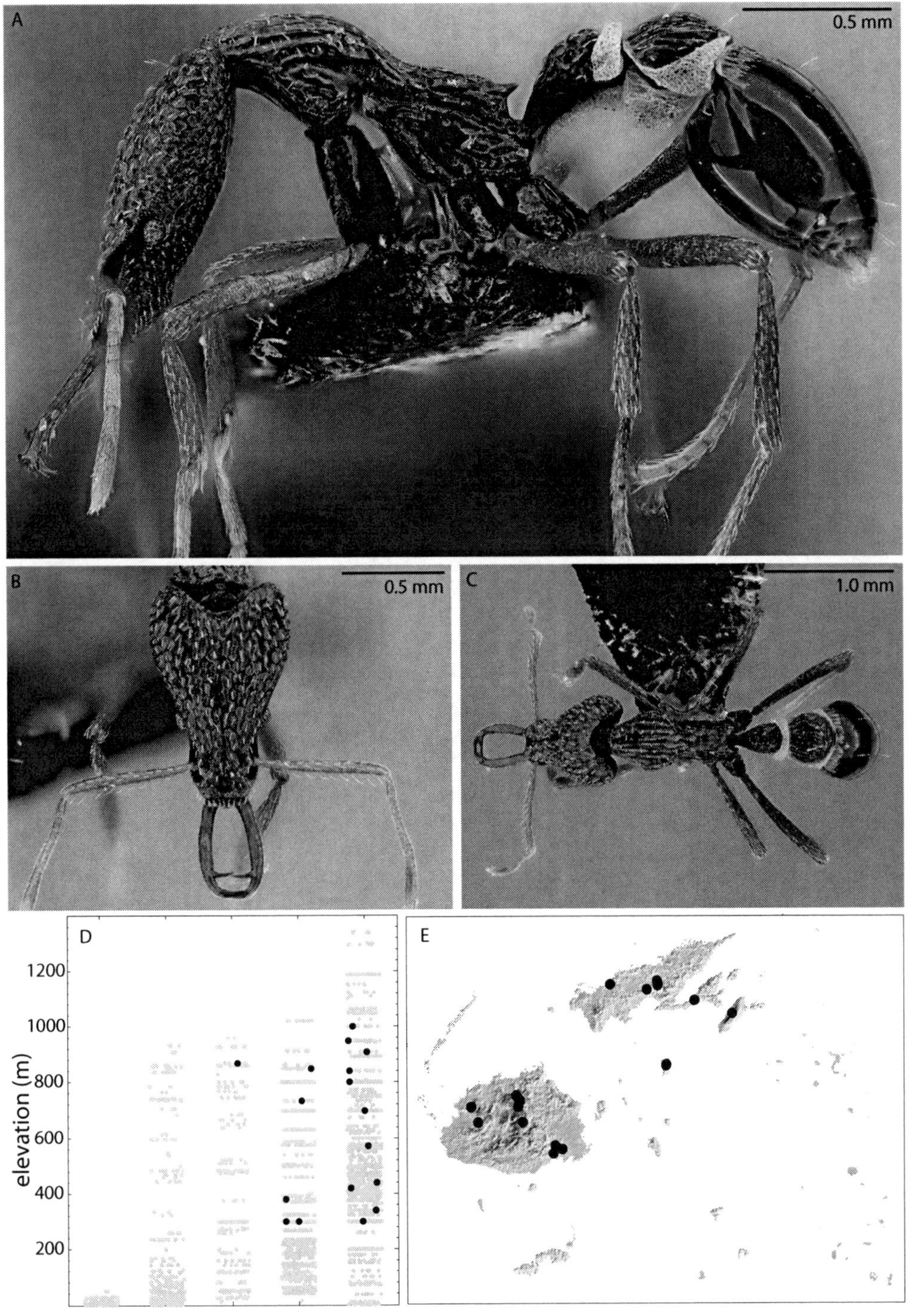

Plate 135. *Strumigenys tumida* worker, CASENT0186506.

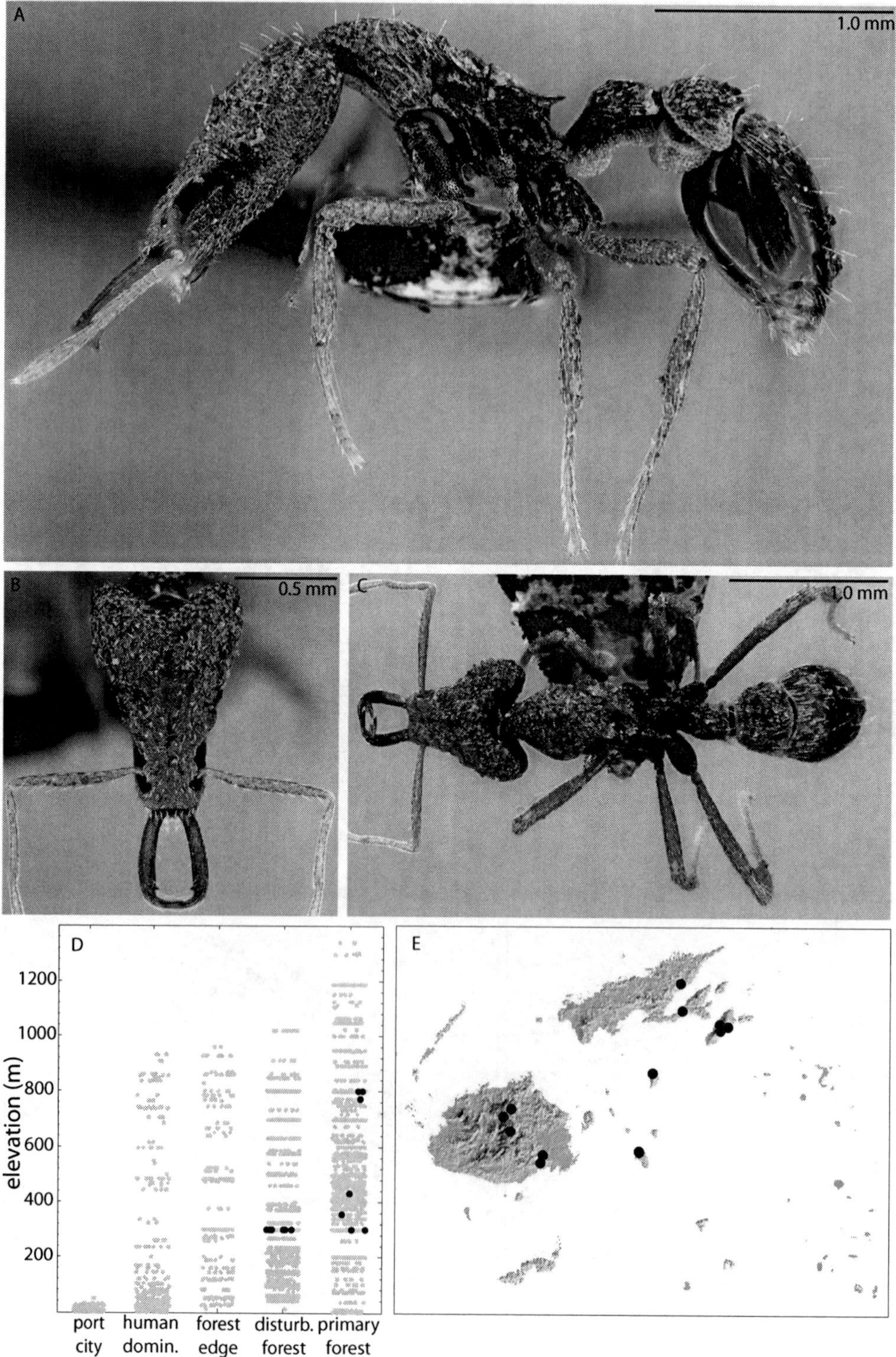

Plate 136. *Strumigenys* sp. FJ01 worker, CASENT0185902.

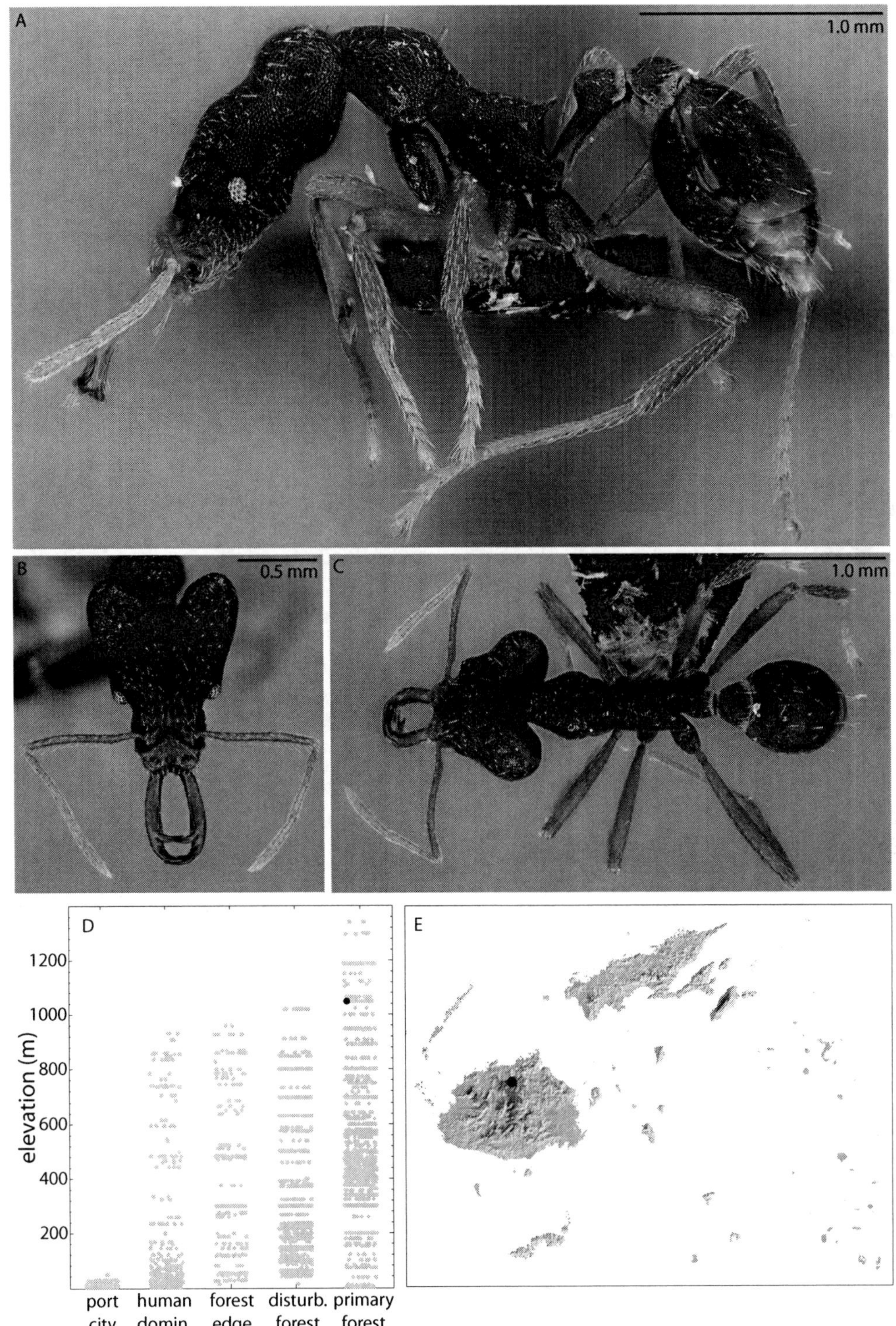

Plate 137. *Strumigenys* sp. FJ13 worker, CASENT0186982.

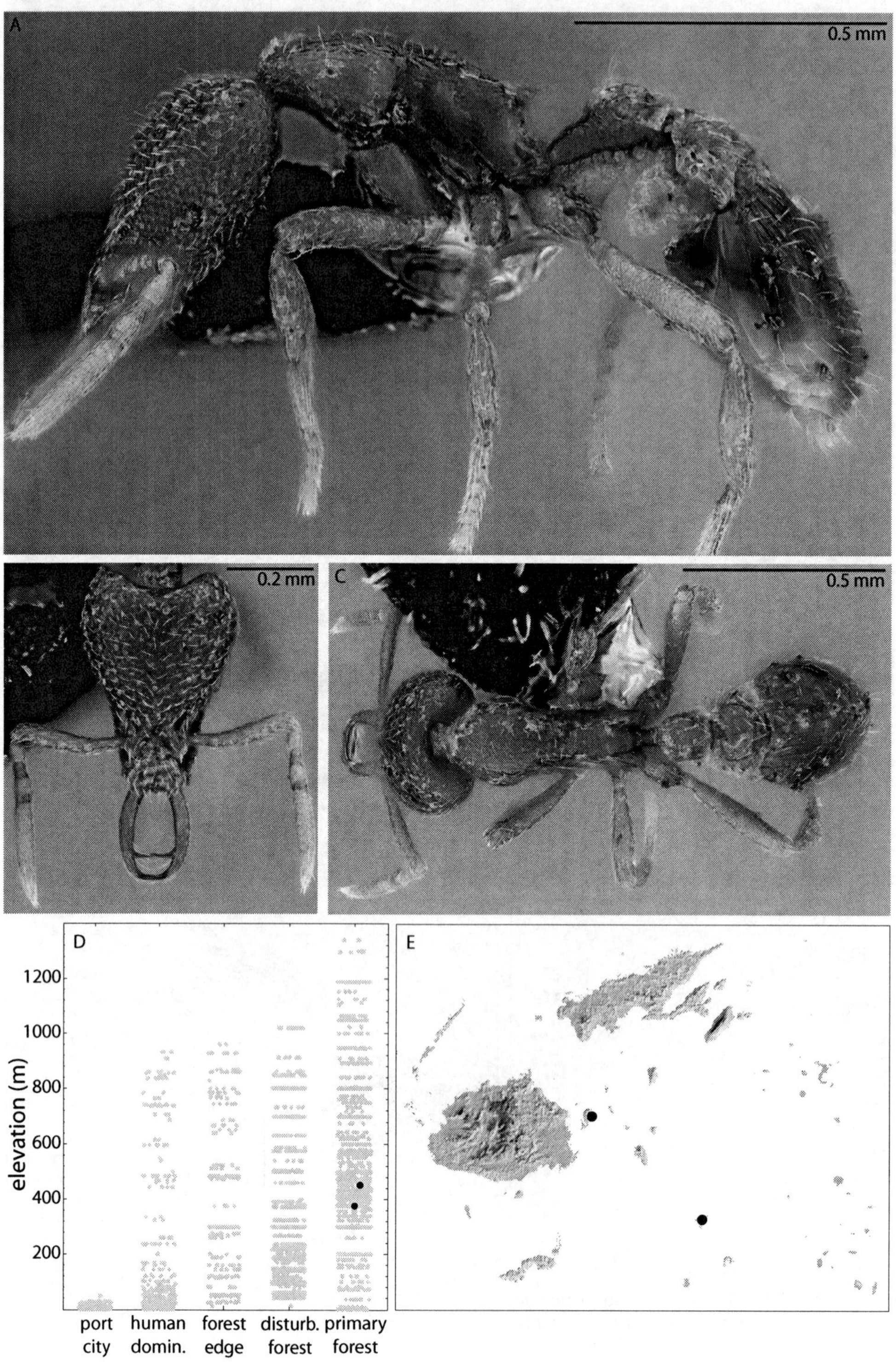

Plate 138. *Strumigenys* sp. FJ14 worker, CASENT0184984.

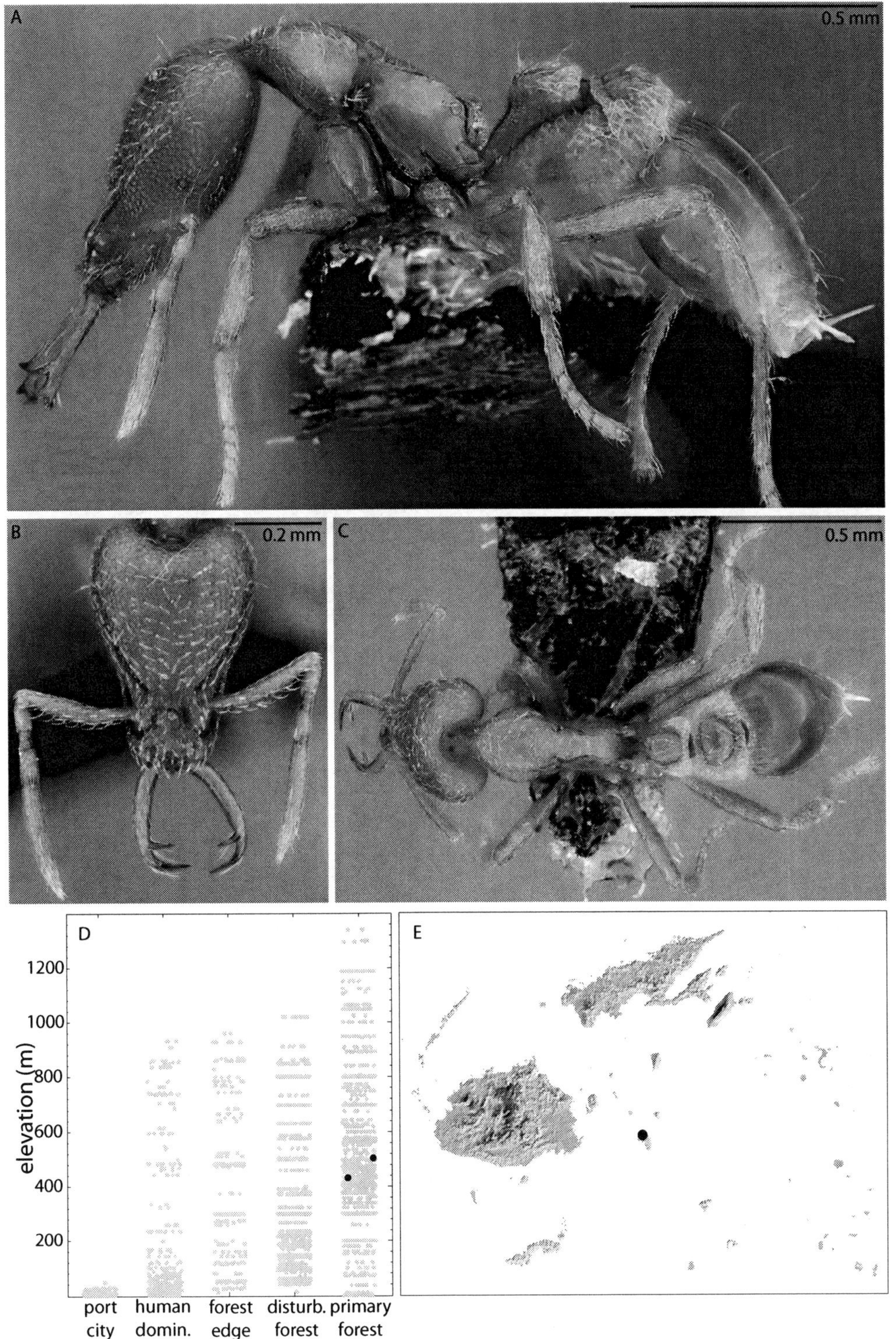

Plate 139. *Strumigenys* sp. FJ17 worker, CASENT0184819.

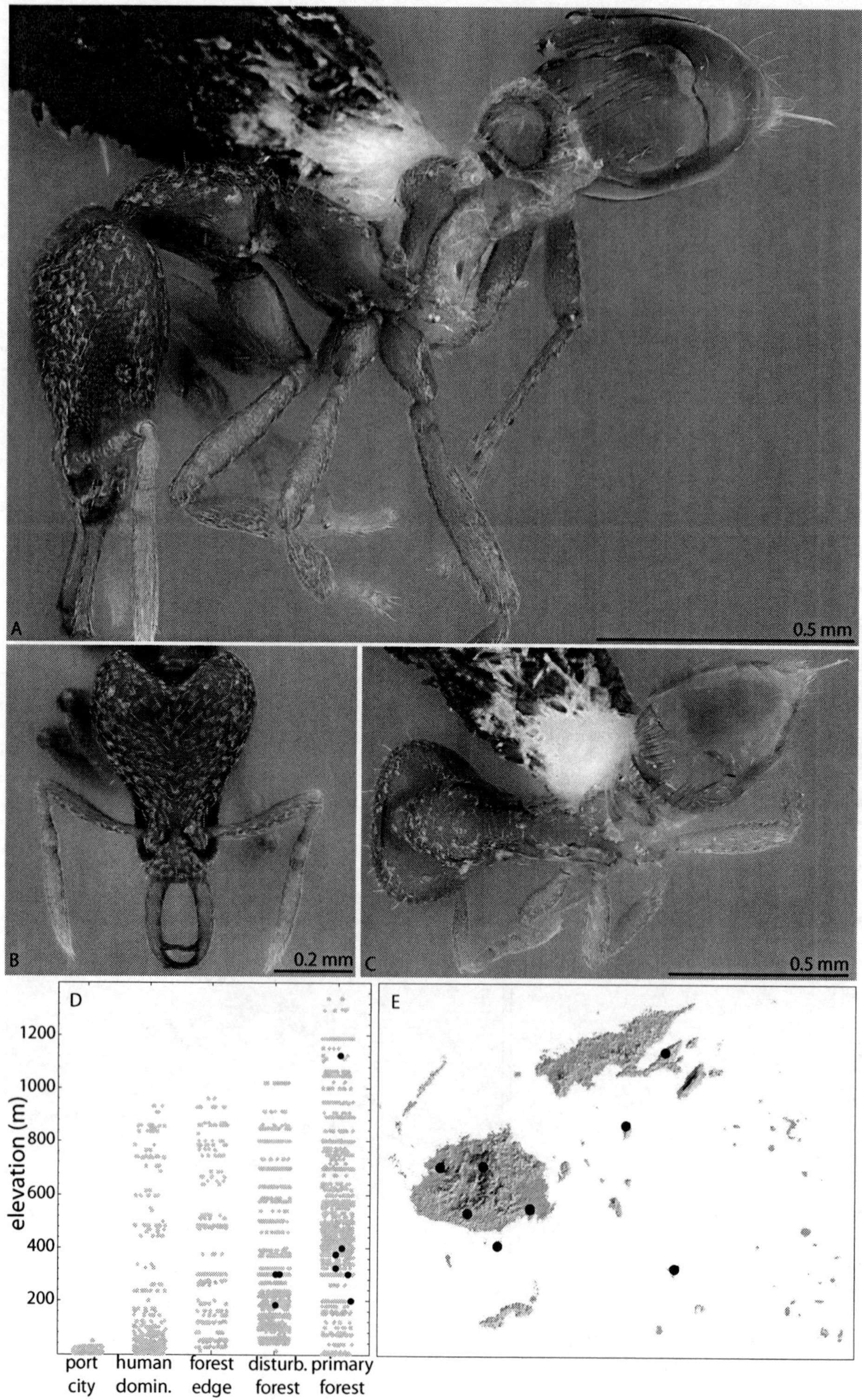

Plate 140. *Strumigenys* sp. FJ18 worker, CASENT0186900.

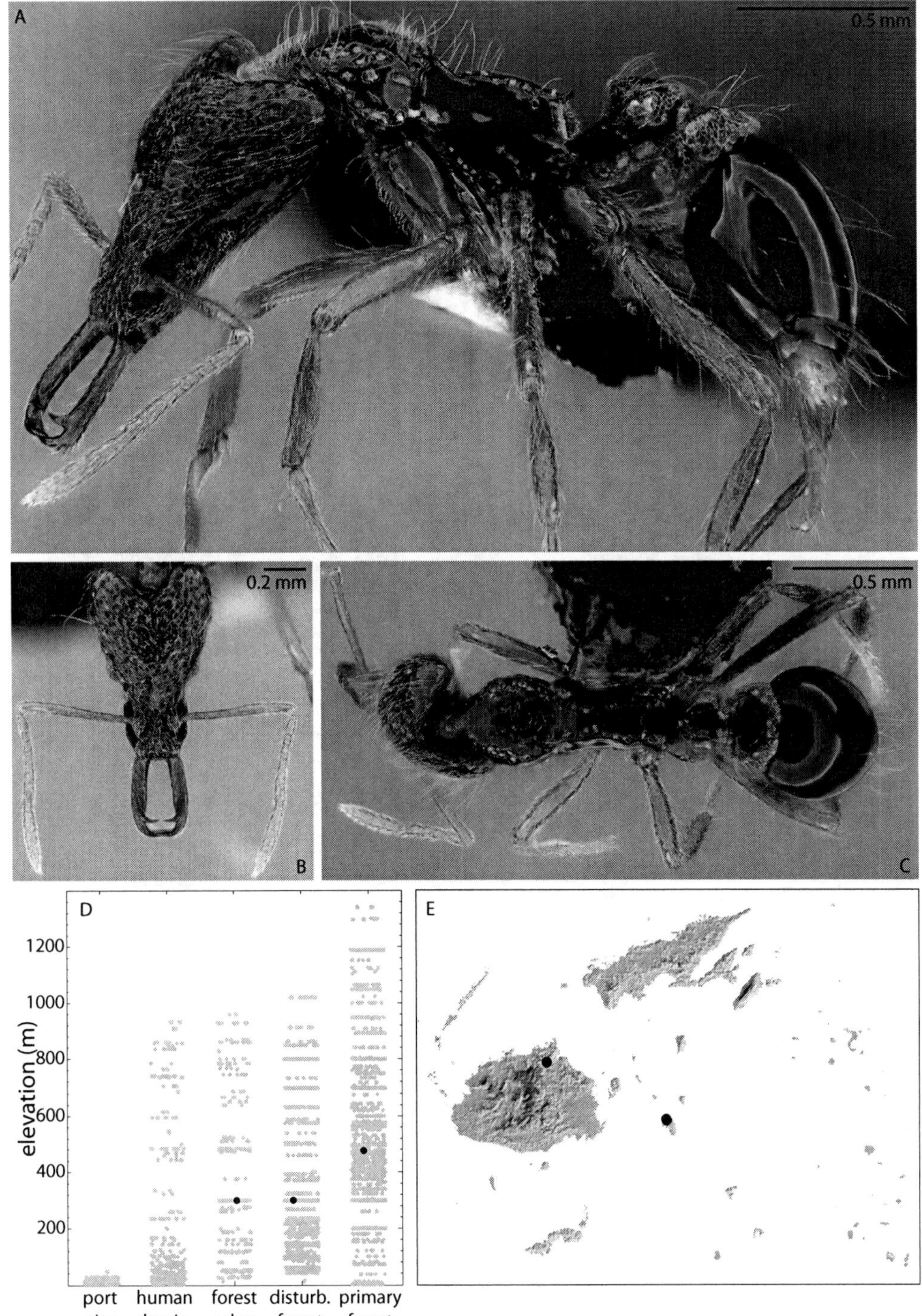

Plate 141. *Strumigenys* sp. FJ19 worker, CASENT0185751.

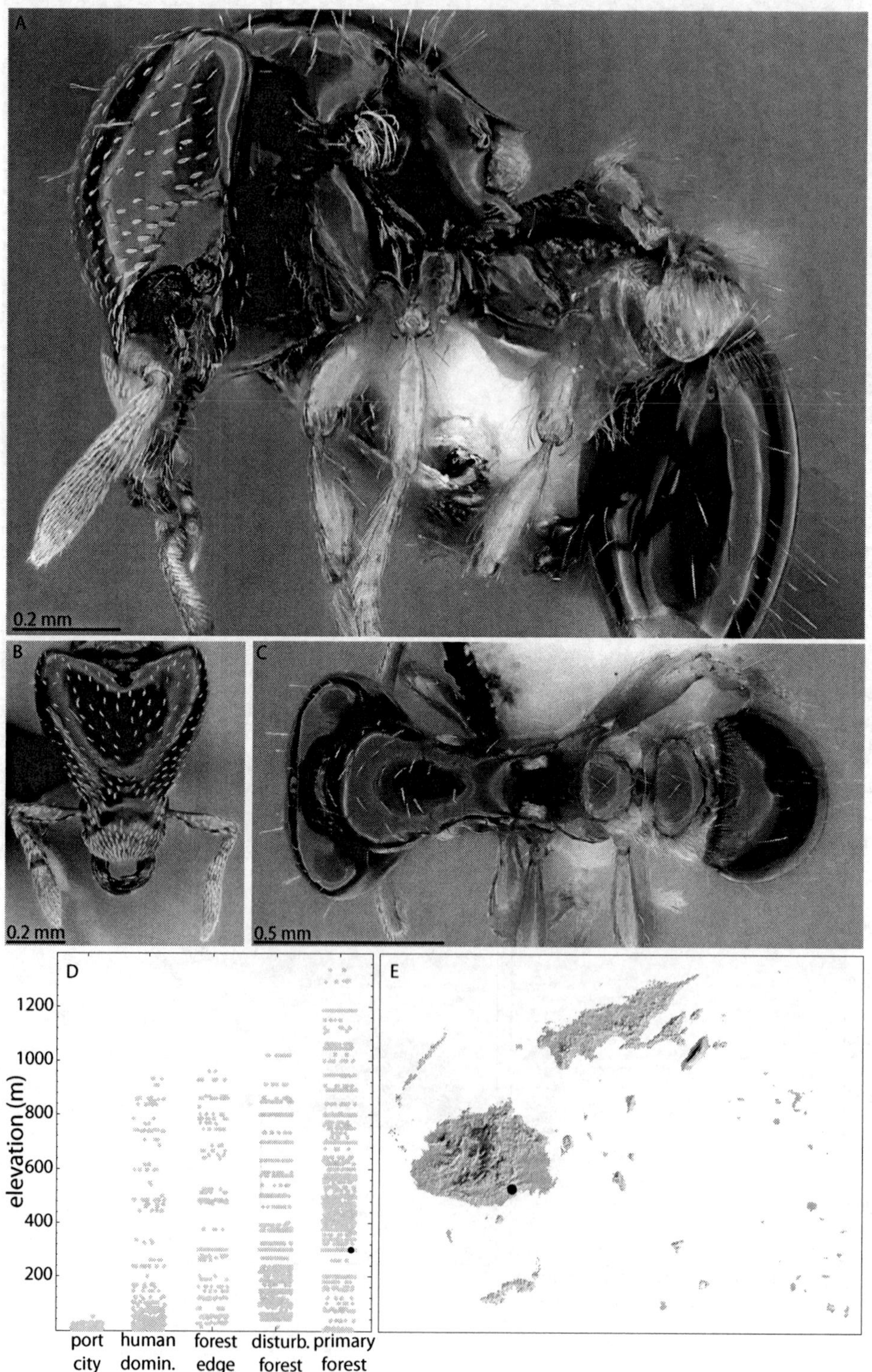

Plate 142. *Tetramorium bicarinatum* worker, CASENT0171032.

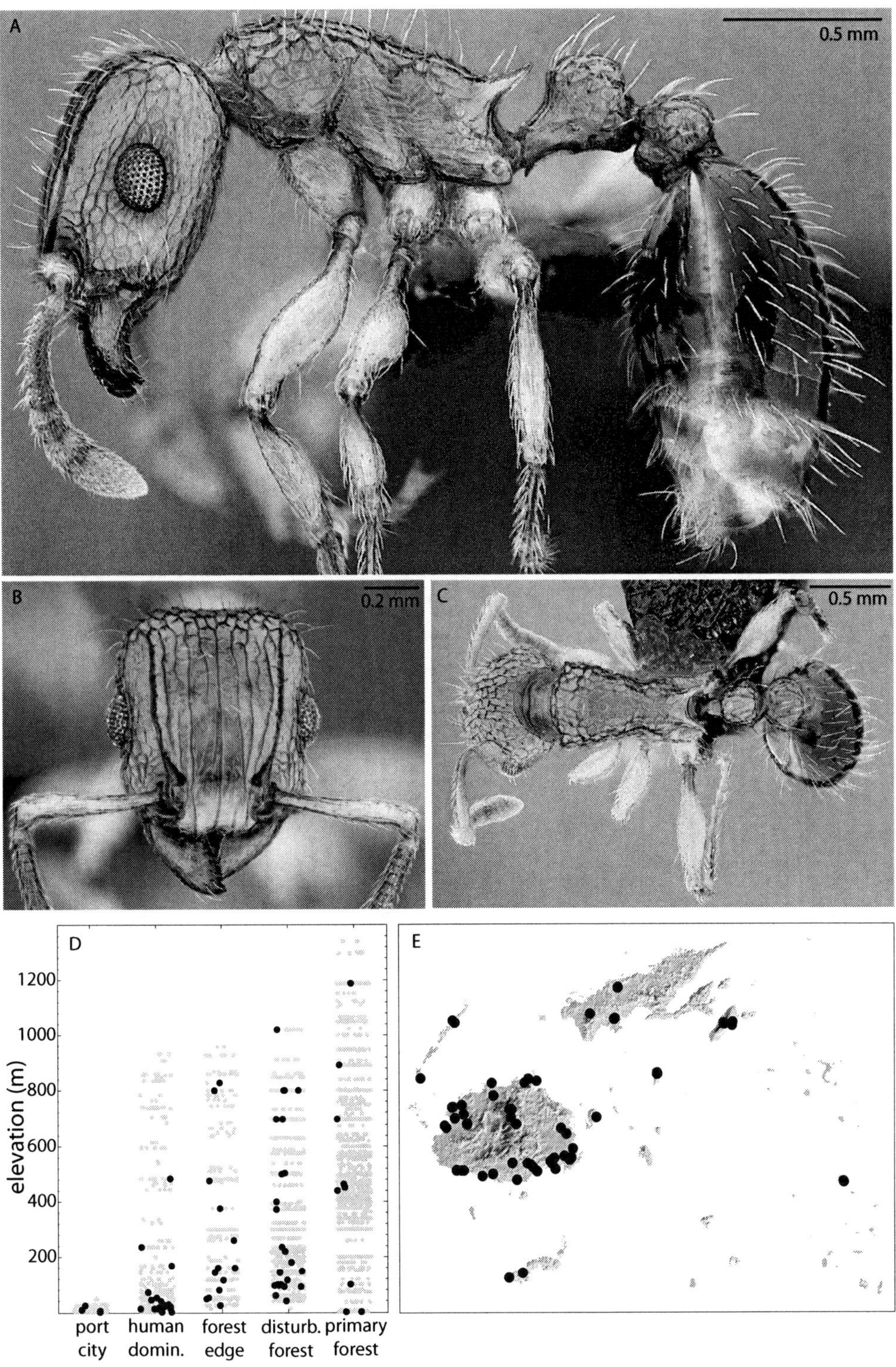

Plate 143. *Tetramorium caldarium* worker, CASENT0171084.

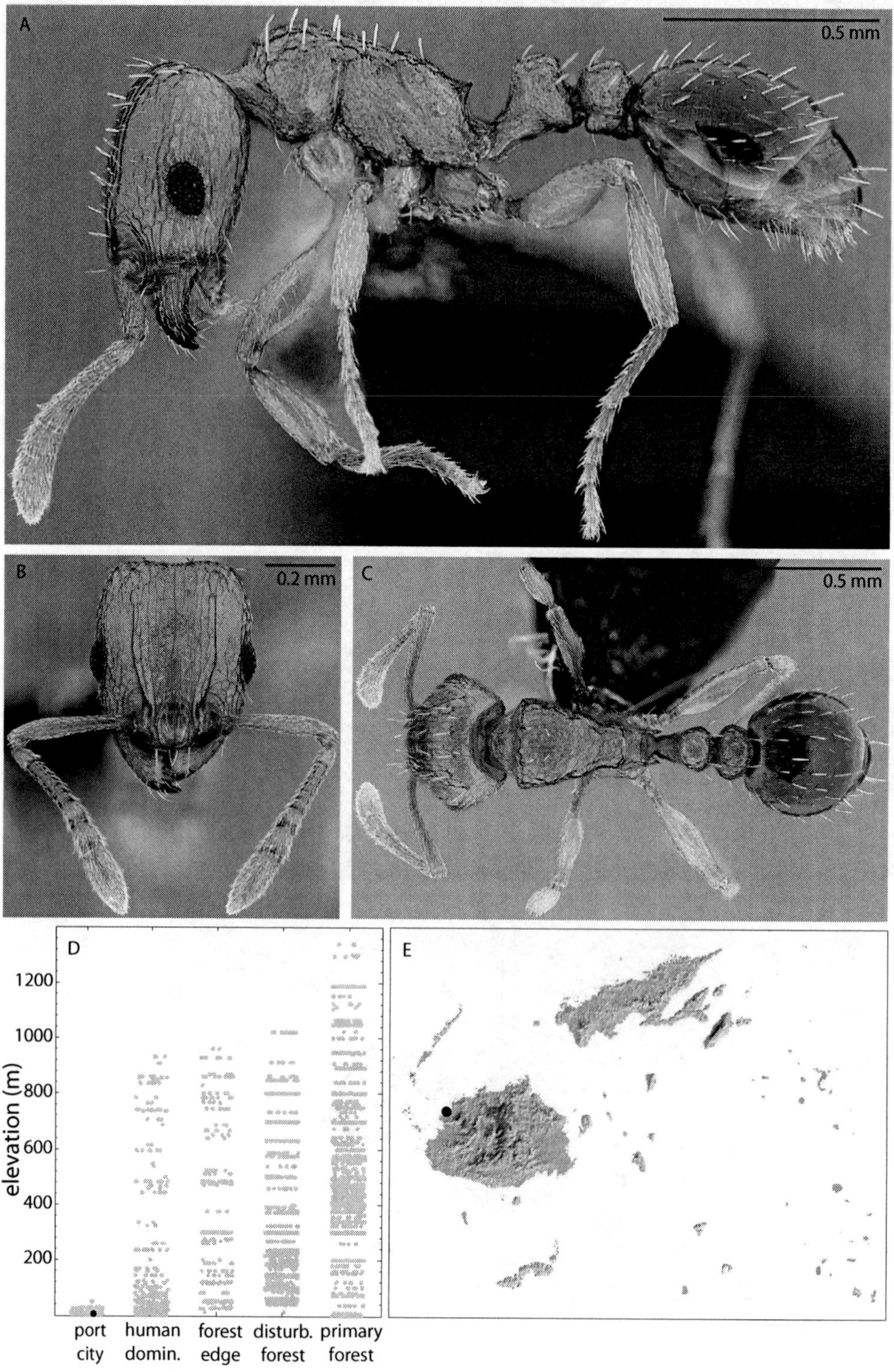

Plate 144. *Tetramorium insolens* worker, CASENT0171138.

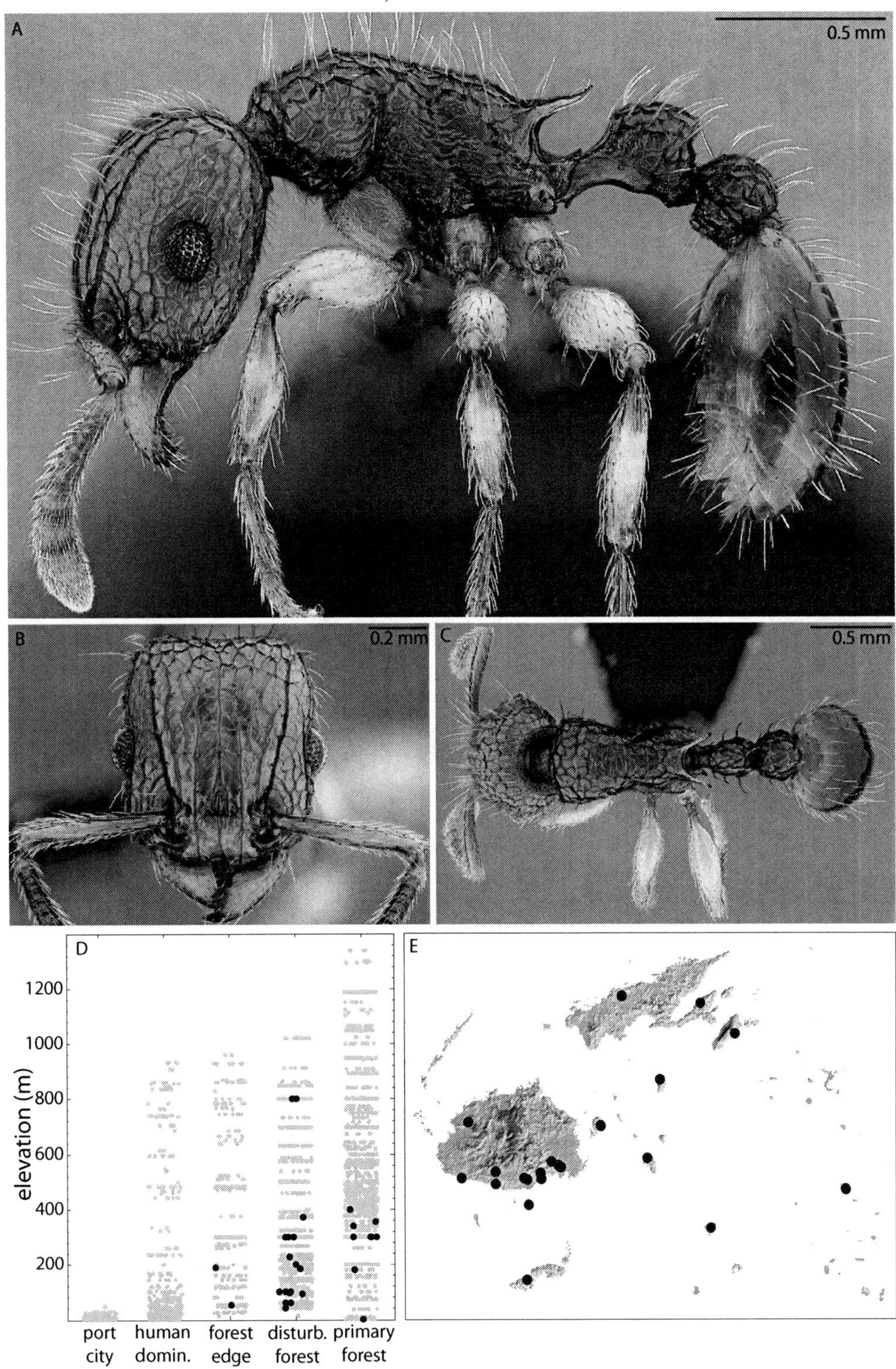

Plate 145. *Tetramorium lanuginosum* worker, CASENT0171128.

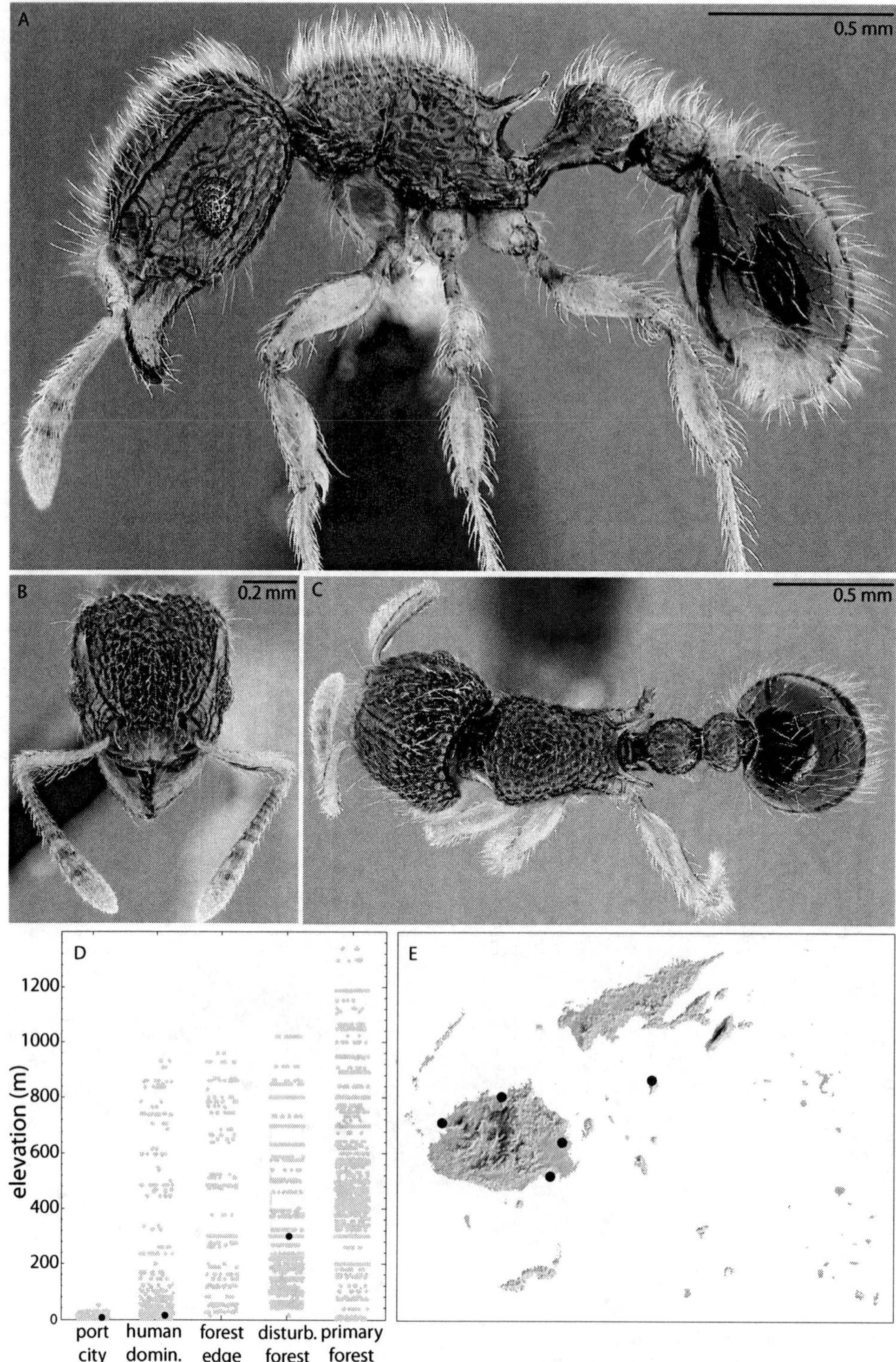

Plate 146. *Tetramorium manni* worker, CASENT0183012.

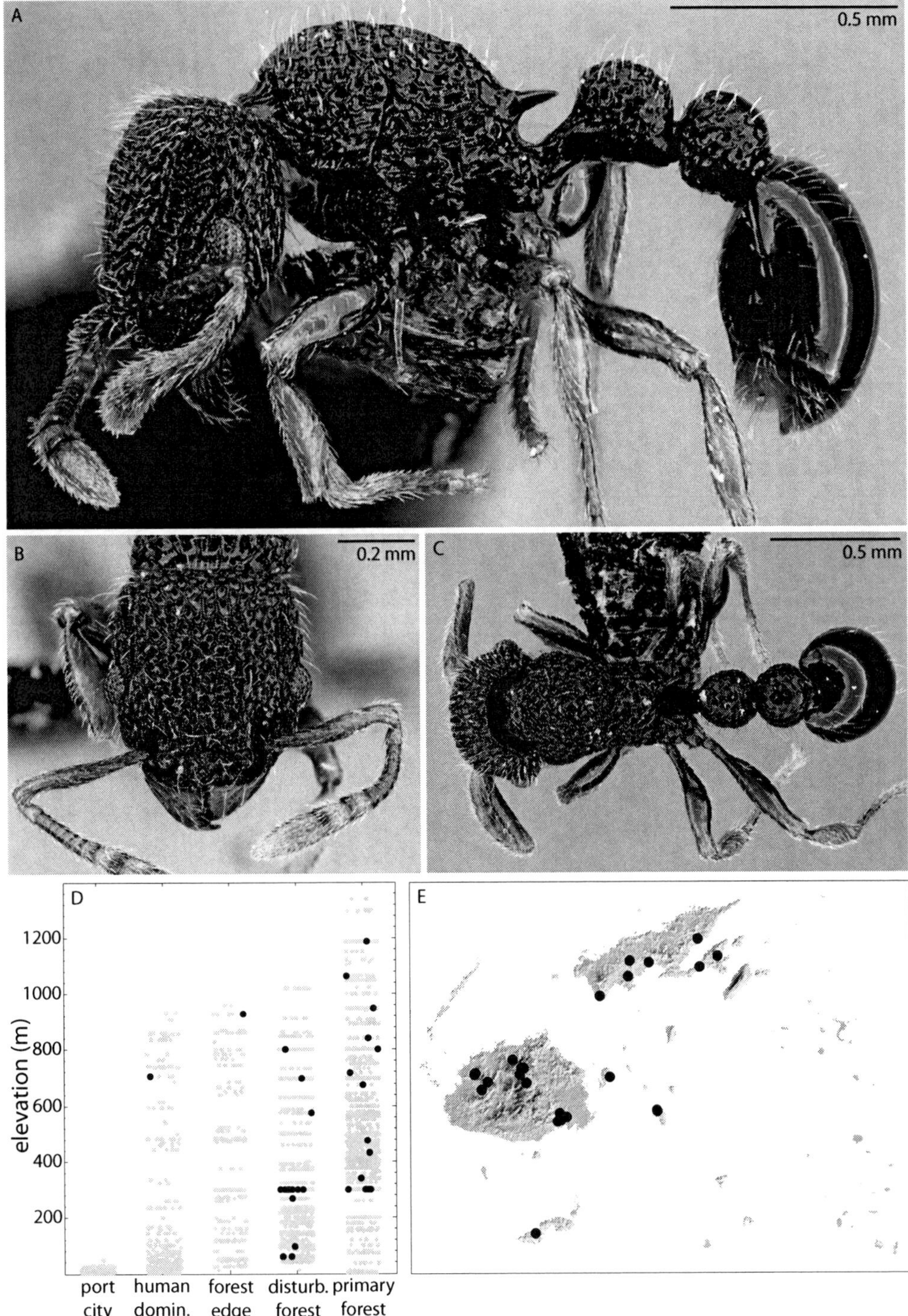

Plate 147. *Tetramorium pacificum* worker, CASENT0171075.

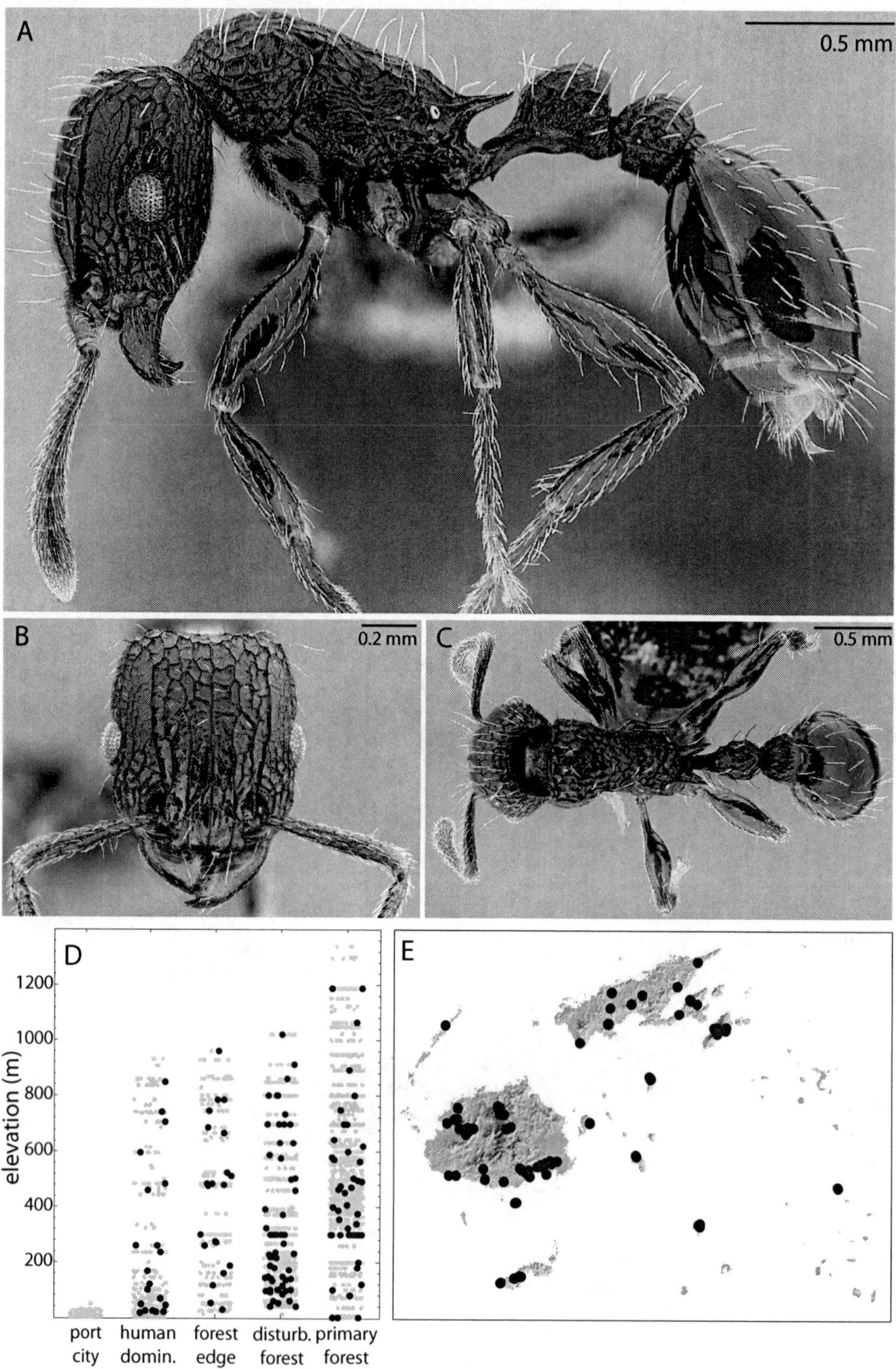

Plate 148. *Tetramorium simillimum* worker, CASENT0171034.

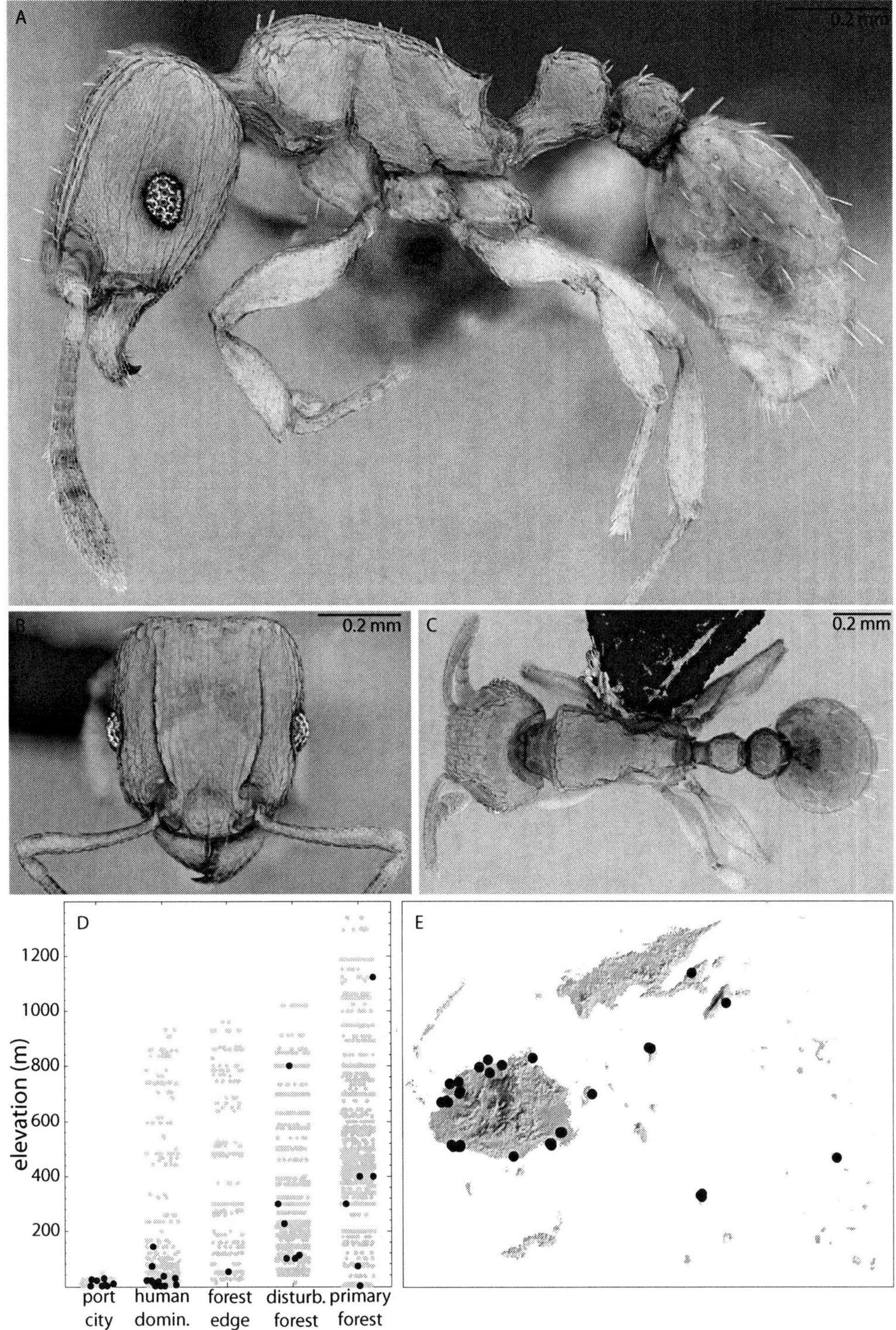

Plate 149. *Tetramorium tonganum* worker, CASENT0171074.

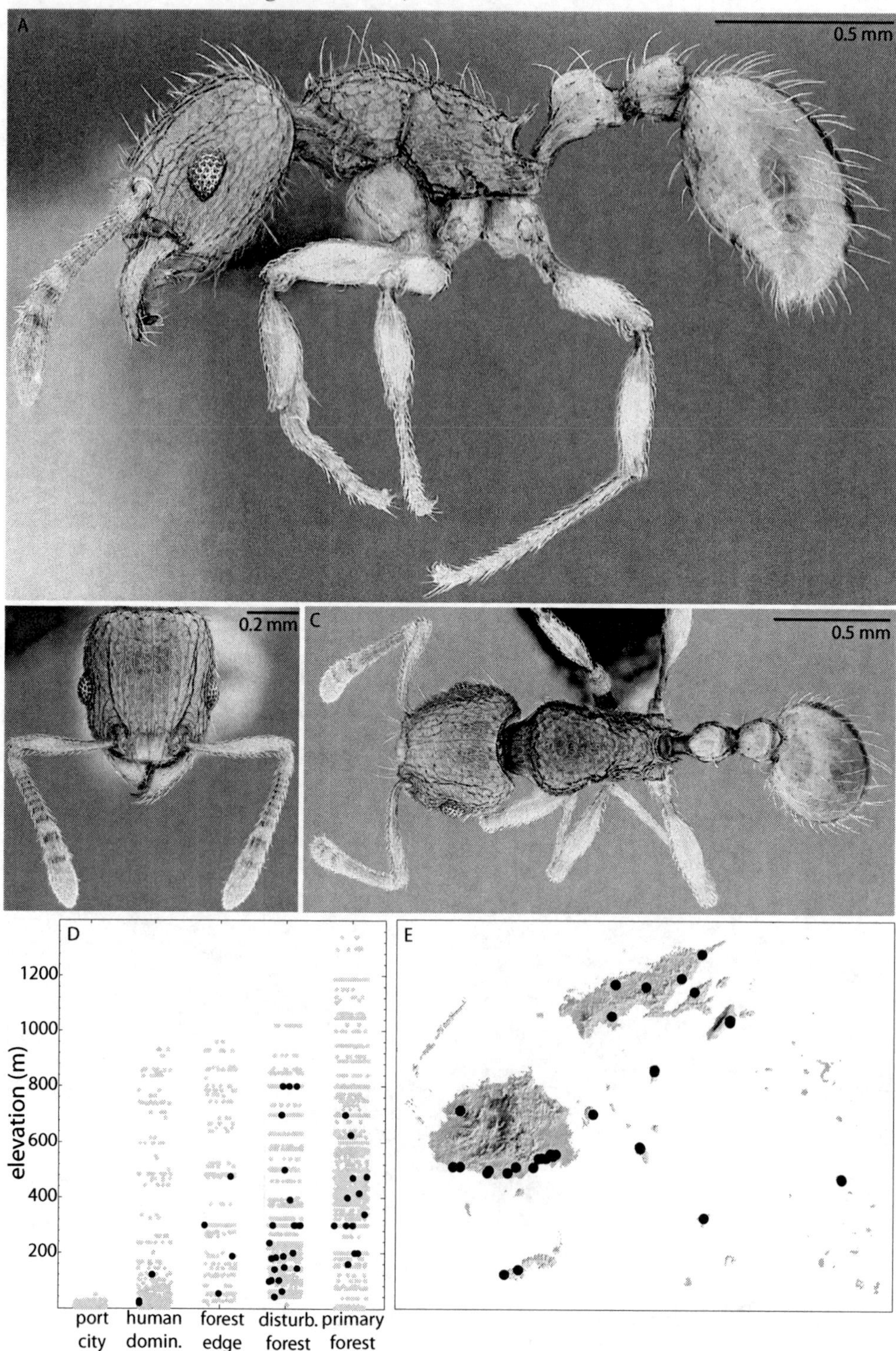

Plate 150. *Vollenhovia denticulata* worker, CASENT0182566.

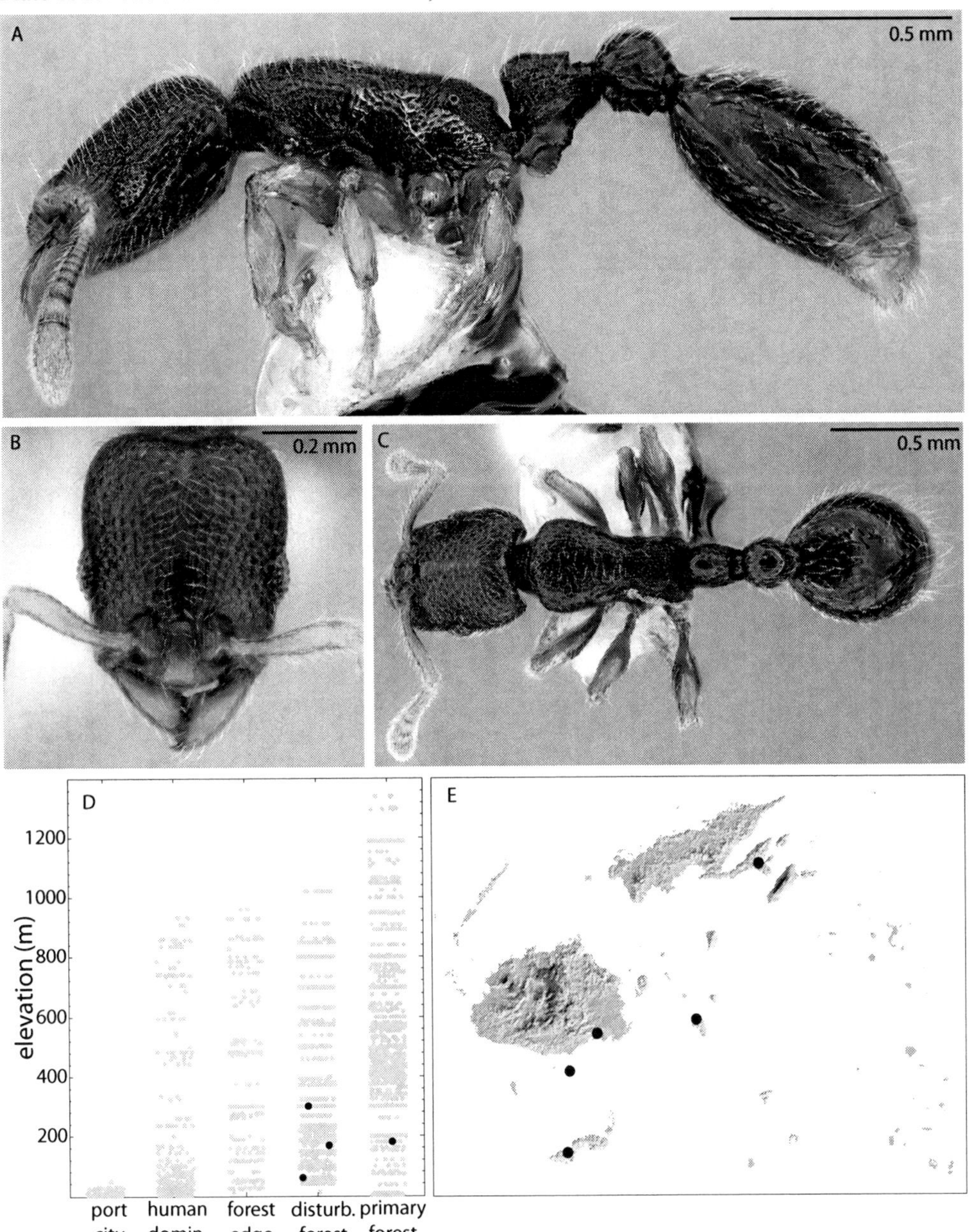

Plate 151. *Vollenhovia* sp. FJ01 worker, CASENT0171049.

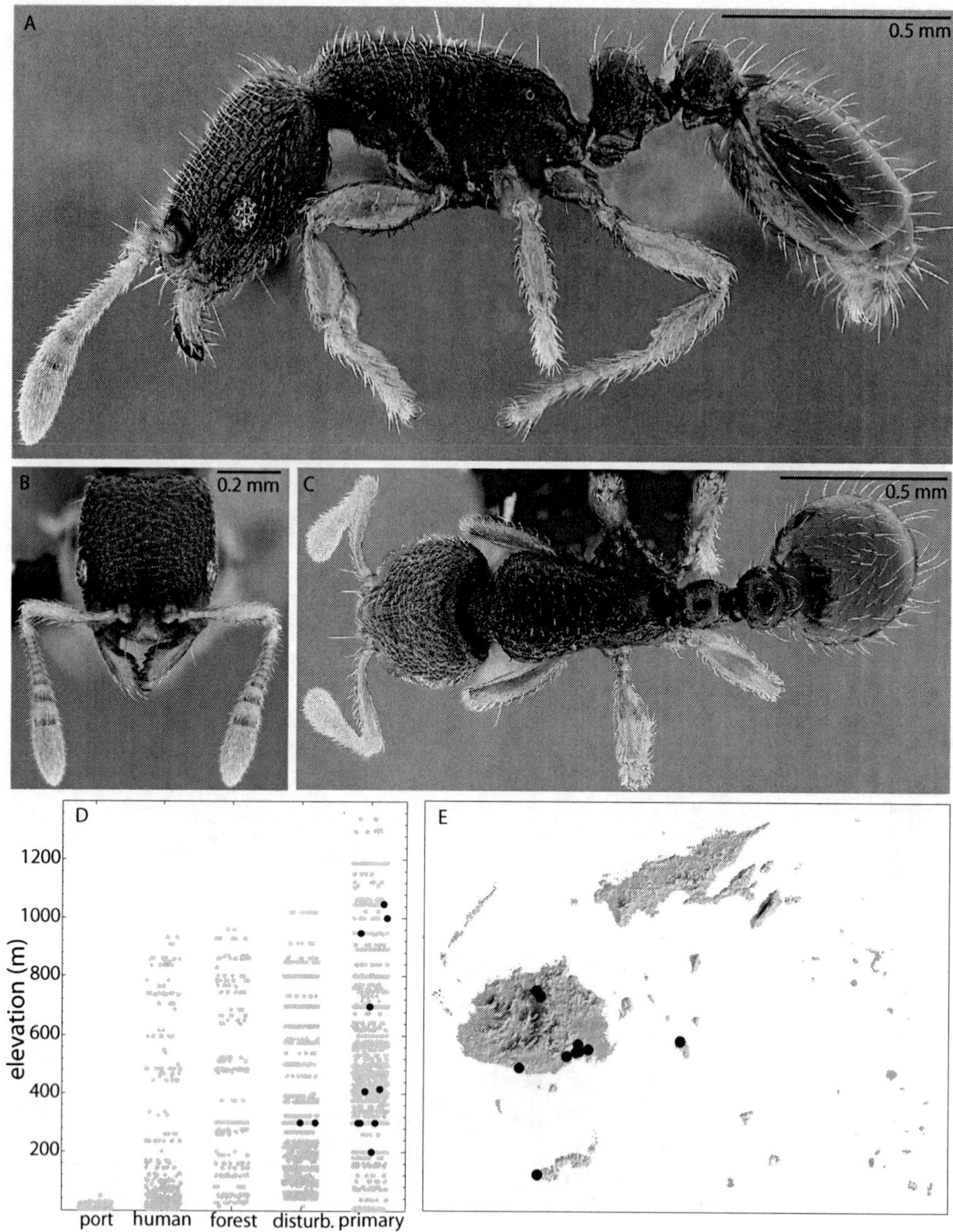

Plate 152. *Vollenhovia* sp. FJ03 worker, CASENT0182605.

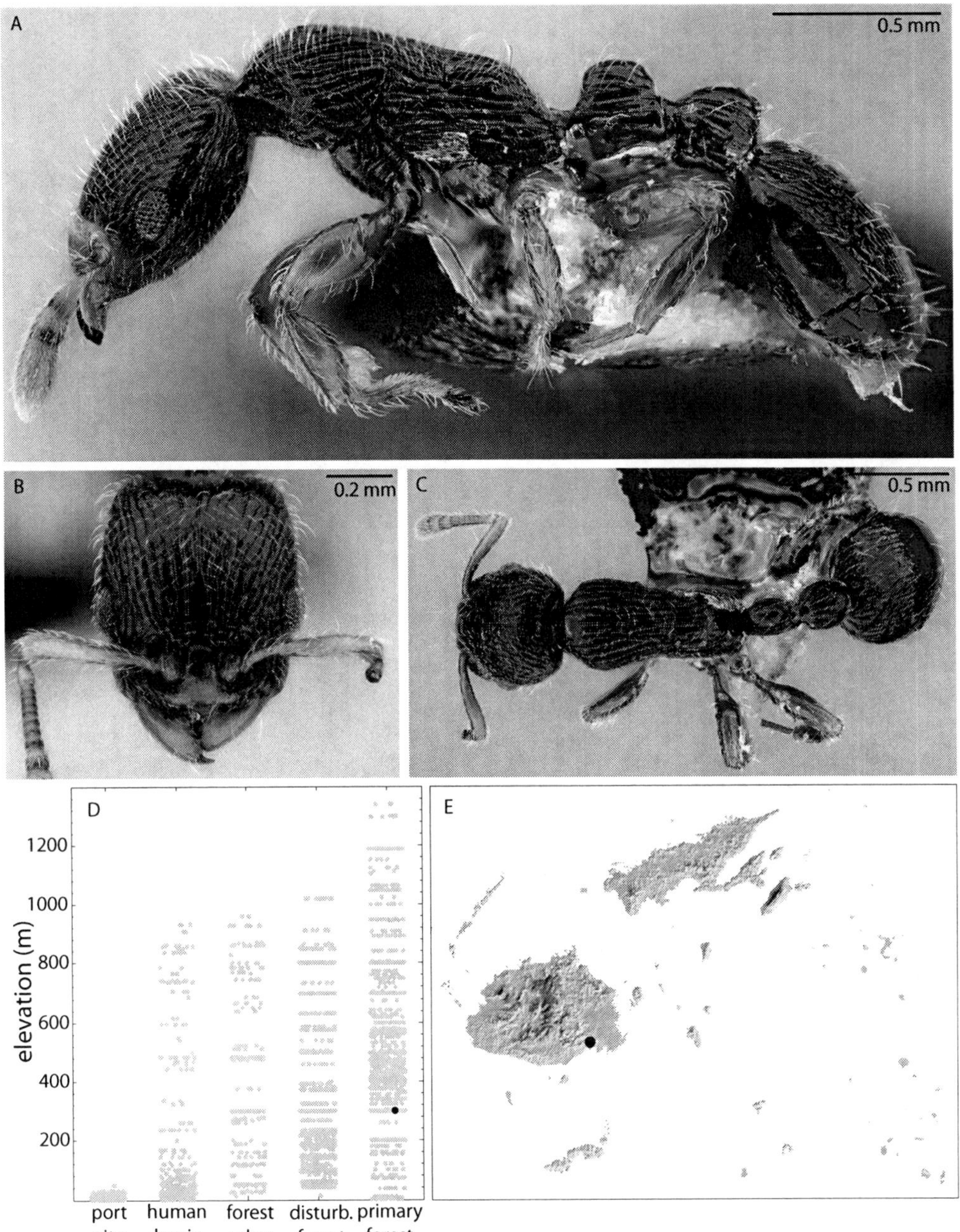

Plate 153. *Vollenhovia* sp. FJ04 worker, CASENT0182576.

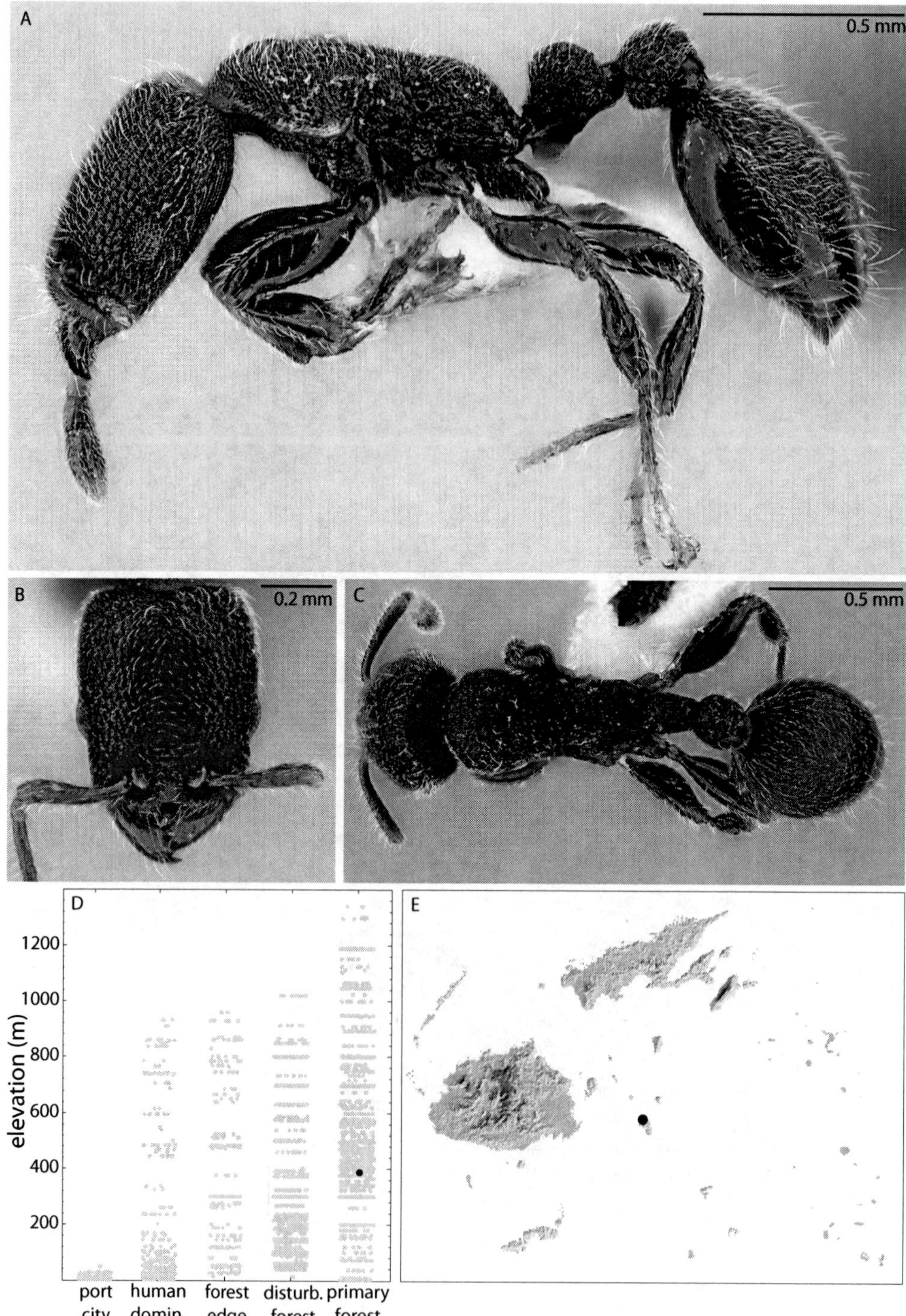

Plate 154. *Vollenhovia* sp. FJ05 worker, CASENT0182577.

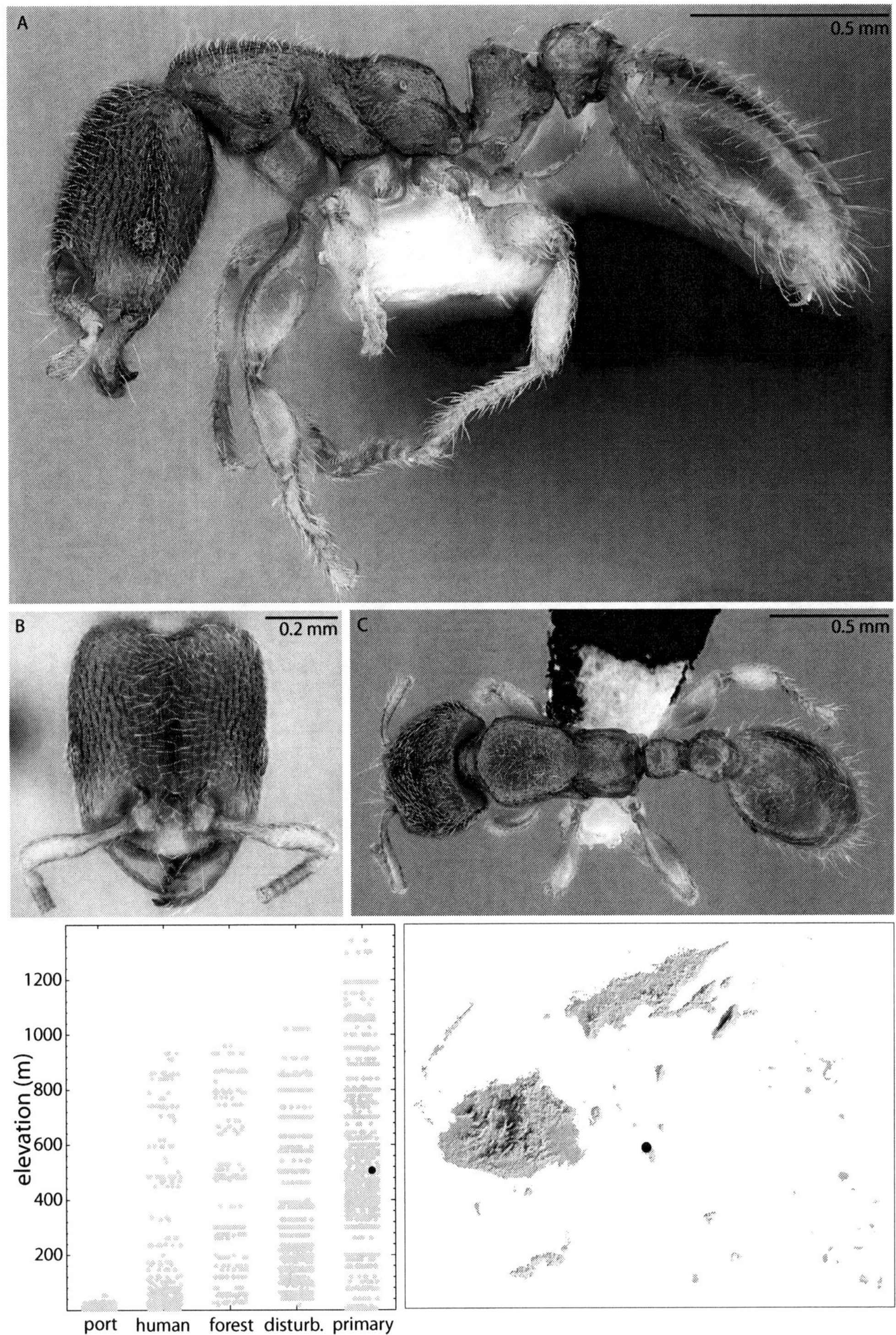

Plate 155. *Anochetus graeffei* worker, CASENT0171067.

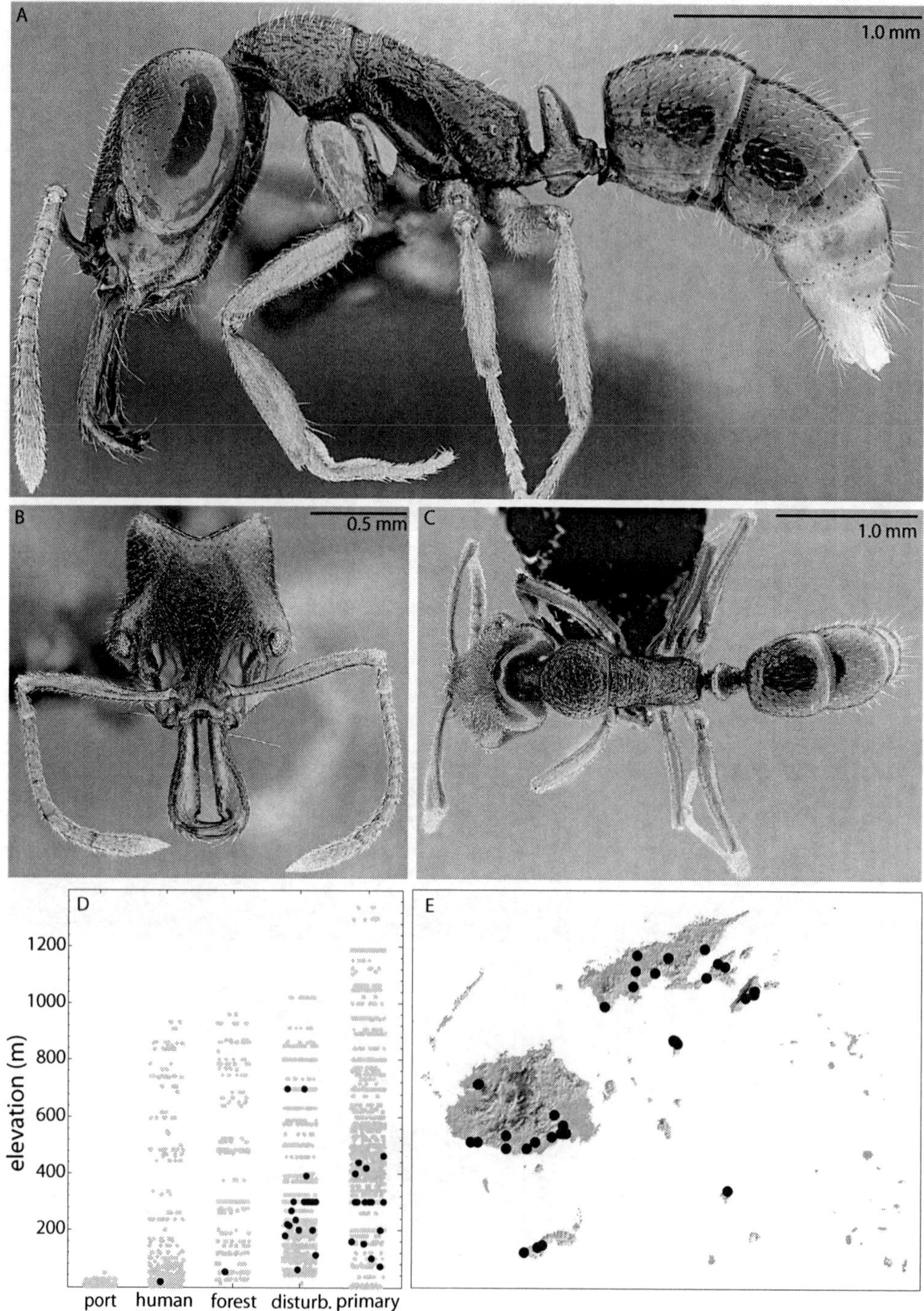

Plate 156. *Hypoponera confinis* worker, CASENT0186525.

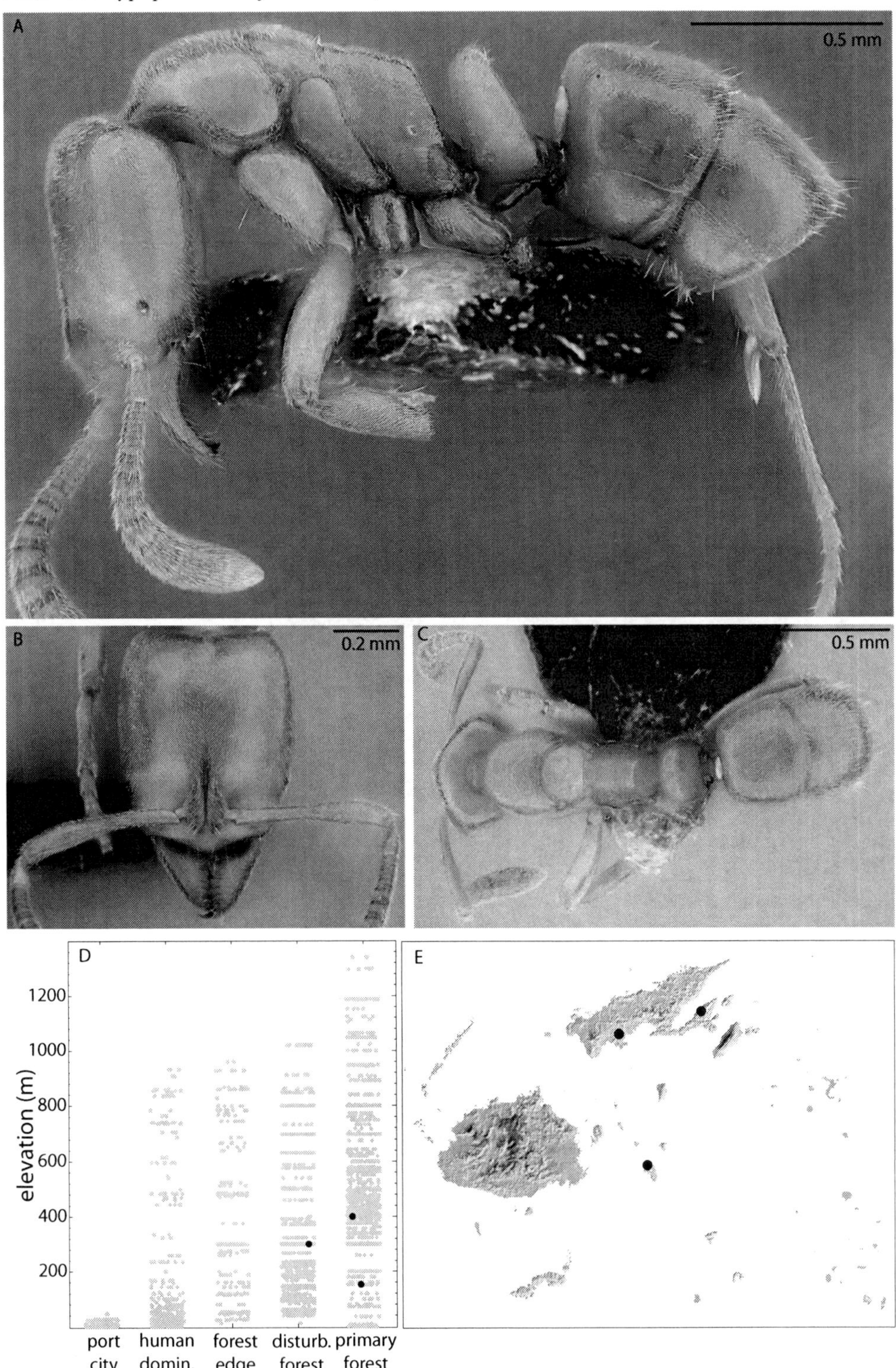

Plate 157. *Hypoponera eutrepta* worker, CASENT0171068.

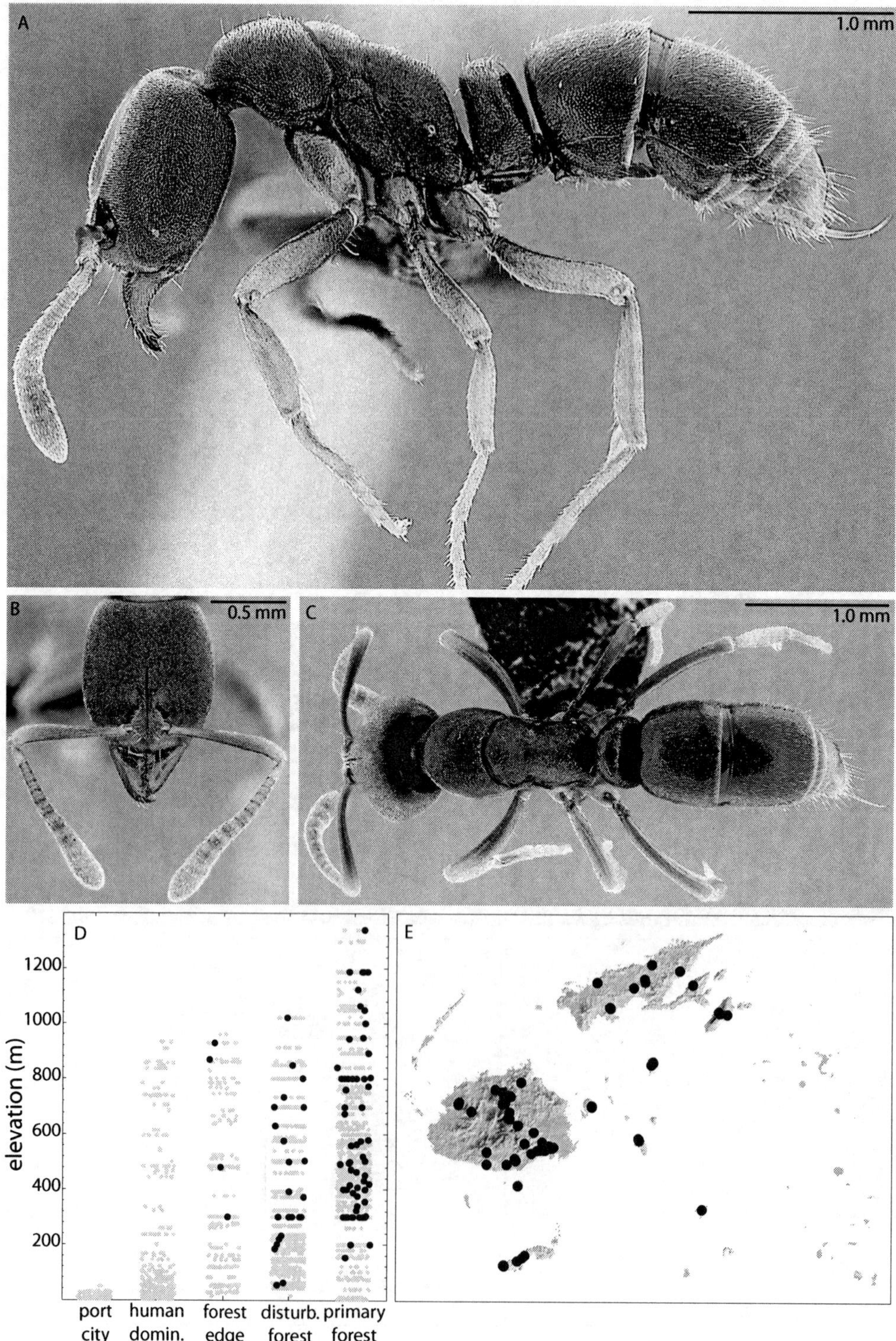

Plate 158. *Hypoponera monticola* worker, CASENT0186141.

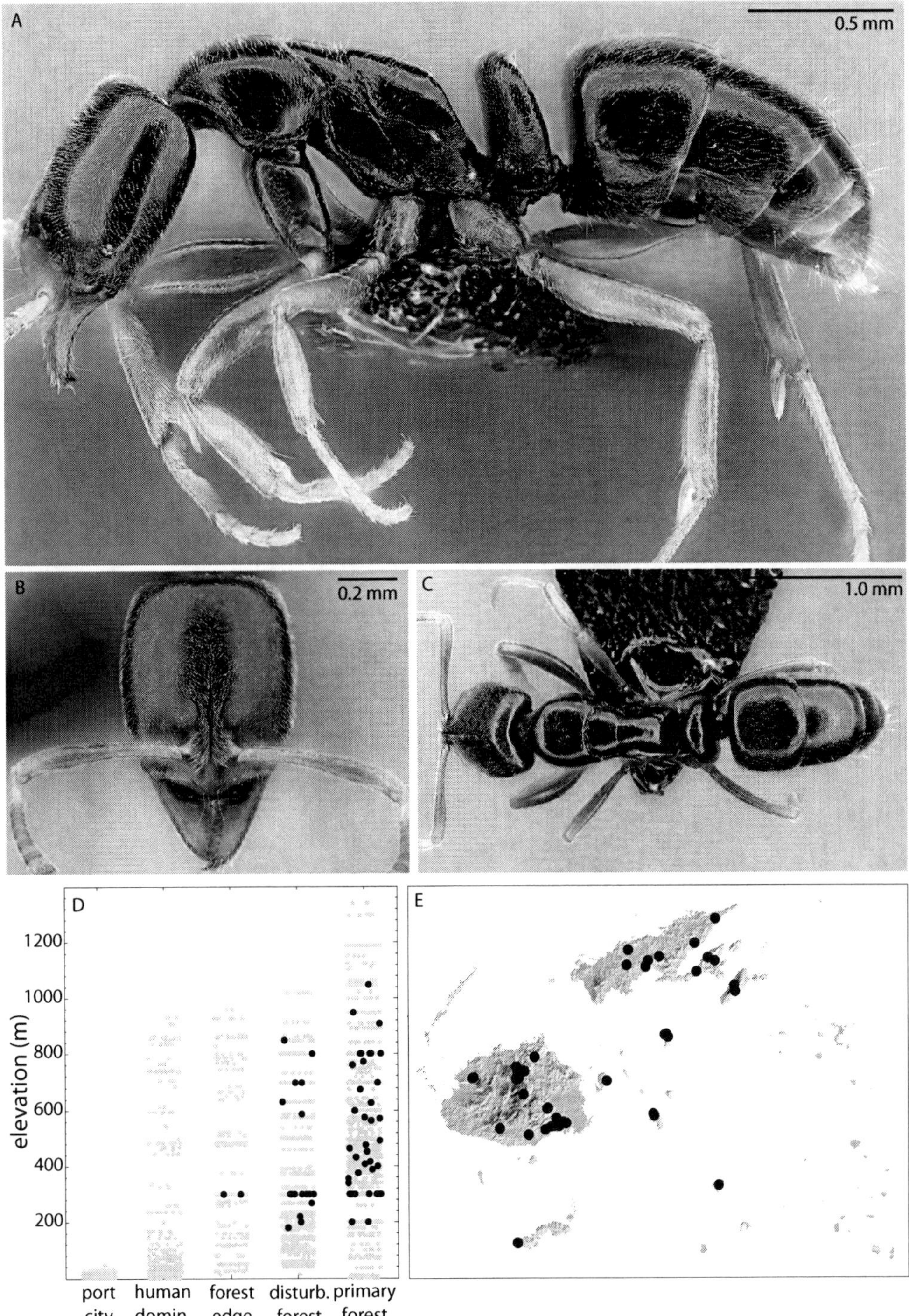

Plate 159. *Hypoponera opaciceps* worker, CASENT0171154.

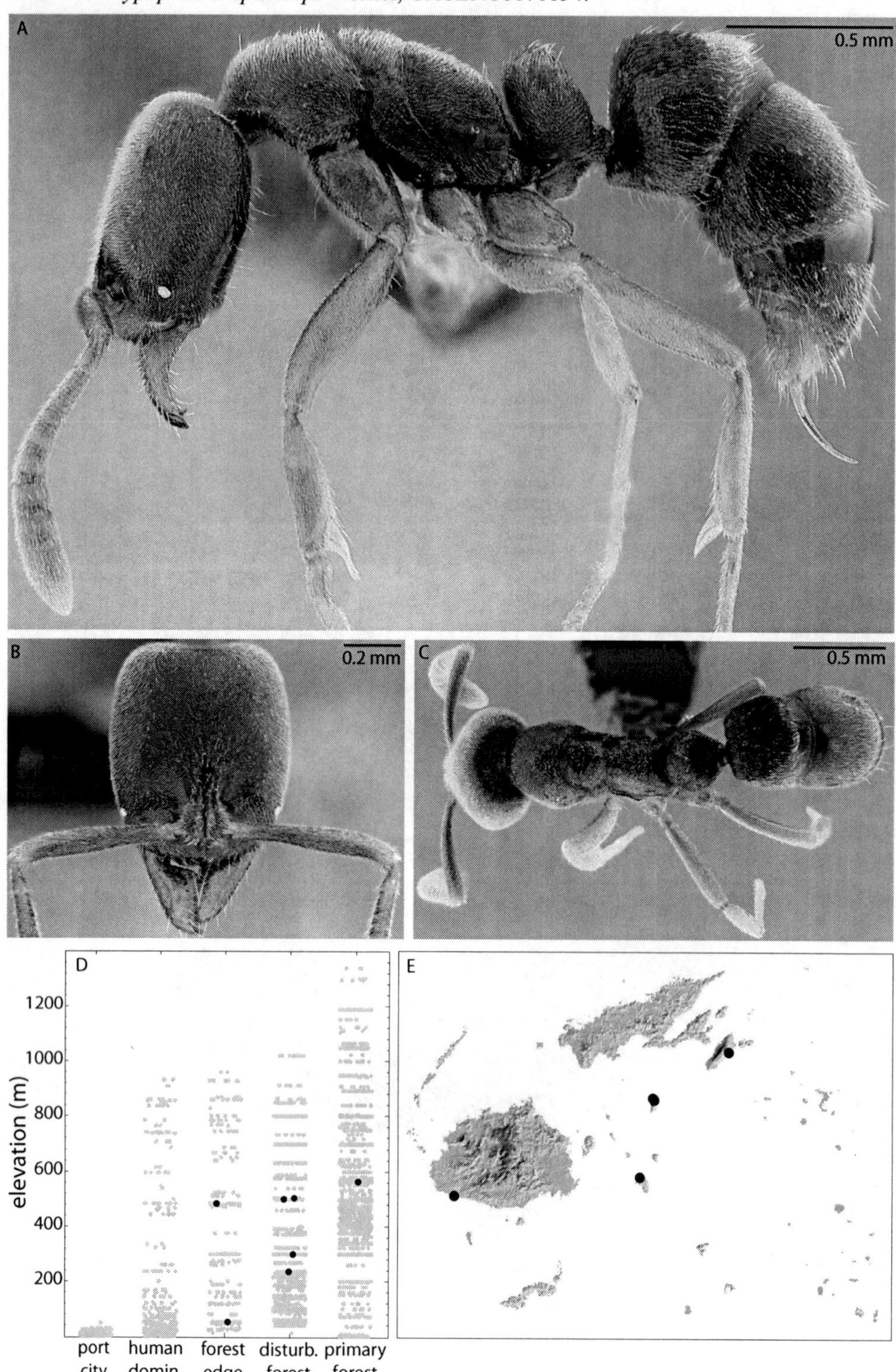

Plate 160. *Hypoponera pruinosa* worker, CASENT0186203.

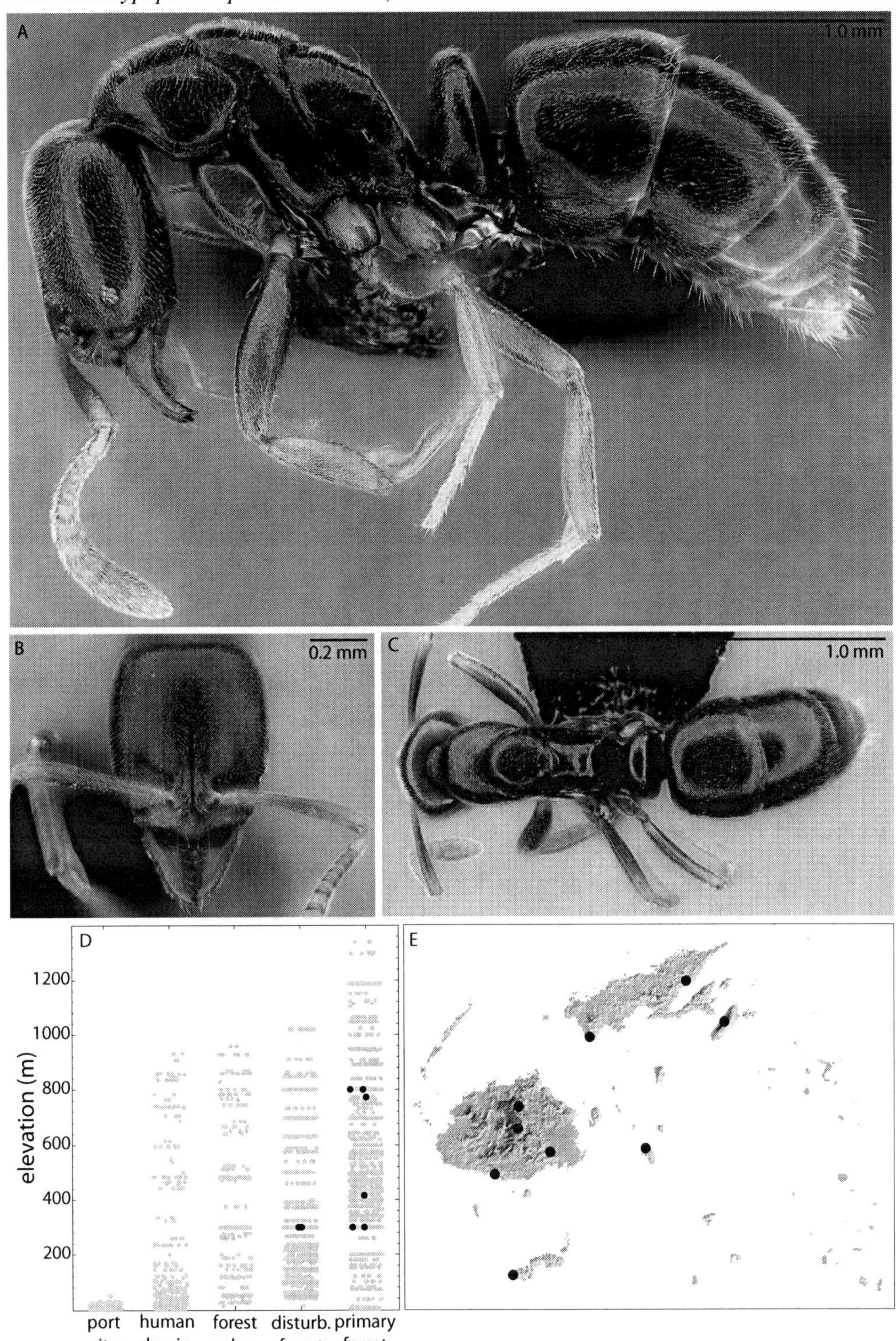

Plate 161. *Hypoponera punctatissima* worker, CASENT0171129.

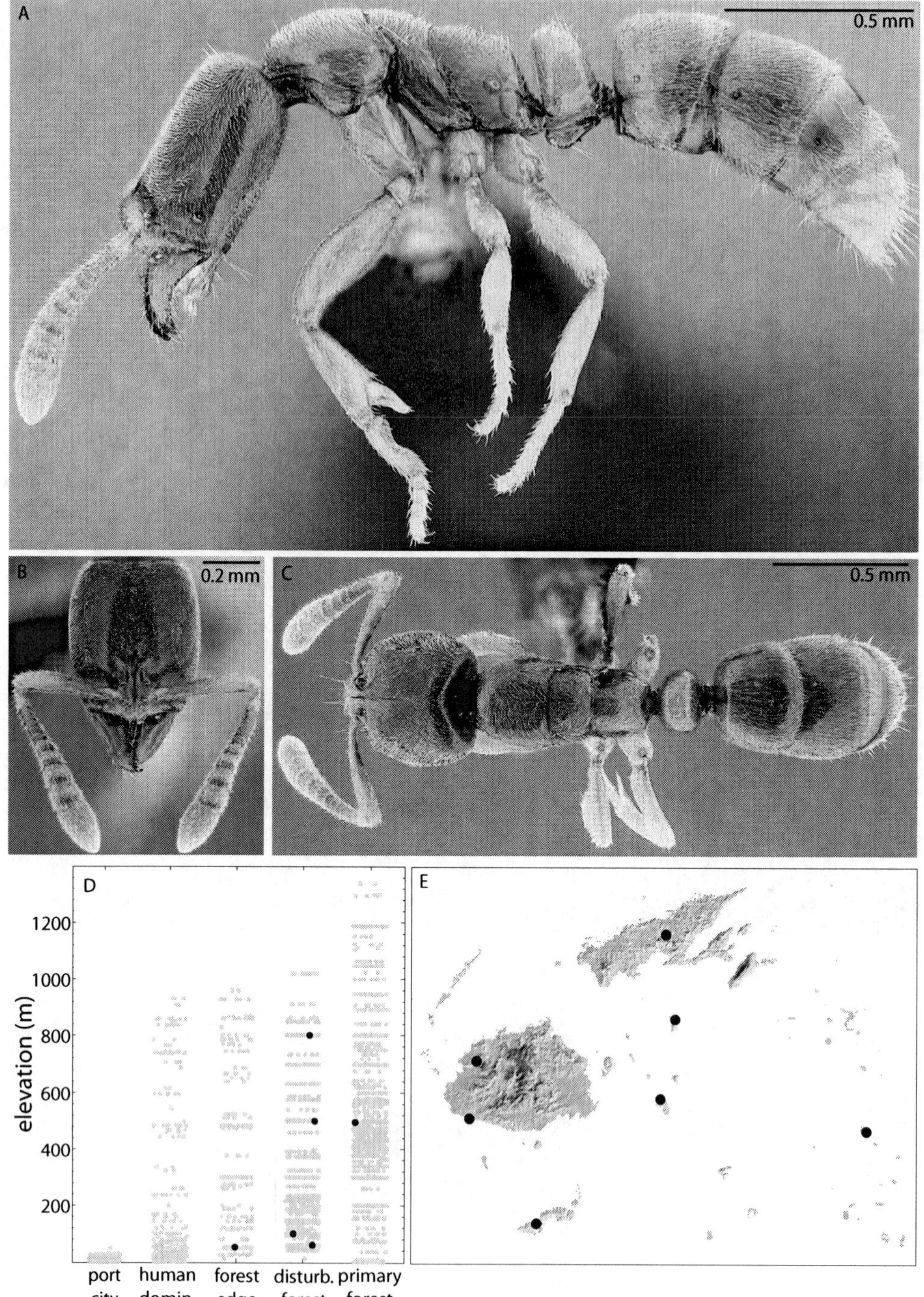

Plate 162. *Hypoponera turaga* worker, CASENT0186517.

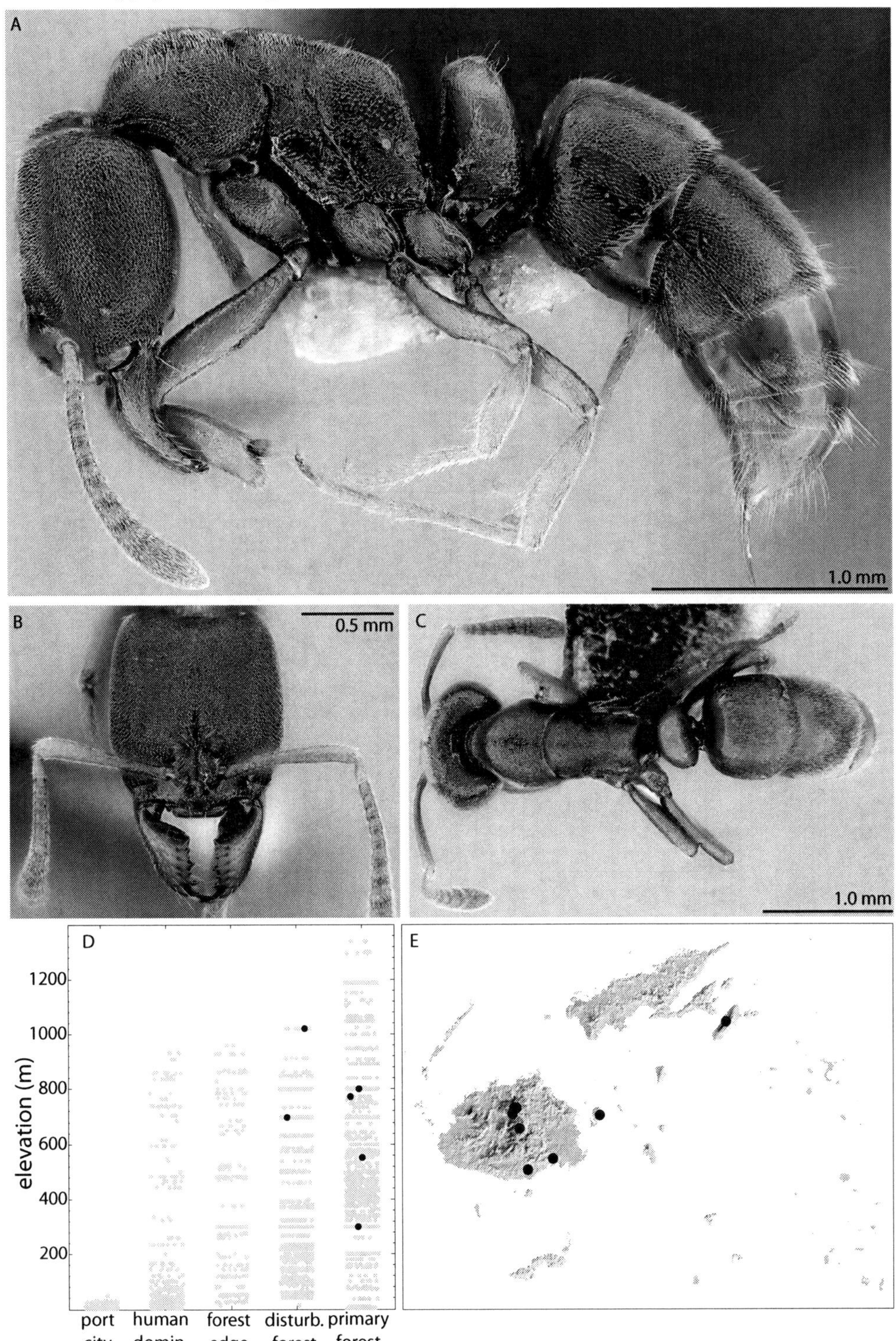

Plate 163. *Hypoponera vitiensis* syntype worker, CASENT0260379.

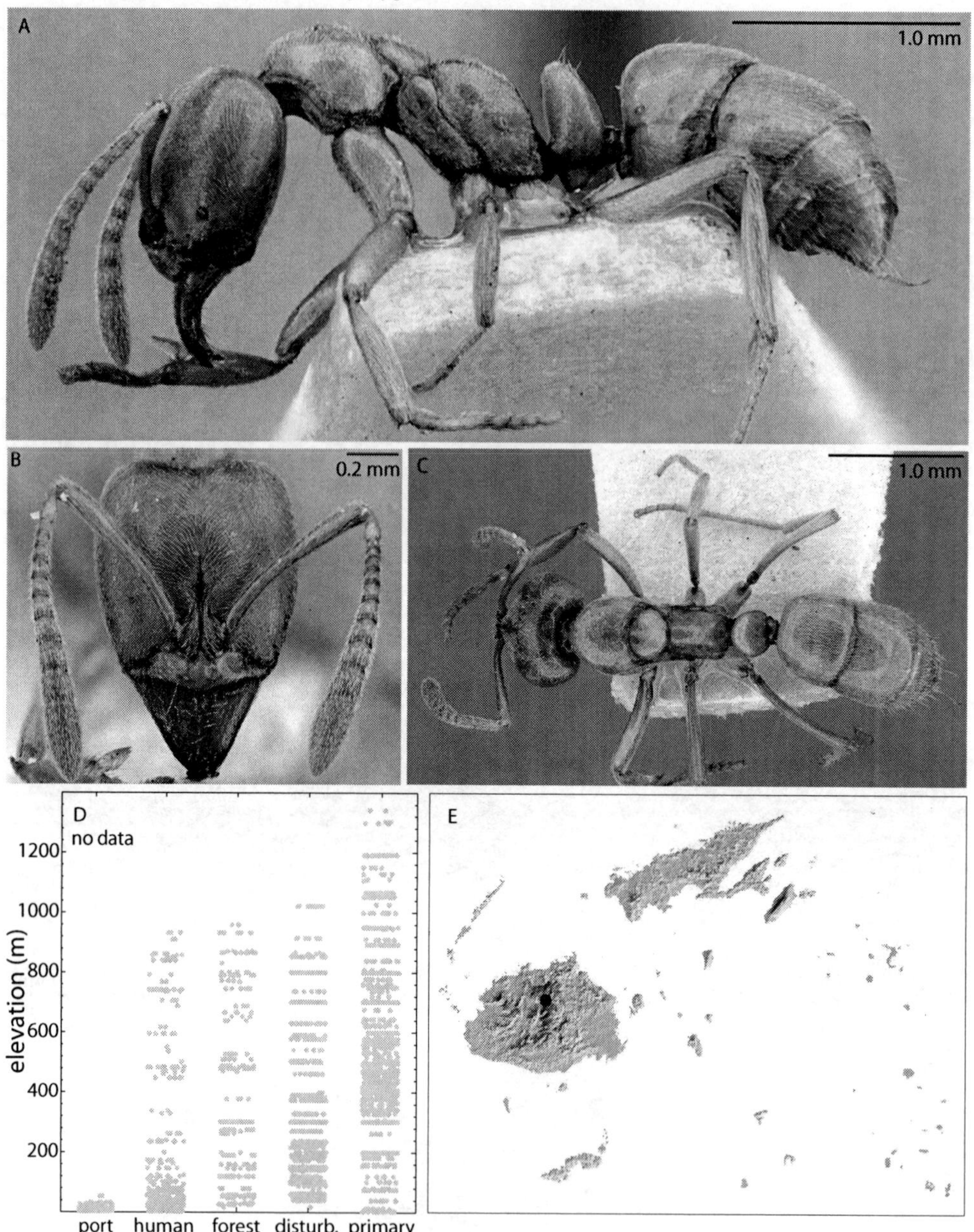

Plate 164. *Hypoponera* sp. FJ16 worker, CASENT0217359.

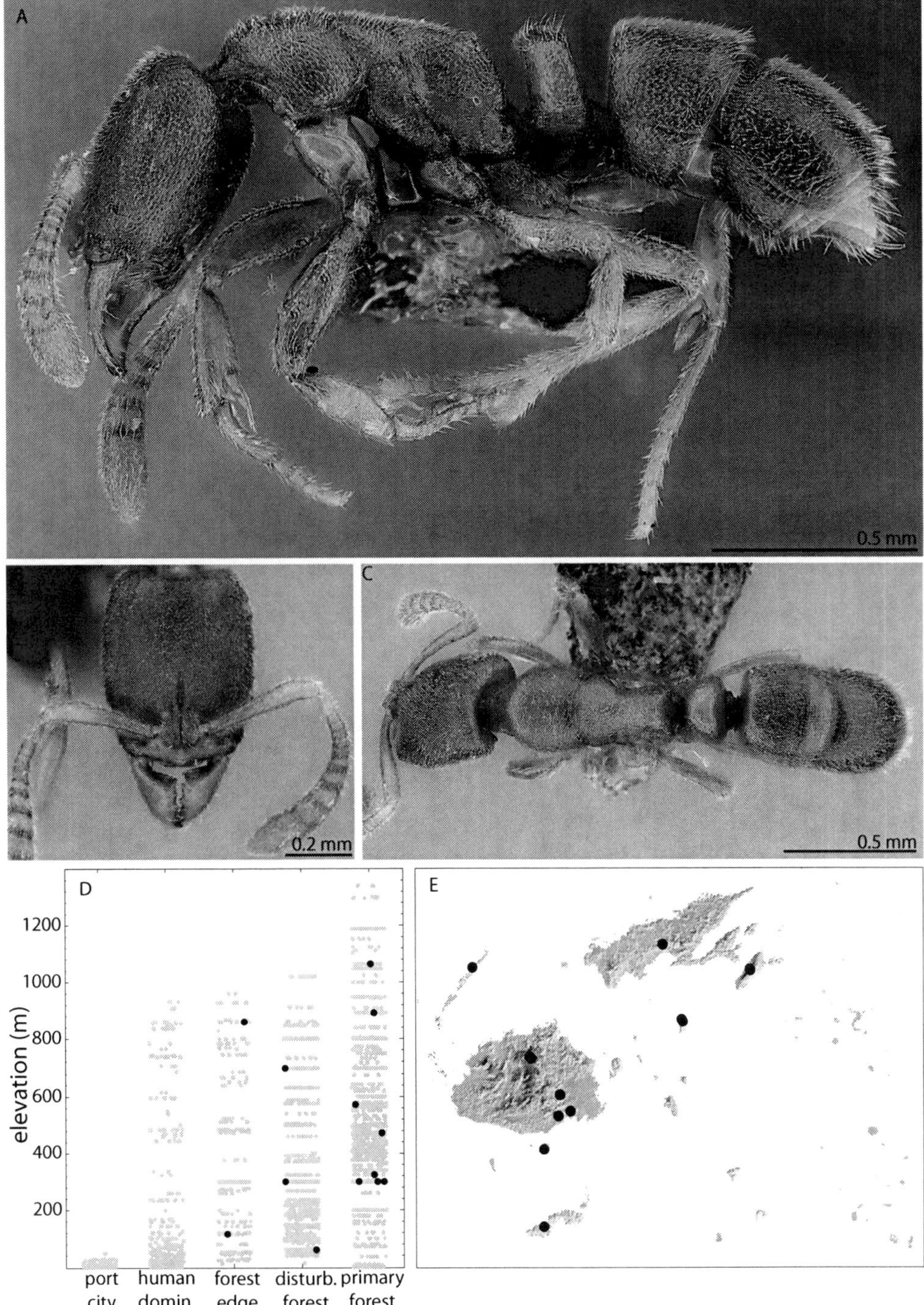

Plate 165. *Leptogenys foveopunctata* syntype worker, CASENT0260380.

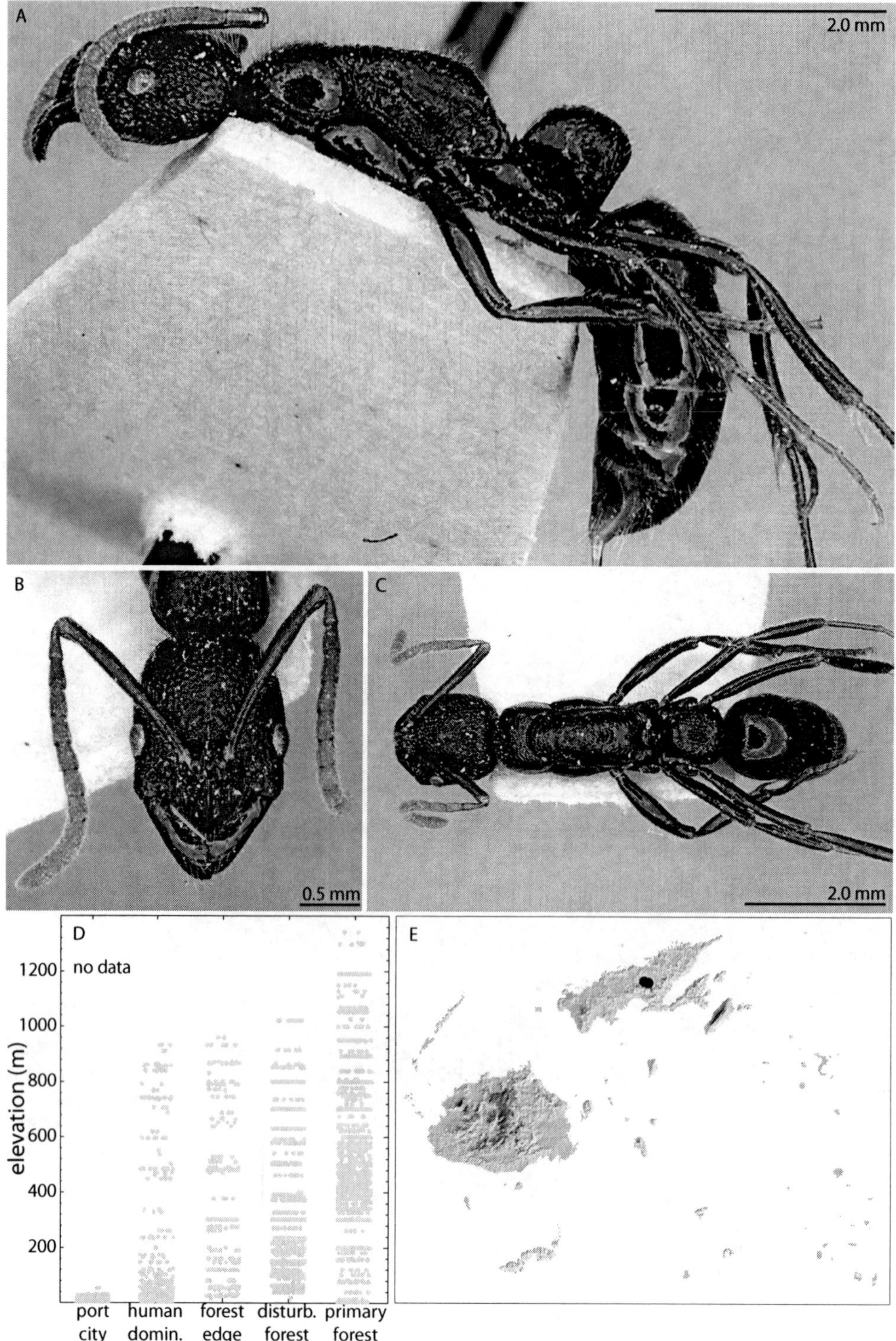

Plate 166. *Leptogenys fugax* worker, CASENT0171055.

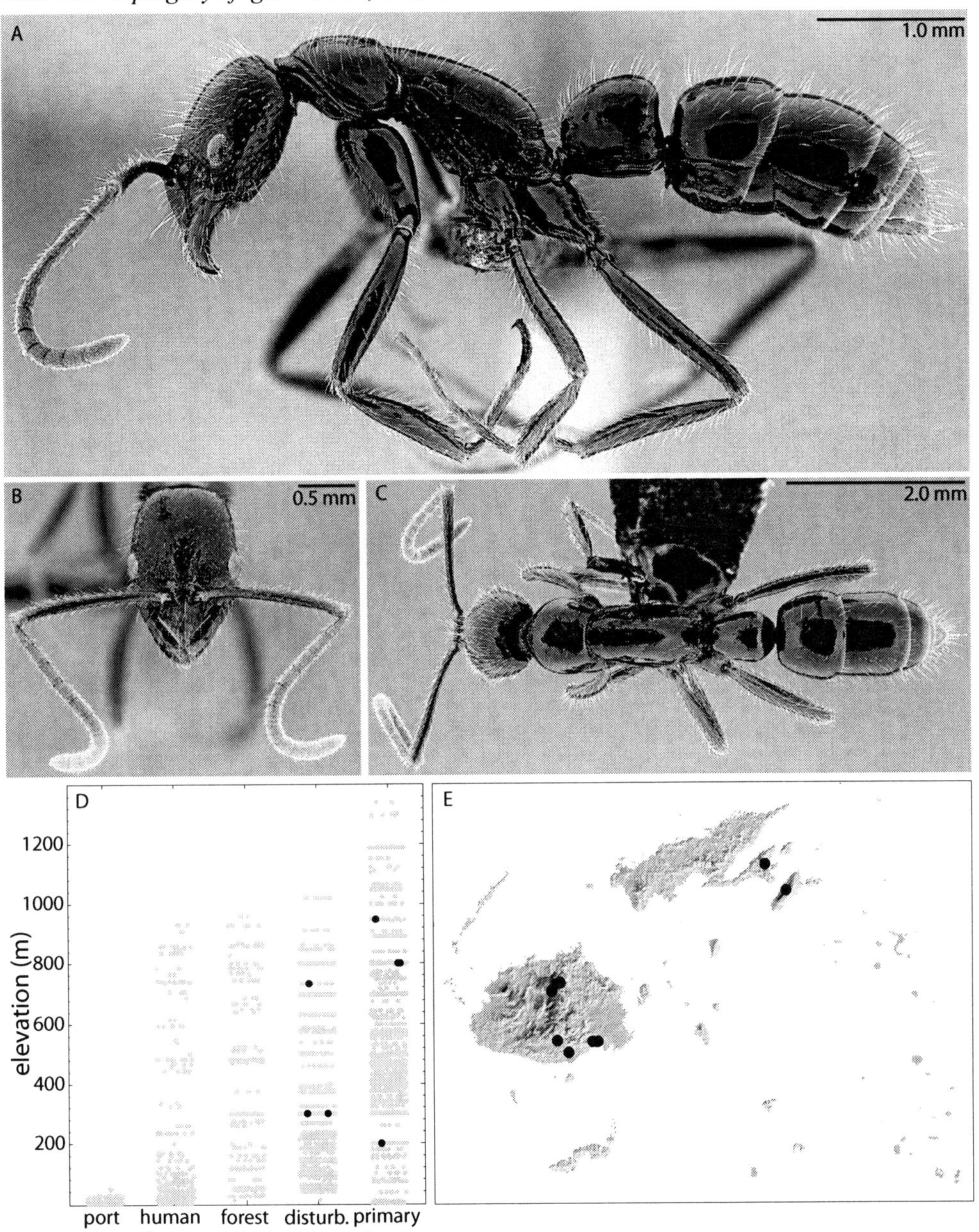

Plate 167. *Leptogenys humiliata* syntype worker, CASENT0260382.

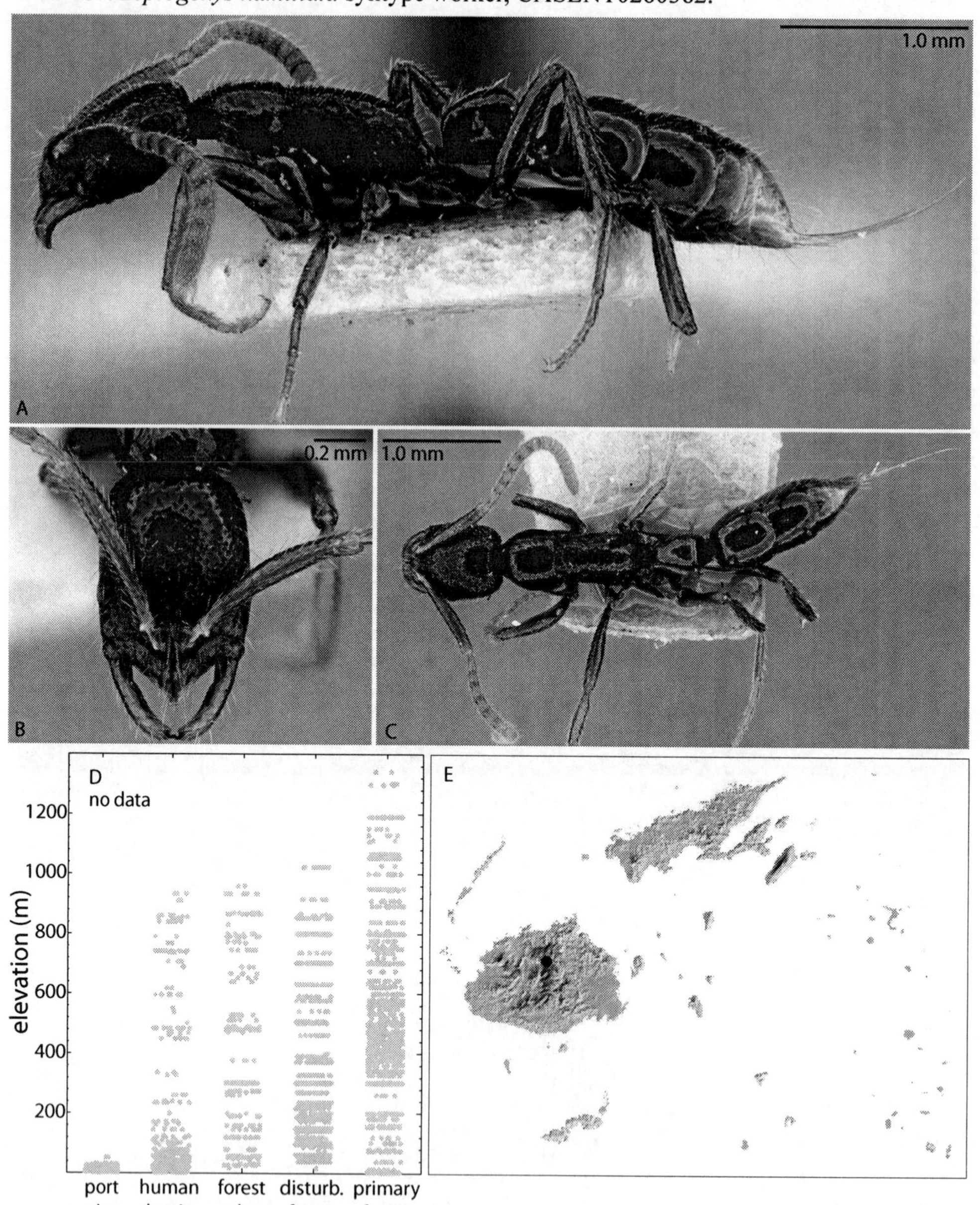

Plate 168. *Leptogenys letilae* worker, CASENT0186196.

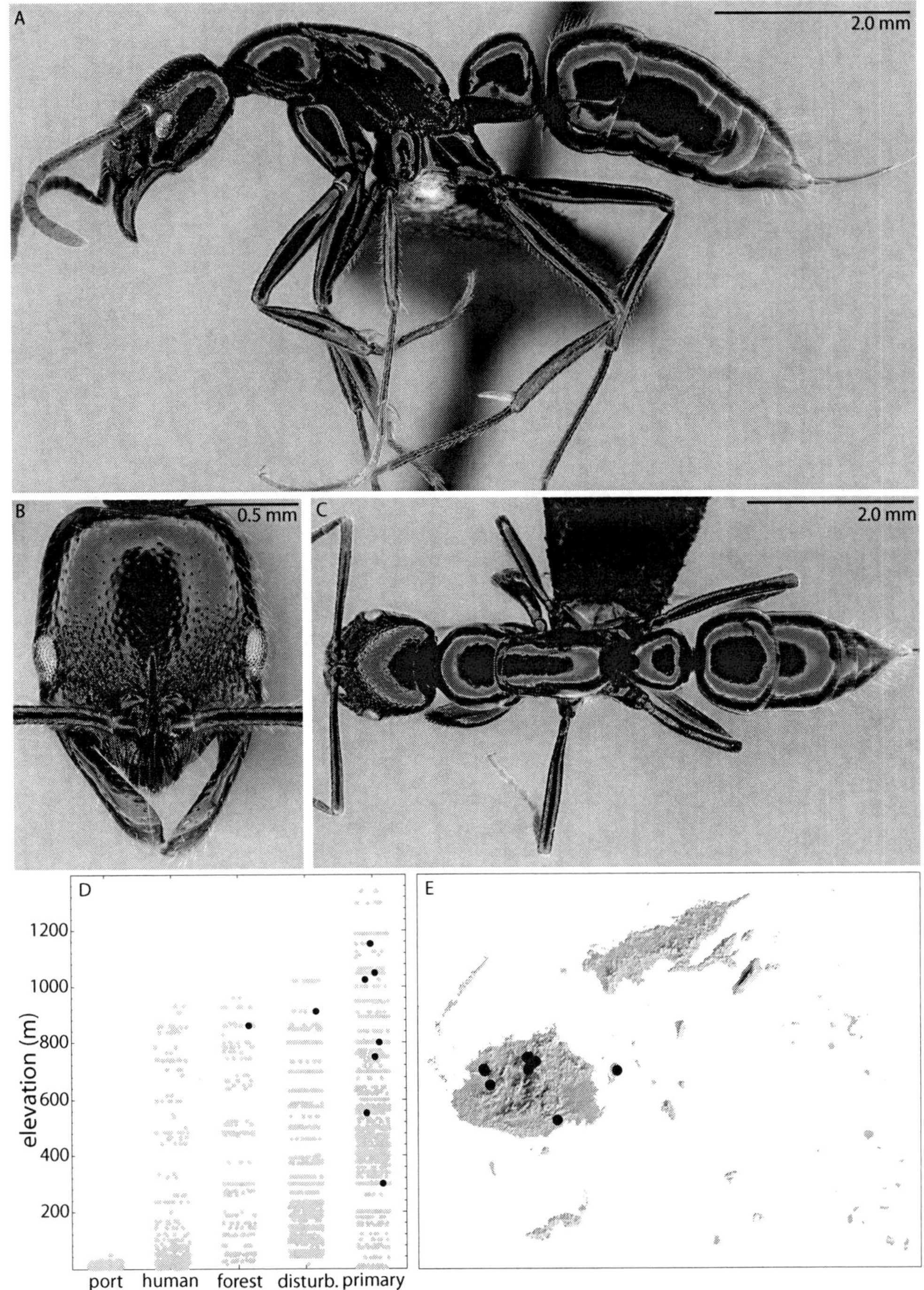

Plate 169. *Leptogenys navua* worker, CASENT0186200.

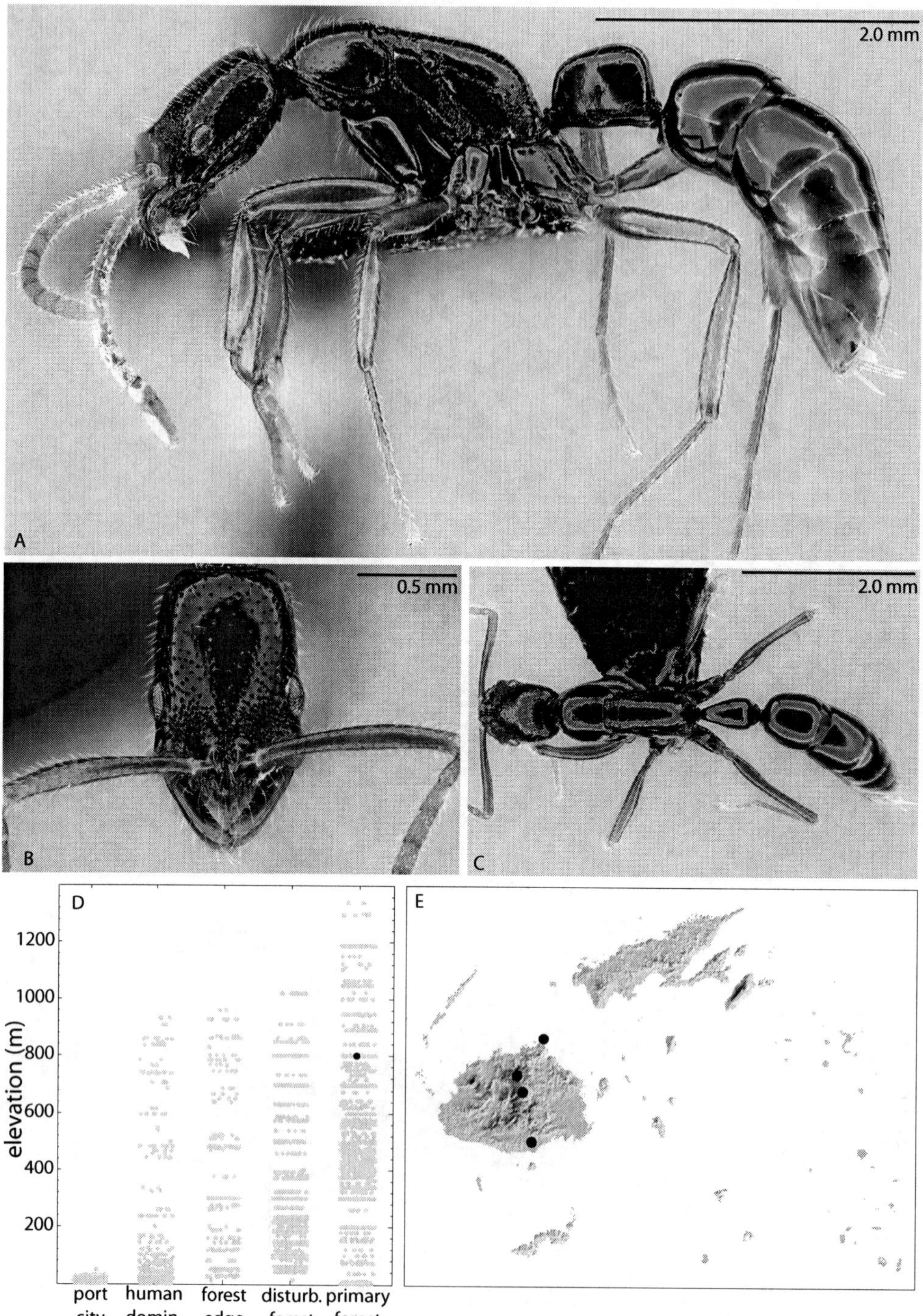

Plate 170. *Leptogenys vitiensis* worker, CASENT0186328.

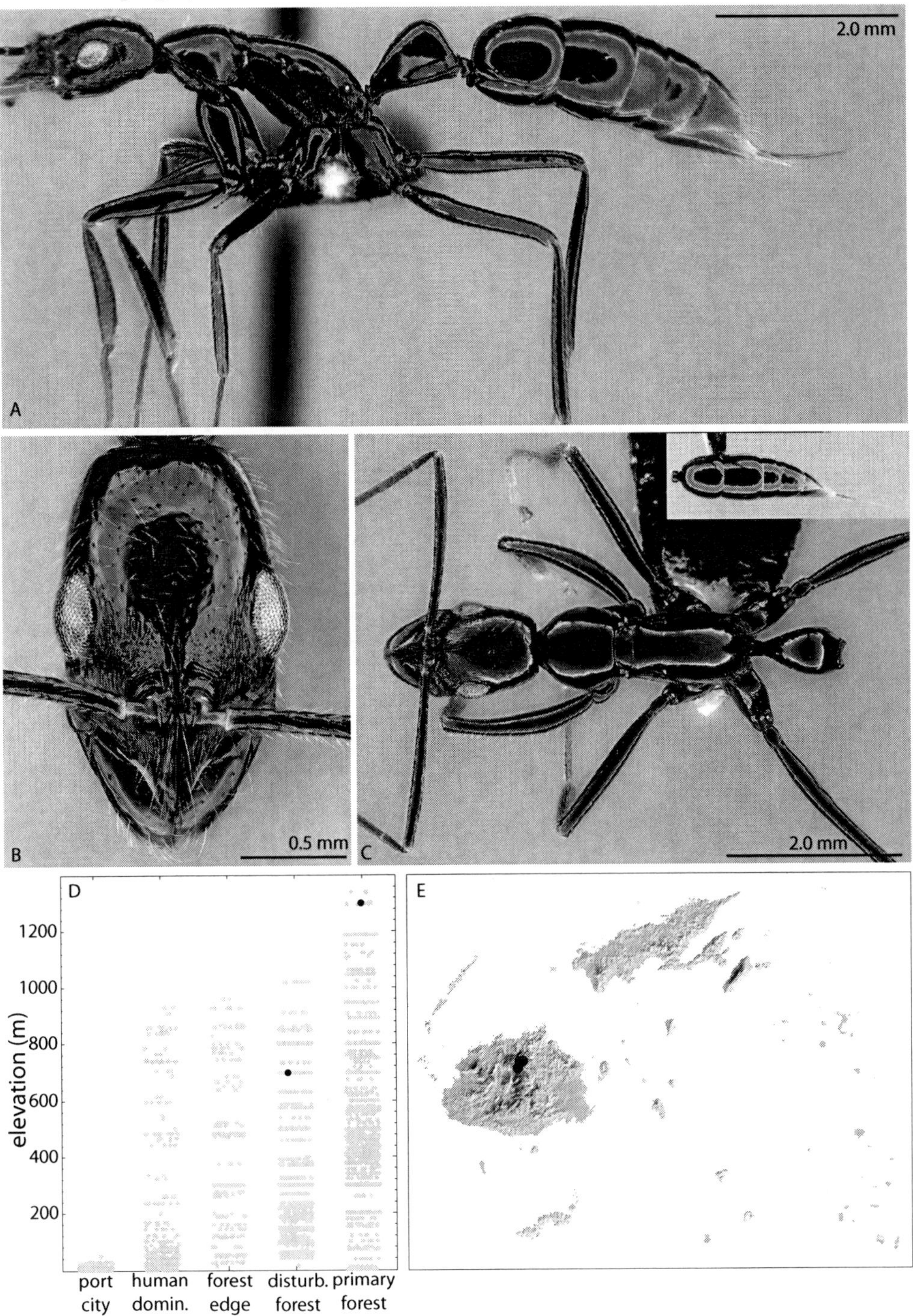

Plate 171. *Leptogenys* sp. FJ01 worker, CASENT0186138.

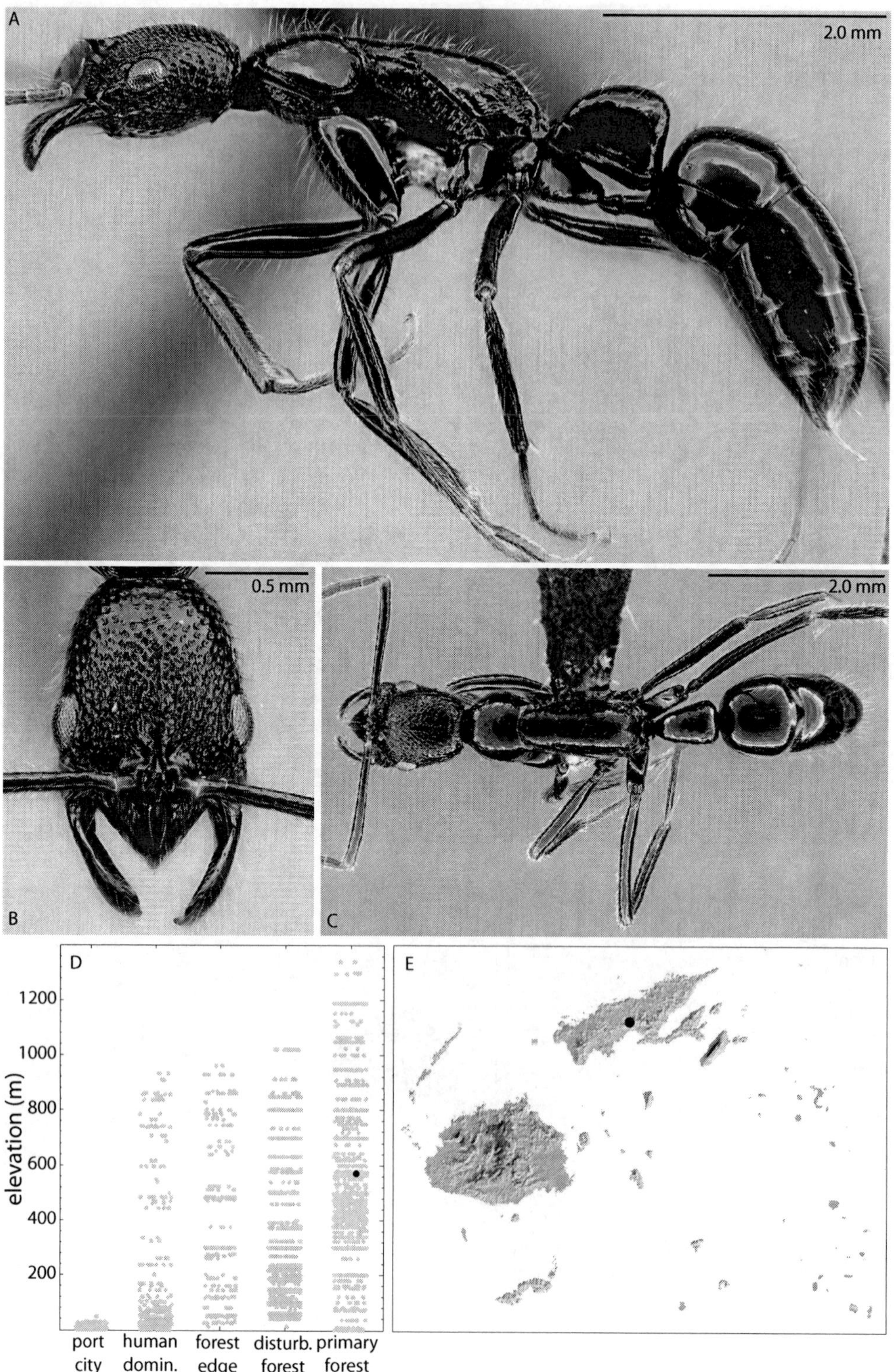

Plate 172. *Odontomachus angulatus* worker, CASENT0184584.

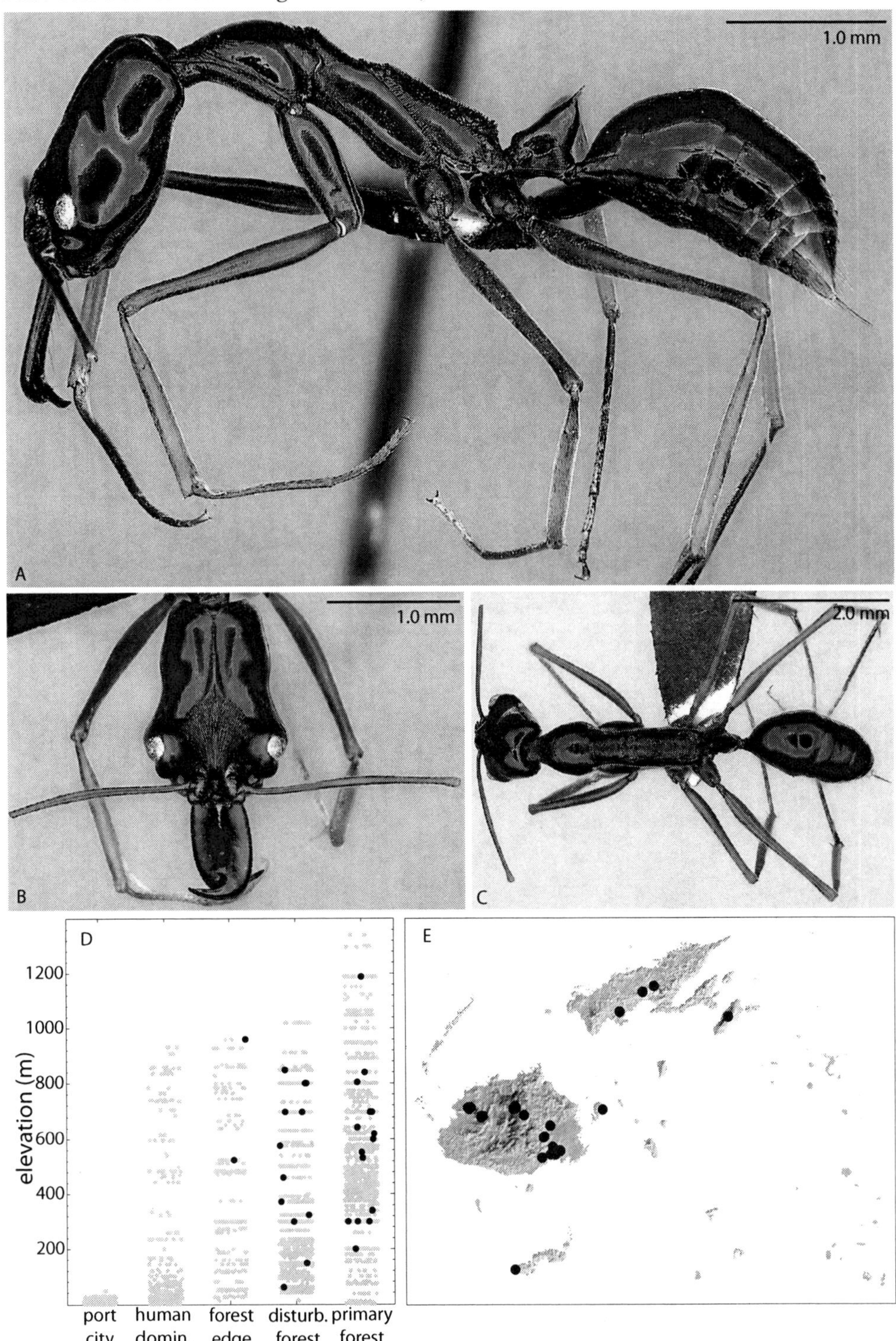

Plate 173. *Odontomachus simillimus* worker, CASENT0171124.

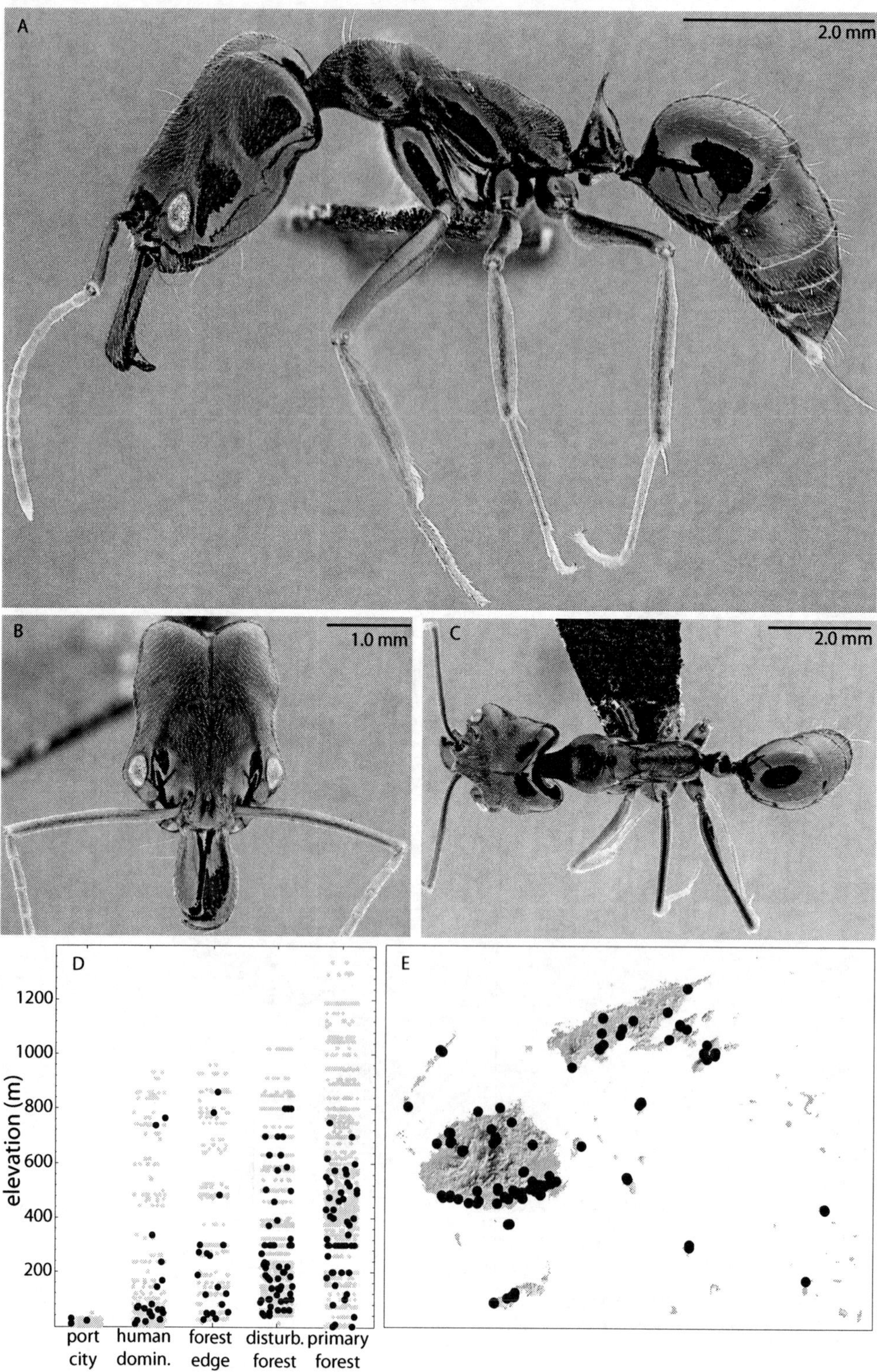

Plate 174. *Pachycondyla stigma* worker, CASENT0171132.

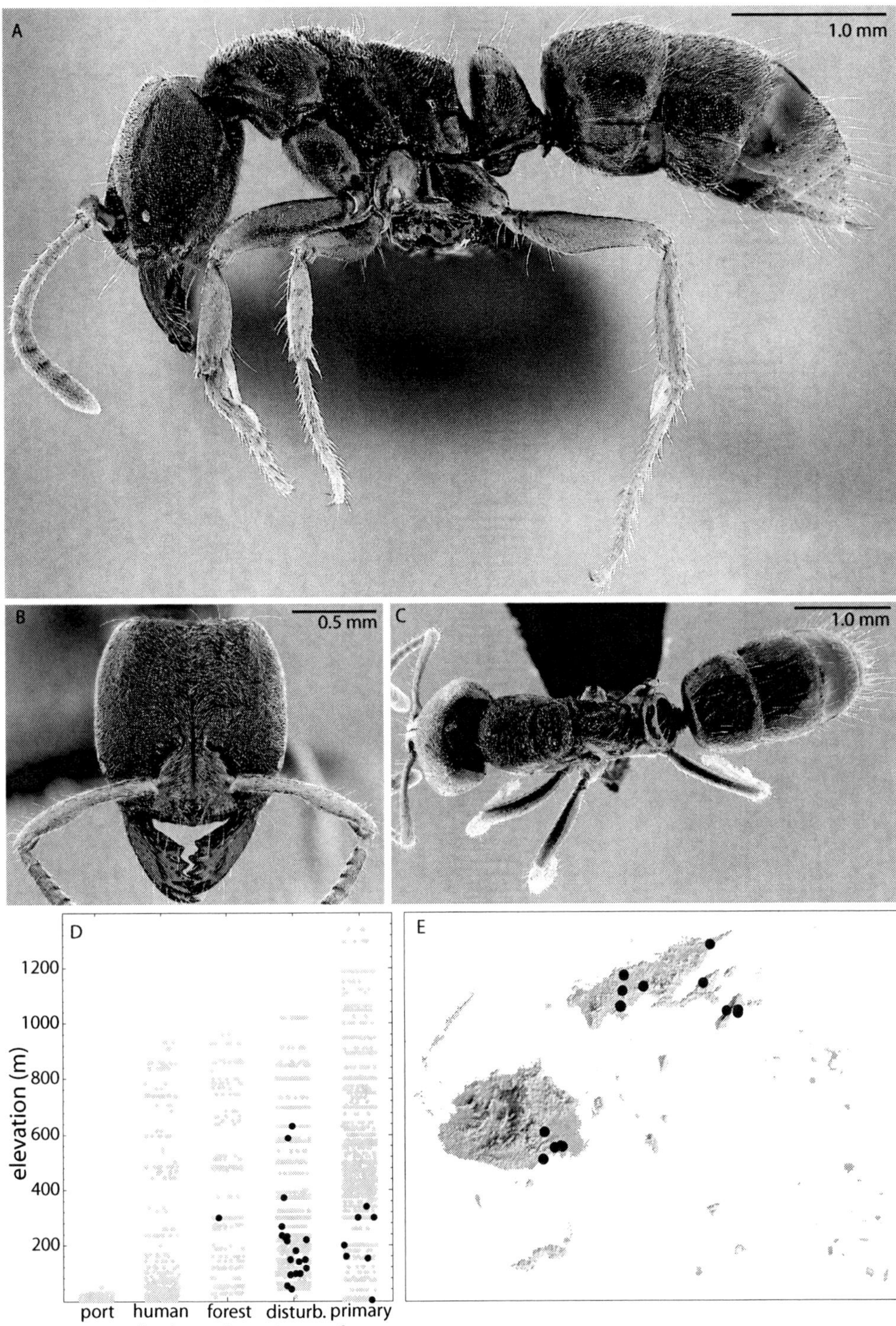

Plate 175. *Platythyrea parallela* queen, CASENT0194735.

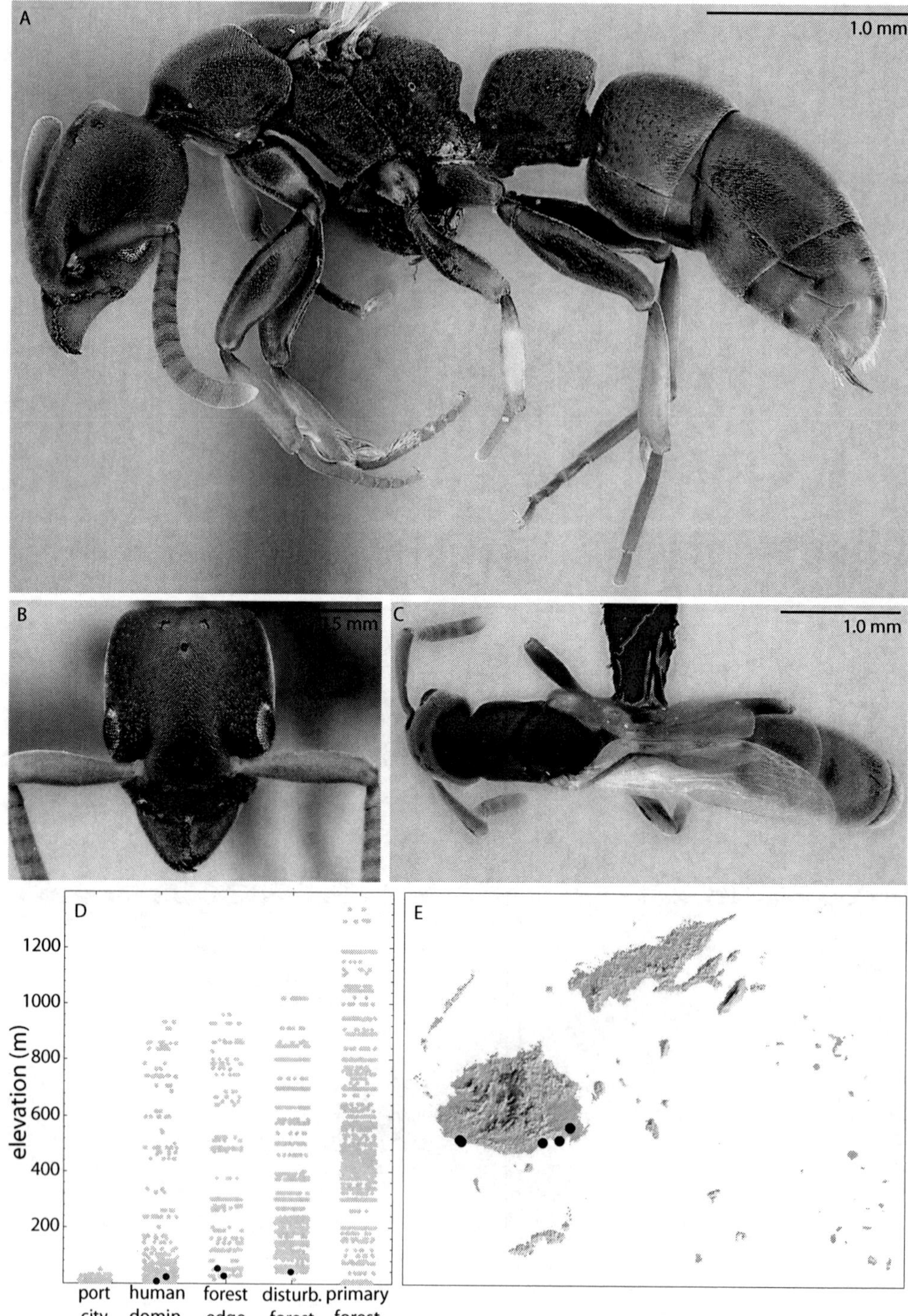

Plate 176. *Ponera colaensis* worker, CASENT0194711.

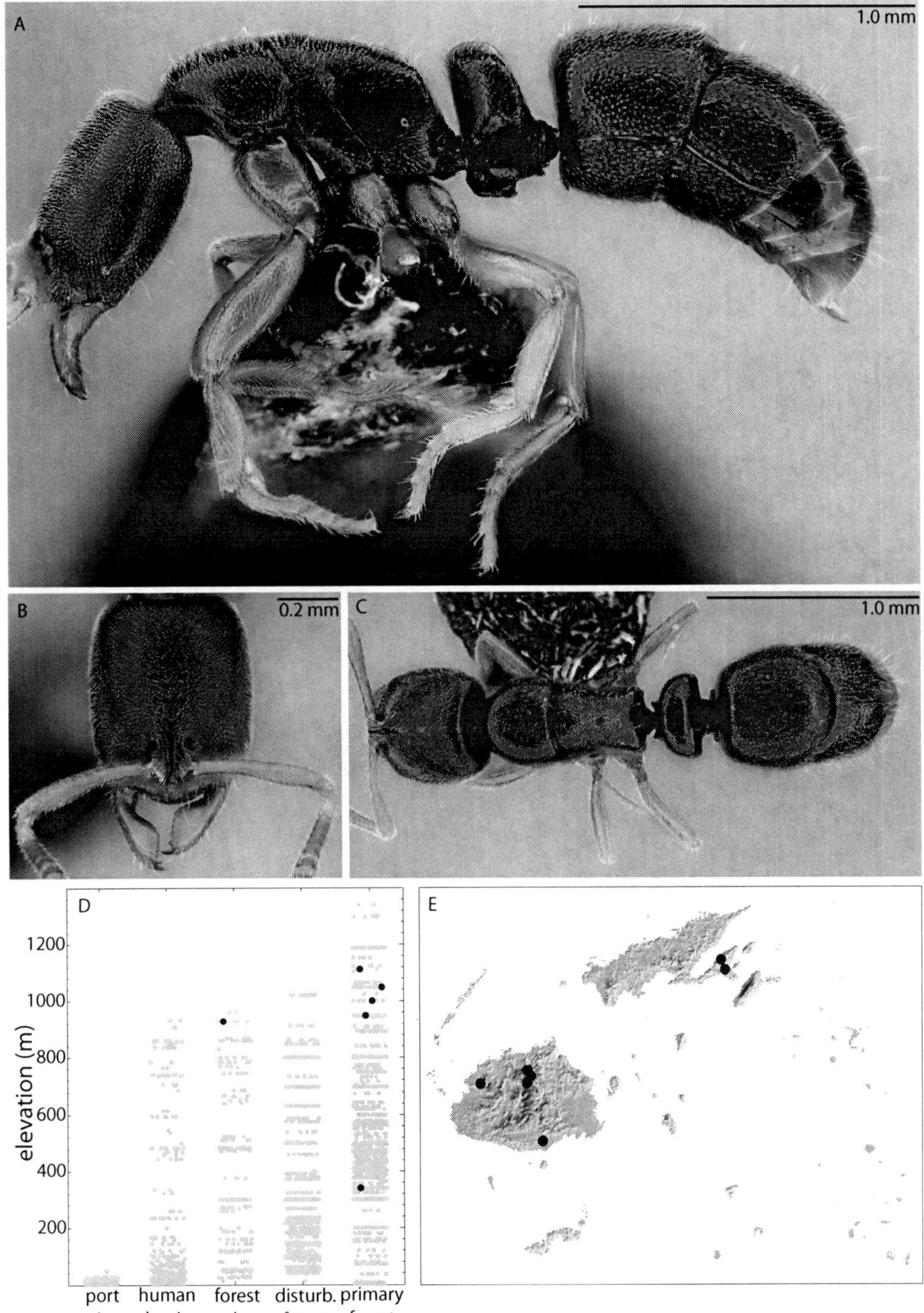

Plate 177. *Ponera manni* queen, CASENT0187758.

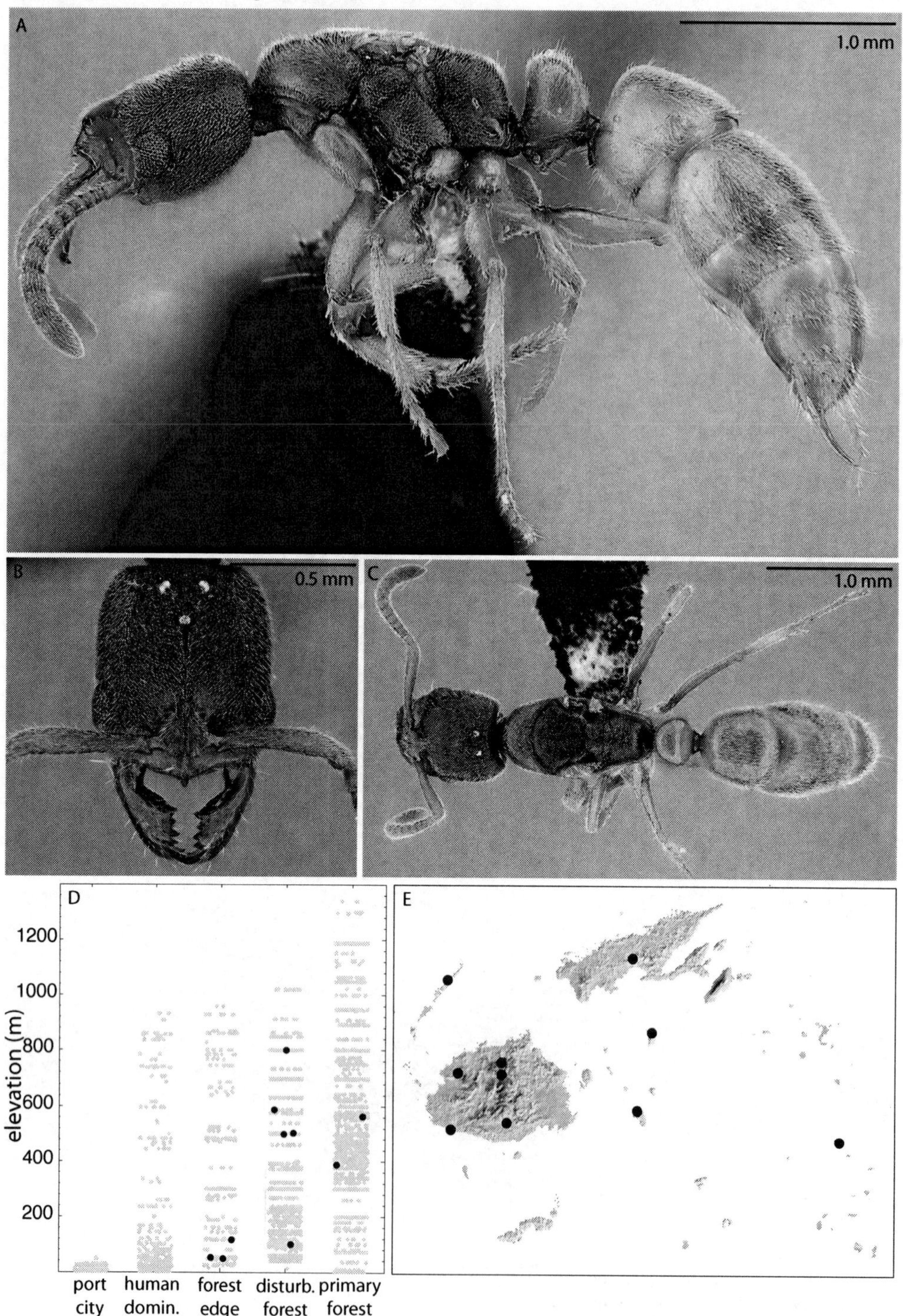

Plate 178. *Ponera swezeyi* worker, CASENT0194716.

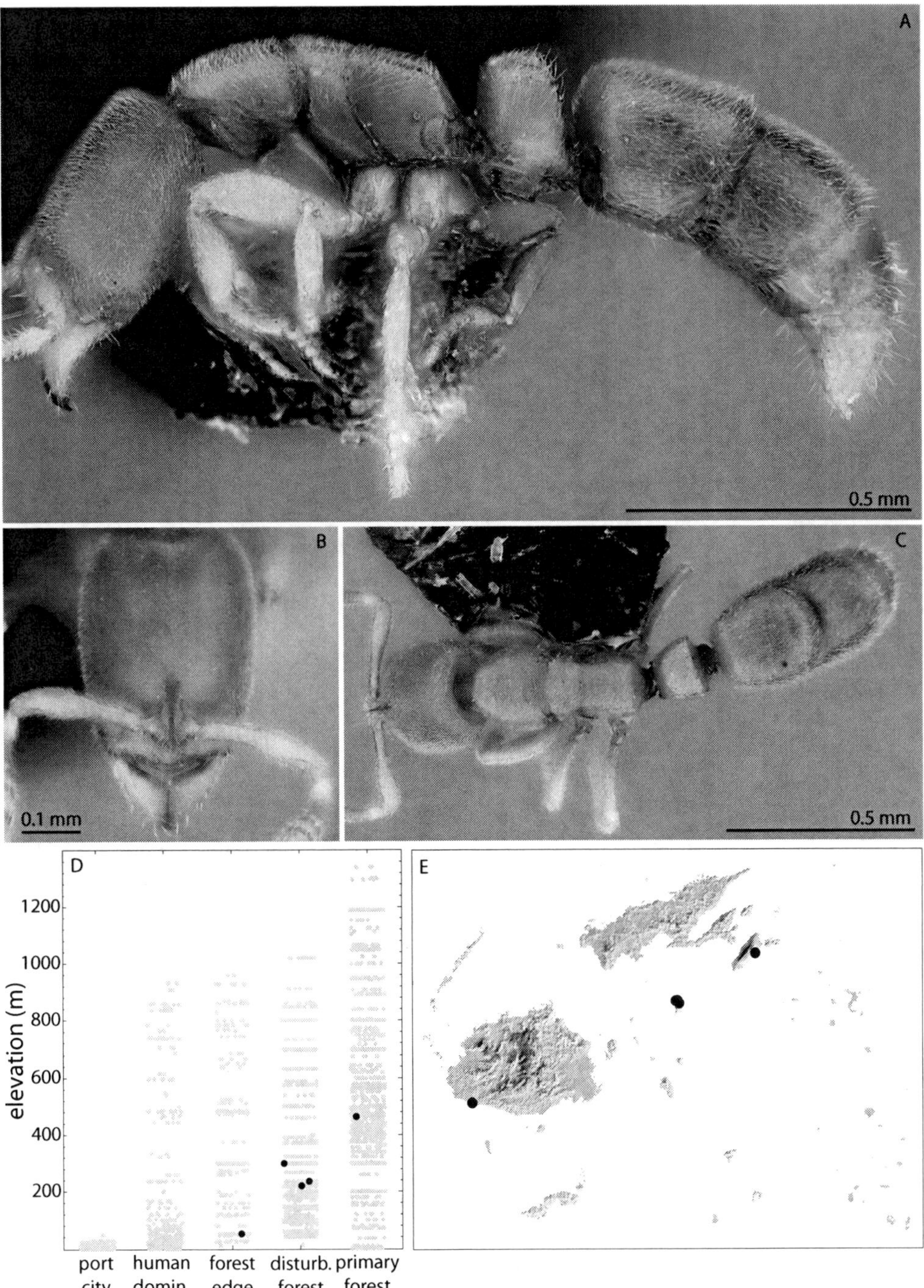

Plate 179. *Ponera* sp. FJ02, CASENT0194723.

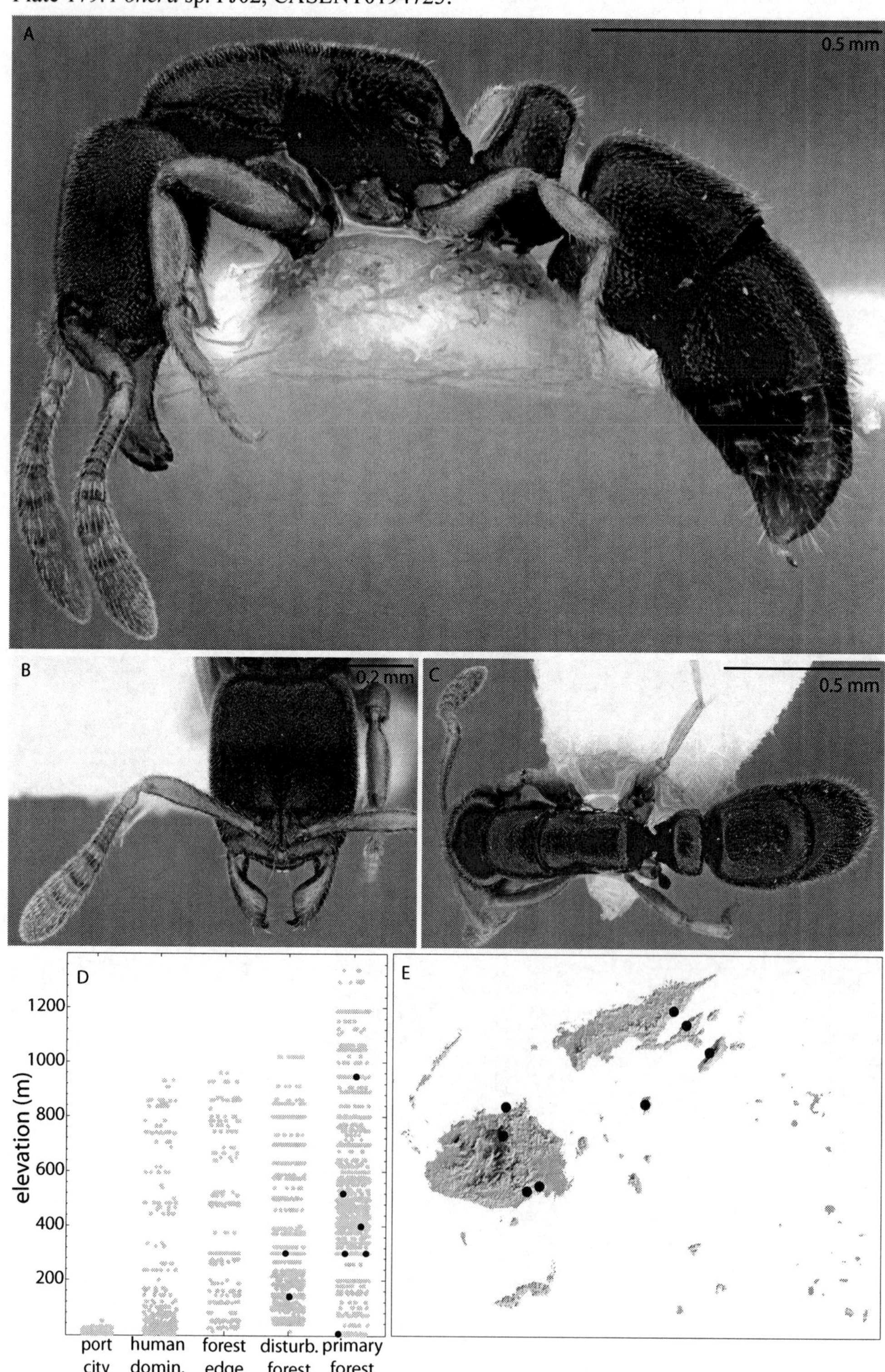

Plate 180. *Discothyrea* sp. FJ01 worker, CASENT0171054.

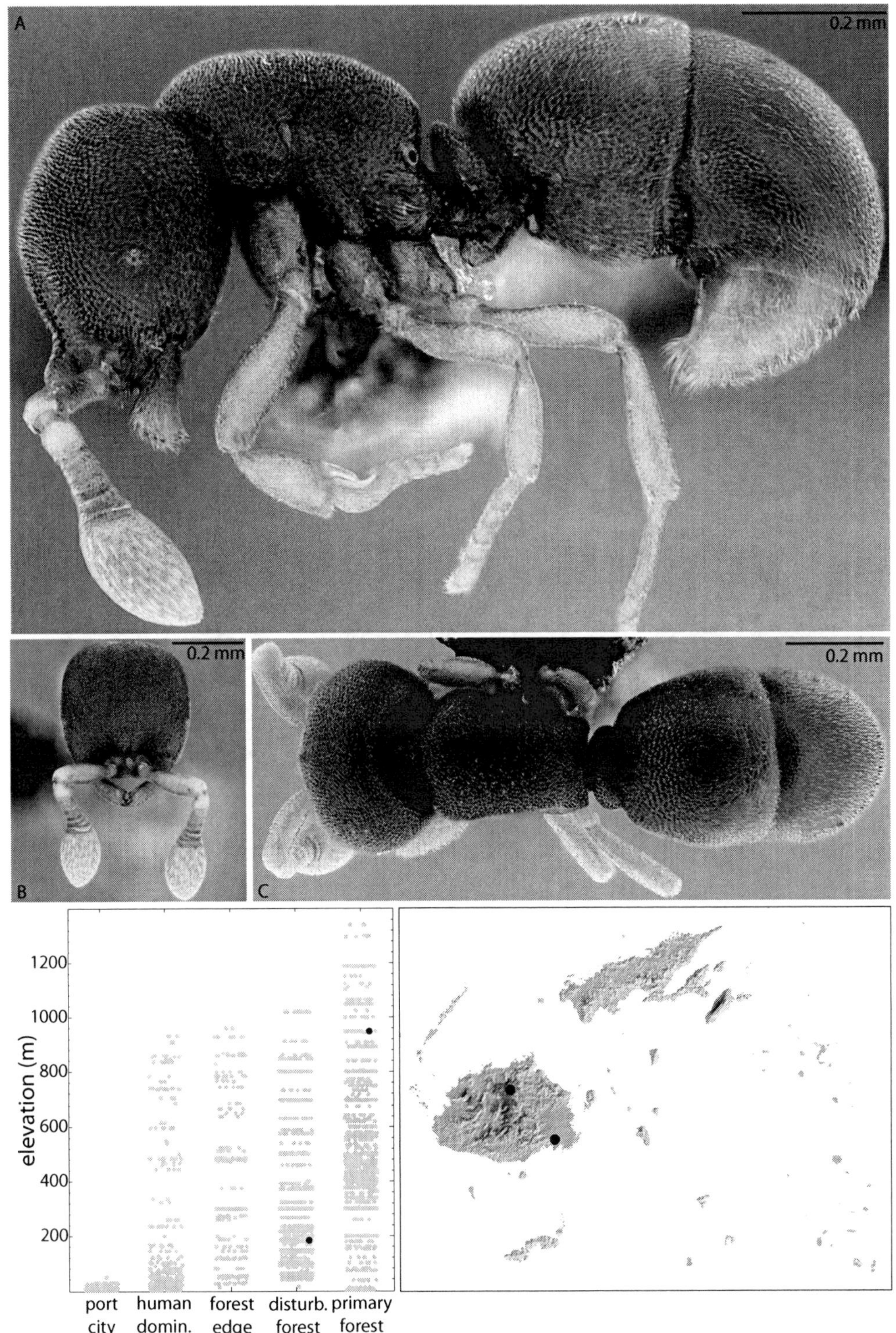

Plate 181. *Discothyrea* sp. FJ02 worker, CASENT0187775.

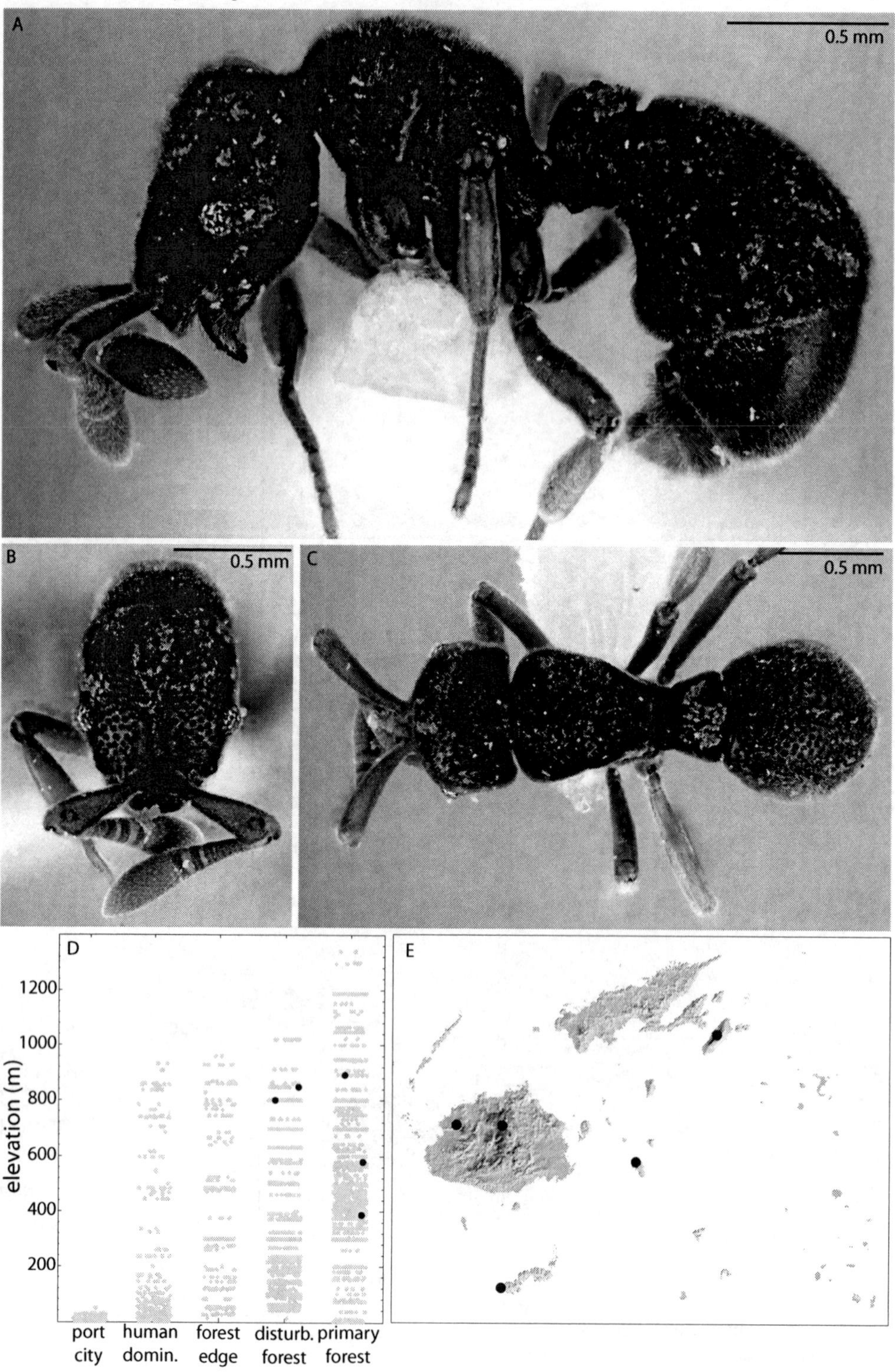

Plate 182. *Discothyrea* sp. FJ04 worker, ANIC32-053451.

Plate 183. *Proceratium oceanicum* worker, CASENT0171053.

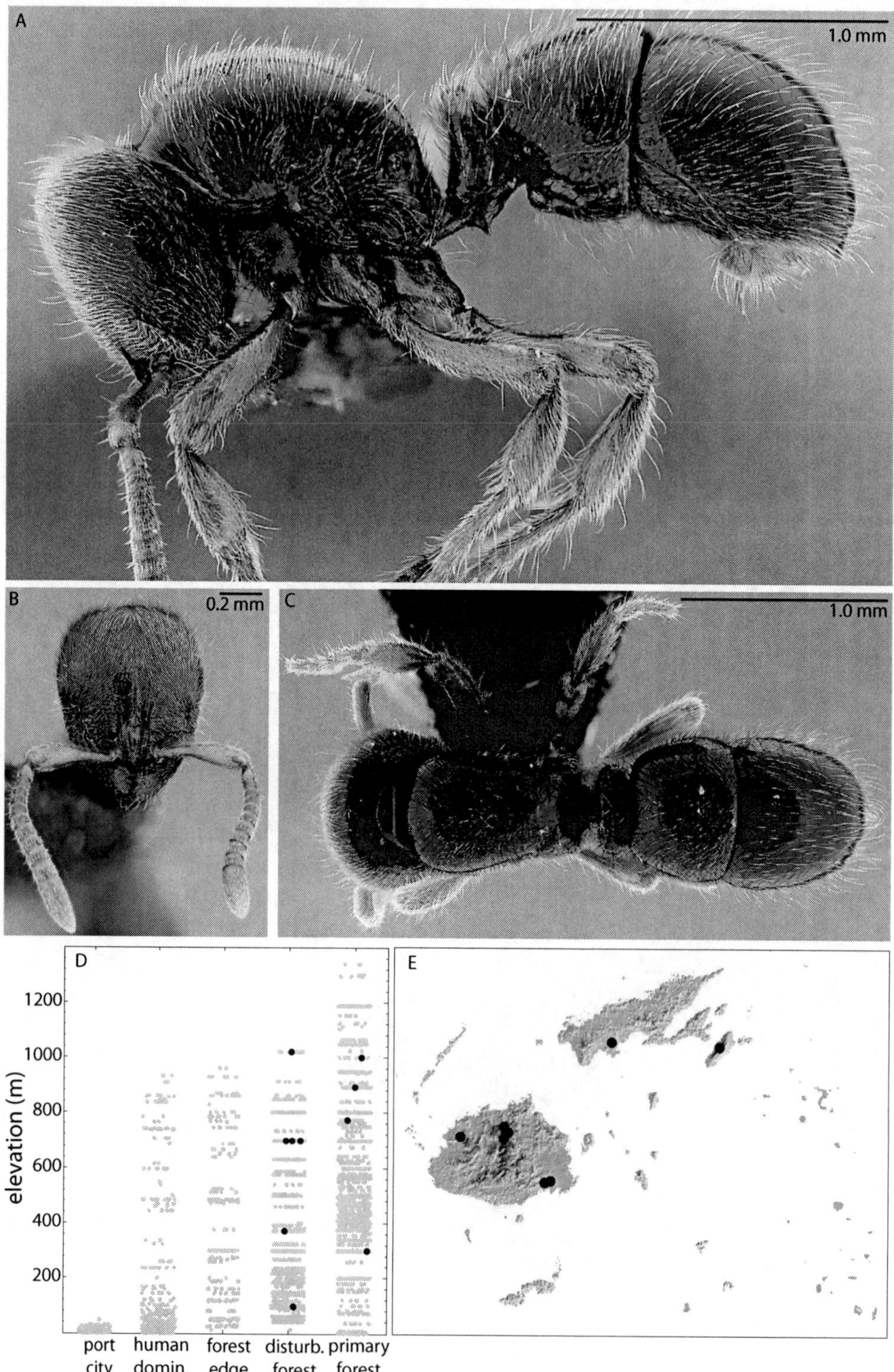

Plate 184. *Proceratium relictum* worker, CASENT0194740.

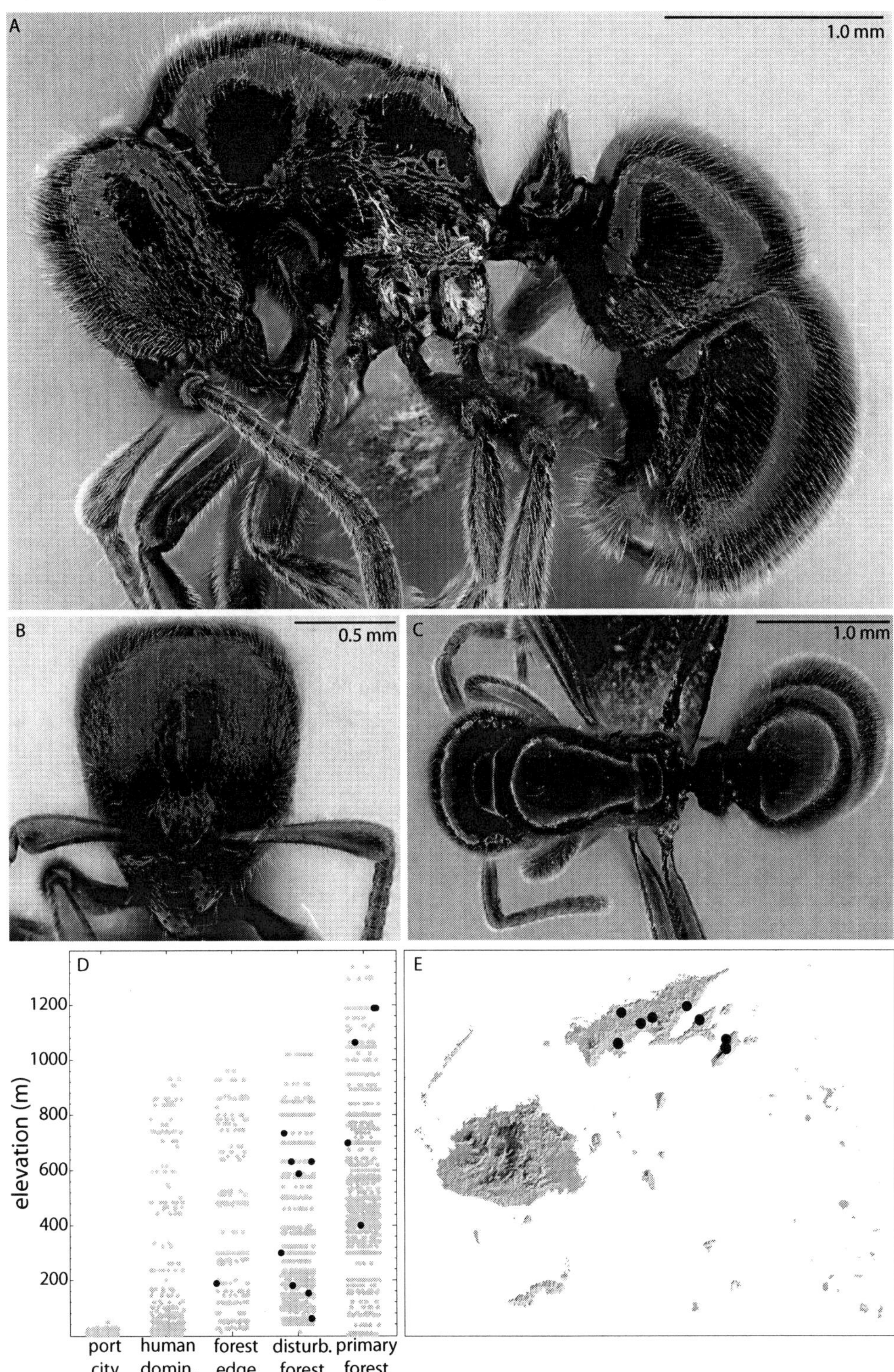

Plate 185. *Proceratium* sp. FJ01 worker, CASENT0187587.

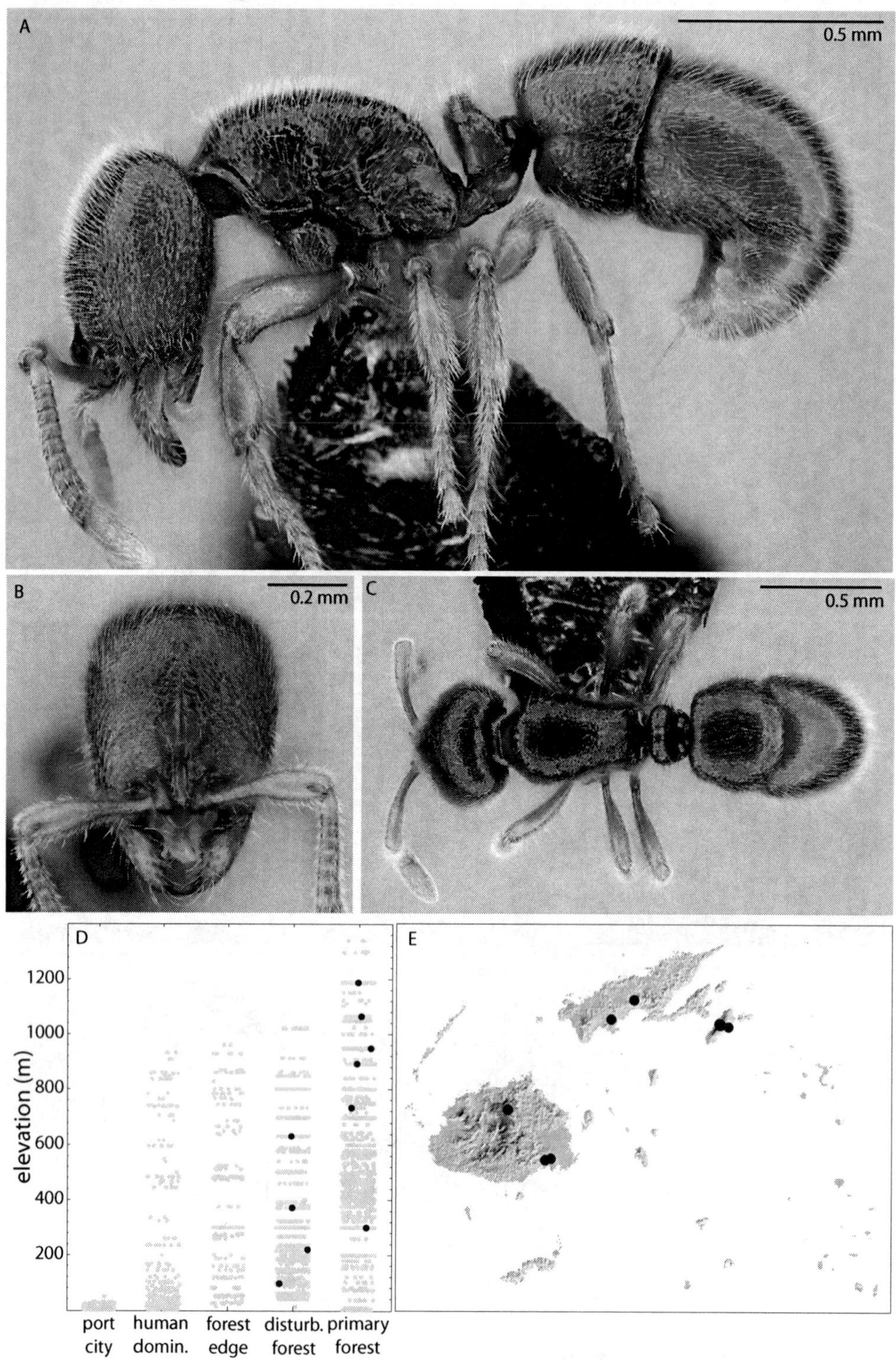

LITERATURE CITED

Ash, J. (1992) Vegetation ecology of Fiji past present and future perspectives. *Pacific Science*, 46, 111-127.

Auzende, J. (1995) Propagating rift west of the Fiji Archipelago (North Fiji Basin, SW Pacific). *Journal of Geophysical Research, B, Solid Earth and Planets*, 100, 17.

Baroni Urbani, C. (1973) Die gattung xenometra, ein objektives synonym (Hymenoptera, Formicidae). *Mitteilungen der Schweizerischen Entomologischen Gessellschaft or Bulletin de la Societe Entomologique Suisse*, 46, 199-201.

Baroni Urbani, C. & De Andrade, M.L. (2003) The ant genus *Proceratium* in the extant and fossil record (Hymenoptera: Formicidae). *Museum Regionale Di Scenze Naturali Monografie*, 36, 1-492.

Bolton, B. (1976) The ant tribe Tetramoriini (Hymenoptera: Formicidae). Constituent genera, review of smaller genera and revision of *Triglyphothrix* Forel. *Bulletin of the British Museum (Natural History) Entomology*, 34, 281-379.

Bolton, B. (1977) The ant tribe Tetramoriini (Hymenoptera: Formicidae). The genus *Tetramorium* Mayr in the Oriental and Indo-Australian regions, and in Australia. *Bulletin of the British Museum (Natural History) Entomology*, 36, 67-151.

Bolton, B. (1979) The ant tribe Tetramoriini (Hymenoptera: Formicidae). The genus *Tetramorium* Mayr in the Malagasy region and in the New World. *Bulletin of the British Museum (Natural History) Entomology*, 38, 129-181.

Bolton, B. (1980) The ant tribe Tetramoriini (Hymenoptera: Formicidae). The genus *Tetramorium* Mayr in the Ethiopian zoogeographical region. *Bulletin of the British Museum (Natural History) Entomology*, 40, 193-384.

Bolton, B. (1982) Afrotropical species of the myrmicine ant genera *Cardiocondyla*, *Leptothorax*, *Melissotarsus*, *Messor* and *Cataulacus* (Formicidae). *Bulletin of the British Museum (Natural History) Entomology*, 45, 307-370.

Bolton, B. (1985) The ant genus *Triglyphothrix* Forel a synonym of *Tetramorium* Mayr. (Hymenoptera: Formicidae). *Journal of Natural History*, 19, 243-248.

Bolton, B. (1987) A review of the *Solenopsis* genus-group and revision of Afrotropical *Monomorium* Mayr (Hymenoptera: Formicidae). *Bulletin of the British Museum (Natural History) Entomology*, 54, 263-452.

Bolton, B. (1995) *A new general catalogue of the ants of the world.* Harvard University Press, Cambridge, Massachusetts, 504 pp.

Bolton, B. (2000) The ant tribe Dacetini. *Memoirs of the American Entomological Institute*, 65, 492-1028.

Bolton, B. (2007) Taxonomy of the Dolichoderine ant genus *Technomyrmex* Mayr (Hymenoptera: Formicidae) based on the worker caste. *Contributions of the American Entomological institute*, 35, 1-150.

Bolton, B., Alpert, G., Ward, P.S. & Nasrecki, P. (2006) *Bolton's Catalogue of ants of the world.* Harvard University Press, Cambridge, Massachusetts, CD-ROM.

Bolton, B. & Fisher, B.L. (2011) Taxonomy of Afrotropical and West Palaearctic ants of the ponerine genus *Hypoponera* Santschi (Hymenoptera: Formicidae). *Zootaxa*, 2843, 1-118.

Brady, S.G., Schultz, T.R., Fisher, B.L. & Ward, P.S. (2006) Evaluating alternative hypotheses for the early evolution and diversification of ants. *Proceedings of the National Academy of Sciences of the United States of America*, 103, 18172-18177.

Brown, W.L., Jr. (1948) A preliminary generic revision of the higher Dacetini (Hymenoptera: Formicidae). *Transactions of the American Entomological Society (Philadelphia)*, 74, 101-129.

Brown, W.L., Jr. (1949) Revision of the ant tribe Dacetini. I. Fauna of Japan, China and Taiwan. *Mushi*, 20, 1-25.

Brown, W.L., Jr. (1954) The ant genus *Strumigenys* Fred. Smith in the Ethiopian and Malagasy regions. *Bulletin of the Museum of Comparative Zoology at Harvard University*, 112, 1-34.

Brown, W.L., Jr. (1958a) Contributions toward a reclassification of the Formicidae. II. Tribe Ectatommini (Hymenoptera). *Bulletin of the Museum of Comparative Zoology*, 118, 173-362.

Brown, W.L., Jr. (1958b) A review of the ants of New Zealand. *Acta Hymenopterologica*, 1, 1-50.

Brown, W.L., Jr. (1960) Contributions toward a reclassification of the Formicidae. III. Tribe Amblyoponini (Hymenoptera). *Bulletin of the Museum of Comparative Zoology*, 122, 143-230.

Brown, W.L., Jr. (1966) [Untitled. Synonymy of *Monomorium floreanum* Stitz under *Monomorium floricola* (Jerdon).] P. 175 in: Linsley, E. G., Usinger, R. L. Insects of the Galápagos Islands. Proceedings of the California Academy of Sciences 33, 113-196

Brown, W.L., Jr. (1971a) Characters and synonymies among the genera of ants. Part IV. Some genera of subfamily Myrmicinae (Hymenoptera: Formicidae). *Breviora*, 365, 1-5.

Brown, W.L., Jr. (1971b) The Indo-Australian species of the ant genus *Strumigenys*: group of *szalayi* (Hymenoptera: Formicidae). *In:* Asahina, S. & al., e. (Eds.) *Entomological essays to commemorate the retirement of Professor K. Yasumatsu*. Hokuryukan Publishing Co., Tokyo. vi + 389 p., pp. 73-86.

Brown, W.L., Jr. (1975) Contributions toward a reclassification of the Formicidae. 5. Ponerinae, tribes Platythyreini, Cerapachyini, Cylindromyrmecini, Acanthostichini, and Aenictogitini. *Search Agriculture*, 5. Entomology (Ithaca) 15, 1-115.

Brown, W.L., Jr. (1976) Contributions toward a reclassification of the Formicidae. Part VI. Ponerinae, tribe Ponerini, subtribe Odontomachiti. Section A. Introduction, subtribal characters. Genus *Odontomachus*. *Studia Entomologica*, 19, 67-171.

Brown, W.L., Jr. (1978) Contributions toward a reclassification of the Formicidae. Part VI. Ponerinae, tribe Ponerini, subtribe Odontomachiti. Section B. Genus *Anochetus* and bibliography. *Studia Entomologica*, 20, 549-638.

Brown, W.L., Jr. (1981) Preliminary contributions toward a revision of the ant genus *Pheidole* (Hymenoptera: Formicidae). Part I. *Journal of the Kansas Entomological Society*, 54, 523-530.

Brown, W.L., Jr. & Kempf, W.W. (1960) A world revision of the ant tribe Basicerotini (Hym. Formicidae). *Studia Entomologica*, (n.s.)3, 161-250.

Brullé, G.A. (1840) Insectes. *In:* Webb, P.B. & Berthelot, S. (Eds.) *Histoire naturelle des Îles Canaries. Vol. 2. (deuxième partie - Entomologie)*. Mellier, Paris, pp. 53-95.

Chase, C.G. (1971) Tectonic history of the Fiji plateau. *Geological Society of America Bulletin*, 82, 3087-3109.

Cheesman, L.E. & Crawley, W.C. (1928) A contribution towards the insect fauna of French Oceania. - Part III. Formicidae. *Annals and Magazine of Natural History*, (10)2, 514-525.

Clouse, R.M. (2007a) The ants of Micronesia (Hymenoptera: Formicidae). *Micronesia*, 39, 171-295.

Clouse, R.M. (2007b) New ants (Hymenoptera: Formicidae) from Micronesia. *Zootaxa*, 1475, 1-19.

Colley, H. & Hindle, W.H. (1984) Volcano-tectonic evolution of Fiji and adjoining marginal basins; Marginal basin geology; volcanic and associated sedimentary and tectonic processes in modern and ancient marginal basins, p. 151-162. *In:* Marginal basin geology. Vol. 16. Kokelaar, B.P. & Howells, M.F. (eds.), Keele, United Kingdom

Dalla Torre, K.W. (1892) Hymenopterologische notizen. *Wiener Entomologische Zeitung*, 11, 89-93.

Dalla Torre, K.W. (1893) *Catalogus Hymenopterorum hucusque descriptorum systematicus et synonymicus. Vol. 7. Formicidae (Heterogyna)*. W. Engelmann, Leipzig, 289 pp.

Davey, F.J. (1982) The structure of the South Fiji Basin; The evolution of the India-Pacific plate boundaries, p. 185-241. *In:* Third Southwest Pacific Workshop symposium, Sydney, N.S.W., Australia, Dec. 1979. Vol. 87. Packham, G.H. (ed.), Netherlands (NLD)

De Andrade, M.L. (2003) *Proceratium oceanicum*, p. 310-313 in: Baroni Urbani, C.; de Andrade, M. L. The ant genus *Proceratium* in the extant and fossil record (Hymenoptera: Formicidae). *Museum Regionale Di Scenze Naturali Monografie*, 36, 1-492.

Delabie, J.H.C. & Blard, F. (2002) The tramp ant *Hypoponera punctatissima* (Roger) (Hymenoptera: Formicidae: Ponerinae): New records from the Southern Hemisphere. *Neotropical Entomology*, 31, 149-151.

Dickinson, W.R. & Shutler, R. (2000) Implications of petrographic temper analysis for Oceanian prehistory. *Journal of World Prehistory*, 14, 203-266.

Dlussky, G.M. (1993a) Ants (Hymenoptera, Formicidae) of Fiji, Tonga, and Samoa, and the problem of island faunas formation. 2. Tribe Dacetini. [in Russian]. *Zoologicheskii Zhurnal [=Zoological Journal, Moscow]*, 72, 52-65.

Dlussky, G.M. (1993b) Ants (Hymenoptera: Formicidae) of Fiji, Tonga, and Samoa, and the problem of island faunas formation. 1. Statement of the problem. [in Russian]. *Zoologicheskii Zhurnal [=Zoological Journal, Moscow]*, 72, 66-76.

Dlussky, G.M. (1994) Zoogeography of southwestern Oceania. [in Russian]. *In:* Puzatchenko, Y.G., Golovatch, S.M., Dlussky, G.M., Diakonov, K.N., Zakharov, A.A. & Korganova, G.A. (Eds.) *Animal populations of the islands of southwestern Oceania (ecogeographic studies). [in Russian]*. Nauka Press, Moscow, pp. 48-93.

Donisthorpe, H. (1915) *British ants, their life-history and classification*. Brendon & Son Ltd., Plymouth, xv + 379 pp.

Donisthorpe, H. (1932) On the identity of Smith's types of Formicidae (Hymenoptera) collected by Alfred Russell Wallace in the Malay Archipelago, with descriptions of two new species. *Annals and Magazine of Natural History*, (10)10, 441-476.

Donisthorpe, H. (1946) Undescribed forms of *Camponotus* (*Colobopsis*) *vitiensis* from the Fiji Islands (Hymenoptera, Formicidae). *Proceedings of the Royal Entomological Society of London Series B Taxonomy*, 15, 69-70.

Eguchi, K. (2001) A revision of the Bornean species of the ant genus *Pheidole* (Insecta: Hymenoptera: Formicidae: Myrmicinae). *Tropics monograph series*, No. 2, 1-154.

Eguchi, K. (2004) Taxonomic revision of two wide-ranging Asian ants, *Pheidole fervens* and *P. indica* (Insects: Hymenoptera), and related species. *Annalen des Naturhistorischen Museums in Wien Serie B Botanik und Zoologie*, 105(B), 189-209.

Emery, C. (1869) Enumerazione dei formicidi che rinvengonsi nei contorni di Napoli con descrizioni di specie nuove o meno conosciute. *Annali dell'Accademia degli Aspiranti Naturalisti Secunda Era*, 2, 1-26.

Emery, C. (1887) Catalogo delle formiche esistenti nelle collezioni del Museo Civico di Genova. Parte terza. Formiche della regione Indo-Malese e dell'Australia. [part b]. *Annali del Museo Civico di Storia Naturale*, 24, 241-256.

Emery, C. (1890) Voyage de M. E. Simon au Venezuela (Décembre 1887 - Avril 1888). Formicides. *Annales de la Société Entomologique de France*, 10, 55-76.

Emery, C. (1891) *Exploration scientifique de la Tunisie. Zoologie. - Hyménoptères. Révision critique des fourmis de la Tunisie*. Imprimerie Nationale, Paris, iii + 21 pp.

Emery, C. (1892) Note sinonimiche sulle formiche. *Bullettino della Società Entomologica Italiana*, 23, 159-167.

Emery, C. (1894a) Mission scientifique de M. Ch. Alluaud aux îles Séchelles (mars, avril, mai 1892). 2e mémoire. Formicides. *Annales de la Société Entomologique de France*, 63, 67-72.

Emery, C. (1894b) Studi sulle formiche della fauna neotropica. VI-XVI. *Bullettino della Società Entomologica Italiana*, 26, 137-241.

Emery, C. (1896) Saggio di un catalogo sistematico dei generi Camponotus, Polyrhachis e affini. *Mem. R. Accad. Sci. Ist. Bologna*, 5, 363-382.

Emery, C. (1897) Formicidarum species novae vel minus cognitae in collectione Musaei Nationalis Hungarici quas in Nova-Guinea, colonia germanica, collegit L. Biró. *Természetrajzi Füzetek*, 20, 571-599.

Emery, C. (1900) Formicidarum species novae vel minus cognitae in collectione Musaei Nationalis Hungarici quas in Nova-Guinea, colonia germanica, collegit L. Biró. Publicatio secunda. *Természetrajzi Füzetek*, 23, 310-338.

Emery, C. (1901) Formiciden von Celebes. *Zoologische Jahrbücher, Abteilung für Systematik, Geographie und Biologie der Tiere*, 14, 565-580.

Emery, C. (1914) Les fourmis de la Nouvelle-Calédonie et des îles Loyalty. *In:* Sarasin, F. & Roux, J. (Eds.) *Nova Caledonia A, Zoologie*. C.W. Kreidels Verlag, Wiesbaden, pp. 393-437.

Emery, C. (1915) Les *Pheidole* du groupe *megacephala* (Formicidae). *Revue de Zoologie Africaine (Bruxelles)*, 4, 223-250.

Emery, C. (1920a) Le genre *Camponotus* Mayr. Nouvel essai de la subdivision en sous-genres. *Revue de Zoologie Africaine (Bruxelles)*, 8, 229-260.

Emery, C. (1920b) Studi sui *Camponotus*. *Bullettino della Società Entomologica Italiana*, 52, 3-48.

Emery, C. (1921) Hymenoptera, family Formicidae, subfamily Myrmicinae, Fasc. *In:* Wytsman, P. (Ed.) *Genera Insectorum*, Bruxelles, pp. 1-94.

Emery, C. (1925) *Hymenoptera. Fam. Formicidae. Subfam. Formicinae*. Brussels, 1-302 p., 4 pls pp.

Evenhuis, N.L. & Bickel, D.J. (2005) The NSF-Fiji terrestrial arthropod survey: overview. *Occasional Papers of the Bernice Pauhahi Bishop Museum*, 82, 3-25.

Ewart, F.T. (1988) Geological history of the Fiji–Tonga–Samoan region of the S.W. Pacific, and some palaeogeographic and biogeographic implications. *In:* Lyneborg, L. (Ed.) *The Cicadas of the Fiji, Samoa and Tonga Islands,their taxonomy and biogeography*. EJ Brill/ Scandinavian Science Press, Leiden, Netherlands, pp. 15-23.

Fabricius, J.C. (1793) *Entomologia systematica emendata et aucta. Secundum classes, ordines, genera, species, adjectis synonimis, locis observationibus, descriptionibus. Tome 2*. C. G. Proft, Hafniae [= Copenhagen], 519 pp.

Fabricius, J.C. (1804) *Systema Piezatorum secundum ordines, genera, species, adjectis synonymis, locis, observationibus, descriptionibus*. C. Reichard, Brunswick, xiv + 15-439 + 30 pp.

Falvey, D.A. (1978) Analysis of palaeomagnetic data from the New Hebrides, p. 117-123. *In:* Second Southwest Pacific earthscience symposium and I.G.C.P. project meeting. Vol. 9. Coleman, P.J. (ed.), Sydney, Australia

Fernández, F. (2003) Revision of the myrmicine ants of the *Adelomyrmex* genus-group (Hymenoptera, Formicidae). *Zootaxa*, 361, 1-52.

Fernández, F. (2004) The American species of the myrmicine ant genus *Carebara* Westwood (Hymenoptera: Formicidae). *Caldasia*, 26, 191-238.

Forel, A. (1881) Die Ameisen der Antille St. Thomas. *Mitteilungen der Münchener Entomologischen Verein*, 5, 1-16.

Forel, A. (1894) Abessinische und andere afrikanische Ameisen, gesammelt von Herrn Ingenieur Alfred Ilg, von Herrn Dr. Liengme, von Herrn Pfarrer Missionar P. Berthoud, Herrn Dr. Arth. Müller etc. *Mitteilungen der Schweizerischen Entomologischen Gessellschaft or Bulletin de la Societe Entomologique Suisse*, 9, 64-100.

Forel, A. (1895) [Untitled. Ponera Gleadowi n. sp.]. Pp. 60-61 [pagination of separate: 292-293] in: Emery, C. Sopre alcune formiche della fauna mediterranea. Mem. R. Accad. Sci. Ist. Bologna (5)5:59-75 [pagination of separate: 291-307]

Forel, A. (1899) Heterogyna (Formicidae). *Fauna Hawaiiensis*, 1, 116-122.

Forel, A. (1901) Formiciden aus dem Bismarck-Archipel, auf Grundlage des von Prof. Dr. F. Dahl gesammelten Materials. *Mitteilungen aus den Zoologischen Museum in Berlin*, 2, 4-37.

Forel, A. (1902) Fourmis nouvelles d'Australie. *Revue Suisse de Zoologie*, 10, 405-548.

Forel, A. (1905) Ameisen aus Java. Gesammelt von Prof. Karl Kraepelin 1904. *Mitteilungen aus dem Naturhistorischen Museum in Hamburg*, 22, 1-26.

Forel, A. (1912a) Einige neue und interessante Ameisenformen aus Sumatra etc. *Zoologische Jahrbücher Supplement*, 15("Erster Band"), 51-78.

Forel, A. (1912b) Formicides néotropiques. Part VI. 5me sous-famille Camponotinae Forel. *Mémoires de la Société Entomologique de Belgique*, 20, 59-92.

Forel, A. (1914) Le genre *Camponotus* Mayr et les genres voisins. *Revue Suisse de Zoologie*, 22, 257-276.

Framenau, V.W. & Thomas, M.L. (2008) Ants (Hymenoptera: Formicidae) of Christmas Island (Indian Ocean): identification and distribution. *Records of the Western Australian Museum*, 25, 45-85.

Gibbons, J.R.H. (1981) The biogeography of *Brachylophus* Iguanidae including the description of *Brachylophus vitiensis* new species from Fiji. *Journal of Herpetology*, 15, 255-274.

Graeffe, E. (1986) Travels in the interior of the island of Vitilevu. *Domodomo*, 4, 98-140.

Hathway, B. & Colley, H. (1994) Eocene to Miocene geology of Southwest Viti Levu, Fiji p. 153-169. *In:* South Pacific Applied Geoscience Commission (SOPAC) Technical Bulletin, vol. 8.

Haupt, A. (1893) *Vermehrung des Kgl. Naturalienkabinets in Bamberg seit 50 Jahren*. Bamberg, Germany, 120 pp.

Heinze, J. (1999) Male polymorphism in the ant *Cardiocondyla minutior* (Hymenoptera: Formicidae). *Entomologia Generalis*, 23, 251-258.

Heterick, B. (2006) A revision of the Malagasy ants belonging to genus *Monomorium* Mayr, 1855 (Hymenoptera: Formicidae). *Proceedings of the California Academy of Sciences*, 57, 69-202.

Heterick, B.E. & Shattuck, S. (2011) Revision of the ant genus *Iridomyrmex* (Hymenoptera: Formicidae). *Zootaxa*, 2845, 1-174.

Hölldobler, B., Oldham, N.J., Alpert, G.D. & Liebig, J. (2002) Predatory behavior and chemical communication in two *Metapone* (Hymenoptera: Formicidae). *Chemoecology*, 12, 147-151.

Imai, H.T., Brown, W.L., Jr, Kubota, M., Yong, H.-S. & Tho, Y.P. (1984) Chromosome observations on tropical ants from western Malaysia. *Annual Report of the National Institute of Genetics*, 34, 66-69.

Ingram, K.K., Bernardello, G., Cover, S. & Wilson, E.O. (2006) The ants of the Juan Fernandez Islands: genesis of an invasive fauna. *Biological Invasions*, 8, 383-387.

Jerdon, T.C. (1851) A catalogue of the species of ants found in Southern India. *Madras Journal of Literature and Science*, 17, 103-127.

Johnson, C., Agosti, D., Delabie, J.H.C., Dumpert, K., Williams, D.J., von Tschirnhaus, M. & Maschwitz, U. (2001) *Acropyga* and *Azteca* ants (Hymenoptera: Formicidae) with scale insects (Sternorhyncha: Coccoidea): 20 Million years of intimate symbiosis. *American Museum Novitates*, 3335, 1-18.

Keogh, J.S., Edwards, D.L., Fisher, R.N. & Harlow, P.S. (2008) Molecular and morphological analysis of the critically endangered Fijian iguanas reveals cryptic diversity and a complex biogeographic history. *Philosophical transactions of the Royal Society of London. Series B, Biological Sciences*, 363, 3413-3426.

Kroenke, L.W. & Yan, C.Y. (1993) An animated plate tectonic reconstruction of the Southwest Pacific, 0-100 Ma, based on the hotspot frame of references, p. 286. *In:* American Geophysical Union, 1993 spring meeting. Vol. 74, Baltimore, MD, United States

Kugler, C. (1994) A revision of the ant genus *Rogeria* with description of the sting apparatus (Hymenoptera: Formicidae). *Journal of Hymenoptera Research*, 3, 17-89.

Kumar, R. (2005) Geology, climate, and landscape of the PABITRA wet-zone transect, Viti Levu Island, Fiji. *Pacific Science*, 59, 141-157.

LaPolla, J.S. (2004) *Acropyga* (Hymenoptera: Formicidae) of the world. *Contributions of the American Entomological institute*, 33, 1-130.

LaPolla, J.S., Brady, S.G. & Shattuck, S.O. (2010) Phylogeny and taxonomy of the *Prenolepis* genus-group of ants (Hymenoptera: Formicidae). *Systematic Entomology*, 35, 118-131.

Latreille, P.A. (1802) *Histoire naturelle générale et particulière des Crustacés et des insectes. Tome 3. Familles naturelles des genres*. F. Dufart, Paris, xii + 467 pp.

Lattke, J.E. (2003) Biogeographic analysis of the ant genus *Gnamptogenys* Roger in South-East Asia-Australasia (Hymenoptera: Formicidae: Ponerinae). *Journal of Natural History*, 37, 1879-1897.

Lattke, J.E. (2004) A taxonomic revision and phylogenetic analysis of the ant genus Gnamptogenys Roger in Southeast Asia and Australasia (Hymenoptera: Formicidae: Ponerinae). *Univ. Calif. Publ. Entomol.*, 122, 1-266.

Linnaeus, C. (1758) *Systema naturae per regna tria naturae, secundum classes, ordines, genera, species, cum characteribus, differentiis, synonymis, locis. Tomus I. Editio decima, reformata*. L. Salvii, Holmiae [= Stockholm], 824 pp.

Lucky, A. & Sarnat, E.M. (2008) New species of *Lordomyrma* (Hymenoptera: Formicidae) from Southeast Asia and Fiji. *Zootaxa*, 1681, 37-46.

Lucky, A. & Sarnat, E.M. (2010) Biogeography and diversification of the Pacific ant genus *Lordomyrma* Emery. *Journal of Biogeography*, 37, 624-634.

MacArthur, R.H. & Wilson, E.O. (1967) *The theory of island biogeography*. Princeton University Press, Princeton, New Jersey,pp.

Mann, W.M. (1919) The ants of the British Solomon Islands. *Bulletin of the Museum of Comparative Zoology*, 63, 273-391.

Mann, W.M. (1920) Ant guests from Fiji and the British Solomon Islands. *Annals of the Entomological Society of America*, 13, 60-69.

Mann, W.M. (1921) The ants of the Fiji Islands. *Bulletin of the Museum of Comparative Zoology*, 64, 401-499.

Mann, W.M. (1925) Ants collected by the University of Iowa Fiji-New Zealand Expedition. *Studies in Natural History, Iowa University*, 11, 5-6.

Mann, W.M. (1948) *Ant hill odyssey*. Little, Brown, Boston, 338 pp.

Mayr, E. (1942) *Systematics and the origin of species*. Columbia University Press, New York, 334 pp.

Mayr, G. (1861) *Die europäischen Formiciden. Nach der analytischen Methode bearbeitet*. C. Gerolds Sohn, Wien, 80 pp.

Mayr, G. (1862) Myrmecologische Studien. *Verhandlungen der Kaiserlich-Königlichen Zoologisch-Botanischen Gesellschaft in Wien*, 12, 649-776.

Mayr, G. (1866) Myrmecologische Beiträge. *Sitzungsberichte der Kaiserlichen Akademie der Wissenschaften. Mathematisch - Naturwissenschaftliche Classe. Abteilung I*, 53, 484-517.

Mayr, G. (1868) Formicidae novae Americanae collectae a Prof. P. de Strobel. *Annu. Soc. Nat. Mat. Modena*, 3, 161-178.

Mayr, G. (1870) Neue Formiciden. *Verhandlungen der Kaiserlich-Königlichen Zoologisch-Botanischen Gesellschaft in Wien*, 20, 939-996.

Mayr, G. (1879) Beiträge zur Ameisen-Fauna Asiens. *Verhandlungen der Kaiserlich-Königlichen Zoologisch-Botanischen Gesellschaft in Wien*, 28, 645-686.

Mayr, G. (1886) Notizen über die Formiciden-Sammlung des British Museum in London. *Verhandlungen der Kaiserlich-Königlichen Zoologisch-Botanischen Gesellschaft in Wien*, 36, 353-368.

Mayr, G. (1887) Südamerikanische Formiciden. *Verhandlungen der Kaiserlich-Königlichen Zoologisch-Botanischen Gesellschaft in Wien*, 37, 511-632.

McArthur, A.J. (2003) New species of *Camponotus* (Hymenoptera: Formicidae) from Australia. *Transactions of the Royal Society of South Australia*, 127, 5-14.

McArthur, A.J. & Leys, R. (2006) A morphological and molecular study of some species in the *Camponotus maculatus* group (Hymenoptera: Formicidae) in Australia and Africa, with a description of a new Australian species. *Myrmecological News*, 8, 99-110.

Moreau, C.S., Bell, C.D., Vila, R., Archibald, S.B. & Pierce, N.E. (2006) Phylogeny of the ants: diversification in the age of angiosperms. *Science (Washington D. C.)*, 312, 101-104.

Musgrave, R.J. & Firth, J.V. (1999) Magnitude and timing of New Hebrides Arc rotation; paleomagnetic evidence from Nendo, Solomon Islands. *Journal of Geophysical Research, B, Solid Earth and Planets*, 104, 2841-2853.

Neall, V.E. & Trewick, S.A. (2008) The age and origin of the Pacific islands: a geological overview. *Philosophical transactions of the Royal Society of London. Series B, Biological Sciences*, 363, 3293-3308.

Nunn, P.D. (1990) Coastal processes and landforms of Fiji their bearing on Holocene sea-level changes in the South and West Pacific. *Journal of Coastal Research*, 6, 279-310.

Nunn, P.D. (1999) Penultimate interglacial emerged reef around Kadavu Island, Southwest Pacific: implications for late Quaternary island-arc tectonics and sea-level history. *New Zealand journal of geology and geophysics*, 42, 219.

Nylander, W. (1846) Additamentum adnotationum in monographiam formicarum borealium Europae. *Acta Societatis Scientiarum Fennicae*, 2, 1041-1062.

O'Dowd, D.J., Green, P.T. & Lake, P.S. (2003) Invasional 'meltdown' on an oceanic island. *Ecology Letters*, 6, 812-817.

Onoyama, K. (1999) A new and a newly recorded species of the ant genus *Amblyopone* (Hymenoptera: Formicidae) from Japan. *Entomological Science*, 2, 157-161.

Özdikmen, H. (2010) New names for the preoccupied specific and subspecific epithets in the genus *Camponotus* Mayr, 1861. *Munis Entomology and Zoology*, 5, 519-537.

Pregill, G.K. & Steadman, D.W. (2004) South Pacific iguanas: Human impacts and a new species. *Journal of Herpetology*, 38, 15-21.

Rodda, P. (1994) Geology of Fiji. *South Pacific Applied Geoscience Commission (SOPAC) Technical Bulletin*, 8, 131-151.

Rodda, P. & Kroenke, L. (1984) Fiji: a fragmented arc, p. 87-110. *In:* Cenozoic Tectonic Development of the Southwest Pacific. Vol. Technical Bulletin No. 6. Kroenke, L. (ed.). U.N. ESCAP, CCOP/SOPAC.

Roger, J. (1857) Einiges über Ameisen. *Berliner Entomologische Zeitschrift*, 1, 10-20.

Roger, J. (1859) Beiträge zur Kenntniss der Ameisenfauna der Mittelmeerländer. I. *Berliner Entomologische Zeitschrift*, 3, 225-259.

Roger, J. (1860) Die Ponera-artigen Ameisen. *Berliner Entomologische Zeitschrift*, 4, 278-312.

Roger, J. (1861) Die Ponera-artigen Ameisen (Schluss). *Berliner Entomologische Zeitschrift*, 5, 1-54.

Roger, J. (1862) Synonymische Bemerkungen. 1. Ueber Formiciden. *Berliner Entomologische Zeitschrift*, 6, 283-297.

Roger, J. (1863a) Die neu aufgeführten Gattungen und Arten meines Formiciden-Verzeichnisses nebst Ergänzung einiger früher gegebenen Beschreibungen. *Berl. Entomol. Z.*, 7, 131-214.

Roger, J. (1863b) Verzeichniss der Formiciden-Gattungen und Arten. *Berliner Entomologische Zeitschrift*, 7[Beilage (Suppl)], 1-65.

Santschi, F. (1928a) Fourmis des iles Fidji. *Revue Suisse de Zoologie*, 35, 67-74.

Santschi, F. (1928b) Nouvelles fourmis d'Australie. *Bulletin de la Société Vaudoise des Sciences Naturelles*, 56, 465-483.

Sarnat, E.M. (2006) *Lordomyrma* (Hymenoptera: Formicidae) of the Fiji Islands. *Occasional Papers of the Bernice Pauhahi Bishop Museum*, 90, 9-42.

Sarnat, E.M. (2008a) Fire ant workshop and surveillance final technical report, p. 1-33. Wildlife Conservation Society, Suva, Fiji

Sarnat, E.M. (2008b) A taxonomic revision of the *Pheidole roosevelti*-group (Hymenoptera: Formicidae) in Fiji. *Zootaxa*, 1767, 1-36.

Sarnat, E.M. & Moreau, C.S. (2011) Biogeography and morphological evolution of a Pacific island ant radiation. *Molecular Ecology*, 20, 114-130.

Schlick-Steiner, B.C., Steiner, F.M. & Zettel, H. (2006) *Tetramorium pacificum* Mayr, 1870, *T. scabrum* Mayr, 1879 sp. rev., *T. manobo* (Calilung, 2000) (Hymenoptera: Formicidae) – three good species. *Myrmecological News*, 8, 181-191.

Schneider, S.A. & LaPolla, J.S. (2011) Systematics of the mealybug tribe Xenococcini (Hemiptera: Coccoidea: Pseudococcidae), with a discussion of trophobiotic associations with *Acropyga* Roger ants. *Systematic Entomology*, 36, 57-82.

Seifert, B. (2003) The ant genus *Cardiocondyla* (Insecta: Hymenoptera: Formicidae) - a taxonomic revision of *C. elegans*, *C. bulgarica*, *C. batesii*, *C. nuda*, *C. shuckardi*, *C. stambuloffii*, *C. wroughtonii*, *C. emeryi* , and *C. minutior* species groups *Annalen des Naturhistorischen Museums in Wien Serie B Botanik und Zoologie*, 104B, 203-338.

Shattuck, S. (2005) Review of the *Camponotus aureopilus* species-group (Hymenoptera, Formicidae), including a second *Camponotus* with a metapleural gland. *Zootaxa*, 1-20.

Shattuck, S. & Janda, M. (2009) A new species of the *Camponotus aureopilus* Viehmeyer, 1914 species group (Hymenoptera: Formicidae) from Papua New Guinea. *Myrmecological News*, 12, 251-253.

Shattuck, S.O. (1992) Generic revision of the ant subfamily Dolichoderinae (Hymenoptera: Formicidae). *Sociobiology*, 21, 1-181.

Shattuck, S.O. (2008) Revision of the ant genus *Prionopelta* (Hymenoptera: Formicidae) in the Indo-Pacific region. *Zootaxa*, 1846, 21-34.

Smith, F. (1851) *List of the specimens of British animals in the collection of the British Museum. Part VI. Hymenoptera, Aculeata*. British Museum, London, 134 pp.

Smith, F. (1857) Catalogue of the hymenopterous insects collected at Sarawak, Borneo; Mount Ophir, Malacca; and at Singapore, by A. R. Wallace. [part]. *Journal and Proceedings of the Linnean Society of London Zoology*, 2, 42-88.

Smith, F. (1858) *Catalogue of hymenopterous insects in the collection of the British Museum. Part VI. Formicidae*. British Museum, London, 216 pp.

Smith, F. (1859) Catalogue of hymenopterous insects collected by Mr. A. R. Wallace at the islands of Aru and Key. [part]. *Journal and Proceedings of the Linnean Society of London Zoology*, 3, 132-158.

Smith, F. (1860) Catalogue of hymenopterous insects collected by Mr. A. R. Wallace in the islands of Bachian, Kaisaa, Amboyna, Gilolo, and at Dory in New Guinea. *Journal and Proceedings of the Linnean Society of London Zoology*, 5(17b) (suppl. to vol. 4), 93-143.

Smith, F. (1861) Catalogue of hymenopterous insects collected by Mr. A. R. Wallace in the islands of Ceram, Celebes, Ternate, and Gilolo. [part]. *Journal and Proceedings of the Linnean Society of London Zoology*, 6, 36-48.

Smith, M.R. (1944) Ants of the genus *Cardiocondyla* Emery in the United States. *Proceedings of the Entomological Society of Washington*, 46, 30-41.

Smith, M.R. (1953) A revision of the genus *Romblonella* W. M. Wheeler (Hymenoptera: Formicidae). *Proceedings of the Hawaiian Entomological Society*, 15, 75-80.

Smith, M.R. (1958) A contribution to the taxonomy, distribution and biology of the vagrant ant, *Plagiolepis alluaudi* Emery (Hymenoptera, Formicidae). *Journal of the New York Entomological Society*, 65, 195-198.

Stratford, J.M.C. & Rodda, P. (2000) Late Miocene to Pliocene palaeogeography of Viti Levu, Fiji Islands. *Palaeogeography, Palaeoclimatology, Palaeoecology*, 162, 137-153.

Taylor, R.W. (1967) *A monographic revision of the ant genus Ponera Latreille (Hymenoptera: Formicidae)*. Entomology Department, Bernice P. Bishop Museum, Honolulu, Hawaii.

Taylor, R.W. (1978) Melanesian ants of the genus *Amblyopone* (Hymenoptera: Formicidae). *Australian Journal of Zoology*, 26, 823-839.

Taylor, R.W. (1980a) Australian and Melanesian ants of the Genus *Eurhopalothrix* Brown and Kempf--notes and new species. *Journal of the Australian Entomological Society*, 19, 229-239.

Taylor, R.W. (1980b) The rare Fijian ant *Myrmecina* (=*Archaeomyrmex*) *cacabau* (Mann) rediscovered (Hymenoptera: Formicidae). *New Zealand Entomologist*, 7, 122-123.

Taylor, R.W. (1990) *Ants*. In: Report: A.N.P.W.S. Consultancy Agreement: C.S.I.R.O. Entomological Survey of Christmas Island. Phase 2 and Final Report. Report of Project No. 8889/13.,

Taylor, R.W. (1991a) Nomenclature and distribution of some Australasian ants of the Myrmicinae (Hymenoptera: Formicidae). *Memoirs of the Queensland Museum*, 30, 599-614.

Taylor, R.W. (1991b) Notes on the ant genera *Romblonella* and *Willowsiella*, with comments on their affinities, and the first descriptions of Australian species (Hymenoptera: Formicidae: Myrmicinae). *Psyche (Cambridge)*, 97, 281-296.

Taylor, R.W. (2009) Ants of the genus *Lordomyrma* Emery (1) Generic synonymy, composition and distribution, with notes on *Ancyridris* Wheeler and *Cyphoidris* Weber (Hymenoptera: Formicidae: Myrmicinae). *Zootaxa*, 1979, 16-28.

Terayama, M. (1999) Taxonomic studies of the Japanese Formicidae, Part 6. Genus *Cardiocondyla* Emery. *Memoirs of The Myrmecological Society of Japan*, 99-107.

Wang, M. (2003) A monographic revision of the ant genus *Pristomyrmex* (Hymenoptera: Formicidae). *Bulletin of the Museum of Comparative Zoology*, 157, 383-542.

Ward, D. & Beggs, J. (2007) Coexistence, habitat patterns and the assembly of ant communities in the Yasawa islands, Fiji. *Acta Oecologica*, 32, 215-223.

Ward, D.F. (2007) The distribution and ecology of invasive ant species in the Pacific region, p. i-xiii, 173. *In:* Biological Sciences. Vol. Ph.D. University of Auckland, Aukland

Ward, D.F. & Wetterer, J.K. (2006) Checklist of the ants of Fiji (Hymenoptera: Formicidae). *Occasional Papers of the Bernice Pauhahi Bishop Museum*, 85, 23-47.

Wetterer, J.K. (2008) Worldwide spread of the longhorn crazy ant, *Paratrechina longicornis* (Hymenoptera: Formicdae). *Myrmecological News*, 11, 137-149.

Wetterer, J.K. (2009a) Worldwide spread of the destroyer ant, *Monomorium destructor* (Hymenoptera: Formicidae). *Myrmecological News*, 12, 97-108.

Wetterer, J.K. (2009b) Worldwide Spread of the Penny Ant, *Tetramorum bicarinatum* (Hymenoptera: Formicidae). *Sociobiology*, 54, 811-830.

Wetterer, J.K. (2010a) Worldwide spread of the flower ant, *Monomorium floricola* (Hymenoptera: Formicidae). *Myrmecological News*, 13, 19-27.

Wetterer, J.K. (2010b) Worldwide spread of the wooly ant, *Tetramorium lanuginosum* (Hymenoptera: Formicidae). *Myrmecological News*, 13, 81-88.

Wetterer, J.K. (2011) Worldwide spread of the stigma ant, Pachycondyla stigma (Hymenoptera: Formicidae). *Myrmecological News*, 16, 39-44.

Wheeler, G.C. (1950) Ant larvae of the subfamily Cerapachyinae. *Psyche (Cambridge)*, 57, 102-113.

Wheeler, W.M. (1913a) The ants of Cuba. *Bulletin of the Museum of Comparative Zoology*, 54, 477-505.

Wheeler, W.M. (1913b) Corrections and additions to "List of type species of the genera and subgenera of Formicidae". *Annals of the New York Academy of Sciences*, 23, 77-83.

Wheeler, W.M. (1922) Ants of the American Museum Congo expedition. A contribution to the myrmecology of Africa. VIII. A synonymic list of the ants of the Ethiopian region. *Bulletin of the American Museum of Natural History*, 45, 711-1004.

Wheeler, W.M. (1929) Ants collected by Professor F. Silvestri in Formosa, the Malay Peninsula and the Philippines. *Bollettino del Laboratorio di Zoologia Generale e Agraria della Reale Scuola Superiore d'Agricoltura, Portici*, 24, 27-64.

Wheeler, W.M. (1933) Three obscure genera of ponerine ants. *American Museum Novitates*, 672, 1-23.

Wheeler, W.M. (1934) Some aberrant species of *Camponotus* (*Colobopsis*) from the Fiji Islands. *Annals of the Entomological Society of America*, 27, 415-424.

Wheeler, W.M. (1935) Check list of the ants of Oceania. *Occasional Papers of the Bernice Pauhahi Bishop Museum*, 11(11), 1-56.

Whelan, P.M., Gill, J.B., Kollman, E., Duncan, R.A. & Drake, R.E. (1985) Radiometric dating of magmatic stages in Fiji. *In:* Scholl, D.W. & Vallier, T.L. (Eds.) *Circum-Pacific Council for Energy and Mineral Resources Earth science series*. Circum-Pacific Council Energy and Mineral Resources, Houston, TX, United States (USA), pp. 415-440.

Wild, A.L. (2007) Taxonomic revision of the ant genus *Linepithema* (Hymenoptera: Formicidae). *University of California Publications in Entomology*, 126, 1-159.

Williams, F.X. (1946) *Stigmatomma* (*Fulakora*) *zwaluwenbergi*, a new species of ponerine ant from Hawaii. *Proceedings of the Hawaiian Entomological Society*, 12, 639-640.

Wilson, E.O. (1953) The ecology of some North American dacetine ants. *Annals of the Entomological Society of America*, 46, 479-495.

Wilson, E.O. (1957) The *tenuis* and *selenophora* groups of the ant genus *Ponera* (Hymenoptera: Formicidae). *Bulletin of the Museum of Comparative Zoology*, 116, 355-386.

Wilson, E.O. (1958a) Studies on the ant fauna of Melanesia. 3. *Rhytidoponera* in western Melanesia and the Moluccas. 4. The tribe Ponerini. *Bulletin of the Museum of Comparative Zoology*, 119, 304-371.

Wilson, E.O. (1958b) Studies on the ant fauna of Melanesia. I. The tribe Leptogenyini. II. The tribes Amblyoponini and Platythyreini. *Bulletin of the Museum of Comparative Zoology*, 118, 101-153.

Wilson, E.O. (1959a) Adaptive shift and dispersal in a tropical ant fauna. *Evolution*, 13, 122-144.

Wilson, E.O. (1959b) Studies on the ant fauna of Melanesia V. The tribe Odontomachini. *Bulletin of the Museum of Comparative Zoology*, 120, 483-510.

Wilson, E.O. (1959c) Studies on the ant fauna of Melanesia. 5. The tribe Cerapachyini. *Pacific Insects*, 1, 39-57.

Wilson, E.O. (1961) The nature of the taxon cycle in the Melanesian ant fauna. *American Naturalist*, 95, 169-193.

Wilson, E.O. (2003) *Pheidole in the new world: a dominant, hyperdiverse ant genus*. Harvard University Press, Cambridge, Mass., 794 pp.

Wilson, E.O. & Taylor, R.W. (1967) The ants of Polynesia (Hymenoptera: Formicidae). *Pacific Insects Monograph*, 14, 1-109.

Zug, G.R. (1991) Lizards of Fiji: natural history and systematics. *Bishop Museum Bulletin in Zoology*, 2, 1-136.

APPENDIX A. COLLECTION LOCALITIES

Each entry represents a unique georeferenced location. Latitude and longitude are given in decimal degree format. Elevation is given in meters. More detailed locality information is available by using the advanced search tool on Antweb.org and searching by locality code. There are several exceptions to the generally phonetic written form of the Fijian language that are useful for pronouncing many of the native locality names:

B (*mb*). e.g., Korobaba is pronounced *Korombamba.*
C (*th*). e.g., Colo-i-Suva is pronounced *Tholo-i-Suva.*
D (*nd*). e.g., Nadala is pronounced *Nandala.*
G (*ng*). e.g., Gau is pronounced *Ngau.*
Q (*ng*). e.g., Beqa is pronounced *Benga.*

Island	*Locality code*	*Locality ID*	*Lat (DD)*	*Long (DD)*	*Elv.*
Beqa	Malovo 182	Malovo peak, 1km NE of Dukuibeqa Vlg.	-18.40417	178.13667	182
Beqa	Mt. Korovou 326	Mt. Korovou, 1.5km WNW of Dukuibeqa Vlg.	-18.40898	178.11943	326
Gau	Navukailagi 300	2.5km SE Navukailagi Vlg.	-17.97139	179.27222	300
Gau	Navukailagi 325	Mt. Delaco, 2.2km SE Navukailagi Vlg.	-17.97203	179.27287	325
Gau	Navukailagi 336	Mt. Delaco, 2.1km SE Navukailagi Vlg.	-17.97197	179.27272	336
Gau	Navukailagi 356	Mt. Delaco, 2.8km SE Navukailagi Vlg.	-17.97803	179.27708	356
Gau	Navukailagi 364	Mt. Delaco, 3.4km SE Navukailagi Vlg.	-17.97852	179.27557	364
Gau	Navukailagi 387	Mt. Delaco, 3.3km SE Navukailagi Vlg.	-17.98000	179.27500	387
Gau	Navukailagi 408	Mt. Delaco, 3km SE Navukailagi Vlg.	-17.97957	179.27633	408
Gau	Navukailagi 415	Mt. Delaco, 2.9km SE Navukailagi Vlg.	-17.97852	179.27557	415
Gau	Navukailagi 432	Mt. Delaco, 3km SE Navukailagi Vlg.	-17.97952	179.27633	432
Gau	Navukailagi 475	Mt. Delaco, 3.4km SE Navukailagi Vlg.	-17.98357	179.27662	475
Gau	Navukailagi 480	Mt. Delaco, 3.4km SE Navukailagi Vlg.	-17.98330	179.27652	480
Gau	Navukailagi 481	Mt. Delaco, 3.25km SE Navukailagi Vlg.	-17.98250	179.27625	481
Gau	Navukailagi 490	Mt. Delaco, 3.5km SE Navukailagi Vlg.	-17.98270	179.27635	490
Gau	Navukailagi 496	Mt. Delaco, 3.8km SE Navukailagi Vlg.	-17.98356	179.27711	496
Gau	Navukailagi 505	Mt. Delaco, 3.5km SE Navukailagi Vlg.	-17.98270	179.27635	505
Gau	Navukailagi 522	Mt. Delaco, 3.4km SE Navukailagi Vlg.	-17.98357	179.27662	522

Gau	Navukailagi 535	Mt. Delaco, 3.3km SE Navukailagi Vlg.	-17.98268	179.27630	535
Gau	Navukailagi 557	Mt. Delaco, 3.6km SE Navukailagi Vlg.	-17.98515	179.27713	557
Gau	Navukailagi 564	Mt. Delaco, 4.0km SE Navukailagi Vlg.	-17.98611	179.27750	564
Gau	Navukailagi 575	Mt. Delaco, 3.9km SE Navukailagi Vlg.	-17.98795	179.27845	575
Gau	Navukailagi 597	Mt. Delaco, 4.4km SE Navukailagi Vlg.	-17.98795	179.27845	597
Gau	Navukailagi 625	Mt. Delaco, 4.4km SE Navukailagi Vlg.	-17.98795	179.27845	625
Gau	Navukailagi 632	Mt. Delaco, 4.34km SE Navukailagi Vlg.	-17.99103	179.28105	632
Gau	Navukailagi 675	Mt. Delaco, 4.4km SE Navukailagi Vlg.	-17.99110	179.28100	675
Gau	Navukailagi 717	Mt. Delaco, 4.4km SE Navukailagi Vlg.	-17.99110	179.28100	717
Gau	Ngau	island record	—	—	—
Kabara	Kabara	island record	-18.87000	-178.95000	—
Kabara	Waquava	Waquava	-18.87000	-178.90000	—
Kadavu	Buka Levu	Buka Levu [Nambukelevuira?]	-19.13000	177.97000	—
Kadavu	Daviqele 300	2.0km N Daviqele Vlg.	-19.12139	177.99556	300
Kadavu	Lomaji 580	Mt. Washington, 1.3km SSW Lomaji	-19.11750	177.99250	580
Kadavu	Moanakaka 60	Moanakaka Bird Sanctuary, 250m SW Soladamu Vlg.	-19.07750	178.12100	60
Kadavu	Moanakaka 60	Moanakaka Bird Sanctuary, 250m SW Soladamu Vlg.	-19.07750	178.12100	60
Kadavu	Mt. Washington 700	Mt. Washington, 1.4km SSW Lomaji Vlg.	-19.11833	177.99028	700
Kadavu	Mt. Washington 760	Mt. Washington, 1.4km SSW Lomaji Vlg.	-19.11806	177.98750	760
Kadavu	Mt. Washington 800	Mt. Washington summit, 1.6km SW Lomaji	-19.11806	177.98750	800
Kadavu	Namalata 100	1.3km E Kadavu Air Strip nr. Namalata Vlg.	-19.05811	178.16892	100
Kadavu	Namalata 120	1.3km E Kadavu Air Strip nr. Namalata Vlg.	-19.05989	178.16892	120
Kadavu	Namalata 139	1.3km E Kadavu Air Strip nr. Namalata Vlg.	-19.05989	178.18739	139
Kadavu	Namalata 50	1.3km E Kadavu Air Strip nr. Namalata Vlg.	-19.05750	178.18556	50
Kadavu	Namalata 75	1.3km E Kadavu Air Strip nr. Namalata Vlg.	-19.06261	178.18739	75
Kadavu	Namara 300	3.0km NW Namara Vlg.	-19.02667	178.19361	300
Kadavu	Solo taviene	Solo taviene (island record)	—	—	—
Kadavu	Vanua Ava b	Vanua Ava	—	—	—
Kadavu	Vunisea	Vunisea	-19.05000	178.17000	—
Kadavu	Vunisea 200	nr Vunisea Vlg.	-19.05700	178.16667	200

Koro	Koro	island record	—	—	—
Koro	Kuitarua 480	Kuitarua Rd, Mt. Kuitarua, 3.7km NW Nasau Vlg.	-17.28806	179.40250	480
Koro	Mt. Kuitarua 380	Mt. Kuitarua 4km WNW Nasau Vlg.	-17.30000	179.40000	380
Koro	Mt. Kuitarua 440 b	Mt. Kuitarua, 3.1km WNW Nasau Vlg.	-17.29050	179.40433	440
Koro	Mt. Kuitarua 485	Mt. Kuitarua, 3.8km NW Nasau Vlg.	-17.28883	179.40400	485
Koro	Mt. Kuitarua 500	Mt. Kuitarua summit, 3.8km NW Nasau Vlg.	-17.28750	179.40233	500
Koro	Mt. Kuitarua 500	Mt. Kuitarua summit, 3.8km NW Nasau Vlg.	-17.28750	179.40389	500
Koro	Mt. Kuitarua 505	Mt. Kuitarua summit, 3.8km NW Nasau Vlg.	-17.28675	179.40436	505
Koro	Mt. Nabukala 500	Mt. Nabukala, 4.7km WSW Nasau Vlg.	-17.31222	179.38767	500
Koro	Mt. Nabukala 520	Mt. Nabukala, 5.0km WSW Nasau Vlg.	-17.31250	179.38617	520
Koro	Nabuna 115	2.1km SW Nabuna Vlg., nr. Wailolo Creek	-17.26583	179.37000	115
Koro	Nasau 420 b	3.0km WNW Nasau Vlg., footrack b/w Mt. Kuitarua & Nasau	-17.29033	179.40500	420
Koro	Nasau 465 a	2.7km NW Nasau Vlg., footrack b/w Mt. Kuitarua & Nasau	-17.29472	179.40650	465
Koro	Nasau 470; 3.7 km	Mt. Kuitarua, 3.7km NW Nasau Vlg.	-17.29083	179.40183	470
Koro	Nasau 470; 4.4 km	4.4km W Nasau Vlg.	-17.30889	179.38883	470
Koro	Nasau 476	3.8km WNW Nasau Vlg.	-17.29667	179.40000	476
Koro	Nasoqoloa 300	2.0km N Nasoqoloa Vlg.	-17.27222	179.38861	300
Koro	Tavua 220	1.6km E Tavua Vlg.	-17.27533	179.37417	220
Lakeba	Lakeba	island record	-18.22944	-178.77892	—
Lakeba	Tubou 100 a	3.2km NE Tubou Vlg.	-18.22139	-178.78445	100
Lakeba	Tubou 100 b	3.2km NE Tubou Vlg.	-18.22083	-178.78430	100
Lakeba	Tubou 100 c	3.2km NE Tubou Vlg.	-18.22944	-178.77892	100
Macuata	Vunitogoloa 10	Vunitogoloa	-17.35361	178.03250	10
Macuata	Vunitogoloa 36	Vunitogoloa	-17.35333	178.03278	36
Macuata	Vunitogoloa 4	Vunitogoloa	-17.35333	178.03250	4
Mago	Mango	island record	-17.44940	179.15720	—
Moala	Maloku 1	Nakorovusa Bay, 1.5km NE Maloku Vlg.	-18.57806	179.88417	1
Moala	Maloku 120	Mt. Natuvu, 2.4km ENE Maloku Vlg.	-18.56833	179.89889	120
Moala	Maloku 80	Mt. Natuvu, 2.4km ENE Maloku Vlg.	-18.56750	179.89806	80
Moala	Mt. Korolevu 300	Mt. Korolevu, 5.5km SW of Naroi Vlg.	-18.59479	179.90000	300

Moala	Mt. Korolevu 375	Mt. Korolevu, 5.3km SW of Naroi Vlg.	-18.59132	179.89930	375
Moala	Naroi	Mangrove malaise site, 5.9km E Naroi Vlg.	-18.57355	179.88380	—
Moala	Naroi	Naroi Vlg.	-18.55920	179.93420	—
Moala	Naroi 75	Naroi coastal forest, 4.13km E of Naroi Vlg.	-18.56600	179.89953	75
Munia	Munia	island record	-17.36860	-178.71220	—
Naroi	McDonald's Resort 10 a	Nanunu-i-Ra I., McDonald"s Resort	-17.29196	178.21709	10
Naroi	McDonald's Resort 15	Nanunu-i-Ra I., McDonald"s Resort	-17.29294	178.22025	15
Naroi	Nanunu-i-Ra Island	Nanunu-i-Ra I.	-17.29000	178.22000	—
Ovalau	Andubagenda	island record (coordinates for Levuka)	-17.68000	178.83000	—
Ovalau	Cawaci	island record (coordinates for Levuka)	-17.68000	178.83000	—
Ovalau	Cawaci	island record (coordinates for Levuka)	-17.68000	178.83000	—
Ovalau	Draiba 300	1.2km NNW Draiba Vlg.	-17.69417	178.82528	300
Ovalau	Levuka	[coordinates for Levuka]	-17.68000	178.83000	—
Ovalau	Levuka 400	1.6km WSW Levuka	-17.68710	178.82350	400
Ovalau	Levuka 450	Mt. Pik, 1.3km SE Levuka	-17.68728	178.82527	450
Ovalau	Levuka 500	2.25km W Levuka	-17.68205	178.81317	500
Ovalau	Levuka 550	2.4km W Levuka	-17.68200	178.81247	550
Ovalau	Lovoni valley	(coords for Lovoni vlg.)	-17.69220	178.79160	—
Ovalau	Ovalau	[coordinates for Levuka]	-17.68000	178.83000	—
Ovalau	Vuma	Vuma (coords for Vuma vlg.)	-17.66470	178.82930	—
Ovalau	Wainiloca	island record (coordinates for Levuka)	-17.68000	178.83000	—
Taveuni	Devo Peak 1187	Devo Peak, 5.6km SE Tavuki Vlg.	-16.84333	-179.95950	1187
Taveuni	Devo Peak 1187 b	Devo Peak, 5.6km SE Tavuki Vlg.	-16.84325	-179.96575	1187
Taveuni	Devo Peak 1187 c	Devo Peak, 5.6km SE Tavuki Vlg.	-16.84306	-179.95967	1187
Taveuni	Devo Peak 1188	Devo Peak, 5.5km SE Tavuki Vlg.	-16.84322	-179.96625	1188
Taveuni	Lake Tagimaucia	nr Lake Tagimaucia (coords for LT)	-16.81870	-179.95320	—
Taveuni	Lavena 217	Mt. Koronibuabua, 3.2km NW Lavena Vlg.	-16.85456	-179.88950	217
Taveuni	Lavena 219	Mt. Koronibuabua, 3.2km NW Lavena Vlg.	-16.85528	-179.88897	219
Taveuni	Lavena 229	Mt. Koronibuabua, 3.2km NW Lavena Vlg.	-16.85553	-179.88850	229
Taveuni	Lavena 234	Mt. Koronibuabua, 3.2km NW Lavena Vlg.	-16.85469	-179.89061	234

Taveuni	Lavena 235	Mt. Koronibuabua, 3.2km NW Lavena Vlg.	-16.85472	-179.89164	235
Taveuni	Lavena 235	4.0km NW Lavena Vlg.	-16.85283	-179.88833	235
Taveuni	Lavena 300	Koronibuabua 3.0km NW Lavena Vlg.	-16.85000	-179.89333	300
Taveuni	Mt. Devo 1064	Mt. Devo, 5.3km SE Tavuki Vlg.	-16.84094	-179.96781	1064
Taveuni	Mt. Devo 734	Mt. Devo, Tavuki Vlg.	-16.83069	-179.97953	734
Taveuni	Mt. Devo 775 a	Mt. Devo, 3.9km SE Tavuki Vlg.	-16.83278	-179.97343	775
Taveuni	Mt. Devo 892	Mt. Devo, Tavuki Vlg.	-16.83722	-179.97303	892
Taveuni	Nagasau	Nagasau	—	—	—
Taveuni	Qacavulo Point 300	1.8km SSE Qacavulo Point	-16.88750	-179.96556	300
Taveuni	Somosomo 200	Somosomo	-16.76000	-179.97000	200
Taveuni	Soqulu Estate 140	Soqulu Estate	-16.83333	-179.99972	140
Taveuni	Tavoro Falls 100	Tavoro Falls, 1.4km WSW Korovou Vlg.	-16.82833	-179.88400	100
Taveuni	Tavoro Falls 160	Tavoro Falls, 2.0km WSW Korovou Vlg.	-16.82972	-179.88717	160
Taveuni	Tavuki 734	Mt. Devo, 3.6km SE Tavuki Vlg.	-16.83056	-179.97433	734
Taveuni	Waiyevo	Waiyevo (coords for Waiyevo vlg.)	-16.78450	-179.98000	—
Vanua Balavu	Lomaloma	Lau Prov.	-17.29280	-178.98700	—
Vanua Levu	Banikea 398	Mt. Koroimari, logging road nr. Banikea Vlg.	-16.76750	178.75694	398
Vanua Levu	Drawa 270	1km NNW Drawa Set.	-16.66848	179.00598	270
Vanua Levu	Dreketi 48	Dreketi Vlg.	-16.58335	178.85995	48
Vanua Levu	Eavatu	island record	—	—	—
Vanua Levu	Kasavu 300	2.0km NNW Kasavu Vlg.	-16.71639	179.66333	300
Vanua Levu	Kilaka 113	Batiqere Range, 6km NW Kilaka Vlg.	-16.73167	178.99972	113
Vanua Levu	Kilaka 146	Batiqere Range, 6km NW Kilaka Vlg.	-16.81528	178.98639	146
Vanua Levu	Kilaka 154	Batiqere Range, 6km NW Kilaka Vlg.	-16.80667	178.98806	154
Vanua Levu	Kilaka 61	Batiqere Range, 6km NW Kilaka Vlg.	-16.81083	178.98806	61
Vanua Levu	Kilaka 98	Batiqere Range, 6km NW Kilaka Vlg.	-16.80667	178.99139	98
Vanua Levu	Labasa	Labasa	-16.42000	179.38000	—
Vanua Levu	Lagi 300	Dogotuki nr Lagi Vlg.	-16.25000	179.83333	300
Vanua Levu	Lasema a	Lasema	-16.68000	179.81000	—

Vanua Levu	Lomaloma 587	Vatudiri, 4km SE Lomaloma Vlg.	-16.63014	179.20750	587
Vanua Levu	Lomaloma 630	Vatudiri, 4km SE Lomaloma Vlg.	-16.62975	179.20789	630
Vanua Levu	Lomaloma 630	Vatudiri, 4km SE Lomaloma Vlg.	-16.62958	179.20806	630
Vanua Levu	Mt. Delaikoro	Mt. Delaikoro	-16.55000	179.31000	—
Vanua Levu	Mt. Delaikoro 391	Mt. Delaikoro, Delaikoro Rd., 2.75km ENE Dogoru Vlg.	-16.55050	179.31237	391
Vanua Levu	Mt. Delaikoro 699	Mt. Delaikoro, Delaikoro Rd., 3.6km SE Dogoru Vlg.	-16.57525	179.31638	699
Vanua Levu	Mt. Delaikoro 734	Mt. Delaikoro, Delaikoro Rd., 3.7km SE Dogoru Vlg.	-16.57827	179.31657	734
Vanua Levu	Mt. Delaikoro 910	Mt. Delaikoro, nr summit, 4.3km SE Dogoru Vlg.	-16.59028	179.31581	910
Vanua Levu	Mt. Kasi Gold Mine 300	0.5km S Mt. Kasi Gold Mine	-16.77056	179.02528	300
Vanua Levu	Mt. Vatudiri 570	Mt. Vatudiri, 3km NW Waisali Vlg.	-16.62905	179.21103	570
Vanua Levu	Mt. Vatudiri 641	Mt. Vatudiri, 3km NW Waisali Vlg.	-16.62847	179.20818	641
Vanua Levu	Mt. Wainibeqa 152 c	Mt. Wainibeqa, 4km NW Kilaka Vlg.	-16.80383	178.98643	152
Vanua Levu	Nakanakana 300	2.0km NW Nakanakana Vlg.	-16.62000	179.83333	300
Vanua Levu	Nakasa 300	2.3km NW Nakasa Vlg.	-16.68139	179.18806	300
Vanua Levu	Rokosalase 118	0.4km S Rokosalase Vlg.	-16.53156	179.01919	118
Vanua Levu	Rokosalase 143	0.5km S Rokosalase Vlg.	-16.53250	179.01733	143
Vanua Levu	Rokosalase 150	0.6km S Rokosalase Vlg.	-16.53294	179.01844	150
Vanua Levu	Rokosalase 180	0.6km S Rokosalase Vlg.	-16.53319	179.01811	180
Vanua Levu	Rokosalase 94	0.3km S Rokosalase Vlg.	-16.53056	179.01817	94
Vanua Levu	Rokosalase 94	0.3km S Rokosalase Vlg.	-16.53050	179.01922	94
Vanua Levu	Rokosalase 97	0.5km S Rokosalase Vlg.	-16.53175	179.01883	97
Vanua Levu	Savusavu	Savusavu	-16.77950	179.33800	5
Vanua Levu	Savusavu	Savusavu (coords for Savusavu town)	-16.77920	179.34000	—
Vanua Levu	Suene	Suene [Sueni]	-16.56000	179.35000	—
Vanua Levu	Vusasivo 50	Natewa Peninsula, Mt. Navatadoi, nr creek, 0.8km SSE Vusasivo Vlg.	-16.57861	179.76361	50
Vanua Levu	Vusasivo Village 190	Natewa Peninsula, Mt. Navatadoi, nr creek, 1.8km SE Vusasivo Vlg.	-16.58611	179.76833	190
Vanua Levu	Vusasivo Village 342 b	Mt. Navatadoi, 2.7km SE Vusasivo Vlg. nr. Matakilawa Crk.	-16.59286	179.77057	342
Vanua Levu	Vusasivo Village 400 b	Natewa Peninsula, Mt. Navatadoi, hilltop, 2.6km SSE Vusasivo Vlg.	-16.59278	179.77167	400
Vanua Levu	Vusasivo Village 400 b	Mt. Navatadoi, 2.6km SE Vusasivo Vlg.	-16.59275	179.76960	400
Vanua Levu	Vuya 300	3.5km NW Vuya Vlg.	-16.98333	178.72528	300

Vanua Levu	Wainibeqa 135	Wainibeqa, 4km NW Kilaka Vlg.	-16.80650	178.98811	135
Vanua Levu	Wainibeqa 150	Wainibeqa, 4km NW Kilaka Vlg.	-16.80597	178.98994	150
Vanua Levu	Wainibeqa 53	Wainibeqa, 4km NW Kilaka Vlg.	-16.81533	178.98986	53
Vanua Levu	Wainibeqa 87	Wainibeqa, 4km NW Kilaka Vlg.	-16.80783	178.98733	87
Vanua Levu	Wainunu	Wainunu [Nasolo?]	-16.95000	178.77000	—
Vanua Levu	Yasawa 300	1.5km N Yasawa Vlg.	-16.46806	179.64361	300
Viti Levu	Abaca 525	Koroyanitu Eco Park, 1km WSW Abaca Vlg.	-17.66861	177.53933	525
Viti Levu	Belt Road	island record	—	—	—
Viti Levu	Colo-i-Suva	Colo-i-Suva area	-18.05889	178.46889	—
Viti Levu	Colo-i-Suva 105 b	Colo-i-Suva, 1.4km SW Forestry Station	-18.06769	178.44212	105
Viti Levu	Colo-i-Suva 186 d	Colo-i-Suva Forest Park, 1km SE Forestry Station	-18.06661	178.44325	186
Viti Levu	Colo-i-Suva 200	Savarua Creek 8.0km W Colo-i-Suva Vlg.	-18.05889	178.44194	200
Viti Levu	Colo-i-Suva 325	Mt. Nakobalevu, 4km WSW Colo-i-Suva Vlg.	-18.05583	178.42222	325
Viti Levu	Colo-i-Suva 372	Mt. Nakobalevu, 4km WSW Colo-i-Suva Vlg.	-18.05528	178.42361	372
Viti Levu	Colo-i-Suva 400	Mt. Nakobalevu, 4.3km WSW, Colo-i-Suva Vlg.	-18.05944	178.40850	400
Viti Levu	Colo-i-Suva 460	Mt. Nakobalevu, TV Tower, 5km WSW Colo-i-Suva Vlg.	-18.05000	178.41667	460
Viti Levu	Colo-i-Suva Forest Park 140	Colo-i-Suva Forest Park	-18.05889	178.47222	140
Viti Levu	Colo-i-Suva Forest Park 220	Colo-i-Suva Forest Park	-18.05917	178.46222	220
Viti Levu	Ellington Wharf 1	Ellington Wharf	-17.33985	178.22185	1
Viti Levu	Galoa 300	Galoa Vlg. nr bridge over creek	-18.25480	177.99092	15
Viti Levu	Galoa 300	4.8km NE Galoa Vlg.	-18.21861	178.01361	300
Viti Levu	Korobaba	Mt. Korobaba	-18.09400	178.38000	—
Viti Levu	Korobaba 300	Mt. Korobaba nr Lami Town	-18.01667	178.35000	300
Viti Levu	Koronivia 10	Koronivia Research Station	-18.03139	178.53333	10
Viti Levu	Korovau	island record	—	—	—
Viti Levu	Lami	Lami (coords for Lami vlg.)	-18.11100	178.41000	—
Viti Levu	Lami 171	Mt. Korobaba, 5km NW Lami Town	-18.09444	178.38467	171
Viti Levu	Lami 200	Mt. Korobaba 1.0km SW Lami Town	-18.08667	178.37861	200
Viti Levu	Lami 260	Mt. Korobaba, 4km NW Lami Town	-18.10417	178.38056	260

Viti Levu	Lami 3	1km NW Lami Queen"s Rd.	-18.11667	178.40000	3
Viti Levu	Lami 304	Mt. Korobaba, 5km NW Lami Town	-18.08803	178.38228	304
Viti Levu	Lami 400	Mt. Korobaba, 4km NW Lami Town	-18.10222	178.38250	400
Viti Levu	Lami 432	Mt. Korobaba summit, 5km NW Lami Town	-18.08578	178.37588	432
Viti Levu	Lautoka Port 5 b	Lautoka Port	-17.60491	177.44002	5
Viti Levu	Lautoka Shirley Park 5	Lautoka, Shirely Park	-17.60447	177.44848	5
Viti Levu	Lomolaki	Mt. Lomolaki [Mt. Lomalagi]	-17.57000	177.97000	—
Viti Levu	McDonald's Resort 10 b	Nanui-i-Ra I., McDonald's Resort	-17.29772	178.21709	10
Viti Levu	Monasavu 800 a	1.5km NE Monasavu Dam	-17.74983	178.05770	800
Viti Levu	Monasavu 800	1.6km NW Monasavu Dam	-17.75000	178.03450	800
Viti Levu	Monasavu Dam 1000	5.8km SSW Monasavu Dam	-17.80167	178.02117	1000
Viti Levu	Monasavu Dam 600	4km NE Monasavu Dam	-17.73417	178.07433	600
Viti Levu	Monasavu Dam 800	7km S Monasavu Dam nr powerstation headquarters	-17.81018	178.03773	800
Viti Levu	Monasavu Dam 830	Rairaimatuku Plateau, Lake Konavatu, 9.6km NW Monasevu Dam	-17.69352	178.00440	830
Viti Levu	Mosquito Island 1	Mosquito I.	-18.11667	178.40000	1
Viti Levu	Mt. Batilamu 1125 b	Koroyanitu Nat'l Heritage Park, Mt. Batilamu, nr summit, 3.2km SE Abaca Vlg.	-17.68658	177.54392	1125
Viti Levu	Mt. Batilamu 840 c	Koroyanitu Nat'l Heritage Park, Mt. Batilamu, 2km SE Abaca Vlg.	-17.67939	177.54194	840
Viti Levu	Mt. Evans 700	Koroyanitu Eco Park 5.0km NE Abaca Vlg.	-17.66667	177.55250	700
Viti Levu	Mt. Evans 700	Mt. Evans Range, Koroyanitu Eco Park, 1.8km NE Abaca Vlg.	-17.66667	177.56333	700
Viti Levu	Mt. Evans 800	Mt. Evans Range, Koroyanitu Eco Park, 0.5km N Abaca Vlg.	-17.66667	177.55000	800
Viti Levu	Mt. Evans 800	Mt. Evans Range, Koroyanitu Eco Park, Kokabula Trail, 1km E Abaca Vlg.	-17.66667	177.55000	800
Viti Levu	Mt. Evans 800	Mt. Evans Range, Koroyanitu Eco Park, Savuione Trail, 1km E Abaca Vlg.	-17.66667	177.55000	800
Viti Levu	Mt. Nakobalevu 200	Mt. Nakobalevu, 1.3km W, Colo-i-Suva Vlg.	-18.04806	178.43767	200
Viti Levu	Mt. Naqarababuluti 864	Mt. Naqarababuluti, 1.1km NE Emperor Gold Mine Rest House	-17.56818	177.95980	864
Viti Levu	Mt. Naqarababuluti 912	Mt. Naqarababuluti, 1.1km NE Emperor Gold Mine Rest House	-17.56973	177.95987	912
Viti Levu	Mt. Naqaranabuluti 1050	Mt. Naqaranibuluti, 1.3km W Emperor Gold Mine Rest House	-17.56944	177.97000	1050
Viti Levu	Mt. Rama 300	Mt. Rama 1.8km NW Naikorokoro Set.	-18.09000	178.30389	300
Viti Levu	Mt. Tomanivi	Mt. Tomanivi	-17.62000	178.00000	—
Viti Levu	Mt. Tomanivi 1105	3km E Navai, trail to Mt. Tomanivi (Mt. Victoria)	-17.61667	178.00680	1105

Viti Levu	Mt. Tomanivi 1294	3.2km E Navai, trail to Mt. Tomanivi (Mt. Victoria)	-17.61583	178.01667	1294
Viti Levu	Mt. Tomanivi 1300	3.4km E Navai, trail to Mt. Tomanivi	-17.61480	178.01825	1300
Viti Levu	Mt. Tomanivi 700	Mt. Tomaniivi, 2km E Navai Vlg.	-17.62111	178.00000	700
Viti Levu	Mt. Tomanivi 700 b	Mt. Tomaniivi, 0.75km E Navai Vlg.	-17.62111	177.98917	700
Viti Levu	Mt. Tomanivi 950	Mt. Tomanivi, 2.4km E Navai Vlg.	-17.61806	178.00550	950
Viti Levu	Naboutini 300	1.8km NW Naboutini Vlg.	-18.22056	177.81667	300
Viti Levu	Nabukavesi 300	nr Nabukavesi Vlg.	-18.11667	178.25000	300
Viti Levu	Nabukavesi 40	Ocean Pacific Resort, 2km SE Nabukavesi Vlg.	-18.17083	178.25806	40
Viti Levu	Nabukelevu 300	2.3km NW Nabukelevu Vlg.	-18.11000	177.81667	300
Viti Levu	Nadakuni 300	Medrausucu Range 5.5km NW Nadakuni Vlg.	-17.93361	178.26278	300
Viti Levu	Nadakuni 300 b	5.5km NNW Nadakuni Vlg.	-17.92694	178.26861	300
Viti Levu	Nadala 300	Nadala, nr. Nadarivatu	-17.55000	177.90000	300
Viti Levu	Nadarivatu 750	Nadarivatu	-17.67570	177.97000	750
Viti Levu	Nadi town	(coords for Nadi town)	-17.80400	177.41490	—
Viti Levu	Naikorokoro 300	2.7km NE Naikorokoro Vlg.	-18.08722	178.33139	300
Viti Levu	Nakavu 200	Navua 3.0km NW Nakavu Vlg.	-18.16472	178.08944	200
Viti Levu	Nakavu 300	3.5km N Nakavu Vlg.	-18.15694	178.08528	300
Viti Levu	Nakobalevu 340	1.5km NE Colo-i-Suva Vlg.	-18.05056	178.41667	340
Viti Levu	Naqaranabuluti 1000	Naqaranibuluti Nature Reserve, nr summit, 0.75km SE Nadarivatu	-17.57278	177.97250	1000
Viti Levu	Naqaranabuluti 860	Naqaranibuluti Nature Reserve, 0.5km SE Nadarivatu	-17.56818	177.96633	860
Viti Levu	Narokorokoyawa 700	1.5km SE Narokorokoyawa	-17.86667	178.11722	700
Viti Levu	Nasoqo	Nasoqo	-18.08000	178.02000	—
Viti Levu	Nasoqo 800 a	8km NWEfrom Nasoqo Vlg.	-17.60458	178.04037	800
Viti Levu	Nasoqo 800 b	8km NWEfrom Nasoqo Vlg.	-17.60886	178.04400	800
Viti Levu	Nasoqo 800 c	8km NWEfrom Nasoqo Vlg.	-17.60788	178.04446	800
Viti Levu	Nasoqo 800 d	8km NWEfrom Nasoqo Vlg.	-17.61058	178.04447	800
Viti Levu	Nausori	Nausori Highlands	—	—	—
Viti Levu	Nausori Highlands 400	Nausori Highlands	-17.81000	177.61000	400
Viti Levu	Navai	(coords for Navai vlg.)	-17.61720	177.98360	—
Viti Levu	Navai 1020	Veilaselase Track, 3.2km E Navai Vlg.	-17.62417	178.00917	1020

Viti Levu	Navai 1023	Mt. Tomanivi, 8km ESE Navai Vlg.	-17.62415	178.00558	1023
Viti Levu	Navai 700	Mt. Tomaniivi, 1.8km E Navai Vlg.	-17.62111	177.99806	700
Viti Levu	Navai 770	0.8km E Navai Vlg.	-17.61952	177.98998	770
Viti Levu	Navai 863	0.9km E Navai, trail to Mt. Tomanivi (Mt. Victoria)	-17.61960	177.99046	863
Viti Levu	Navai 870	0.2km E Navai, trail to Mt. Tomanivi (Mt. Victoria)	-17.61900	178.00275	870
Viti Levu	Navai 930	2.4km E Navai, trail to Mt. Tomanivi (Mt. Victoria)	-17.61833	178.00500	930
Viti Levu	Navai Forestry Camp	Navai Forestry Camp [coordinates for Navai Vlg.]	-17.61000	177.98000	—
Viti Levu	Nuku 50	Navua River, at Sauniwaqa, nr Nuku Vlg.	-18.10172	178.01927	50
Viti Levu	Ocean Pacific 1	Ocean Pacific	-18.17361	178.25722	1
Viti Levu	Ocean Pacific 2	Ocean Pacific	-18.17333	178.25944	1
Viti Levu	Ratu Sukuna Park 5	Ratu Sukuna Park	-18.14002	178.42230	5
Viti Levu	Saiaro	Saiaro (unknown locality, forisland record)	—	—	—
Viti Levu	Savione 750 a	Koroyanitu National Heritage Park, Savione Falls, 2km ESE Abaca Vlg.	-17.67593	177.55015	750
Viti Levu	Sigatoka	8mi. N Sigatoka	—	—	—
Viti Levu	Sigatoka 30 a	Sigatoka Sand Dunes, 0.5km. SW visitor center	-18.16842	177.48667	30
Viti Levu	Suva	Suva	-18.13000	178.42000	—
Viti Levu	Suva 10	Suva, nr. USP campus [Lacala Bay]	-18.15000	178.44000	10
Viti Levu	Tailevu	Tailevu (coords for Tailevu town)	—	—	—
Viti Levu	Tamavua	Tamavua	-18.10000	178.45000	—
Viti Levu	Vatubalavu 300	Vaturu nr Vatubalavu Vlg.	-18.18333	178.10000	300
Viti Levu	Vaturu Dam 530	1.3km SW Vaturu Dam	-17.74778	177.67722	530
Viti Levu	Vaturu Dam 540	Mt. Lomalagi, Heavens Edge Lodge, nr. Vaturu Dam	-17.74472	177.66500	540
Viti Levu	Vaturu Dam 550	1.5km SW Vaturu Dam	-17.74389	177.67556	550
Viti Levu	Vaturu Dam 575 b	1km NNE Vaturu Dam	-17.74333	177.66817	575
Viti Levu	Vaturu Dam 620	1.0km SW Vaturu Dam	-17.75417	177.66500	620
Viti Levu	Vaturu Dam 700	2.0km SW Vaturu Dam	-17.75583	177.66028	700
Viti Levu	Veisari 300; 3.5 km N	Waivudawa, 3.5km N Veisari Set.	-18.06806	178.36694	300
Viti Levu	Veisari 300; 3.8 km N	Waivudawa, 3.8km N Veisari Set.	-18.07917	178.36250	300
Viti Levu	Vesari	Vesari (island record)	—	—	—
Viti Levu	Volivoli 25	Sigatoka Sand Dunes, 0.8km SSW Volivoli Vlg.	-18.16667	177.48514	25

Viti Levu	Volivoli 50	Sigatoka Sand Dunes, 1.1km SSW Volivoli Vlg.	-18.17158	177.48422	50
Viti Levu	Volivoli 55	Sigatoka Sand Dunes, 1.1km SSW Volivoli Vlg.	-18.16936	177.48469	55
Viti Levu	Vunidawa	Vinidawa (coords for Vunidawa vlg.)	-17.82800	178.32670	—
Viti Levu	Vunisea 300	7.5km NE Vunisea Vlg.	-17.48333	178.14333	300
Viti Levu	Waimoque 850	Monasavu Rd., 1.75km SE Waimoque Set.	-17.67035	177.99375	850
Viti Levu	Waisoi 300	Waisoi Forestry Camp, nr Namosi Vlg.	-18.03000	178.13000	300
Viti Levu	Waivaka	nr. Waivaka	-18.03000	178.18000	—
Viti Levu	Waivudawa 300	Waivudawa Creek 6.0km NNW Lami Town	-18.07605	178.36278	300
Viti Levu	Waiyanitu	Waiyanitu	-18.18000	178.12000	—
Yasawa	Nabukeru 120	Yasawa i Lau Cave, 1km SE Nabukeru Vlg.	-16.85222	177.46889	120
Yasawa	Nabukeru 144	Yasawa i Lau Cave, 1km SE Nabukeru Vlg.	-16.85278	177.46733	144
Yasawa	Tamusua 118	Taucake, 700m NW Tamusua Vlg.	-16.83706	177.44500	118
Yasawa	Wayalailai Resort 55	Wayasewa I., Wayalailai Resort	-17.35117	177.13803	55
Yasawa	Wayalailai Resort 55 b	Wayasewa I., Wayalailai Resort	-17.34892	177.13867	55

APPENDIX B. CHECKLIST OF THE ANTS OF FIJI

The following checklist is sorted by subfamily, genus and species. Distribution: E = endemic to Fiji, N = native to the Pacific region, I = introduced to the Pacific region. Voucher specimens are listed where available and can be entered into Antweb for most up to date taxonomy.

	Page	*Plate*	*Dist.*	*Voucher specimen*	*Subgenus/species group*
Subfamily Amblyoponinae					
Amblyopone zwaluwenburgi	22	1	N	CASENT0187702	
Prionopelta kraepelini	23	2	N	—	
Subfamily Cerapachyinae					
Cerapachys cryptus	26	3	E	—	*typhlus* group
Cerapachys fuscior	26	4	E	CASENT0171152	*typhlus* group
Cerapachys lindrothi	28	6	E	CASENT0171147	*dohertyi* group
Cerapachys majusculus	30	9	E	—	*dohertyi* group
Cerapachys sculpturatus	30	10	E	—	*dohertyi* group
Cerapachys vitiensis	30	11	E	CASENT0175795	*dohertyi* group
Cerapachys zimmermani	28	7	E	CASENT0175759	*dohertyi* group
Cerapachys sp. FJ01	29	8	E	CASENT0175817	*dohertyi* group
Cerapachys sp. FJ04	32	14	E	CASENT0171150	
Cerapachys sp. FJ05	32	13	E	CASENT0175808	*dohertyi* group
Cerapachys sp. FJ06	27	5	E	CASENT0171145	*typhlus* group
Cerapachys sp. FJ07	31	12	E	CASENT0175790	*dohertyi* group
Cerapachys sp. FJ08	32	15	E	CASENT0175805	
Cerapachys sp. FJ10	32	16	E	CASENT0177223	
Cerapachys sp. FJ52	25	—	E	—	
Subfamily Dolichoderinae					
Iridomyrmex anceps	33	17	N	CASENT0171061	*anceps* group
Ochetellus sororis	34	18	E	CASENT0171060	*glaber* group
Philidris nagasau	36	19	E	CASENT0171058	*cordatus* group
Tapinoma melanocephalum	37	20	I	CASENT0171078	
Tapinoma minutum	38	21	N	CASENT0177146	
Tapinoma sp. FJ01	38	22	E	CASENT0177206	
Tapinoma sp. FJ02	39	23	E	CASENT0177104	
Technomyrmex vitiensis	39	24	N	CASENT0171057	
Subfamily Ectatomminae					
Gnamptogenys aterrima	41	25	E	CASENT0171062	*albiclava* group
Subfamily Formicinae					
Acropyga lauta	42	26	N	CASENT0171066	*myops* group
Acropyga sp. FJ02	43	27	E	CASENT0127488	*myops* group
Anoplolepis gracilipes	43	28	I	CASENT0171031	
Camponotus bryani	54	35	E	CASENT0177563	(*Colobopsis*) *dentatus* group.
Camponotus chloroticus	46	29	N	CASENT0171140	*maculatus* group
Camponotus cristatus	57	40	E	CASENT0180230	(*Myrmogonia*)
Camponotus dentatus	53	34	E	CASENT0177557	(*Colobopsis*) *dentatus* group.

	Page	*Plate*	*Dist.*	*Voucher specimen*	*Subgenus/species group*
Camponotus fijianus	52	33	E	CASENT0177590	(*Colobopsis*) *dentatus* group
Camponotus kadi	61	45	E	CASENT0180467	(*Myrmogonia*)
Camponotus laminatus	58	41	E	CASENT0180461	(*Myrmogonia*)
Camponotus lauensis	62	46	E	CASENT0235853	(*Myrmogonia*)
Camponotus levuanus	59	42	E	CASENT0180271	(*Myrmogonia*)
Camponotus maafui	60	43	E	CASENT0180410	(*Myrmogonia*)
Camponotus manni	54	36	E	CASENT0177564	(*Colobopsis*) *dentatus* group.
Camponotus oceanicus	47	—	E	—	(*Colobopsis*)
Camponotus polynesicus	48	30	E	CASENT0187263	(*Colobopsis*)
Camponotus sadinus	60	44	E	CASENT0180304	(*Myrmogonia*)
Camponotus schmeltzi	63	47	E	CASENT0187073	(*Myrmogonia*)
Camponotus umbratilis	55	37	E	CASENT0177580	(*Colobopsis*) *dentatus* group.
Camponotus vitiensis	50	31	E	CASENT0186859	(*Colobopsis*)
Camponotus sp. FJ02	56	38	E	CASENT0177581	(*Colobopsis*) *dentatus* group.
Camponotus sp. FJ03	56	39	E	CASENT0177667	(*Colobopsis*) *dentatus* group.
Camponotus sp. FJ04	51	32	E	CASENT0177727	(*Colobopsis*)
Nylanderia glabrior	64	48	N	CASENT0181474	
Nylanderia vaga	64	49	N	CASENT0171069	
Nylanderia vitiensis	65	50	E	CASENT0181499	
Nylanderia sp. FJ03	66	51	N	CASENT0171141	
Paraparatrechina oceanica	67	52	E	CASENT0181303	
Paratrechina longicornis	68	53	I	CASENT0171073	
Plagiolepis alluaudi	69	54	I	CASENT0171065	
Subfamily Myrmicinae					
Adelomyrmex hirsutus	70	55	E	CASENT0181559	
Adelomyrmex samoanus	71	56	N	CASENT0171037	
Cardiocondyla emeryi	72	57	I	CASENT0171087	*emeryi* group
Cardiocondyla kagutsuchi	72	58	N	CASENT0171071	*nuda* group
Cardiocondyla minutior	73	59	N	CASENT0171077	*minutior* group
Cardiocondyla nuda	73	60	N	CASENT0181806	*nuda* group
Cardiocondyla obscurior	74	61	I	CASENT0171038	*wroughtonii* group
Carebara atoma	74	62	N	CASENT0171039	
Eurhopalothrix emeryi	76	63	E	CASENT0181833	*procera* group
Eurhopalothrix insidiatrix	77	64	E	CASENT0171052	*procera* group
Eurhopalothrix sp. FJ52	77	65	E	CASENT0194603	
Lordomyrma curvata	79	66	E	CASENT0171008	
Lordomyrma desupra	80	67	E	CASENT0171002	
Lordomyrma levifrons	80	68	E	CASENT0171004	
Lordomyrma polita	81	69	E	CASENT0171007	
Lordomyrma rugosa	81	70	E	CASENT0171009	
Lordomyrma stoneri	82	71	E	CASENT0171014	
Lordomyrma striatella	82	72	E	CASENT0171010	
Lordomyrma sukuna	83	73	E	CASENT0171011	
Lordomyrma tortuosa	83	74	E	CASENT0171000	
Lordomyrma vanua	84	75	E	CASENT0171051	

	Page	*Plate*	*Dist.*	*Voucher specimen*	*Subgenus/species group*
Lordomyrma vuda	84	76	E	CASENT0171018	
Metapone sp. FJ01	85	77	E	CASENT0181802	
Monomorium destructor	87	78	I	CASENT0171088	*destructor* group
Monomorium floricola	88	79	I	CASENT0171072	*monomorium* group
Monomorium pharaonis	88	80	I	CASENT0171086	*salomonis* group
Monomorium sechellense	89	81	I	CASENT0181877	*hildebrandti* group
Monomorium vitiense	89	82	E	CASENT0181872	*fossulatum* group
Monomorium sp. FJ02	90	83	I?	CASENT0181926	*monomorium* group
Myrmecina cacabau	91	84	E	CASENT0260381	
Myrmecina sp. FJ01	92	85	E	CASENT0182571	
Pheidole bula	106	98	E	CASENT0171113	*roosevelti* group
Pheidole caldwelli	96	86	E	CASENT0185147	
Pheidole colaensis	106	99	E	CASENT0171103	
Pheidole fervens	97	87	N	CASENT0171099	
Pheidole furcata	107	100	E	CASENT0171111	*roosevelti* group
Pheidole knowlesi	98	88	E	CASENT0171097	*knowlesi* group
Pheidole megacephala	99	89	I	CASENT0171036	
Pheidole oceanica	100	90	N	CASENT0171126	
Pheidole onifera	101	91	E	CASENT0185253	
Pheidole pegasus	107	101	E	CASENT0171108	*roosevelti* group
Pheidole roosevelti	108	102	E	CASENT0171027	*roosevelti* group
Pheidole sexspinosa	101	92	N	CASENT0194651	*quadrispinosa* group
Pheidole simplispinosa	109	103	E	CASENT0171106	*roosevelti* group
Pheidole umbonata	102	93	N	CASENT0171135	
Pheidole uncagena	109	104	E	CASENT0171110	*roosevelti* group
Pheidole vatu	102	94	E	CASENT0185525	
Pheidole wilsoni	103	95	E	CASENT0183622	*knowlesi* group
Pheidole sp. FJ05	104	96	E	CASENT0183626	
Pheidole sp. FJ09	105	97	E	CASENT0183927	*knowlesi* group
Poecilomyrma myrmecodiae	112	105	E	CASENT0217355	
Poecilomyrma senirewae	113	106	E	CASENT0181574	
Poecilomyrma sp. FJ03	113	107	E	CASENT0181454	
Poecilomyrma sp. FJ05	113	108	E	CASENT0217355	
Poecilomyrma sp. FJ06	114	109	E	CASENT0181607	
Poecilomyrma sp. FJ07	114	110	E	CASENT0181539	
Poecilomyrma sp. FJ08	114	111	E	CASENT0181455	
Pristomyrmex mandibularis	115	112	E	CASENT0171044	*laevigatus* group
Pristomyrmex sp. FJ02	116	113	E	CASENT0171144	*laevigatus* group
Pyramica membranifera	118	114	I	CASENT0171134	
Pyramica trauma	119	115	E	CASENT0184729	*capitata* group
Pyramica sp. FJ02	119	116	E	CASENT0184976	*capitata* group
Rogeria stigmatica	120	117	N	CASENT0171046	
Romblonella liogaster	121	118	E	CASENT0013155	
Solenopsis geminata	122	119	I	CASENT0171029	*geminata* group
Solenopsis papuana	123	120	N	CASENT0171089	

	Page	*Plate*	*Dist.*	*Voucher specimen*	*Subgenus/species group*
Strumigenys basiliska	127	121	E	CASENT0185864	*biroi* group
Strumigenys chernovi	128	122	E	CASENT0184914	*godeffroyi* group
Strumigenys daithma	128	123	E	—	*caniophanes* group
Strumigenys ekasura	129	124	E	CASENT0184678	*godeffroyi* group
Strumigenys frivola	129	125	E	CASENT0186989	*godeffroyi* group
Strumigenys godeffroyi	130	126	I	CASENT0171155	*godeffroyi* group
Strumigenys jepsoni	130	127	N	CASENT0235852	*godeffroyi* group
Strumigenys mailei	130	128	N	CASENT0186741	*godeffroyi* group
Strumigenys nidifex	131	129	N	CASENT0171047	*szalayi* group
Strumigenys panaulax	132	130	E	CASENT0186960	*godeffroyi* group
Strumigenys praefecta	133	131	E	CASENT0186986	*godeffroyi* group
Strumigenys rogeri	133	132	I	CASENT0171133	
Strumigenys scelesta	133	133	E	CASENT0186833	*godeffroyi* group
Strumigenys sulcata	134	134	E	CASENT0185962	*godeffroyi* group
Strumigenys tumida	135	135	E	CASENT0186506	*godeffroyi* group
Strumigenys sp. FJ01	135	136	E	CASENT0185902	*szalayi* group
Strumigenys sp. FJ13	135	137	E	CASENT0186982	*godeffroyi* group
Strumigenys sp. FJ14	136	138	E	CASENT0184984	
Strumigenys sp. FJ17	136	139	E	CASENT0184819	
Strumigenys sp. FJ18	136	140	E	CASENT0186900	
Strumigenys sp. FJ19	136	141	E	CASENT0185751	
Tetramorium bicarinatum	138	142	I	CASENT0171032	*bicarinatum* group
Tetramorium caldarium	139	143	I	CASENT0171084	*simillimum* group
Tetramorium insolens	140	144	N	CASENT0171138	*bicarinatum* group
Tetramorium lanuginosum	140	145	I	CASENT0171128	
Tetramorium manni	141	146	E	CASENT0183012	*szalayi* group
Tetramorium pacificum	141	147	N	CASENT0171075	*bicarinatum* group
Tetramorium simillimum	142	148	I	CASENT0171034	*simillimum* group
Tetramorium tonganum	143	149	N	CASENT0171074	*tonganum* group
Vollenhovia denticulata	144	150	N	CASENT0182566	
Vollenhovia sp. FJ01	145	151	E	CASENT0171049	
Vollenhovia sp. FJ03	145	152	E	CASENT0182605	
Vollenhovia sp. FJ04	146	153	E	CASENT0182576	
Vollenhovia sp. FJ05	146	154	E	CASENT0182577	
Subfamily Ponerinae					
Anochetus graeffei	147	155	N	CASENT0171067	
Hypoponera confinis	149	156		CASENT0186525	
Hypoponera eutrepta	149	157	E	CASENT0171068	*biroi* group
Hypoponera monticola	150	158	E	CASENT0186141	*pruinosa* group
Hypoponera opaciceps	150	159	I	CASENT0171154	
Hypoponera pruinosa	151	160	N	CASENT0186203	
Hypoponera punctatissima	152	161	I	CASENT0171129	
Hypoponera turaga	152	162	E	CASENT0186517	
Hypoponera vitiensis	153	163	E	CASENT0260379	
Hypoponera sp. FJ16	153	164	E	CASENT0217359	

	Page	*Plate*	*Dist.*	*Voucher specimen*	*Subgenus/species group*
Leptogenys foveopunctata	155	165	E	CASENT0260380	*chinensis* group
Leptogenys fugax	155	166	E	CASENT0171055	*chinensis* group
Leptogenys humiliata	156	167	E	CASENT0260382	*chinensis* group
Leptogenys letilae	156	168	E	CASENT0186196	*chinensis* group
Leptogenys navua	157	169	E	CASENT0186200	*chinensis* group
Leptogenys vitiensis	157	170	E	CASENT0186328	*chinensis* group
Leptogenys sp. FJ01	157	171	E	CASENT0186138	*chinensis* group
Odontomachus angulatus	158	172	E	CASENT0184584	*saevissimus* group
Odontomachus simillimus	159	173	N	CASENT0171124	
Pachycondyla stigma	160	174	I	CASENT0171132	
Platythyrea parallela	161	175	N	CASENT0194735	
Ponera colaensis	162	176	E	CASENT0194711	*taipingensis* group
Ponera manni	163	177	E	CASENT0187758	*japonica* group
Ponera swezeyi	163	178	I	CASENT0194716	
Ponera sp. FJ02	163	179	E	CASENT0194723	
Subfamily Proceratiinae					
Discothyrea sp. FJ01	165	180	E	CASENT0171054	
Discothyrea sp. FJ02	165	181	E	CASENT0187775	
Discothyrea sp. FJ04	165	182	E	ANIC32-053451	
Proceratium oceanicum	166	183	E	CASENT0171053	*silaceum* group
Proceratium relictum	167	184	E	CASENT0194740	*silaceum* group
Proceratium sp. FJ01	167	185	E	CASENT0187587	*silaceum* group

Lightning Source UK Ltd.
Milton Keynes UK
UKOW011837110113

204769UK00002B/12/P